Handbook of Botany

Handbook of Botany

Edited by Edward Ackles

MURPHY & MOORE
www.murphy-moorepublishing.com

Published by Murphy & Moore Publishing,
1 Rockefeller Plaza,
New York City, NY 10020, USA
www.murphy-moorepublishing.com

Handbook of Botany
Edited by Edward Ackles

International Standard Book Number: 978-1-63987-275-6 (Hardback)

Cataloging-in-Publication Data

Handbook of botany / edited by Edward Ackles.
p. cm.
Includes bibliographical references and index.
ISBN 978-1-63987-275-6
1. Botany. 2. Plants. I. Ackles, Edward.
QK48 .H36 2022
580--dc23

TABLE OF CONTENTS

PREFACE

Botany, also known as plant biology, is the study of plant life. Plants have been used as food, medicine and in science, which makes their study extremely important in today's time. This field studies the internal processes and functions of various parts of plants such as tissues, cells, the whole plant and their communities. The various methods used to study the life of plants are live cell imaging, optical microscopy and electron microscopy. Botany also focuses on mechanisms of gene expressions using methods like molecular genetics and epigenetics. The main area of study in this field includes plant structure, growth, biochemistry, reproduction and plant taxonomy. This book traces the progress of this field and highlights some of its key concepts and applications. While understanding the long-term perspectives of these topics, the book makes an effort in highlighting their impact as a modern tool for the growth of the discipline. For all those who are interested in this area of study, this book can prove to be an essential guide.

The information shared in this book is based on empirical researches made by veterans in this field of study. The elaborative information provided in this book will help the readers further their scope of knowledge leading to advancements in this field.

Finally, I would like to thank my fellow researchers who gave constructive feedback and my family members who supported me at every step of my research.

Editor

PREFACE

[illegible]

Editor

1

Squalene synthase cloning and functional identification in wintersweet plant (*Chimonanthus zhejiangensis*)

Guanhua Liu[1,2,3] and Jianyu Fu[1,2*]

Abstract

Background: Three species of wintersweets: *Chimonanthus salicifolius* S. Y. Hu, *Chimonanthus zhejiangensis* M. C. Liu and *Chimonanthus grammalus* M. C. Liu are widely distributed in China. The three wintersweets belonging to the genus of *Chimonanthus* that can synthesize abundant terpenoids that are beneficial to human health. Their buds and leaves are traditional Chinese herb applied by the 'She' ethnic minority in southeast of China. Squalene is a multi-functional and ubiquitous triterpene in plants, which is biosynthesized by squalene synthase (SQS) using farnesyl diphosphate (FPP) as the substrate. The synthesis of squalene in wintersweet was not clearly. This work would provide us much help to further understand the terpene metabolism in wintersweet and its health function to people at phytochemistry and molecular levels.

Results: In this study, we identified squalene component in the extractions of leaves of three wintersweets and isolated SQS genes from leaf transcriptomes. The three SQSs were highly conservative, so CzSQS from *C. zhejiangensis* was just determined the enzymatic activity. The in vitro expressed CzSQS that deleted two transmembrane domains could catalyze FPP to generate squalene with the presence of NADPH and Mg^{2+}.

Conclusions: The squalene was one of wintersweet leaves phytochemicals. The squalene synthases of three wintersweet plants were highly conserved. The CzSQS was capable to catalyze two FPP molecules to squalene.

Keywords: Wintersweet, Squalene, Squalene synthase

Introduction

Wintersweet (*Calycanthaceae*) contains ten species belonging to three genera: *Calycanthus* L. contains three species distributed in America and Eastern Asia, *Chimonanthus* contains six species originated in China, and *Idiospermum Blake* contains a single species distributed in Australia (Christenhusz and Byng 2016). The six species of *Chimonanthus* are further classified into aromatic type and nonaromatic type according to whether their leaves release aroma or not. (Zhang and Shen 1999). Three species of *Chimonanthus salicifolius* S. Y. Hu, *Chimonanthus Zhejiangensis* M. C. Liu and *Chimonanthus grammalus* M. C. Liu are aromatic type (Zhang and Shen 1999). *C. salicifolius* S. Y. Hu, an endemic species of China, has been used as Chinese traditional medicine to therapy cough, vomiting, heatstroke, rheumatic arthritis and measles for thousands of years (Ma et al. 2017). *C. zhejiangensis* M. C. Liu, only found in Zhejiang, Jiangxi and Fujian provinces, is also a Chinese herb and a courtyard ornamental plant, which is widely planted in the scenic area of Zhejiang province (Ouyang and Mai 2010). *C. grammalus* M. C. Liu is an ancient angiosperm and a transitional species between *Chimonanthus* and *Neochimonanthus*, and is an endangered and national-protected medicinal plant in China and mainly distributed in Anyuan and Huichang counties of Jiangxi Province (Jiang et al. 2015a, b). The three *Chimonanthus* plants mostly distributed in the area where 'She' ethnic minority lived in, so the tender stems were firstly applied by them as a traditional herb to treat cough and rheumatism for a long

*Correspondence: mybatigoal@mail.tricaas.com
[1] Tea Research Institute, Chinese Academy of Agricultural Sciences, Hangzhou 310008, People's Republic of China
Full list of author information is available at the end of the article

time, and the Chinese name of the galenical is Shiliang-cha (herb tea) (Ma et al. 2017). To date, the three species of *Chimonanthus* have been made into herbal beverage named as 'Gold tea', a type of healthy decoction in the counties around Zhejiang, Fujian and Jiangxi provinces.

In order to identify the bioactive components, both volatiles and non-volatiles of wintersweet plants were taken into account. The non-volatiles of *Calycanthaceae* include alkaloid (Ma et al. 2015a, b), terpene (Li et al. 2016), flavone (Zhang et al. 2017), coumarin (Li et al. 2013; Wang et al. 2016) and so forth. The volatiles of wintersweet plants were essential oil, which can be used as cosmetics, perfumery and aromatherapy. The aromatic type species of *Chimonanthus* usually release large quantity of volatiles and abundant terpenes (terpenoids) as β-caryophyllene, β-elemene, γ-elemene, germacrene-D, trans-β-ocimene, sabinene, δ-cadinene were found in essential oils (Lv et al. 2012; Farsam et al. 2007). Phytochemical analyses have revealed that they were not only rich in essential oils but also in semi-volatile fractions. Recently, squalene was identified in the volatile oil of *C. salicifolius* by GC–MS (Zhou et al. 2013). We also found squalene in the extractions of leaves of *C. zhejiangensis* detected by GC × GC-TOFMS method in this work. Squalene, a triterpene with a formula of $C_{30}H_{50}$ including six isoprene units, is an important intermediate of the endogenous synthesis of cholesterol in animals or sterol in plants. It is ubiquitous and has been detected in bacteria, alga and mammals. In 1935, the first higher plant-derived squalene was gained from olive oil (Thorbjarnarson and Drummond 1935). Up to now, squalene has been isolated in more plants as *Arabidopsis thaliana*, *Vitis vinifera* (Nakashima et al. 1995; Kribii et al. 1997; Genova et al. 2014; Uchida et al. 2015).

Terpene or terpenoid is the largest class of secondary metabolites in plants, and it is synthesized by either the mevalonate pathway (MVA) in the cytosol or the methylerythritol phosphate pathway (MEP) in plastids (Jiang et al. 2015a, b). Squalene is synthesized in the triterpene biosynthesis pathway, a branch of the terpene biosynthetic network of MVA pathway in plants (Chappell et al. 1995, 2002). It is generated from two molecules of farnesyl diphosphate (FPP) in a two-step reductive dimerization reaction catalyzed by squalene synthase (SQS) (Pandit et al. 2000). The next metabolism step is that squalene is cyclized to lanosterol followed by many enzymatic steps for the conversion to cholesterol (Smith 2000). SQS is the key enzyme of squalene metabolism, and the genes have been isolated in animals, plants and bacteria (Pandit et al. 2000; Jiang et al. 2015a, b; Dale et al. 2013; Okada et al. 2000). SQSs from eukaryotic organisms are associated with the ER (endoplasmic reticulum) membranes through a hydrophobic membrane-spanning α-helix located at the C-terminus of the protein. In contrast, bacterial SQSs lack of the C-terminal α-helix, and are not affiliated with the membrane (Stamellos et al. 1993a, b). The SQSs have been studied deeply in many organisms including *Arabidopsis thaliana*, tobacco and human. In contrast, squalene was found in *C. zhejiangensis*, a traditional Chinese herb, but our knowledge on the molecular basis of squalene biosynthesis in which is absent. Here, we further confirmed squalene generation in the leaves of wintersweet, isolated the squalene synthase gene of *C. zhejiangensis* and identified its enzymatic activity, to give a complete explanation on the squalene biosynthesis in wintersweet.

Methods

Plant materials

Three species of wintersweet: *Chimonanthus salicifolius* S. Y. Hu, *Chimonanthus Zhejiangensis* M. C. Liu and *Chimonanthus grammalus* M. C. Liu were collected from Yushan and Wuyuan, Jiangxi province and Kaihua, Zhejiang Province. The leaves were stored at − 80 °C after being frozen in liquid N_2.

Squalene collection and detection in *C. zhejiangensis* leaves

Fresh leaves of *C. zhejiangensis* (3 g) was cut into pieces and sealed in a petri plate and then SPME (solid-phase micro-extraction, SUPELCO, Sigma-Aldrich Co.) column was injected into the plate to collect the volatiles. For more accurate analysis, the volatiles emitted from *C. zhejiangensis* leaves pieces were collected in an open headspace sampling system (Analytical Research System, Gainesville, FL) as previously reported (Li et al. 2012). Volatiles were collected for 24 h by pumping air from the chamber through a Super-Q volatile collection trap and eluted using 300 μL n-hexane containing 0.003% (w/v) 1-octanol as an internal standard. The leaves (1 g) was further ground into powder in liquid nitrogen and extracted by chromatographic *n*-hexane, dichloromethane and ethyl acetate at 37 °C and shaking at 200 rpm for 24 h. Finally, the extract was centrifuged at 12,000 rpm for 30 min and the supernatant was used for analyzing.

The non-volatile compounds of *C. zhejiangensis* were analyzed by GC × GC-TOFMS (multidimensional gas chromatography coupled to time-of-flight mass spectrometry LECO Pegasus 4D GC × GC-TOFMS instrument, LECO Corporation, St. Joseph, MI, USA) and Gas Chromatography-Mass Spectrometer (GC–MS, Agilent Technologies 7890B GC–MS system). The chromatographic column of GC × GC-TOFMS was Rxi-5MS column (30 m × 250 μm, Restek, Bellefonte, PA, USA) coupled to Rxi-17Sil MS column (1.9 m × 100 μm × 0.10 μm, Restek, Bellefonte, PA, USA).

The GC–MS column was HP-5MS (30 m × 0.250 mm, 1909IS-433UI, Agilent Technologies. Inc. California, USA). For GC–MS analysis (Jiang et al. 2015a, b), the injection temperature was 250 °C and a temperature gradient of 5 °C/min from 50 °C (hold 3 min) to 250 °C. The electronic impact (EI) mode was at 70 eV. For GC × GC-TOFMS analysis (Yoshimi et al. 2018), samples were injected at 50 °C in the splitless mode. After holding the samples for 5 min at 50 °C, the oven temperature was increased at 10 °C/min to 300 °C and hold for 5 min. The flow rate of helium carrier gas was set at 1.9 mL/min. All MS data was collected from 40 to 400 m/z.

Cloning of squalene synthase genes from wintersweet

The transcriptome of the *C. zhejiangensis* foliage was sequenced by Illumina. The data was assembled by the Trinity software (Grabherr et al. 2011) and the result was de-redundant and re-spliced by the Tgicl program to obtain the final unigene. The unigene was Blast with other organism SQSs by Blastp program. Total RNAs were extracted from leaves using Polysaccharide and Polyphenol Total RNA Isolation Kit (Tiangen Biotech Co., Ltd., Beijing, China). The quality and concentration were checked by NanoDrop 1000 spectrophotometer (Thermo Fisher Scientific, Waltham, MA, USA) and formaldehyde agarose gel electrophoresis. Total RNA (100 μg/μL) was reverse-transcribed into first strand cDNA in a reaction volume (80 μL) using PrimeScript™ RT reagent Kit with gDNA Eraser (Perfect Real Time) (TaKaRa, Co. Ltd, Dalian, China) following the manufacture's protocol and diluted to 60 μL. The primers of 5′-ATGGGGAAG TTAGGAACAATGCTG-3′ and 5′-TCATTTAATGGA AAGAAATGCAAAC-3′ was designed based on the unigene sequence on primer 6.0. The PCR reaction was set as follows: pre-denature at 94 °C for 2 min; 35 cycles of denature at 98 °C for 10 s, annealing at 56 °C for 30 s, extension at 68 °C for 2 min; a final extension at 72 °C for 10 min. The PCR products were separated by agarose gel and purified using TIANgel Midi Purification kit (TIANGEN, Beijing, China), and then were cloned into pGEM-T Easy vector (Promega Corporation, Madison, WI, USA) to sequence bidirectionally with the universal primers of M13.

Squalene synthase sequences analysis

The SQS genes from three species of wintersweet were translated into protein sequences by ExPASy online (http://web.expasy.org/translate/). Multiple sequences of other plant SQSs were retrieved by Blastp from NCBI (https://blast.ncbi.nlm.nih.gov/Blast.cgi), and the detail information of all SQSs was in Additional file 1: Table S3. Sequence alignment for SQSs was done by ClustalW program embedded in MAGE 7.0, and the phylogenetic tree was also constructed by MAGE 7.0 based on Neighbor Joining (NJ) method (Kumar et al. 2016). The protein structure model was constructed by Phyre2 Fold Recognition online (http://www.sbg.bio.ic.ac.uk/phyre2). The signal peptide and transmembrane region of SQSs were analyzed by SignalP 4.1 Server (http://www.cbs.dtu.dk/services/SignalP/) and TMHMM program online (http://www.cbs.dtu.dk/services/TMHMM/), respectively.

Protein expression and enzymatic activity determination of SQS

The *C. zhejiangensis* squalene synthase (CzSQS) gene removed hydrophobic region was introduced into pEXP5-CT/TOPO vector and the recombinant plasmids were transformed into *E. coli* (BL21 Condon plus) by heat shock at 42 °C for 1 min. After the cells were rejuvenation in LB liquid medium at 37 °C, 225 rpm for an hour, they were cultured to an OD_{600} of 0.6 in 50 mL LB liquid medium with 50 μg/mL ampicillin and 25 μg/mL chloramphenicol at the same condition. Gene expression was induced by 1 mM IPTG (Isopropyl-β-D-thiogalactopyranoside, 50 mg/mL, TIANGEN, Ltd., Beijing, China) at 16 °C, 150 rpm for 20 h. The cells were harvested by centrifugation, sonicated for 6 × 30 s in chilled 1 × PBS buffer and desalted through Econo-Pac 10DG columns (Bio-Rad). The enzymatic activity was measured on the basis of the conversion of farnesyl diphosphate (FPP) (Echelon Biosciences) to squalene in the presence of NADPH and Mg^{2+}: 250 μL of the reaction mixture included 10 mM MOPSO (pH 7.0), 25 μM FPP (Echelon Biosciences), 25 mM $MgCl_2$, 25 mM mercaptoethanol, 5 mM NADPH and 125 μL enzyme. The reaction was conducted at 37 °C for 5 h. After incubation, the mixture was extracted three times with 500 μL *n*-hexane. The *n*-hexane solution was concentrated by the blowing of nitrogen at room temperature until the total volume reached approximately 120 μL (Jiang et al. 2015a, b). The catalytic product was analyzed by GC–MS (Agilent 7890B-7000C, Agilent Technologies. Inc. California, USA) and confirmed by authentic standard of squalene.

Results

Identification of squalene in leaves of wintersweet

Squalene was not found in the volatiles of *C. zhejiangensis* leaves collected by the SPME and Super-Q using static and dynamic headspace methods, while it was detected by GC × GC-TOFMS in the leaf extractions of *n*-hexane, dichloromethane and ethyl acetate. Both the 1st and 2nd dimension time (minutes/s) and the unique mass were equivalent with the authentic standard of squalene (Fig. 1). The chromatogram peak area of the extractions of dichloromethane was the largest, which indicated that although squalene was identified in all the extractions of

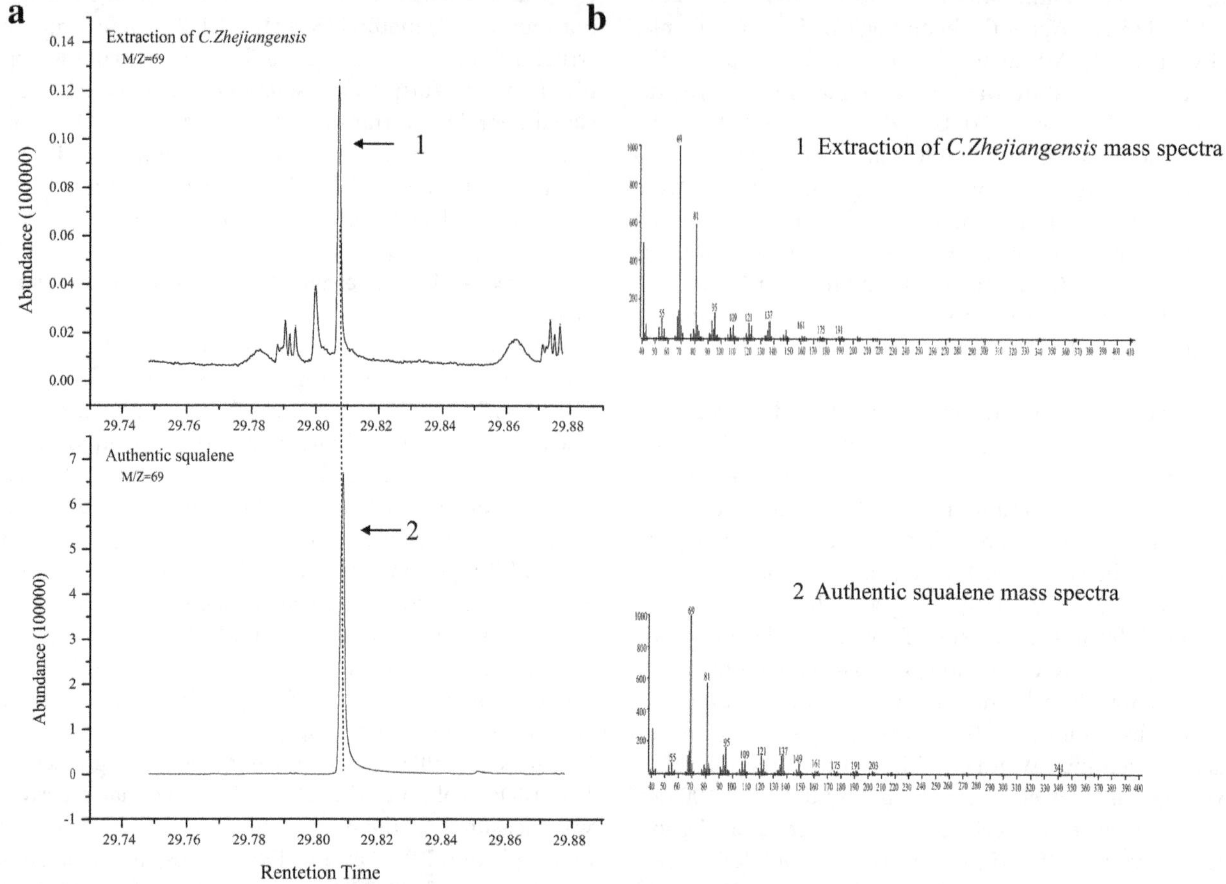

Fig. 1 A partial gas spectrum of the *C. zhejiangensis* foliage extraction with dichloromethane

three solvent, the extraction efficiency of dichloromethane was the highest (Fig. 1).

Isolation of squalene synthase genes from wintersweet

The full-length sequences of squalene synthase from *C. salicifolius, C. zhejiangensis, C. grammalus* were first identified by transcriptomic sequencing and further confirmed by PCR amplification and the NCBI accession numbers are MH277638, MH277639 and MH277637, respectively. Although the sequences of wintersweet plants were all 1227 bp long and encoded 408 amino acids, there were five different sites among the three SQSs (Additional file 1: Figure S1). At 42nd, 52nd and 313rd sites, the SQSs of *C. zhejiangensis, C. grammalus* and were Ile, Ser and Met, but the three sites were Thr, Gly and Leu in *C. Salicifolius* squalene synthase (CsSQS). At the positions of 172nd and 208th, the sequences of CsSQS, CzSQS were Val and Ala, while the same sites replaced by Ala and Thr in *C. grammalus* squalene synthase (CgSQS), respectively.

Multiple sequences alignment was made by MAGE7.0 (Fig. 2) to further confirm the SQS sequences of three wintersweet plants. The sequences, including *Arabidopsis thaliana* SQS1 (Nakashima et al. 1995), *Rattus norvegicus* SQS (McKenzie et al. 1992), *Selaginella moellendorffii* SQS (Jiang et al. 2015a, b) and *Populus trichocarpa* SQS (Tuskan et al. 2006) were searched from the NCBI Blastp, and the function of these enzymes were conducted. The TcSQS (*Trypanosoma cruzi* SQS, Shang et al. 2014), elucidated senior structure and obtained on the Phyre2 on Internet, was included for comparison in this work. The two aspartate rich domains 'DXXXDD' and 'DXXXD' that mediate the binding of prenyl diphosphate are highly conserved among all the SQSs. The five different sites were out of the active domain compared with other SQSs.

Homology-based structural modeling and phylogenetic analysis of SQSs

The homology-based structural modeling on Phyre2 indicated that the model of three wintersweet SQSs were similar to *Trypanosoma cruzi* squalene synthase (TcSQS, Fig. 3, Shang et al. 2014), and they were 100% confidence with TcSQS. The different amino acids of three

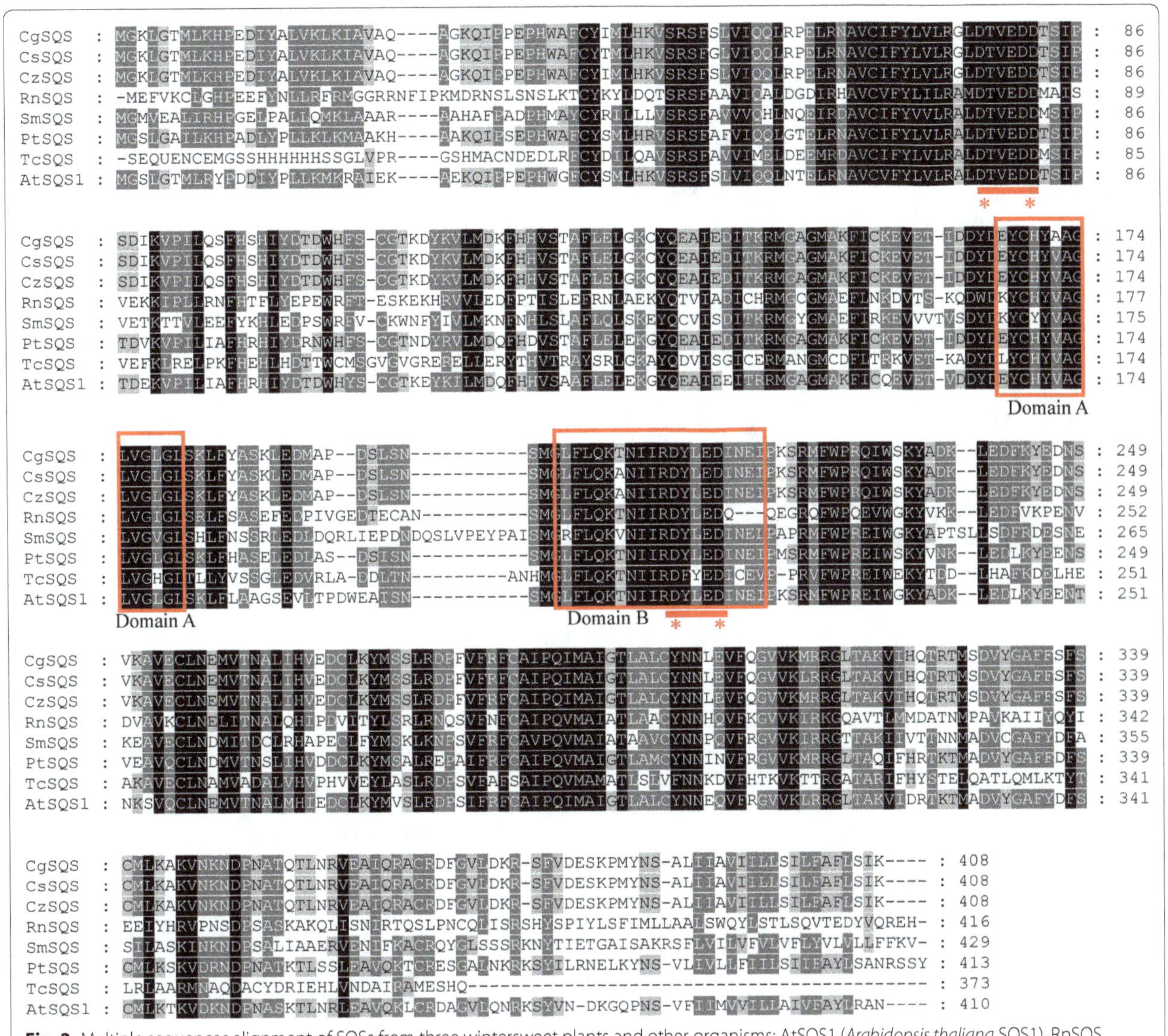

Fig. 2 Multiple sequences alignment of SQSs from three wintersweet plants and other organisms: AtSQS1 (*Arabidopsis thaliana* SQS1), RnSQS (*Rattus norvegicus* SQS), SmSQS (*Selaginella moellendorffii* SQS), PtSQS (*Populus trichocarpa* SQS) and TcSQS (*Trypanosoma cruzi* SQS). The two conserved domains, DXXXDD and DXXXD, which mediate the binding prenyl diphosphate are underlined among SQSs from different species

wintersweet SQSs, located at the 42nd, 52nd, 172nd, 208th and 313th sites, which didn't affect the active center structure formation of α-helix on the base of the homology-based structural modeling of TcSQS in the same sites. The analyses of TMHHM and SingalP showed that CsSQS, CzSQS and CgSQS contained no any N-terminal signal peptide but one C-terminal transmembrane region located at 390th to 407th amino acids (Additional file 1: Figure S2).

Other twenty-eight SQSs of fungi, alga, lichen and higher plants were retrieved from NCBI to run the phylogenetic analysis with three SQSs of wintersweet. The phylogenetic tree was divided into four large groups based on the four phyla: the SmSQS (Jiang et al. 2015a, b), BbSQS (Okada et al. 2000) and CalSQS (Dale et al. 2013) are on behalf of Lycopodiophyta, Chlorophyta and Ascomycota, and the rest SQSs clustered into another branch as the spemaiophyla (Fig. 4). The SQSs of three Chimonanthus plants clustered into one clade and then they prior clustered with MoSQS (*Magnolia officinalis* squalene synthases, Zha et al. 2016) (Fig. 4). It indicated that MoSQS had close relationship with the SQSs from wintersweet, which perhaps because both *Magnolia officinalis* and Chimonanthus were Magnoliids plants. The detail information of the SQSs is available in Additional file 1: Table S3.

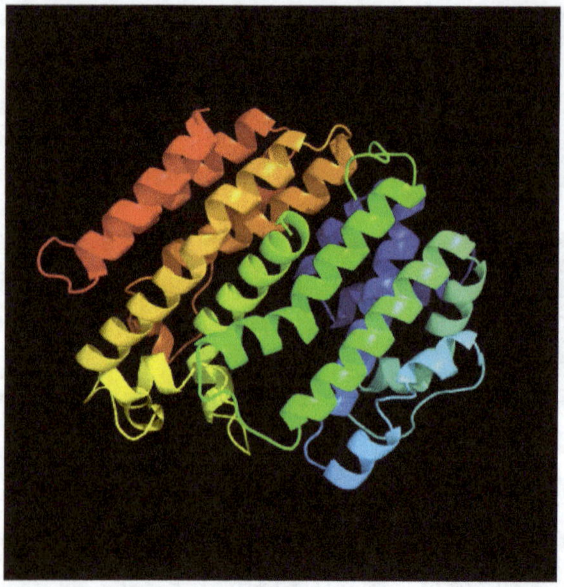

Fig. 3 The homology-based structural modeling of c3wccC (TcSQS)

E. coli expressed SQS catalyzes the formation of squalene using FPP as substrate

Eukaryotic squalene synthases localized in the ER membrane through its C-terminus, which usually lead to insoluble question when expression eukaryotic SQSs in *E. coli*. The deletion of residues from the C-terminus of tobacco SQS (Hanley et al. 1996; Devarenne et al. 2002), capsicum SQS (Lee et al. 2002) and yeast SQS (LoGrasso et al. 1993) resulted in functionally soluble enzymes. The coding sequence of CzSQS without C-terminal member-anchoring peptide was cloned and expressed in *E. coli* (strain: BL21 Condon plus) to determine the enzymatic activity. The catalyzed product in vitro was detected in GC–MS, and its peak at 39.998 min was the same as that of authentic squalene (Fig. 5a). Nevertheless, no any peak was detected at uniform retention time in the empty vector as control. The major mass fragments of m/z = 69 and m/z = 81 at 39.998 min of the peak of CzSQS is consistent with full-scan mass spectrum of authentic squalene (Fig. 5b). The results indicated that CzSQS encoded a

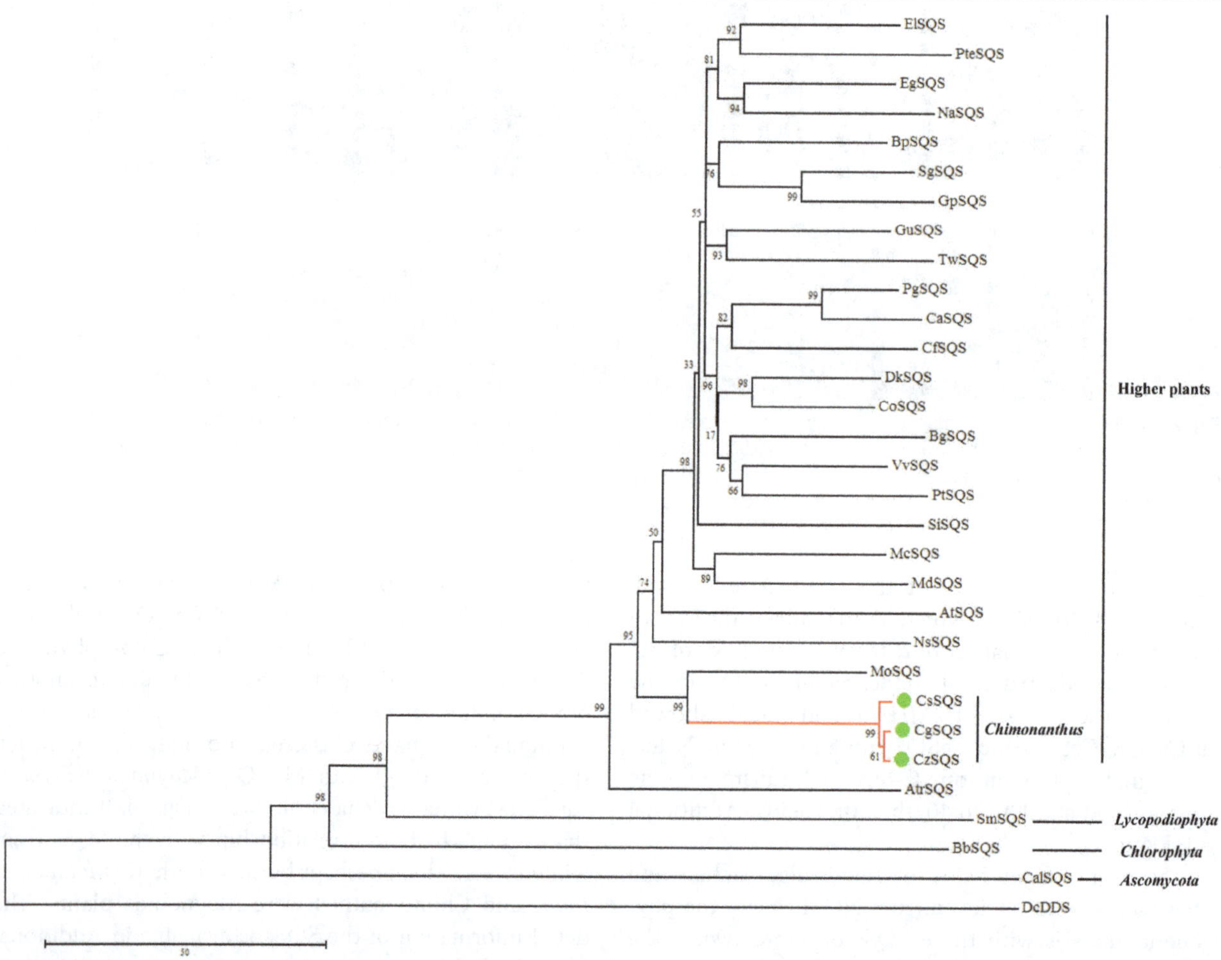

Fig. 4 Phylogenetic tree of SQSs. The NCBI accession number of the sequences was in Additional file 1: Table S3

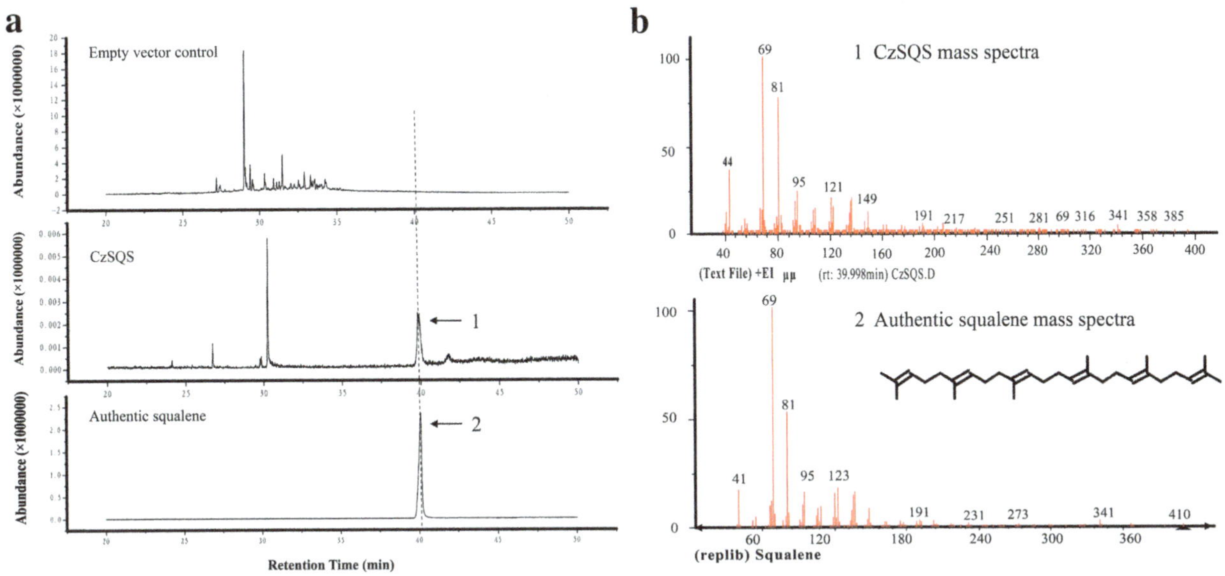

Fig. 5 Partial gas chromatogram of products of squalene synthase enzyme assays. **a** Gas chromatogram of the empty vector control (pET32a), CzSQS reflects squalene synthase enzyme reaction catalyzed by recombinant CzSQS expressed in *E. coli* and the authentic squalene from SIGMA. **b** Mass spectrum of the 39.998-min peak in the gas chromatogram of the CzSQS and the authentic squalene at the same time

functional SQS enzyme which converted two farnesyl diphosphate molecules into a squalene in vitro without the C-terminal member-anchoring peptide.

Discussion

Wintersweet has attracted much attention not only because of their ornamentality and rarity, but also due to their affluent pharmacologically active substances. Their buds and leaves have been applied as a traditional herb by the 'She' ethnic minority people that distributed in the southeast of China. The pharmacology studies of wintersweet have revealed that its volatile or non-volatile compounds has phytochemical functions as anticonvulsion, abirritation, inhibiting cancer cells and anti-mycobacteria (Chebib et al. 2003; Amador et al. 2000; Verotta et al. 2002; Ma et al. 2015a, b; Wang et al. 2011; Yu et al. 2014). It was reported that 35.97% of the volatile compounds of wintersweet was sesquiterpenoids, and so the essential oil has anti-bacteria and free radical scavenging functions (Yu et al. 2014). In this paper, we analyzed the volatiles and non-volatiles compounds in leaves of wintersweet by SPME and GC × GC-TOFMS methods, respectively. Totally, 54 volatile terpenoids including 25 monoterpenoids and 29 sesquiterpenoids were detected in the leaves of *C. zhejiangensis* by SPME. The contents of eucalyptol, β-myrcene and pinene were the highest in monoterpenoids, and germacrene D, cadina-1(10), 4-diene and copaene were the highest in sesquiterpenoids. In the concentrated extractions of *n*-hexane, dichloromethane and ethyl acetate, up to 104 terpenes or terpenoids were identified and some of them were also included in the volatiles detected by the SPME (Additional file 1: Tables S1 and S2). Our result was similar to other studies that the terpenes or terpenoids were the major compounds of the wintersweet plants. Besides large quantity of monoterpenoids and sesquiterpenoids, we also found squalene in the extractions of leaves of *C. zhejiangensis* and which has been confirmed in one species of wintersweet (Zhou et al. 2013).

Squalene is taken part in the ergosterol biosynthetic process, sterol biosynthetic process and terpene metabolism. Squalene synthase (SQS), the farnesyl-diphosphate farnesyl transferase, catalyzes two molecules farnesyl diphosphate to form squalene. SQS is the key enzyme of the ergosterol, sterol and terpene metabolism. Here, we cloned SQSs from three wintersweet plants and identified the function in vitro for the first time. The subcellular location of AtSQS1 is the endoplasmic reticulum (ER) (Busquets et al. 2008). The CzSQS, CgSQS, CsSQS transmembrane regions were similar to AtSQS1 based on the sequences alignment and the TMHHM analysis (Additional file 1: Figure S2). The subcellular of the wintersweet SQS was like the AtSQS1, they locate on the ER. The active site of domain A, combining the two FPP into PSPP and loosen one pyrophosphate (Gu et al. 1998), is Try. These sites are the same among the three wintersweet plants SQS. The different five sites of the three wintersweet plants SQS was absent of the active site. In domain B, binding the diphosphate moiety of the allylic substrates via an Mg^{2+} bridge (Gu et al. 1998), the active domain is DXXXD that binding prenyl diphosphate. The different amino acid residues of the three wintersweet plants SQSs are absent in DXXXD. The other different sites were not in the active domain (Additional file 1: Figure S1).

In view of highly conserved sequences and homology-based structural modeling result, we just performed the enzymatic analysis for CzSQS. Prokaryotic expression demonstrated that CzSQS-pEXP5-CT/TOPO was expressed in its soluble form in *E.coli* (strain: *BL21 Condon plus*) induced with 1 mM IPTG. In vitro enzyme reactions manifested that CzSQS-pEXP5-CT/TOPO encoded a functional squalene synthase. Squalene is spotlighted by its biofunction in clinic usage, such as antioxidant, antitumor, and cyto-protective effects (Kim and Karadeniz 2012). While the function of squalene from wintersweet plants was deficiency. On the other hand, the botanical squalene was restricted to olive oil, camellia oleifera, amaranth and palm (Murkovica et al. 2004; Mlakar et al. 2010; Lehmann 1996; Sun et al. 1995). We isolated the squalene synthase and authenticated squalene in the leaves of *C. zhejiangensis*. The results provide a theoretical basis for the development and utilization of Shiliangcha, buds and leaves of the three wintersweet plants, and its pharmacological analysis.

Conclusions

In a word, squalene was detected in the extraction of *C. zhejiangensis* leaves. The three subspecies wintersweet plants squalene synthase were cloned and analyzed results indicated that the three squalene synthase were highly conversed. The CzSQS possessed biological activity in catalyzed the squalene synthesis.

Abbreviations

CzSQS: *Chimonanthus zhejiangensis* squalene synthase; CsSQS: *Chimonanthus salicifolius* squalene synthase; CgSQS: *Chimonanthus grammalus* squalene synthase; TcSQS: *Trypanosoma cruzi* squalene synthase.

Authors' contributions

JF initiated the project and designed the experiment. GL carried out the experiments and analyzed the data. Both authors wrote and revised the manuscript. Both authors read and approved the final manuscript.

Author details

[1] Tea Research Institute, Chinese Academy of Agricultural Sciences, Hangzhou 310008, People's Republic of China. [2] Key Laboratory of Tea Quality and Safety Control, Ministry of Agriculture and Rural Affairs, Hangzhou, 310008, People's Republic of China. [3] Graduate School of Chinese Academy of Agricultural Sciences, Beijing 100081, People's Republic of China.

Acknowledgements

Not applicable.

Competing interests

The authors declare that they have no competing interests.

Funding

This work was supported by the National Natural Science Foundation of China (31470693, 31100503), the Natural Science Foundation of Zhejiang Province (LY18C160006), the Central Public-interest Scientific Institution Basal Research Fund (1610212018004, 1610212016017) and the Science and Technology Innovation Project of the Chinese Academy of Agricultural Sciences (CAAS-ASTIP-2014-TRICAAS).

References

Amador TA, Verotta L, Nunes DS, Elisabetsky E (2000) Antinociceptive profile of hodgkinsine. Planta Med 66(2):770. https://doi.org/10.1055/s-2000-9604

Busquets A, Keim V, Closa M, del Arco A, Boronat A, Arró M, Ferrer A (2008) *Arabidopsis thaliana* contains a single gene encoding squalene synthase. Plant Mol Biol 67(1–2):25–36. https://doi.org/10.1007/s11103-008-9299-3

Chappell J (2002) The genetics and molecular genetics of terpene and sterol origami. Curr Opin Plant Biol 5:151–157. https://doi.org/10.1016/S1369-5266(02)00241-8

Chappell MJ, Proulx MJ, Saunders RC, Cuellar CR, Wolf MTF (1995) Is the reaction catalyzed by 3-hydroxy-3-methylglutaryl coenzyme a reductase a rate-limiting step for isoprenoid biosynthesis in plants? Plant Physiol 109(4):1337–1343

Chebib M, Duke RK, Duke CC, Connor M, Mewett KN, Johnston GAR (2003) Convulsant actions of Calycanthine. Toxicol Appl Pharmacol 190:58. https://doi.org/10.1016/S0041-008X(03)00149-2

Christenhusz MJM, Byng JW (2016) The number of known plants species in the world and its annual increase. Phytotaxa 261(3):201–217. https://doi.org/10.11646/phytotaxa.261.3.1

Dale M, Katja S, Jonathan SW, Gavin S (2013) Assembly of a phased diploid *Candida albicans* genome facilitates allele-specific measurements and provides a simple model for repeat and indel structure. Genome Biol R10:1038. https://doi.org/10.1186/gb-2013-14-9-r97

Devarenne TP, Ghosh A, Chappell J (2002) Regulation of squalene synthase, a key enzyme of sterol biosynthesis in tobacco. Plant Physiol 129:1095e1106. https://doi.org/10.1104/pp.001438

Farsam H, Amanlou M, Taghi-Cheetsaz N, Amin GR, Salehi-Sormaghi MH (2007) Essential oil constituents of Chimonanthus fragrans flowers population of tehran. Daru J Pharm Sci 15(3):129–131

Genova AD, Almeida AM, Muñoz-Espinoza C, Vizoso P, Travisany D, C C, Pinto M, Hinrichsen P, Orellana A, Maass A (2014) Whole genome comparison between table and wine grapes reveals a comprehensive catalog of structural variants. BMC Plant Biol 14:7. https://doi.org/10.1186/1471-2229-14-7

Grabherr MG, Haas BJ, Yassour M, Levin JZ, Thompson DA, Amit I, Adiconis X, Fan L, Raychowdhury R, Zeng Q, Chen Z, Mauceli E, Hacohen N, Gnirke A, Rhind N, Palma F, Birren BW, Nusbaum C, Lindblad-Toh K, Friedman N, Regev A (2011) Full-length transcriptome assembly from RNA-Seq data without a reference genome. Nat Biotechnol 29(7):644–652. https://doi.org/10.1038/nbt.1883

Gu P, Ishii Y, Spencer AT, Shechter I (1998) Function-structure studies and identification of three enzyme domains involved in the catalytic activity in rat hepatic squalene synthase. J Biol Chem 273(20):12515–12525. https://doi.org/10.1074/jbc.273.20.12515

Hanley KM, Nicolas O, Donaldson TB, Smith-Monroy C, Robinson GW, Hellmann GM (1996) Molecular cloning, in vitro expression and characterization of a plant squalene synthetase cDNA. Plant Mol Biol 30(6):1139–1151. https://doi.org/10.1007/BF00019548

Jiang YM, Wang J, Liu YX, Zhang ZB, Cao G, Lu S, Zhu D (2015a) Genetic patterns investigation of wild *Chimonanthus grammatus* M.C. Liu by using SSR markers. Acta Ecol Sin 35(6):203–209. https://doi.org/10.1016/j.chnaes.2015.07.007

Jiang Y, Chen H, Chen X, Köllner TG, Jia Q, Wymore TW, Wang F, Chen F (2015b) Volatile squalene from a nonseed plant *Selaginella moellendorffii*: emission and biosynthesis. Plant Physiol Biochem 96(1–8):0981–9428. https://doi.org/10.1016/j.plaphy.2015.07.010

Kim SK, Karadeniz F (2012) Biological importance and applications of squalene and squalane. In: SK Kim, Karadeniz F (eds) Advances in food and nutrition research. Academic Press, Cambridge. https://doi.org/10.1016/b978-0-12-416003-3.00014-7

Kribii R, Arró M, Del Arco A, González V, Balcells L, Delourme D, Ferrer A, Karst F, Boronat A (1997) Cloning and characterization of the *Arabidopsis thaliana* SQS1 gene encoding squalene synthase-involvement of the C-terminal region of the enzyme in the channeling of squalene through the sterol. Eur J Biochem 249:61e69. https://doi.org/10.1111/j.1432-1033.1997.00061

Kumar S, Stecher G, Tamura K (2016) MEGA7:molecular evolutionary genetics analysis version 7.0 for bigger datasets. Mol Biol Evol 33:1870–1874. https://doi.org/10.1093/molbev/msw054%20

Lee JH, Yoon YH, Kim HY, Shin DH, Kim DU, Lee IJ (2002) Cloning and expression of squalene synthase cDNA from hot pepper (*Capsicum annuum* L.). Mol Cells 13:436e443

Lehmann JW (1996) Case history of grain amaranth as an alternative crop. Cereal Foods World 41(5):72–74

Li G, Köllner TG, Yin Y, Jiang Y, Chen H, Xu Y, Gershenzon J, Pichersky E, Chen F (2012) Nonseed plant *Selaginella moellendorffii* has both seed plant and microbial types of terpene synthases. Proc Natl Acad Sci USA 109:14711e14715. https://doi.org/10.1073/pnas.1204300109

Li QJ, Wang ML, Yang XS, Ma L, Hao XJ (2013) Two new coumarin glycosides from *Chimonanthus nitens*. Asian Nat Prod Res 15(3):270. https://doi.org/10.1080/10286020.2012.762766

Li D, Jiang YY, Jin ZM, Li HY, Xie HJ, Wu B, Wang KW (2016) Isolation and absolute configurations of diastereomers of 8-hydroxy-Tmuurolol and (1,6,7)-cadinane-4-en-8,10-diol from *Chimonanthus salicifolius*. Phytochemistry 122:294

LoGrasso PV, Soltis DA, Boettcher BR (1993) Overexpression, purificatin, and kinetic characterization of a carboxyl-terminal-truncated yeast squalene synthetase. Arch Biochem Biophys 307:193e199

Lv JS, Zhang LL, Chu XZ, Zhou JF (2012) Chemical composition, antioxidant and antimicrobial activity of the extracts of the flowers of the Chinese plant *Chimonanthus praecox*. Nat Prod Res 26(14):1363–1367. https://doi.org/10.1080/14786419.2011.602828

Ma GL, Yang GX, Xiong J, Chen WL, Cheng KJ, Hu JF (2015a) Salicifoxazines A and B, new cytotoxic tetrahydro-1,2-oxazine-containing tryptamine-derived alkaloids from the leaves of *Chimonanthus salicifolius*. Tetrahedron Lett 56:4071. https://doi.org/10.1016/j.tetlet.2015.05.008

Ma G, Yang G, Xiong J, Cheng W, Cheng K, Hu J (2015b) Salicifoxazines A and B, new cytotoxic tetrahydro-1,2-oxazine-containing tryptamine-derived alkaloids from the leaves of *Chimonanthus salicifolius*. Tetrahedron Lett 56(27):4071–4075. https://doi.org/10.1016/j.tetlet.2015.05.008

Ma S, Lv Q, Zhou H, Fang J, Cheng W, Jiang C, Cheng K, Yao H (2017) Identification of traditional she medicine shi-liang tea species and closely related species using the ITS2 barcode. Appl Sci 195(7):2076–3417. https://doi.org/10.3390/app7030195

McKenzie TL, Jiang G, Straubhaar JR, Conrad DG, Shechter I (1992) Molecular cloning, expression, and characterization of the cDNA for the rat hepatic squalene synthase. J Biol Chem 267(30):21368–21374

Mlakar SG, Turinek M, Jakop M, Bavec M, Bavec F (2010) Grain amaranth as an alternative and perspective crop in temperate climate. J Geogr 5(1):140

Murkovica M, Lechner S, Pietzkaa A, Bratacos M, Katzogiannos E (2004) Analysis of minor components in olive oil. J Biochem Biophys Methods 61(1/2):155–160. https://doi.org/10.1016/j.jbbm.2004.04.002

Nakashima T, Inoue T, Oka A, Nishino T, Osumi T, Hata S (1995) Cloning, expression, and characterization of cDNAs encoding *Arabidopsis thaliana* squalene synthase. Proc Natl Acad Sci USA 92(6):2328–2332. https://doi.org/10.1073/pnas.92.6.2328

Okada S, Devarenne TP, Chappell J (2000) Molecular characterization of squalene synthase from the green microalga *Botryococcus braunii*, Race B. Arch Biochem Biophys 373:307–317. https://doi.org/10.1006/abbi.1999.1568

Ouyang T, Mai X (2010) Analysis of the chemical constituents of essential oil from *Chimonanthus zhejiangensis* by GC–MS. J Chin Med Mater 33(3):385–387

Pandit J, Danley DE, Schulte GK (2000) Crystal structure of human squalene synthase. A key enzyme in cholesterol biosynthesis. J Biol Chem 275:30610e30617. https://doi.org/10.1074/jbc.m004132200

Shang N, Li Q, Ko TP, Chan HC, Li J (2014) Squalene synthase as a target for chagas disease therapeutics. PLoS Pathog 10(5):e1004114. https://doi.org/10.1371/journal.ppat.1004114

Smith TJ (2000) Squalene: potential chemopreventive agent. Expert Opin Investig Drugs 9(8):1841–1848. https://doi.org/10.1517/13543784.9.8.1841

Stamellos KD, Shackelford JE, Shechter I (1993a) Subcellular localization of squalene synthase in rat hepatic cells. Biochemical and immunochemical evidence. Biol Chem 268:12825e12836

Stamellos KD, Shackelford JE, Shechter I, Jiang G, Conrad D, Keller GA, Krisans SK (1993b) Subcellular localization of squalene synthase in rat hepatic cells Biochemical and immunochemical evidence. Biol Chem 268:12825

Sun H, Wiesenborn D, Rayasduarte P, Mohamed A, Hagen K (1995) Bench-scale processing of amaranth seed for oil. J Am Oil Chem Soc 72(12):1551–1555. https://doi.org/10.1007/BF02577853

Thorbjarnarson T, Drummond JC (1935) Occurrence of an unsaturated hydrocarbon in olive oil. Analyst 60(706):23–29. https://doi.org/10.1039/an9356000023

Tuskan GA, Difazio S, Jansson S, Bohlmann J, Grigoriev I, Hellsten U, Putnam N (2006) The genome of black cottonwood, *Populus trichocarpa* (Torr. & Gray). Science 313(5793):1596–1604. https://doi.org/10.1126/science.1128691

Uchida H, Sumimoto K, Ferriols VME, Imou K, Saga K, Furuhashi K, Matsunaga S, Okada S (2015) Isolation and characterization of two squalene epoxidase genes from *Botryococcus braunii*, Race B. PLoS ONE 10(4):e0122649. https://doi.org/10.1371/journal.pone.0122649

Verotta L, Orsini F, Sbacchi M, Scheildler MA, Amador TA, Elisabetsky E (2002) Synthesis and antinociceptive activity of chimonanthines and pyrrolidinoindoline-type alkaloids. Bioorg Med Chem 10:2133. https://doi.org/10.1016/S0968-0896(02)00078-0

Wang W, Cao L, Xiong J, Xia G, Hu J (2011) Constituents from *Chimonanthus praecox* (wintersweet). Phytochem Lett 4:271. https://doi.org/10.1016/j.phytol.2011.04.012

Wang KW, Li D, Wu B, Cao XJ (2016) New cytotoxic dimeric and trimeric coumarins from *Chimonanthus salicifolius*. Phytochem Lett 16:115. https://doi.org/10.1016/j.phytol.2016.03.009

Yoshimi Y, Yukari T, Kei I, Io U, Ayako O, Fumiya K, Jung-Bum L (2018) Characterization of ent-kaurene synthase and kaurene oxidase involved in gibberellin biosynthesis from *Scoparia dulcis*. J Nat Med 72:456–463. https://doi.org/10.1016/j.plaphy.2015.07.010

Yu CL, Kuang Y, Yang SX, Liu L, Liu CG (2014) Chemical composition, antifungal activity and toxicity of essential oils from leaves of *Chimonanthus praecox* and *Chimonanthus zhejiangensis*. Asian J Chem 26(1):254–256. https://doi.org/10.14233/ajchem.2014.15860

Zha L, Liu S, Su P, Yuan Y, Huang L (2016) Cloning, prokaryotic expression and functional analysis of squalene synthase (SQS) in *Magnolia officinalis*. Protein Expr Purif 120:28–34. https://doi.org/10.1016/j.pep.2015.12.008

Zhang R, Shen X (1999) Taxonomy geographical distribution and evolution of *Calycanthaceae*. Chin J Beijing For Univ 21:7–11

Zhang X, Xu M, Zhang J, Wu L, Liu J, Si J (2017) Identification and evaluation of antioxidant components in the flowers of five *Chimonanthus* species. Ind Crops Prod 102:164–172. https://doi.org/10.1016/j.indcrop.2017.03.014

Zhou J, Qian C, Song Y, Zhang Y, Bi S (2013) Study on chemical constituents of volatile oil from leaf of wild *Chimonanthus salicifolicus* with GC–MS analysis and their antioxidant activity. Chin West China J Pharm Sci 3:008

2

Identification among morphologically similar *Argyreia* (Convolvulaceae) based on leaf anatomy and phenetic analyses

Paweena Traiperm[1*], Janene Chow[1,2], Possathorn Nopun[1,2], G. Staples[3] and Sasivimon C. Swangpol[1]

Abstract

Background: The genus *Argyreia* Lour. is one of the species-rich Asian genera in the family Convolvulaceae. Several species complexes were recognized in which taxon delimitation was imprecise, especially when examining herbarium materials without fully developed open flowers. The main goal of this study is to investigate and describe leaf anatomy for some morphologically similar *Argyreia* using epidermal peeling, leaf and petiole transverse sections, and scanning electron microscopy. Phenetic analyses including cluster analysis and principal component analysis were used to investigate the similarity of these morpho-types.

Results: Anatomical differences observed between the morpho-types include epidermal cell walls and the trichome types on the leaf epidermis. Additional differences in the leaf and petiole transverse sections include the epidermal cell shape of the adaxial leaf blade, the leaf margins, and the petiole transverse sectional outline. The phenogram from cluster analysis using the UPGMA method represented four groups with an R value of 0.87. Moreover, the important quantitative and qualitative leaf anatomical traits of the four groups were confirmed by the principal component analysis of the first two components. The results from phenetic analyses confirmed the anatomical differentiation between the morpho-types.

Conclusions: Leaf anatomical features regarded as particularly informative for morpho-type differentiation can be used to supplement macro morphological identification.

Keywords: Leaf epidermis, Leaf transverse section, PCA, SEM, Species complex, Stomatal index, UPGMA

Background

The genus *Argyreia* Lour., one of the larger genera in Convolvulaceae with approximately 135 species, is mainly distributed in tropical Asia (Staples and Traiperm 2017). During preparation of a nomenclatural review of the genus, we found that some species of *Argyreia* are quite similar in their vegetative morphology leading to confusion in identification when only leaf and stem characters are available. Species delimitation in Convolvulaceae generally, including *Argyreia*, is heavily dependent on floral characters; when fully developed flowers are lacking it can be difficult or impossible to identify species precisely. Thus, when specimens with only leaves, or young flower buds, or fruits are all that is available it may not be possible to do more than narrow down the identification to a group of similar species. We have identified several such cases; these seem to represent species-complexes of morphologically similar species, or perhaps single highly polymorphic species. One such group of species includes *Argyreia breviscapa* (Kerr) Ooststr., *Argyreia kerrii* Craib, *Argyreia suddeeana* Traiperm & Staples, *Argyreia variabilis* Traiperm & Staples and two unnamed *Argyreia* that appear morphologically distinct. Previously we pointed out that some species complexes have similar macro morphological leaf characters (Staples and Traiperm 2008; Traiperm and Staples 2014). In order to explore other characters that might be useful for identification,

*Correspondence: paweena.tra@mahidol.edu
[1] Department of Plant Science, Faculty of Science, Mahidol University, Rama VI Road, Ratchathewi, Bangkok 10400, Thailand
Full list of author information is available at the end of the article

this research investigates foliar micromorphology as a means to identify plants or specimens that lack flowers.

Leaf anatomy of some species in *Argyreia* had been investigated by Metcalfe and Chalk (1950), Sayeedud-Din (1953) and Pant and Bhatnagar (1975). Leaf epidermis and stem transverse sections of four *Argyreia* species [*Argyreia nervosa* (Burm.f.) Bojer, *Argyreia onilahiensis* Deroin, *Argyreia sericea* Dalzell & A. Gibson and *Argyreia splendens* (Hornem.) Sweet] were studied by Tayade and Patil (2003, 2012a). A key to species was constructed based on the epidermal characters such as cuticular striations, anticlinal cell walls, and basal cells of trichomes (Tayade and Patil 2003). Recent publication of a new species of *Argyreia* suggested that leaf epidermal characters, such as the pattern of cutin, can be used to distinguish *Argyreia albiflora* Staples & Traiperm from the morphologically similar species *Argyreia wallichii* Choisy (Staples et al. 2015). Anatomical characters strongly supported species identification in an investigation of *Merremia* section *Xanthips* (Pisuttimarn et al. 2013).

Anatomical information has been used for this analysis because it has proven useful and informative in assisting with taxonomic identification and classification in other plants (Thadeo et al. 2014). Leaf anatomical characters from several families also revealed similarities between plant species when using phenetic analysis [cluster analysis (CA) and principal component analysis (PCA)] (Nikolić and Mitić 1991; Breitwieser and Ward 1993; De Faria et al. 2010; Aghababaeyan et al. 2014; Moraes et al. 2011; Thadeo et al. 2014; Arthan et al. 2016; Jayarathna et al. 2016). In Convolvulaceae, particularly the genus *Argyreia*, phenetic analysis based on leaf anatomy has not been done before. The aim of this research was to study the leaf anatomy of *A. breviscapa* and similar-looking morpho-types and to conduct phenetic analyses using leaf anatomical characters to determine their utility for taxon discrimination and identification.

Methods

Plant collections

Plant materials were collected from their natural habitats for six morpho-types that are similar to *A. breviscapa* (e.g., *A. kerrii*, *A. suddeeana*, *A. variabilis*, *Argyreia* 1 and *Argyreia* 2). For each morpho-type, 1–5 populations were sampled. At least five mature leaves in each population were fixed in a mixture of 70% ethyl alcohol and 0.05% glycerol. Voucher specimens were prepared for each morpho-type and deposited at BKF (Table 1). Investigation of the micro-morphological characters was divided into three parts: epidermal surfaces examined by the peeling method; epidermal surfaces viewed with scanning electron microscopy (SEM); and leaf transverse sections prepared by modified paraffin method (Johansen 1940).

Leaf epidermal surfaces

Mature leaves were selected and peeled by a razor blade. The epidermis was stained with Safranin-O for 10–15 min, and then dehydrated in an ethyl alcohol series, finally mounted with DePeX for a permanent slide. The stomatal index was calculated according to the method described by Salisbury (1927).

Leaf surfaces were also studied intensively by the scanning electron microscope (SEM). Small pieces of leaf (5×5 mm^2) were dehydrated in ethanol series and sonicated to remove unwanted parts from the surfaces. After that, dried samples were coated with platinum-palladium in a sputter coater (Hitachi E-102 Ion Sputter). The cuticular patterns were observed and imaged under a Hitachi S-2500 scanning electron microscope.

Table 1 List of the plant materials used in this study

No. of accession	Taxa	Voucher specimens	Locality
1	*Argyreia breviscapa* (Kerr) Ooststr.	J. Chow and P. Traiperm s.n.	Khon Kaen, Thailand
2		Staples et al. 1550	Kanchanaburi, Thailand
3	*A. kerrii* Craib	J. Chow and P. Traiperm 002	Chiang Mai, Thailand
4	*A. suddeeana* Staples & Traiperm	J. Chow and P. Traiperm 017	Ratchaburi, Thailand
5		J. Chow and P. Traiperm 019	Ratchaburi, Thailand
6	*A. variabilis* Traiperm & Staples	S. Suddee s.n.	Phetchaburi, Thailand
7		J. Chow and P. Traiperm s.n.	Phetchaburi, Thailand
8	*Argyreia* 1	J. Chow and P. Traiperm 030	Kanchanaburi, Thailand
9	*Argyreia* 2	J. Chow and P. Traiperm 031	Kanchanaburi, Thailand
10		J. Chow and P. Traiperm 032	Kanchanaburi, Thailand
11		J. Chow and P. Traiperm 033	Kanchanaburi, Thailand
12		J. Chow and P. Traiperm 034	Kanchanaburi, Thailand
13		J. Chow and P. Traiperm 035	Kanchanaburi, Thailand

Leaf transverse section

The paraffin method used in this study was modified from Johansen (1940). Leaf samples were cut using a sliding microtome, stained with Safranin-O and counterstained with Fast Green. The tissues were dehydrated using ethanol series and mounted with DePeX.

The leaf characters from the epidermis and the transverse sections were observed using a light microscope, Olympus BX43, and photo-micrographed using an Olympus DP21 camera.

Phenetic analyses

Thirteen accessions (Table 1) were used in the analysis. Twenty-four leaf anatomical characters (six qualitative and 18 quantitative) were analyzed from leaf surfaces, leaf and petiole transverse sections (Table 2). The qualitative traits were repeatedly observed under light microscopy with five different leaf sample slides per one accession to confirm character stability. For the quantitative characters, each character was measured and means with standard deviations were calculated in ten replicates. For reducing bias in quantitative characters, most of them were expressed in terms of ratios (except blade thickness).

All data were transformed into discrete numbers to score and form the data matrix. The phenetic analyses including CA and PCA were performed. The cluster analysis was conducted based on Unweighted Pair Group Method with Arithmetic Mean (UPGMA) clustering in NTSYS-pc version 2.1 (Rohlf 2000). The similarity between morpho-types resulted in phenogram was estimated by simple matching coefficient (SM). The PCA was performed using Minitab version 16 (2010). The first and second components from PCA were presented for identification of the important anatomical traits.

Table 2 The leaf anatomical characters and character states used in phenetic analyses

Anatomical characters	Character states	Score
Qualitative characters		
1. Epidermal cell wall on adaxial leaf surface	Straight to slightly curved/sinuate	0/1
2. Epidermal cell wall on abaxial leaf surface	Straight to slightly curved/sinuate	0/1
3. Types of non-glandular trichome on abaxial leaf surface	Large and small straight trichome/large straight and small curly trichome	0/1
4. Shape of petiole	Slightly concave/convex	0/1
5. Midrib outline and adaxial curvature	Rounded/acute	0/1
6. Leaf margin shape	Rounded/acute	0/1
Quantitative characters		
7. Ratio of adaxial leaf epidermal cell size (width: length)	Ratio is less than or equal to 0.7/ratio is more than 0.7	0/1
8. Stomatal index (SI) on adaxial leaf surface	SI is less than or equal to 6/SI is more than 6	0/1
9. Ratio of stomatal size on adaxial leaf surface (width: length)	Ratio is less than or equal to 0.6/ratio is more than 0.6	0/1
10. Ratio of glandular trichome size on adaxial leaf surface (width: length)	Ratio is less than or equal to 0.92/ratio is more than 0.92	0/1
11. Ratio of non-glandular trichome size on adaxial leaf surface (width: length)	Ratio is less than or equal to 0.08/ratio is more than 0.08	0/1
12. Ratio of abaxial leaf epidermal cell size (width: length)	Ratio is less than or equal to 0.65/ratio is more than 0.65	0/1
13. Stomatal index (SI) on abaxial leaf surface	SI is less than or equal to 15/SI is more than 15, but less than or equal to 30/SI is more than 30	0/1/2
14. Ratio of stomatal size on abaxial leaf surface (width: length)	Ratio is less than or equal to 0.75/ratio is more than 0.75	0/1
15. Ratio of glandular trichome size on abaxial leaf surface (width: length)	Ratio is less than or equal to 0.9/ratio is more than 0.9	0/1
16. Ratio of non-glandular trichome size on abaxial leaf surface (width: length)	Ratio is less than or equal to 0.03/ratio is more than 0.03, but less than or equal to 0.04/ratio is more than 0.04	0/1/2
17. Ratio of epidermal cells size in petiole (width: length)	Ratio is less than or equal to 0.7/ratio is more than 0.7	0/1
18. Ratio of petiole transverse section (width: length)	Ratio is less than or equal to 0.82/ratio is more than 0.82, but less than or equal to 0.95/ratio is more than 0.95	0/1/2
19. Ratio of palisade mesophyll cell size in leaf transverse section (width: length)	Ratio is less than or equal to 0.2/ratio is more than 0.2	0/1
20. Ratio of spongy mesophyll cell size in leaf transverse section (width: length)	Ratio is less than or equal to 0.7/ratio is more than 0.7	0/1
21. Ratio of adaxial epidermal cell size in leaf transverse section (width: length)	Ratio is less than or equal to 0.7/ratio is more than 0.7	0/1
22. Ratio of abaxial epidermal cell size in leaf transverse section (width: length)	Ratio is less than or equal to 0.7/ratio is more than 0.7	0/1
23. Blade thickness (µm)	Thickness is less than or equal to 208/thickness is more than 208, but less than or equal to 248/thickness is more than 248	0/1/2
24. Ratio of midrib thickness (width: length)	Ratio is less than or equal to 1/ratio is more than 1	0/1

Results

Epidermal cell wall and cuticle

The patterns of anticlinal cell walls as seen under light microscopy appear straight to slightly curved and sinuate (Table 3). Almost all morpho-types have straight to slightly curved cell walls on adaxial surfaces, except in *Argyreia* 1, which shows a sinuate pattern of anticlinal cell walls (Fig. 1). Straight to slightly curved cell walls are present on the abaxial side of *A. breviscapa* and *A. variabilis* (Fig. 1g, j), whereas the others have sinuate patterns of anticlinal cell walls (Fig. 1h, i, k, l). The cuticular striation is shown on both sides of the lamina in all morpho-types (Fig. 2). However, *A. kerrii* possesses numerous trichomes on the abaxial leaf surface which interfere with viewing the surface (Fig. 2h).

Type of stomata and stomatal index

All of the morpho-types have amphistomatous leaves. The stomata on the adaxial leaf surface are anisocytic and paracytic in most of the morpho-types, excluding *A. kerrii* and *A. variabilis,* in which only the paracytic stomatal type was found. Conversely, a paracytic stomatal type presents on the abaxial leaf surface of every morpho-type, whereas *A. breviscapa* possesses both anisocytic and paracytic stomata (Table 3). On the adaxial epidermis the highest stomatal index was observed in *A. breviscapa* and the lowest in *A. variabilis.* On the other hand, the highest stomatal index on the abaxial epidermis was found in *A. variabilis* and the lowest in *Argyreia* 1 (Table 3).

Type, density, size and length of trichomes

Two types of trichomes are present in all morpho-types (Table 3): peltate glandular and simple non-glandular trichomes. Three types of non-glandular trichomes were recognized based on their length and direction: i.e. large straight; small straight; and small curly trichomes were found (Table 3). On their adaxial leaf surfaces, all of the morpho-types possess large and small, straight trichomes (Fig. 3a–f). Similarly on the abaxial leaf surfaces, most morpho-types have large and small straight trichomes excluding *A. kerrii,* which has large straight alternating with small curly trichomes (Fig. 3g–l). The longest trichomes on the adaxial epidermis occur in *A. variabilis,* while the shortest ones are in *A. suddeeana* (Table 3). Meanwhile, the longest trichomes on abaxial epidermis are in *A. breviscapa* and the shortest in *A. suddeeana* (Table 3).

Transverse section of leaf blade and petiole

Leaf blades of all six morpho-types have one layer of epidermis with thin cuticle (Fig. 4). Adaxial epidermal cells are irregular to round in shape excluding *A. variabilis,* which has long and narrow shaped cells (Fig. 4d). *Argyreia breviscapa, A. kerrii,* and *A. suddeeana* have dark-staining deposits in some of the cells on the adaxial surface (Fig. 5a–c). Palisade and spongy cells are different in shape. Palisade parenchyma is composed of 1–2 cell layers. Most morpho-types have rounded leaf margins (Fig. 4g–i, k, l), except *A. variabilis,* which has acute margins (Fig. 4j) (Table 4).

Outlines of the midrib are convex on the adaxial surface with a single acute to rounded adaxial curvature, while abaxial side are rounded or flattened. Almost all have rounded adaxial curvature (Fig. 5b–f), except *A. breviscapa* (Fig. 5a). Vascular tissue in the midrib is bicollateral type (Fig. 5a–f). Secretory cavities occur in all morpho-types, especially in *A. kerrii* (Fig. 5b). The largest cross-sectional midrib size is in *A. variabilis* and the smallest in *Argyreia* 1 (Table 4). Druse crystals are commonly found in the petioles, midribs, and leaf blades of all.

Petiole transverse section outlines are convex in *A. breviscapa, A. variabilis* and *Argyreia* 2 (Fig. 5g, j, l), while *A. kerrii, A. suddeeana* and *Argyreia* 1 show slightly concave outlines on the adaxial surface (Fig. 5h, i, k). Chlorenchyma comprises three to five cell layers in the cortex. Secretory cavities with thin walls are present in the ground parenchyma of all the morpho-types, particularly in *A. kerrii* (Fig. 5h). The vascular tissue of the petioles in all morpho-types consists of three bicollateral bundles. The largest cross-sectional petiole size is shown in *A. variabilis* and the smallest in *Argyreia* 1 (Table 4).

Cluster analysis

The phenogram of *Argyreia* morpho-types (Fig. 6) using 24 anatomical traits (Table 2) was constructed by UPGMA algorithm, for which the cophenetic correlation coefficient (or R value) was equal to 0.87. The high R value (>0.8) indicated the suitability of morpho-types clustering. From the phenogram, four major clusters (A, B, C, and D) were distinctly recognized by similarity coefficient from at least 0.60 and up to 0.70 (Fig. 6). The first cluster A comprises two accessions of *A. breviscapa;* it had a high similarity coefficient around 0.70. Cluster B contains only one morpho-type, *A. kerrii,* whereas the adjacent cluster C was separated from *A. kerrii* at similarity coefficient around 0.60. The cluster C contains two samples of *A. variabilis* that had a similarity coefficient about 0.71. The last cluster D was separated from the other clusters by a similarity coefficient about 0.72, and formed two subgroups. The subgroup E comprises two accessions of *A. suddeeana* that had a high similarity coefficient around 0.88. The second subgroup, F, is composed of two *Argyreia* morpho-types which are a single accession of *Argyreia* 1, and a group of five samples

Table 3 Leaf epidermal characters of all morpho-types in summary

Name	Adaxial epidermis						Abaxial epidermis					
	Stomatal index (S.I.)[a]	Epidermal cell wall	Cuticle	Stomatal type	Non glandular trichome and length (μm)	Density of trichome	Stomatal index (S.I.)[a]	Epidermal cell wall	Cuticle	Stomatal type	Non glandular trichome and length (μm)	Density of trichome
A. breviscapa	8.92 ± 2.76	Str-cur	Striated	Ani-para	LS = 546–820 SS = 311–440	+	24.92 ± 3.35	Str-cur	Striated	Ani-para	LS = 569–1291 SS = 249–487	++
A. breviscapa	13.68 ± 4.45	Str-cur	Striated	Ani-para	LS = 563–981 SS = 296–444	+	25.33 ± 6.87	Str-cur	Striated	Para	LS = 1053–1499 SS = 217–462	++
A. kerrii	2.42 ± 1.45	Str-cur	Striated	Para	LS = 512–738 SS = 280–402	++	22.87 ± 3.63	Sinuate	Striated	Para	LS and SC NT	+++
A. suddeeana	6.04 ± 2.23	Str-cur	Striated	Ani-para	LS = 515–945 SS = 128–410	+	26.84 ± 6.77	Sinuate	Striated	Para	LS = 583–1599 SS = 110–451	+++
A. suddeeana	7.19 ± 3.14	Str-cur	Striated	Ani-para	LS = 525–1037 SS = 149–313	+	25.70 ± 1.39	Sinuate	Striated	Para	LS = 667–1478 SS = 172–446	+++
A. variabilis	3.54 ± 1.05	Str-cur	Striated	Para	LS = 565–1100 SS = 155–478	+	33.21 ± 6.65	Str-cur	Striated	Para	LS = 796–1616 SS = 259–440	+++
A. variabilis	0.58 ± 1.32	Str-cur	Striated	Para	LS = 665–1295 SS = 145–420	+	25.46 ± 4.10	Str-cur	Striated	Para	LS = 948–1625 SS = 206–432	+++
Argyreia 1	4.37 ± 2.05	Sinuate	Striated	Ani-para	LS = 572–1042 SS = 142–391	+	22.22 ± 3.88	Sinuate	Striated	Para	LS = 617–1583 SS = 207–434	+++
Argyreia 2	5.21 ± 2.18	Str-cur	Striated	Ani-para	LS = 528–1073 SS = 197–428	+	27.68 ± 3.85	Sinuate	Striated	Para	LS = 708–1496 SS = 154–310	+++
Argyreia 2	10.71 ± 4.13	Str-cur	Striated	Ani-para	LS = 548–903 SS = 183–363	+	29.56 ± 3.08	Sinuate	Striated	Para	LS = 713–1673 SS = 173–305	+++
Argyreia 2	9.56 ± 1.32	Str-cur	Striated	Ani-para	LS = 551–1304 SS = 131–306	+	25.00 ± 2.77	Sinuate	Striated	Para	LS = 884–1994 SS = 144–424	+++
Argyreia 2	8.03 ± 2.47	Str-cur	Striated	Ani-para	LS = 544–824 SS = 173–390	+	24.42 ± 6.45	Sinuate	Striated	Para	LS = 685–1452 SS = 144–359	++
Argyreia 2	9.49 ± 4.75	Str-cur	Striated	Ani-para	LS = 523–945 SS = 160–318	+	25.00 ± 3.46	Sinuate	Striated	Para	LS = 661–1497 SS = 148–386	++

NT numerous trichome, *Str-cur* straight to slightly curved, *Ani-para* anisocytic and paracytic, *Para* paracytic, *10X* trichome density, + 1–5 trichomes, ++ 6–10 trichomes, +++ more than 10 trichomes, *LS* large straight trichome, *SS* small straight trichome, *SC* small curly trichome

[a] Mean ± SD

of *Argyreia* 2 that mutually diverged from each other with the similarity coefficient of 0.74. Five accessions of *Argyreia* 2 were obviously separated from *Argyreia* 1 at around 0.81 of similarity coefficient. Moreover, *Argyreia* 2 was subdivided into one subgroup (accession numbers 9 and 11) with high similarity coefficient of 0.88, and a second subgroup (accession numbers 10, 12 and 13) with the same value of similarity (0.88).

Principal component analysis

The PCA resulting from examination of 24 anatomical characters (Table 5) from 13 *Argyreia* accessions representing six morpho-types showed the scatter plots of four clusters (Fig. 7). The two accessions of *A. breviscapa* in cluster A, and also both samples of *A. variabilis* in cluster C were quite loosely grouped, assuming that they were affected by the first component, while a single accession of *A. kerrii* was isolated from them. The large cluster, D, was formed into two subgroups, E and F. The morpho-types of subgroup E were closely grouped together, whereas those in subgroup F were widely separated by the first component.

The first two components accounted respectively for 28 and 21.9% of the total variance (Table 5). In the first component, the one character with high positive factor loading was stomatal index on the abaxial leaf surface (13), while high negative factor loading were the ratio of non-glandular trichome size on abaxial leaf surface (16); ratio of petiole transverse section (18); and blade thickness (23). The second component was weighted heavily positive for epidermal cell walls on abaxial leaf surface (2), and ratio of epidermal cell size in petiole (17), whereas it was negative for ratio of glandular trichome size on abaxial leaf surface (15), and ratio of non-glandular trichome size on abaxial leaf surface (16).

Discussion and conclusions

The anatomical characters that are informative for identifying these six morpho-types are the epidermal cell walls and trichome types on the leaf surfaces. Moreover, the characters from leaf transverse section were also useful to distinguish each morpho-type such as the shape of adaxial epidermal cells of the leaf blades, the shape of leaf margins and the outline of petioles.

Epidermal cell wall

Two types of epidermal cell walls are found on both surfaces: straight to slightly curved, and sinuate. This finding is in agreement with the investigations by Metcalfe and Chalk (1950) and Tayade and Patil (2003, 2012c). *Argyreia* 1 is the only one that possesses sinuate epidermal cell walls on the adaxial surface, whereas most of them have this type of epidermal cell wall on the abaxial side, except *A. breviscapa* and *A. variabilis*. *Argyreia breviscapa* is quite similar to *Argyreia* 1 in terms of morphological characters, especially when compared as pressed and dried herbarium specimen. This study shows that observation of the epidermal cell walls can separate *A. breviscapa* from *Argyreia* 1, and cluster analysis (CA) further revealed that these two morpho-types are placed in different clusters with similarity coefficient of 0.53 (Fig. 6). This suggests that the two morpho-types are not the same taxon. Further investigations, using different techniques such as molecular genetics, are required for more accurate characterization.

Type of stomata and stomatal index

Plants in the family Convolvulaceae generally have amphistomatic leaves with paracytic type of stomata, rarely anisocytic (Metcalfe and Chalk 1950; Tayade and Patil 2003, 2012b). The results from this study agreed with the previous researches; nevertheless, anisocytic stomata were present in four morpho-types. The stomatal frequency on both leaf surfaces varies in each morpho-type; this is considered systematically significant and may be connected with the ecology of the plants (Stace 1965). Commonly, the stomatal index (SI) might vary depending on light intensity, humidity, and level of air pollution (Nikolić and Mitić 1991). However, both the stomatal frequency and stomatal index were treated as useful characters for diagnostic purposes (Metcalfe and Chalk 1979).

Trichomes and cuticular pattern on leaf surfaces

Trichomes are known to play essential roles in protecting plants from chilling, drought, UV radiation, disease infection, and particularly against insect herbivory (Liakoura et al. 1997; Pfeiffer et al. 2003). Furthermore, trichome density may vary due to genetic as well as environmental factors, including water and mineral nutrients or the accumulation of water which alters leaf size (Johnson 1975; Wilkens et al. 1996; Roy et al. 1999).

Both glandular and non-glandular trichomes are present in *Argyreia*. Trichomes are normally present on both leaf surfaces in the six morpho-types studied. Finding glandular peltate trichomes in all morpho-types corresponded to reports in the previous studies by Tayade and Patil (2012a, c). Non-glandular trichomes, both large and small straight unicellular types, are found in all the morpho-types, but the large straight and small curly trichomes occur only on the abaxial surface of *A. kerrii*. The CA result also confirmed that *A. kerrii* is separated from cluster C with a 0.602 similarity coefficient (Fig. 6).

Cuticular striations occur in all six morpho-types studied. The diversity of patterning on the cuticular layers can be conveniently employed in taxonomic delineations in this family, as suggested by Tayade and Patil (2003). Our

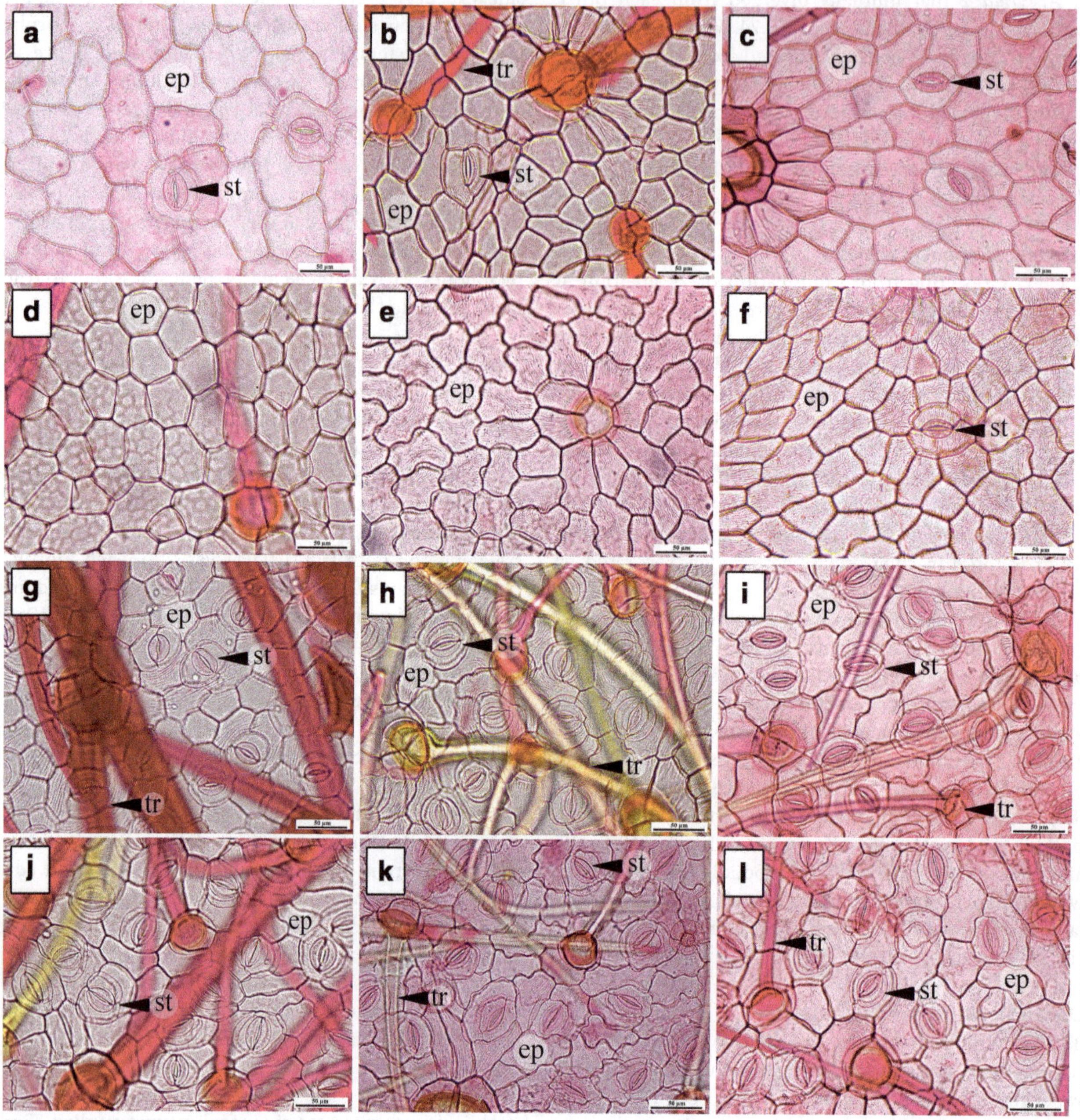

Fig. 1 The epidermal cell walls on adaxial (**a**–**f**) and abaxial leaf surfaces (**g**–**l**) under light microscope (LM). **a** *A. breviscapa*, **b** *A. kerrii*, **c** *A. suddeeana*, **d** *A. variabilis* and **f** *Argyreia* 2, straight to slightly curved walls. **e** *Argyreia* 1, sinuate walls. **g** *A. breviscapa* and **j** *A. variabilis*, straight to slightly curved. **h** *A. kerrii*, **i** *A. suddeeana*, **k** *Argyreia* 1 and **l** *Argyreia* 2, sinuate (*ep* epidermal cell, *st* stomata, *tr* trichome). *Scale bar* 50 μm

recent paper showed that the cutin layer can distinguish two morphologically similar species, *A. albiflora* and *A. wallichii* (Staples et al. 2015).

Transverse sections of leaf blade and petiole

Characters obtained from leaf transverse sections in this research are in agreement with previous studies (Metcalfe and Chalk 1950; Tayade and Patil 2012b, c) that demonstrated all *Argyreia* leaves are dorsiventral; the leaf blades of every morpho-type have one layer of epidermis; and the mesophyll is composed of palisade and spongy cells. Additionally, there are some dark-staining deposits, which could be flavonoids or alkaloids, in some of the epidermal cells on the adaxial surface. This occurrence corresponded to the study of loline alkaloids in *Argyreia mollis* (Burm.f.) Choisy, which also can be found in roots

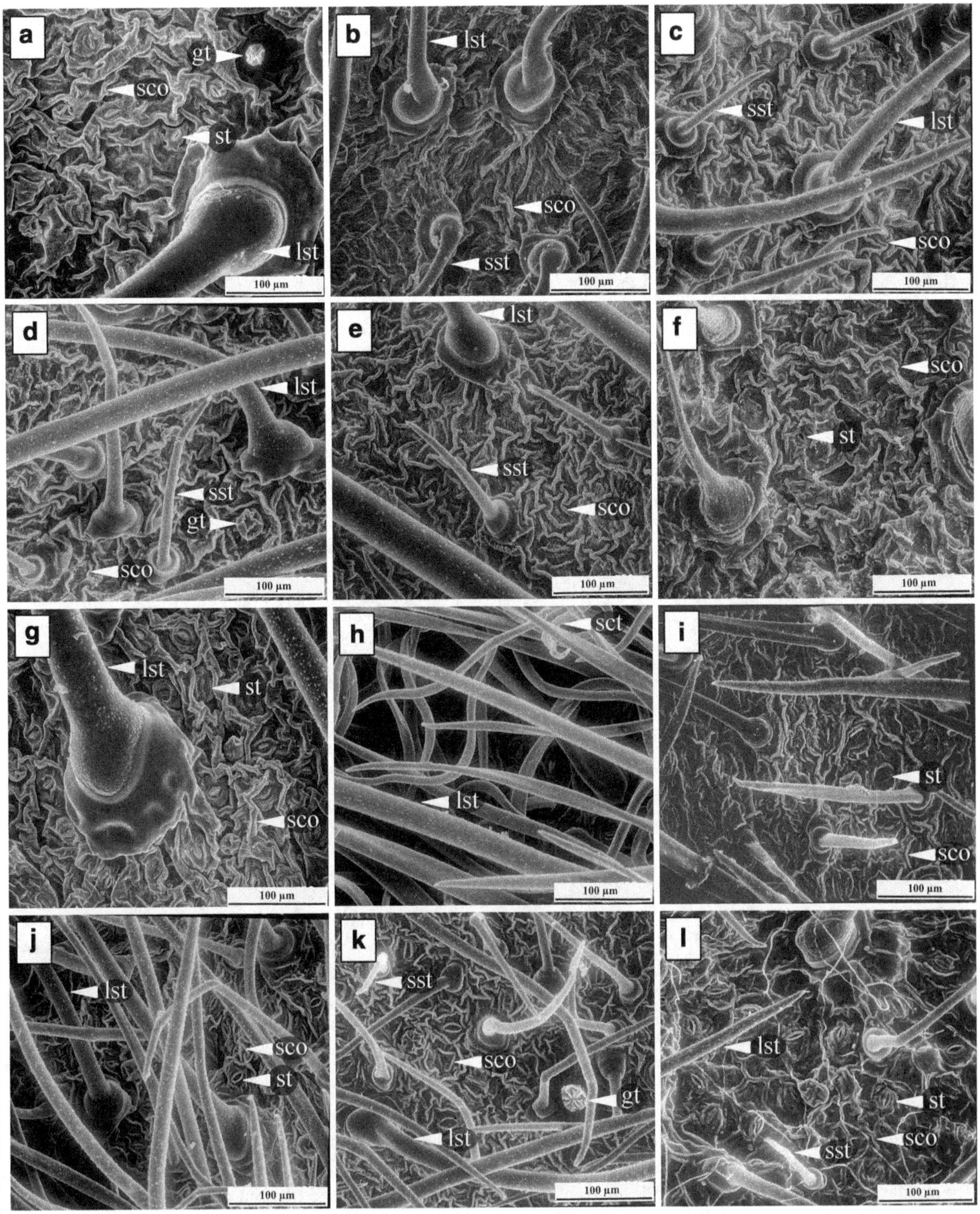

Fig. 2 Cuticle ornamentation on adaxial (**a–f**) and abaxial leaf surfaces (**g–l**) from SEM. **a**, **g** *A. breviscapa*, **b**, **h** *A. kerrii*, **c**, **i** *A. suddeeana*, **d**, **j** *A. variabilis*, **e**, **k** *Argyreia* 1 and **f**, **l** *Argyreia* 2 (*gt* glandular trichome, *lst* large straight trichome, *sst* small straight trichome, *sct* small curly trichome, *st* stomata, *sco* striate cuticular ornamentation). *Scale bar* 100 µm

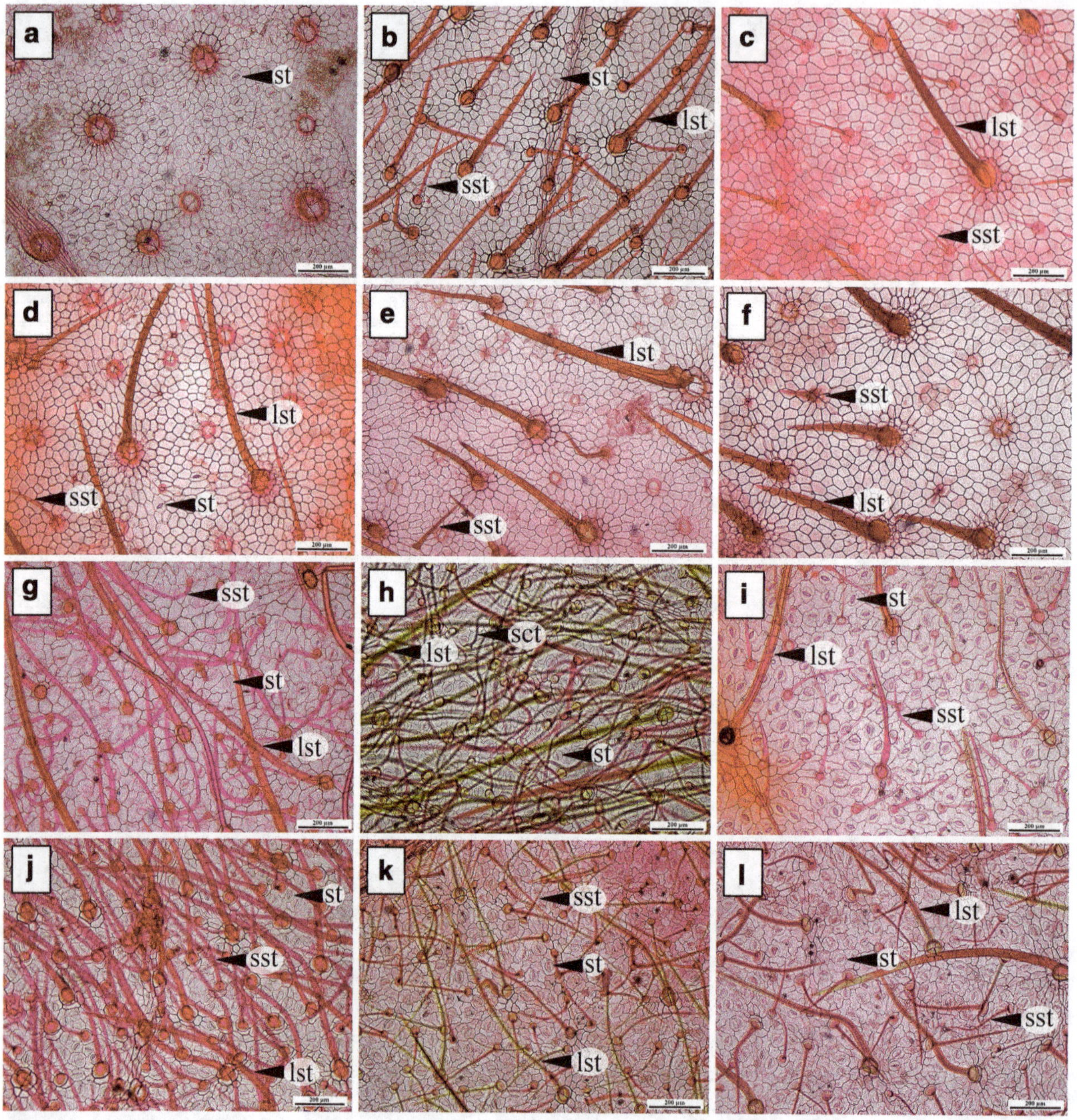

Fig. 3 Trichomes on adaxial (**a–f**) and abaxial leaf surfaces (**g–l**) under light microscopy (LM). **a**, **g** *A. breviscapa*, **b**, **h** *A. kerrii*, **c**, **i** *A. suddeeana*, **d**, **j** *A. variabilis*, **e**, **k** *Argyreia* 1 and **f**, **l** *Argyreia* 2 (*lst* large straight trichome, *sst* small straight trichome, *sct* small curly trichome, *st* stomata). *Scale bar* 200 μm

and aerial vegetative parts of plants (Tofern et al. 1999; Mann et al. 1999; Tayade and Patil 2012b).

The epidermal cell shapes of leaf blade in transverse section of *A. variabilis* show long and narrow shapes and its leaf margins are acute, which is different from the other morpho-types. This coincides with phenetic result that show two accessions of *A. variabilis* separated from cluster B (Fig. 6) in cluster analysis by similarity coefficient of 0.602 and forming a distinct cluster C. The epidermal cell shape character and results from phenetic analyses can be used to point out that *Argyreia* 2 is different from *A. variabilis*.

The vascular tissue of the petiole in the family Convolvulaceae commonly exhibits bicollateral bundles (Metcalfe and Chalk 1950; Tayade and Patil 2012a, c). The results from this research correspond to the prior studies.

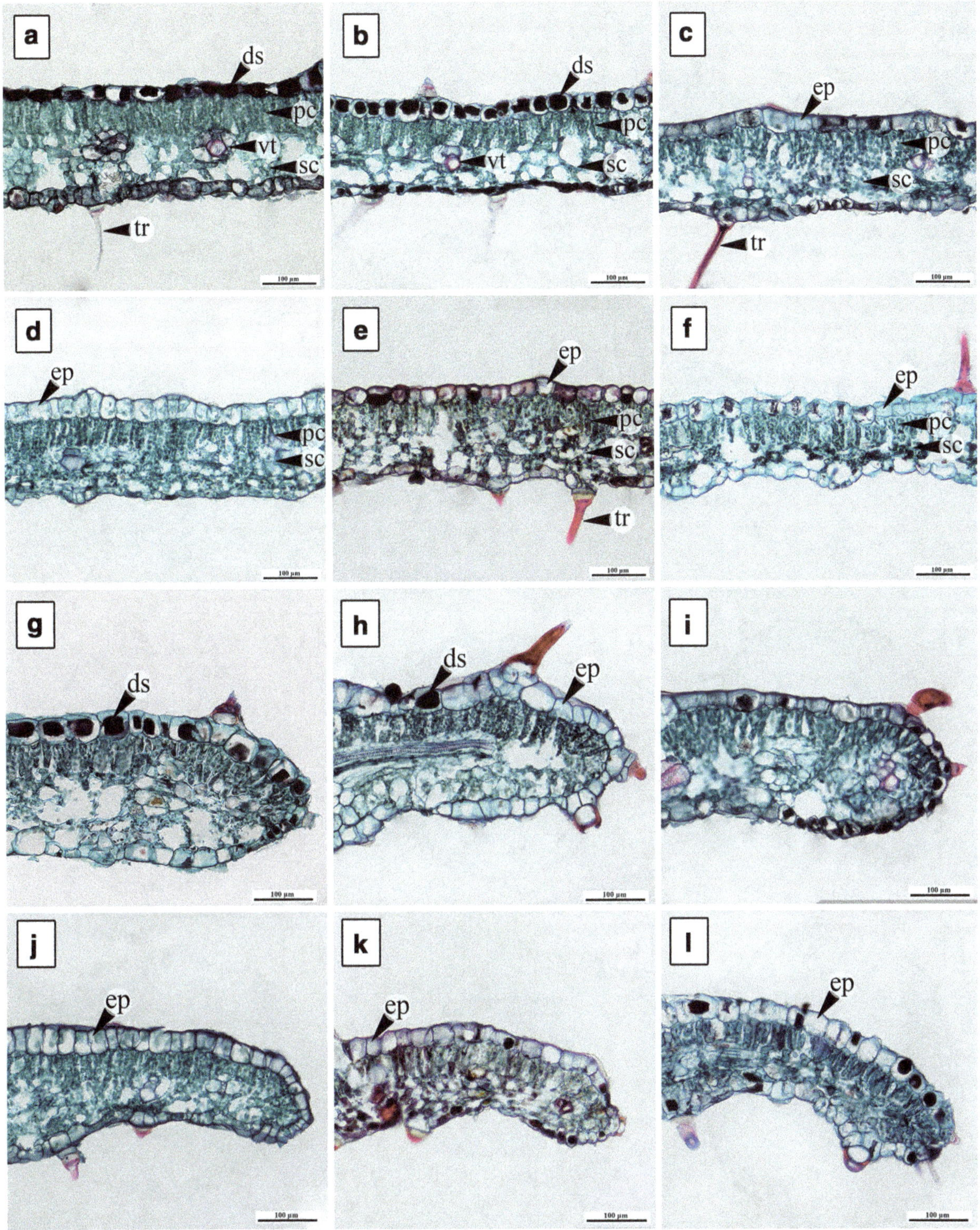

Fig. 4 Transverse sections of leaf blade (**a**–**f**) and leaf margin (**g**–**l**). **a**, **g** *A. breviscapa*, **b**, **h** *A. kerrii*, **c**, **i** *A. suddeeana*, **d**, **j** *A. variabilis*, **e**, **k** *Argyreia* 1 and **f**, **l** *Argyreia* 2 (*ds* dark-staining deposit, *ep* epidermal cell, *pc* palisade cell, *sc* spongy cell, *tr* trichome, *vt* vascular tissue). *Scale bar* 100 µm

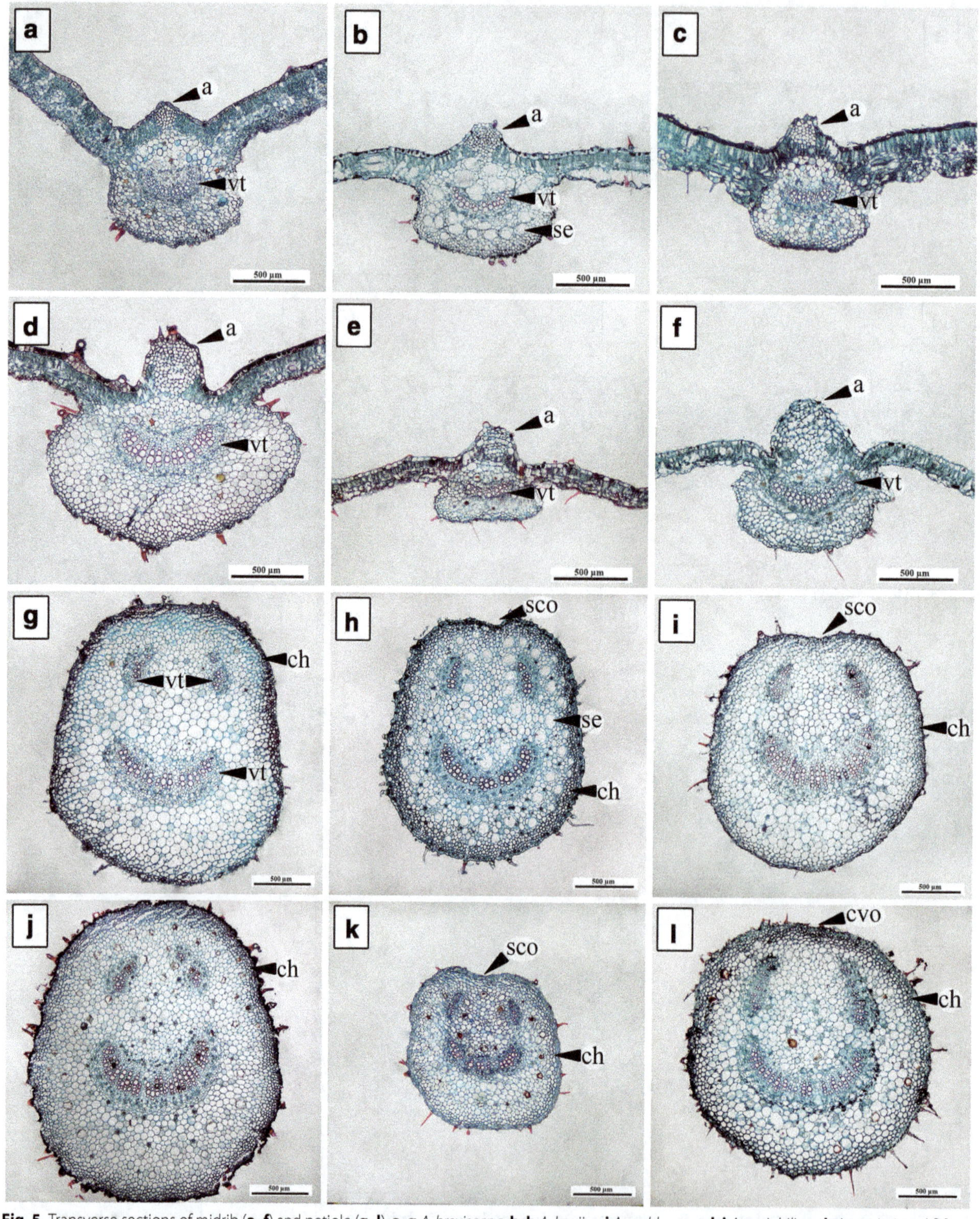

Fig. 5 Transverse sections of midrib (**a**–**f**) and petiole (**g**–**l**). **a**, **g** *A. breviscapa*, **b**, **h** *A. kerrii*, **c**, **i** *A. suddeeana*, **d**, **j** *A. variabilis*, **e**, **k** *Argyreia* 1 and **f**, **l** *Argyreia* 2 (*a* adaxial curvature, *ch* chlorenchyma cell, *cvo* convex outline, *sco* slightly concave outline, *se* secretory cavity, *vt* vascular tissue). *Scale bar* 500 µm

Table 4 Leaf and petiole transverse section characters in summary

Name	Petiole outline	Cross-sectional petiole size (mm)		Midrib outline and adaxial curvature	Cross-sectional midrib size (mm)		Leaf blade width (μm)	Leaf margins
		Width	Height		Width	Height		
A. breviscapa	Convex	2.02 ± 0.05	2.10 ± 0.04	Convex with acute adaxial curvature	1.17 ± 0.24	1.03 ± 0.08	269.40 ± 35.64	Round
A. breviscapa	Convex	1.98 ± 0.02	2.28 ± 0.11	Convex with acute adaxial curvature	1.19 ± 0.08	1.14 ± 0.07	211.77 ± 7.84	Round
A. kerrii	Concave	1.73 ± 0.03	1.76 ± 0.23	Convex with round adaxial curvature	1.01 ± 0.06	0.95 ± 0.05	198.70 ± 45.96	Round
A. suddeeana	Concave	2.01 ± 0.17	2.06 ± 0.12	Convex with round adaxial curvature	1.40 ± 0.29	1.21 ± 0.15	183.03 ± 3.46	Round
A. suddeeana	Concave	1.51 ± 0.15	1.65 ± 0.12	Convex with round adaxial curvature	0.96 ± 0.31	0.92 ± 0.08	291.23 ± 28.90	Round
A. variabilis	Convex	2.24 ± 0.09	2.77 ± 0.39	Convex with round adaxial curvature	1.33 ± 0.12	1.34 ± 0.12	177.60 ± 29.23	Acute
A. variabilis	Convex	1.98 ± 0.58	2.34 ± 0.59	Convex with round adaxial curvature	1.52 ± 0.20	1.26 ± 0.15	244.87 ± 44.14	Acute
Argyreia 1	Concave	1.47 ± 0.18	1.24 ± 0.15	Convex with round adaxial curvature	0.85 ± 0.26	0.71 ± 0.10	162.53 ± 11.59	Round
Argyreia 2	Convex	1.99 ± 0.19	2.20 ± 0.02	Convex with round adaxial curvature	1.34 ± 0.17	1.26 ± 0.13	191.53 ± 9.76	Round
Argyreia 2	Convex	1.44 ± 0.09	1.82 ± 0.02	Convex with round adaxial curvature	0.91 ± 0.09	0.94 ± 0.06	171.83 ± 22.91	Round
Argyreia 2	Convex	1.52 ± 0.12	1.84 ± 0.04	Convex with round adaxial curvature	0.98 ± 0.12	1.05 ± 0.07	174.57 ± 33.96	Round
Argyreia 2	Convex	1.43 ± 0.09	1.83 ± 0.07	Convex with round adaxial curvature	0.91 ± 0.10	1.05 ± 0.01	156.73 ± 14.62	Round
Argyreia 2	Convex	1.73 ± 0.03	1.73 ± 0.23	Convex with round adaxial curvature	0.95 ± 0.09	0.92 ± 0.04	220.97 ± 16.15	Round

Shape of the *Argyreia* petiole in transverse section is convex (Fig. 5g, j, l) or slightly concave (Fig. 5h, i, k) on the adaxial side, as also seen in the study of genus *Merremia* Dennst. ex. Endl. (Pisuttimarn et al. 2013). In our results, half the taxa have convex petiole shape (*A. breviscapa, A. variabilis,* and *Argyreia* 2), whereas in *A. kerrii, A. suddeeana* and *Argyreia* 1, petioles have a concave shape (about 1/6 of the length of the petiole). The result from CA indicated that *A. suddeeana, Argyreia* 1, and *Argyreia* 2 were grouped together in the same cluster (D), however, *A. suddeeana* was separated from this group forming a subgroup E, at similarity coefficient

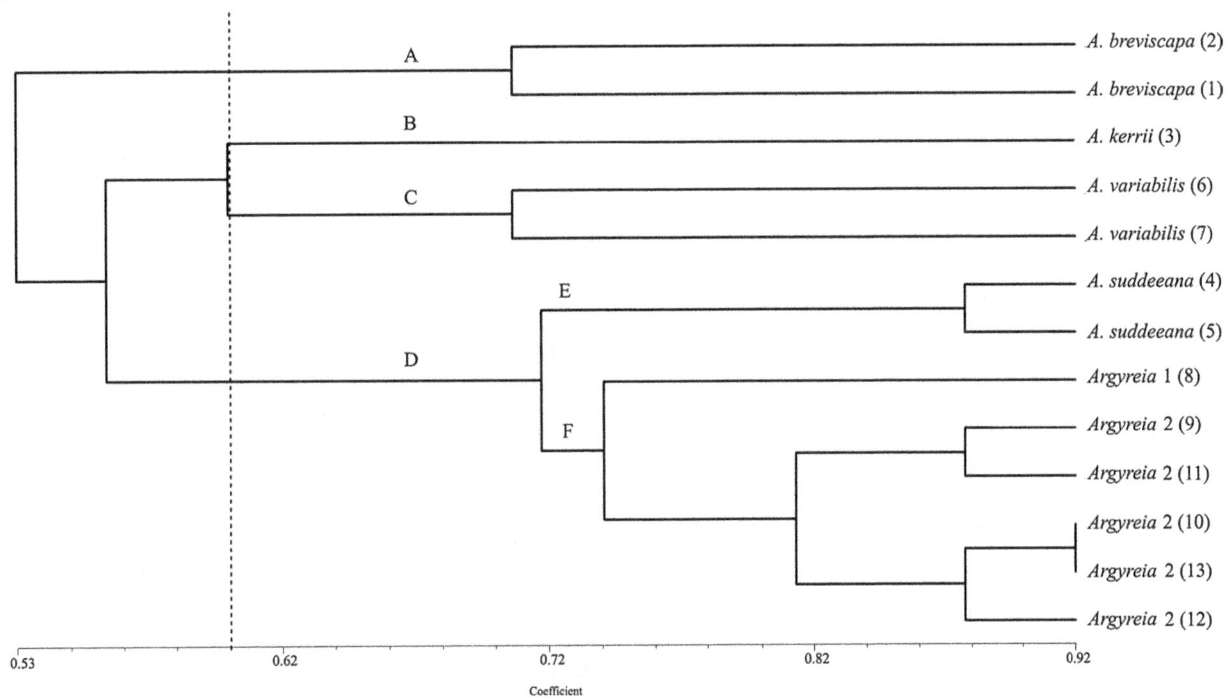

Fig. 6 Phenogram for six *Argyreia* morpho-types from cluster analysis based on UPGMA method using 24 anatomical traits; the *vertical dashed line* indicates similarity coefficient; parenthetical numbers (*1*)–(*13*) indicate accession number of plant materials cited in Table 1

about 0.718 (Fig. 6). Therefore, the outline of petiole can be used to separate *A. suddeeana* from *Argyreia* 2.

Druse crystals (calcium oxalate) are commonly found in petiole, midrib, and leaf blade of all six morpho-types, which agrees with the previous investigations by Metcalfe and Chalk (1950), Sayeedud-Din (1953) and Tayade and Patil (2012a, c). The druse crystal has many functions in plants including calcium regulation, ion balance, tissue support, plant protection, detoxification and light assembly and reflection (Franceschi and Horner 1980; Nakata 2003). Recent research about druse crystals in a C_4 plant, *Amaranthus hybridus* L., revealed that the plant used calcium oxalate crystals as a non-atmospheric carbon source in photosynthesis assimilation during the day, especially in drought condition (Tooulakou et al. 2016). In the daytime, druse crystals were degraded and provided subsidiary carbon used in photosynthesis under ambient conditions that caused stomatal closure (for preventing water loss). Moreover, these crystals were recovered during the night as a biochemical carbon reservoir.

Table 5 Factor loading and cumulative eigenvalue percentage received from 24 leaf anatomical characters

Anatomical characters	Component	
	1	2
1. Epidermal cell wall on adaxial leaf surface	0.024	−0.086
2. Epidermal cell wall on abaxial leaf surface	0.071	0.397
3. Types of non-glandular trichome on abaxial leaf surface	0.051	0.059
4. Shape of petiole	0.076	0.131
5. Midrib outline and adaxial curvature	−0.217	0.153
6. Leaf margin shape	0.146	0.245
7. Ratio of adaxial leaf epidermal cell size	0.051	0.059
8. Stomatal index (SI) on adaxial leaf surface	−0.193	0.067
9. Ratio of stomatal size on adaxial leaf surface	−0.124	0.281
10. Ratio of glandular trichome size on adaxial leaf surface	−0.202	0.118
11. Ratio of non-glandular trichome size on adaxial leaf surface	0.000	−0.163
12. Ratio of abaxial leaf epidermal cell size	0.093	0.172
13. Stomatal index (SI) on abaxial leaf surface	0.222	0.159
14. Ratio of stomatal size on abaxial leaf surface	0.043	0.113
15. Ratio of glandular trichome size on abaxial leaf surface	−0.197	−0.304
16. Ratio of non-glandular trichome size on abaxial leaf surface	−0.448	−0.229
17. Ratio of epidermal cell size in petiole	0.122	0.390
18. Ratio of petiole transverse section	−0.430	0.026
19. Ratio of palisade mesophyll cell size in leaf transverse section	0.076	0.063
20. Ratio of spongy mesophyll cell size in leaf transverse section	−0.114	0.284
21. Ratio of adaxial epidermal cell size in leaf transverse section	0.177	0.103
22. Ratio of abaxial epidermal cell size in leaf transverse section	0.108	−0.245
23. Blade thickness	−0.504	0.243
24. Ratio of midrib thickness	0.083	−0.156
Eigenvalue	1.3776	1.0784
Cumulative % of eigenvalue	28	49.9

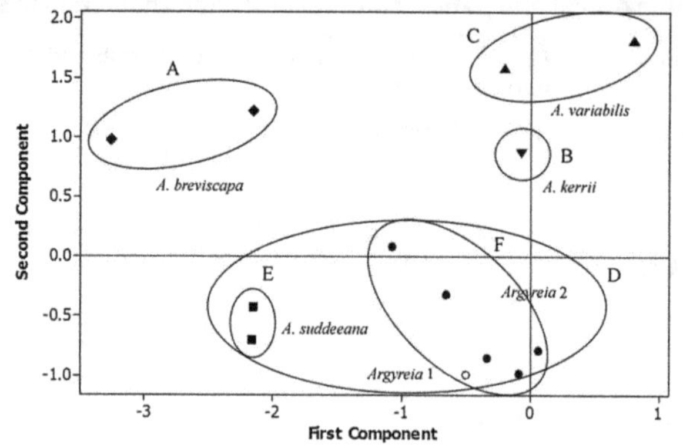

Fig. 7 The scattered plot of first and second components obtained from principal component analysis (PCA) based on 24 anatomical traits. Group *A, B, C, D* and subgroups *E* and *F* corresponding to the phenogram in Fig. 6

Numerous secretory cavities in the ground tissue of the petiole and midrib are also observed in *A. kerrii*, in agreement with previous researches, which reported that plants in this family possess secretory cavities in the ground tissue of leaf, petiole, and stem (Metcalfe and Chalk 1950; Sayeedud-Din 1953; Tayade and Patil 2012a, c).

Although the results from this study were consistent with earlier anatomical studies they incrementally increase the fundamental knowledge base about the genus *Argyreia* and point the way for further study. Rather few plant samples in genus *Argyreia* were examined previously; in this study, we investigated several more *Argyreia* taxa that have not been studied before. Furthermore, our data demonstrate that anatomical traits can provide informative characters that have a useful role for plant identification. *Argyreia* has proven to be an extraordinarily complex genus and species delimitation is difficult, especially in the complexes of vegetatively similar species where floral characters have been the only reliable method to discriminate taxa. Accordingly, evidence from methods and techniques outside the usual morphological ones must be explored to determine whether they afford taxonomically informative characters. Since the genus *Argyreia* is one of the larger genera in the family Convolvulaceae comprising around 135 species (Staples and Traiperm 2017) the results from our study expand the baseline for understanding morphological and anatomical diversity within and between species.

Authors' contributions

PT created the study topic, collected the plants, planned the study, guided all lab work, drafted, revised and developed the manuscript, prepared the manuscript for submission. JC analyzed the anatomical data and drafted some parts of the manuscript. PN analyzed the CA and PCA data and drafted some parts of the manuscript. GWS planned the study, guided taxonomic identifications, provided important taxonomic literatures, drafted, revised, developed the manuscript and polished the English. SCS guided lab work. All authors read and approved the final manuscript.

Author details

[1] Department of Plant Science, Faculty of Science, Mahidol University, Rama VI Road, Ratchathewi, Bangkok 10400, Thailand. [2] Department of Pharmaceutical Botany, Faculty of Pharmacy, Mahidol University, Bangkok 10400, Thailand. [3] HRA, Herbarium, Royal Botanic Gardens, Kew, Richmond, Surrey TW9 3AE, UK.

Acknowledgements

We would like to thank the Department of Plant Science, Faculty of Science, Mahidol University for laboratory facilities. Thanks also to Wirata Rujichaipimon for help in preparing the colour photos. Traiperm wishes to acknowledge the Thailand Research Fund (TRF), RSA5880022 and the Faculty of Science, Mahidol University for financial support. GS gratefully acknowledges funding support from Faculty of Graduate Studies, Mahidol University, which facilitated our collaboration.

Competing interests

The authors declare that they have no competing interests.

Funding

The Thailand Research Fund (TRF), RSA5880022, the Faculty of Science and Faculty of Graduate Studies, Mahidol University.

References

Aghababaeyan E, Pakravan M, Ghahremaninejad F (2014) Anatomical study and pollen micromorphology of *Onopordum* L. in Iran. Notulae Scientia Biologicae 6(1):66–76

Arthan W, Traiperm P, Gale SW, Norsaengsri M, Kethirun L (2016) Re-evaluation of the taxonomic status of Hackelochloa (Poaceae) based on anatomical and phenetic analyses. Bot J Linn Soc 181(2):224–245

Breitwieser I, Ward JM (1993) Systematics of New Zealand Inuleae (Compositae-Asteraceae)-3 numerical phenetic analysis of leaf anatomy and flavonoids. N Z J Bot 31(1):43–58

De Faria AP, Wendt T, Brown GK (2010) A revision of *Aechmea* subgenus *Macrochordion* (Bromeliaceae) based on phenetic analyses of morphological variation. Bot J Linn Soc 162(1):1–27

Franceschi VR, Horner HT (1980) Calcium oxalate crystals in plants. Bot Rev 46(4):361–427

Jayarathna SP, Senanayake SP, Rajapakse S, Jayasekera LR (2016) Phenetic variation and preliminary phytochemical screening of *Piper* species in Sri Lanka. J Agric Sci 11(3):155–163

Johansen DA (1940) Plant microtechnique, 1st edn. McGraw-Hill Book, New York

Johnson HB (1975) Plant pubescence: an ecological perspective. Bot Rev 41:233–258

Liakoura V, Stefanou M, Manetas Y, Cholevas C, Karabourniotis G (1997) Trichome density and its UV-B protective potential are affected by shading and leaf position on the canopy. Environ Exp Bot 38(3):223–229

Mann P, Tofern B, Kaloga M, Eich E (1999) Flavonoid sulfates from the Convolvulaceae. Phytochemistry 50(2):267–271

Metcalfe CR, Chalk L (1950) Convolvulaceae. In: Metcalfe CR, Chalk L (eds) Anatomy of the dicotyledons, vol 2. Clarendon Press, Oxford, pp 954–964

Metcalfe CR, Chalk L (1979) Anatomy of the dicotyledons, vol 1, 2nd edn. Clarendon Press, Oxford

Minitab (2010) Minitab 16 statistical software. Minitab Inc, State College

Moraes TM, Rabelo GR, Alexandrino CR, Silva Neto SJ, Da Cunha M (2011) Comparative leaf anatomy and micromorphology of *Psychotria* species (Rubiaceae) from the Atlantic Rainforest. Acta Bot Bras 25(1):178–190

Nakata PA (2003) Advances in our understanding of calcium oxalate crystal formation and function in plants. Plant Sci 164(6):901–909

Nikolić T, Mitić B (1991) Phenetic relationships within populations of *Iris illyrica*, *I. pallida* and *I. pseudopallida* (Iridaceae) with regard to morphological characteristics of epidermis. Acta Bot Croat 50(1):99–106

Pant DD, Bhatnagar S (1975) Morphological studies in *Argyreia* Lour. (Convolvulaceae). Bot J Linn Soc 70(1):45–69

Pfeiffer TW, Peyyala R, Ren Q, Ghabrial SA (2003) Increased soybean pubescence density: yield and soybean mosaic virus resistance effects. Crop Sci 43:2071–2076

Pisuttimarn P, Traiperm P, Pornpongrungrueng P (2013) Comparative anatomy of *Merremia* section *Xanthips* (Convolvulaceae) in Thailand. In: Proceedings in the national graduate research conference 29th (NGRC29) (in Thai), pp 93–101

Rohlf FJ (2000) NTSYS-pc: numerical taxonomy and multivariate analysis system, version 2.1. Exeter Software, New York

Roy BA, Stanton ML, Eppley SM (1999) Effects of environmental stress on leaf hair density and consequences for selection. J Evol Biol 12(6):1089–1103

Salisbury EJ (1927) On the causes and ecological significance of stomatal frequency, with special reference to the woodland flora. Philos Trans R Soc Lond 216:1–65

Sayeedud-Din M (1953) Observations on the anatomy of some of the Convolvulaceae. Bot J Linn Soc 37:106–109

Stace CA (1965) Cuticular studies as an aid in plant taxonomy. Bull Brit Mus (Nat Hist) Bot 4:1–78

Staples GW, Traiperm P (2008) New species, new combinations, and new records in Convolvulaceae for the Flora of Thailand. Thai Forest Bull Bot 36:86–108

Staples GW, Traiperm P (2017) A nomenclatural review of *Argyreia* (Convolvulaceae). Taxon 66:445–477

Staples GW, Traiperm P, Chow J (2015) Another new Thai *Argyreia* species (Convolvulaceae). Phytotaxa 204(3):223–229

Tayade SK, Patil DA (2003) Foliar epidermal features and their taxonomic significance in the genus *Argyreia* Lour. (Convolvulaceae). J Swam Bot Club 20:15–18

Tayade SK, Patil DA (2012a) Anatomical studies of stem in some species of Convolvulaceae. Int J Sci Pharm Educ Res 1(2):42–46

Tayade SK, Patil DA (2012b) Leaf anatomical studies in some species of Convolvulaceae. Life Sci Leafl 3:64–74

Tayade SK, Patil DA (2012c) The family Convolvulaceae: anatomical conspectus. World J Sci Technol 2(6):42–50

Thadeo M, Azevedo AA, Meira RM (2014) Foliar anatomy of neotropical Salicaceae: potentially useful characters for taxonomy. Plant Syst Evol 300(9):2073–2089

Tofern B, Kaloga M, Witte L, Hartmann T, Eich E (1999) Occurrence of loline alkaloids in *Argyreia mollis* (Convolvulaceae). Phytochemistry 51(8):1177–1180

Tooulakou G, Giannopoulos A, Nikolopoulos D, Bresta P, Dotsika E, Orkoula MG, Kontoyiannis CG, Fasseas C, Liakopoulos G, Klapa MI, Karabourniotis G (2016) Alarm photosynthesis: calcium oxalate crystals as an internal CO_2 source in plants. Plant Physiol 171(4):2577–2585

Traiperm P, Staples GW (2014) A new endemic Thai species of *Argyreia* (Convolvulaceae). Phytotaxa 164(4):281–285

Wilkens RT, Shea GO, Halbreich S, Stamp NE (1996) Resource availability and the trichome defense of tomato plants. Oecologia 106(2):181–191

3

Genome-wide characterization of the *Rab* gene family in *Gossypium* by comparative analysis

Peng Li and Wangzhen Guo*

Abstract

Background: *Rab* protein family is the largest subfamily of small G protein family. As one of the most important families in plant, *Rab* family plays an important role in the process of plant growth and development. So far, the identification of 57 members of the *Rab* family in *Arabidopsis* has been completed. In cotton, the relevant family has not been reported.

Results: Here, we identified 87, 169, 136, 80 *Rabs* in the four sequenced cotton species, *G. raimondii* (D_5), *G. hirsutum* acc. TM-1 (AD_1), *G. barbadense* acc. 3-79 (AD_2) and *G. arboreum* (A_2), respectively. Biological information analysis showed that the number of amino acid is 200–300 aa among *Rab* family members in *G. raimondii* and the protein molecular weight is between 20 and 30 kDa, which is consistent with the characterization of the *Rab* protein itself. 87 *GrRabs* in *G. raimondii* are divided into eight groups. In each group, intron numbers and subcellular localization of *Rab* protein are basically the same. We mapped the distribution of *GrRab* genes on 13 chromosomes of *G. raimondii* except two genes. Among the 87 *GrRabs* in *G. raimondii*, we identified 60 pairs of *GrRabs* formed in whole genome duplication. Among all the gene pairs, the *Ka/Ks* values were less than 1. This indicates that it is the results of the purification selection and will help maintain the conservation of gene in structure and function. Further, 4 of the 87 *GrRabs* showed tandem duplication. They were *GrRabA2a* vs *GrRabD1a* and *GrRabA2h* vs *GrRabD1b* respectively. Expression patterns analysis of 169 *GhRabs* in *G. hirsutum* acc. TM-1 indicates that most *Rab* family members play a certain role in different tissues/organs and different growth stages of cotton, implying their potential function in the polar growth of pollen tube, root hair and fiber cell, as well as improving stress and disease tolerance.

Conclusion: The systematic investigation of *Rab* genes in cotton will lay a foundation for understanding the functional roles of different *Rab* members in the polar growth and stress tolerance.

Keywords: *Rab* gene family, Genome-wide identification, Gene duplication, Phylogenetic relationship, Gene structure, Expression pattern

Background

As the most important cash crop in the world, cotton provides a very important natural fiber for human beings. Therefore, we tried to explore the important genes related to abiotic stress, biotic stress and fiber development from the whole genome of cotton, hoping to improve cotton quality and yield. *Rab* protein family in cotton is a very large family. According to previous reports, *Rab* family is not only closely related to plant growth and development, but also plays an important role in stress and disease resistance.

The world's first *Rab* gene (*YPT1*) was discovered in yeast in 1983 (Gallwitz et al. 1983). Four years later, Salminen and Novick demonstrated that *Ras* (*Sec4*) is involved in vesicle trafficking (Salminen and Novick 1987). In the same year, Tavitian and his colleagues cloned the homologous gene similar to the *SEC4/YPT* gene for the first time through cDNA library in rat brain and named

*Correspondence: moelab@njau.edu.cn
State Key Laboratory of Crop Genetics and Germplasm Enhancement, Nanjing Agricultural University, Nanjing 210095, Jiangsu Province, People's Republic of China

ras-like in rat brain (*Rab*) (Martinez and Goud 1998). In *Arabidopsis thaliana*, 57 Rab proteins were found and divided into *RabA* to *RabH* (Hill and Sylvester 2007). *RabA* to *RabH* in *A. thaliana* correspond to *Rab11, Rab2, Rab1, Rab18, Rab8, Rab5, Rab7* and *Rab6* in animal, indicating that *Rab* family has further differentiation in plant (Pereira-Leal and Seabra 2001; Brighouse et al. 2010).

Rab protein family members have about 200 amino acids. Their sequences are conservative with high sequence similarity. All *Rab* proteins have five typical conserved domains, including four guanine nucleotide binding domains (G1, G3, G4 and G5) and an effector binding domain (G2) (Takai et al. 2001; Agarwal et al. 2009). Four domains of G1, G3, G4 and G5 participate in the binding and hydrolysis of nucleotides. Among them, G1 is the binding site of phosphate or Mg^{2+}, G4 and G5 are key sites involved in GTP-GDP binding and hydrolysis (Stenmark and Olkkonen 2001). Mutations in the important amino acid sites of these conserved domains will cause *Rab* proteins to produce some constitutive inhibiting or activating mutant proteins. Normally, the amino acid sequence at the C end of the *Rab* protein is highly variable, but it ends with two conserved cysteine residues (CC) ultimately. These two highly conserved cysteine residues play an important role in membrane localization and protein function (Rutherford and Moore 2002). Although five G domains of *Rab* protein is discrete distribution in amino acid sequence, they are close to each other to form a special catalytic domain in the three-dimensional conformation of proteins, so as to better exercise the function of a protein (Rutherford and Moore 2002). The main function of the *Rab* protein family member is responsible for intracellular protein transport and they are essential regulator of vesicle trafficking way (Novick and Zerial 1997; Brennwald 2000). Newly synthesized secretory proteins are usually transported from one compartment of the organelle to another membrane through vesicles (Gurkan et al. 2005). They are transported to the endoplasmic reticulum at first, then transported to the plasma membrane through the Golgi apparatus, and some are delivered to the lysosome. In general, *Rab* proteins were involved in various cellular physiological functions of vesicular transport, such as cell polarity, cytokinesis, cell plate formation and so on (Barr 2009). Therefore, vesicular transport includes four steps at least. Vesicles budded from the donor membrane, moved to the receptor, anchored in the membrane receptor, and fused with membrane receptors. These processes all need the active involvement of *Rab* proteins (Tuvim et al. 2001; Yang 2002).

Up to now, the identification of *Rab* protein gene family in *Arabidopsis* (Rutherford and Moore 2002) has been very clear. But the related gene family research has not been reported in cotton. As is known to all, *Rab* gene family is a vital family in plant (Hill and Sylvester 2007). As one of the most important families in cotton, the *Rab* gene family plays a key role in the process of fiber development and biotic and abiotic stresses. With the great progress of genome sequence information in four different cotton species (*G. raimondii, G. arboreum, G. hirsutum* acc. TM-1 and *G. barbadense* acc. 3-79) (Wang et al. 2012; Li et al. 2014; Zhang et al. 2015; Yuan et al. 2015), we tried to mine the important genes related to fiber development and biotic and abiotic stresses in the whole genomic level for their functional analysis.

In this study, combined with the released genome information in four sequenced cotton species, the cotton *Rab* gene family members were systematically studied. We analyzed the characteristics on the gene structures, classification, chromosomal locations, and expression patterns of *Rab* gene family members. Our studies will lay a foundation for understanding the functional roles of different *Rab* members in the polar growth and stress tolerance in cotton.

Methods

Identification on *Rab* genes

Using the known *Rab* amino acid sequences (Accessions NP_187823.1 and NP_199607.1), we searched the seed file of *Rab* structure in the Pfam database (http://www.phytozome.net/) website. The number of seed file is PF00071. After that, the seed file was used to further retrieve the whole genome database of *G. raimondii* (Paterson et al. 2012) by HMMER 3.0 software (Finn et al. 2011). Finally, all obtained *Rab* proteins were used the blastp program on the NCBI site to further validate the conserved domains of proteins. The gene information of the other three cotton species, *G. arboreum, G. hirsutum* acc. TM-1, and *G. barbadense* acc. 3-79, were downloaded from http://cgp.genomics.org.cn, http://mascotton.njau.edu.cn and http://cotton.cropdb.org, respectively.

Chromosomal mapping and gene duplication

We used Mapchart 2.2 software to anchor the *Rab* genes to all 13 chromosomes of *G. raimondii*. The syntenic information of *G. raimondii* was downloaded from the Plant Genome Duplication Database (PGDD; http://chibba.agtec.uga.edu/duplication/). *Rabs* were mapped to the syntenic blocks for intra- and inter-genomic comparison (Xu et al. 2016a, b). The chromosome numbers were integrated with the interspecific genetic map (D1 to D13) in allotetraploid cultivated cotton species (Wang et al. 2015) and the scaffolds (Chr.1 to Chr.13) in the genomic data of *G. raimondii* (Paterson et al. 2012).

The timing of segmental duplication events can be estimated by computing mean *Ks* values for all anchor points located in the corresponding syntenic block (Paterson et al. 2012; Wang et al. 2012), and all the *Ks* values were parsed from PGDD syntenic data. Genes separated by five or fewer genes within a 100-kb region on a chromosome may have resulted from tandem duplication (Wang et al. 2010).

Calculating *Ka* and *Ks*

Coding sequences were aligned using the PRANK codon model with the default options, and alignment gaps were deleted manually (Loytynoja and Goldman 2005). On the basis of the aligned coding sequences, the non-synonymous substitutions per non-synonymous site (*Ka*) and the synonymous substitutions per synonymous site (*Ks*) of homologous gene pairs were computed by the maximum likelihood method in Codeml from the PAML package v4.7 (Yang 2007).

Basic bioinformatics analysis

We use ExPASy online program (http://web.expasy.org/compute_pi/) to predict molecular weight and isoelectric point of all the *Rab* genes. Then we use Softberry online website (http://www.softberry.com) for subcellular localization prediction (Zhang et al. 2016).

Phylogenetic and exon–intron structural analysis

Through the ClustalX software (http://www-igbmc.u-strasbg.fr/BioInfo/), all amino acid sequences of *Rab* proteins obtained were used for alignment analysis (Thompson et al. 1994). Then the results were saved with the MSF format. Later, the results were edited with the Genedoc software (http://www.psc.edu/biomed/genedoc/). We used MEGA5.0 software (http://www.megasoftware.net/index.html) to construct the phylogenetic tree of *Rab* proteins in cotton. Checking parameter Bootstrap was set to 1000 with NJ method for operation (Saitou and Nei 1987). We compared the cDNA of *Rabs* in cotton with DNA sequence and analyzed the gene structure by using online site (http://wheat.pw.usda.gov/piece/GSDraw.php).

RNA-seq data analysis

The RNA-seq data for *Rab*s expression profile analysis was derived from TM-1 transcriptome data of Zhang et al. (2015). The data used for spatial–temporal expression analysis mainly includes the expression data of tissues and organs at different stages of growth and development.

Using RNA-seq data in the different tissues of TM-1, we filtered out the reads of multiple loci mapping, and only gene-specific read counts with unique mapping were remained to calculate the gene FPKM value, where FPKM referred to fragments per kilobase of exon model per million mapped reads with Cufflinks software (http://cufflinks.cbcb.umd.edu/). Analysis method of data model expression as follows: if the value of $\log_2$ FPKM is more than 1, we consider it to be expressed and if the value of $\log_2$ FPKM is less than or equal to 1, we consider it not to be expressed (Xu et al. 2016a, b).

Results

Genome-wide identification of the *Rab* gene family in *Gossypium*

Using the four sequenced cotton species (*G. raimondii, G. hirsutum* acc. TM-1, *G. barbadense* acc. 3-79 and *G. arboreum*), genome-wide identification of the *Rab* gene family were carried out in *Gossypium*. With several known amino acid sequences of the *Rab* gene family members, seed file (PF00071) about the *Rab* domain was queried on the website of Pfam data base. Then the protein databases of *Gossypium* were searched using HMMER 3.0 software. At last, the *Rab* domain was confirmed by a blastp program, and 87 *GrRabs*, 169 *GhRabs*, 136 *GbRabs* and 80 *GaRabs* were obtained from the four cotton species, respectively (Additional file 1).

In allotetraploid cotton *G. hirsutum* acc. TM-1, the number of *Rab* genes is about twice as many as in diploid cotton *G. raimondii* and *G. arboreum*. By comparison, in allotetraploid cotton *G. barbadense* acc. 3-79, the number of *Rab* genes is much less than that of *G. hirsutum* acc. TM-1. This might result from different sequencing methods, assembly error in partial chromosomal regions, or different degrees of colonization during the evolutionary process of *Gossypium*, and need to be further confirmed. Especially in *RabE* group, the number of *Rab* genes in *G. barbadense* has decreased drastically, implying that *Rab* members have further differentiation of function in the group *RabE* of *G. barbadense*.

The basic bioinformatics of 87 *Rabs* in *G. raimondii* were analyzed by using various softwares and online tools (Table 1). We found that the number of amino acids about 87 *GrRabs* ranged from 200 to 258 aa, which was consistent with the previous reports that the *Rab* protein usually has about more than 200 amino acids. It was also reported that the molecular weight of *Rab* proteins is generally at 20–30 kDa. In the present study, we also found that all members follow this rule, in which the minimum *Rab* protein has a molecular weight of 21.77 kDa and the maximum 28.65 kDa.

The nomenclature of the *Rab* gene family in *Gossypium*

In total, 57 *Rabs* in *A. thaliana* were identified. Then they were divided into eight groups and eighteen subgroups. These groups were named as *AtRabA, AtRabB, AtRabC,*

Table 1 Basic bioinformatics analysis of *Rab* gene family in *G. raimondii*

Gene name	Gene location	Number of amino acid (aa)	pI	Mw (kDa)	Chromosome	Subcellular location
RabA1a	Gorai.002G065100.1	216	5.87	24.20	Chr02	Golgi
RabA1b	Gorai.009G160700.1	218	5.49	24.09	Chr09	Golgi
RabA1c	Gorai.009G374600.1	217	6.53	24.24	Chr09	Golgi
RabA1d	Gorai.010G156800.1	215	6.12	23.89	Chr10	Golgi
RabA1e	Gorai.010G161100.1	218	5.50	24.25	Chr10	Golgi
RabA1f	Gorai.010G243000.1	217	5.98	24.01	Chr10	Golgi
RabA1g	Gorai.011G031200.1	217	5.70	23.90	Chr11	Golgi
RabA1h	Gorai.011G037400.1	218	5.36	24.01	Chr11	Golgi
RabA1i	Gorai.013G172400.1	217	5.98	24.09	Chr13	Golgi
RabA2a	Gorai.005G265900.1	215	6.76	23.87	Chr05	Golgi
RabA2b	Gorai.003G128100.1	216	6.22	23.89	Chr03	Golgi
RabA2c	Gorai.012G073300.1	216	6.91	24.01	Chr12	Golgi
RabA2d	Gorai.009G320100.1	216	6.32	24.04	Chr09	Golgi
RabA2e	Gorai.001G113000.1	216	7.01	23.85	Chr01	Golgi
RabA2f	Gorai.004G006300.1	216	5.89	23.81	Chr04	Golgi
RabA2g	Gorai.011G271400.1	214	6.32	23.67	Chr11	Golgi
RabA2h	Gorai.008G046400.1	215	6.76	23.89	Chr08	Golgi
RabA3a	Gorai.001G152600.1	232	5.00	25.49	Chr01	Golgi
RabA3b	Gorai.007G110800.1	236	5.35	26.04	Chr07	Golgi
RabA4a	Gorai.002G157000.1	225	6.62	24.96	Chr02	Golgi
RabA4b	Gorai.003G040900.1	223	9.00	24.64	Chr03	Golgi
RabA4c	Gorai.004G174200.1	229	6.84	25.16	Chr04	Golgi
RabA4d	Gorai.006G140300.1	222	5.61	24.92	Chr06	Golgi
RabA4e	Gorai.007G060000.1	225	6.84	24.80	Chr07	Golgi
RabA4f	Gorai.007G234300.1	233	6.84	25.83	Chr07	Golgi
RabA5a	Gorai.005G017600.1	219	5.23	24.53	Chr05	Golgi
RabA5b	Gorai.005G022300.1	215	5.03	24.10	Chr05	Golgi
RabA5c	Gorai.010G188100.1	224	5.12	24.69	Chr10	Golgi
RabA5d	Gorai.001G229200.1	217	4.99	24.02	Chr01	Golgi
RabA5e	Gorai.011G009000.1	224	5.01	24.69	Chr11	Golgi
RabA5f	Gorai.011G144500.1	258	6.24	28.65	Chr11	Golgi
RabA5g	Gorai.N003900.1	217	4.83	24.29	scaffold_14	Golgi
RabA6a	Gorai.007G004100.1	221	5.77	25.28	Chr07	Golgi
RabA6b	Gorai.008G289500.1	216	5.88	24.54	Chr08	Golgi
RabB1a	Gorai.009G185200.1	211	6.95	23.20	Chr09	Golgi
RabB1b	Gorai.001G224400.1	218	6.59	23.73	Chr01	Golgi
RabB1c	Gorai.001G267000.1	210	6.31	23.07	Chr01	Golgi
RabB1d	Gorai.006G035100.1	211	6.90	23.19	Chr06	Golgi
RabB1e	Gorai.011G093900.1	211	6.91	23.22	Chr11	Golgi
RabB1f	Gorai.N000100.1	211	7.01	23.42	scaffold_14	Golgi
RabC1a	Gorai.005G084300.1	212	5.39	23.60	Chr05	Plasma membrane
RabC1b	Gorai.012G035500.1	204	7.55	22.68	Chr12	Plasma membrane
RabC1c	Gorai.011G119600.1	208	6.41	23.25	Chr11	Plasma membrane
RabC2a	Gorai.002G160900.1	212	8.47	23.33	Chr02	Plasma membrane
RabC2b	Gorai.009G412800.1	212	5.45	23.18	Chr09	Plasma membrane
RabC2c	Gorai.006G161100.1	221	6.62	24.12	Chr06	Golgi
RabC2d	Gorai.011G234300.1	218	8.93	24.15	Chr11	Golgi
RabC2e	Gorai.007G372500.1	217	5.62	23.69	Chr07	Golgi
RabD1a	Gorai.005G266000.1	201	5.93	22.35	Chr05	Golgi

Table 1 continued

Gene name	Gene location	Number of amino acid (aa)	pI	Mw (kDa)	Chromosome	Subcellular location
RabD1b	Gorai.008G046500.1	201	5.15	22.44	Chr08	Cytoplasmic
RabD2a	Gorai.003G111900.1	203	7.59	22.56	Chr03	Golgi
RabD2b	Gorai.003G176500.1	203	5.52	22.60	Chr03	Golgi
RabD2c	Gorai.004G070000.1	203	5.27	22.49	Chr04	Golgi
RabD2d	Gorai.004G194200.1	203	5.14	22.60	Chr04	Golgi
RabD2e	Gorai.007G085600.1	203	5.09	22.46	Chr07	Golgi
RabE1a	Gorai.002G253800.1	215	7.64	23.71	Chr02	Plasma membrane
RabE1b	Gorai.009G101400.1	217	8.37	24.01	Chr09	Cytoplasmic
RabE1c	Gorai.009G424200.1	215	7.64	23.72	Chr09	Plasma membrane
RabE1d	Gorai.010G140100.1	215	6.61	23.76	Chr10	Cytoplasmic
RabE1e	Gorai.006G175600.1	216	7.64	23.80	Chr06	Plasma membrane
RabE1f	Gorai.011G255400.1	206	5.93	22.70	Chr11	Plasma membrane
RabE1g	Gorai.013G023100.1	214	7.64	23.64	Chr13	Plasma membrane
RabE1h	Gorai.013G265100.1	215	7.64	23.62	Chr13	Golgi
RabF1	Gorai.011G237500.1	213	6.6	23.28	Chr11	Golgi
RabF2a	Gorai.002G059500.1	225	6.44	24.92	Chr02	Golgi
RabF2b	Gorai.009G024300.1	200	9.34	22.32	Chr09	Cytoplasmic
RabF2c	Gorai.009G150100.1	200	7.71	21.77	Chr09	Golgi
RabF2d	Gorai.010G070200.1	203	9.54	22.80	Chr10	Cytoplasmic
RabF2e	Gorai.010G171600.1	228	5.00	25.35	Chr10	Golgi
RabF2f	Gorai.010G236900.1	200	7.74	21.81	Chr10	Golgi
RabF2g	Gorai.001G005000.1	200	9.52	22.69	Chr01	Cytoplasmic
RabF2h	Gorai.006G091700.1	203	4.99	22.55	Chr06	Golgi
RabG2	Gorai.008G101600.1	208	5.08	23.46	Chr08	Vacuole
RabG3a	Gorai.005G076500.1	213	6.30	24.57	Chr05	Vacuole
RabG3b	Gorai.005G242800.1	207	5.06	23.05	Chr05	Vacuole
RabG3c	Gorai.003G064900.1	207	5.42	23.19	Chr03	Vacuole
RabG3d	Gorai.009G259500.1	205	5.49	23.03	Chr09	Vacuole
RabG3e	Gorai.006G098100.1	207	4.80	23.27	Chr06	Vacuole
RabG3f	Gorai.011G074700.1	208	5.37	23.27	Chr11	Vacuole
RabG3g	Gorai.008G028300.1	206	5.09	23.01	Chr08	Vacuole
RabG3h	Gorai.013G037800.1	207	5.10	23.22	Chr13	Vacuole
RabH1a	Gorai.004G037900.1	208	6.38	23.18	Chr04	Golgi
RabH1b	Gorai.004G067600.1	208	7.70	23.20	Chr04	Golgi
RabH1c	Gorai.007G060700.1	208	7.67	23.09	Chr07	Golgi
RabH1d	Gorai.007G238500.1	208	6.38	23.18	Chr07	Golgi
RabH1e	Gorai.008G090100.1	207	7.67	23.17	Chr08	Golgi
RabH1f	Gorai.008G111500.1	208	5.94	23.10	Chr08	Golgi

AtRabD, AtRabE, AtRabF, AtRabG and *AtRabH*, respectively. According to cluster analysis of 57 members in *A. thaliana* and 87 in *G. raimondii*, we found that 87 *GrRabs* were also divided into eight groups, and named as *GrRabA, GrRabB, GrRabC, GrRabD, GrRabE, GrRabF, GrRabG* and *GrRabH*, correspondingly. There were 17 subgroups in *G. raimondii* compared with 18 subgroups in *A. thaliana* (Fig. 1). The number of each subgroup was named by combining their chromosome order (D1 to D13) with their location on the chromosome. Based on the nomenclature of *G. raimondii*, the corresponding *Rab* orthologs in *G. arboreum, G. hirsutum* acc. TM-1, and *G. barbadense* acc. 3-79 were named, respectively, with the same number for orthologs as in *G. raimondii* (Additional file 1).

We counted the numbers and proportion of *Rab* genes of the eight groups between *Arabidopsis* and *G. raimondii* (Tables 2 and 3). As we can see that *RabA* is the largest group of the family and the largest proportion. The proportion of the same group was similar between

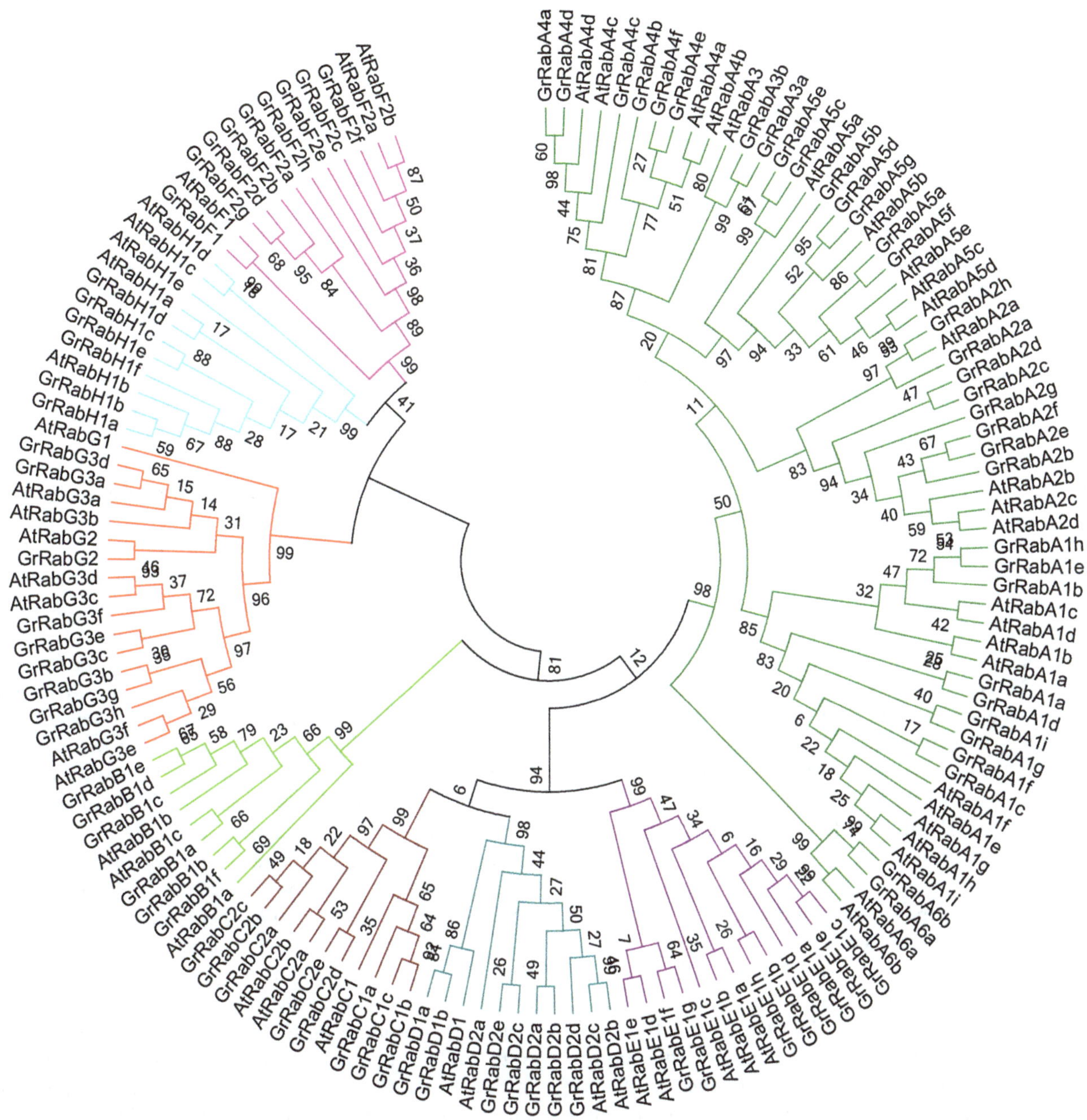

Fig. 1 Phylogenetic relationships of *Rab* family members between *G. raimondii* and *Arabidopsis*. The multiple alignment of the conserved *Rab* family domain between *G. raimondii* and *Arabidopsis* were constructed with Clustal X (version 2.0). Phylogenetic tree was generated using the maximum likelihood method under WAG model in MEGA v5.2, and the reliability of interior branches was assessed with 1000 bootstrap resamplings

Table 2 The numbers of *Rab* genes of the eight groups between *Arabidopsis* and *G. raimondii*

	RabA	*RabB*	*RabC*	*RabD*	*RabE*	*RabF*	*RabG*	*RabH*
Arabidopsis	26	3	3	4	5	3	8	5
G. raimondii	34	6	8	7	8	9	9	6

Table 3 The proportion of *Rab* genes in the eight groups between *Arabidopsis* and *G. raimondii*

	RabA (%)	*RabB* (%)	*RabC* (%)	*RabD* (%)	*RabE* (%)	*RabF* (%)	*RabG* (%)	*RabH* (%)
Arabidopsis	45.61	5.26	5.26	7.02	8.77	5.26	14.04	8.77
G. raimondii	39.08	6.90	9.20	8.05	9.20	10.34	10.34	6.90

Arabidopsis and *G. raimondii*. According to the position of each member in the phylogenetic tree, combining with the function of reported genes in the same group, we speculated that there were the similar function in the same group.

The chromosome distribution of the *Rab* gene family in *G. raimondii*

To elucidate the chromosomal distribution of these *Rab* genes, we integrated 13 scaffolds of the *G. raimondii* genome (named as Chr. 1 to Chr. 13) with the reported high-density interspecific genetic map of allotetraploid cultivated cotton species (Wang et al. 2015). Using Mapchart 2.2 software and the genomic localization information, we mapped the distribution of 87 *GrRab* genes on 13 chromosomes of *G. raimondii* (Fig. 2). 85 *GrRab* genes were tagged on the corresponding physical location of all the 13 chromosomes except for two genes, indicating that the scaffolds involved in the two genes had not been integrated in the chromosome in the present draft genome. The chromosomal distribution patterns of these *GrRabs* were uneven. Chr.11 (D10) contained the most *Rabs* (11 genes), while Chr.12 (D4) contained the fewest (two genes).

Rab gene family expansion in *G. raimondii*

To investigate the *Rab* gene family expansion pattern in cotton, we download the syntenic data of *G. raimondii* from the Plant Genome Duplication Database (PGDD). We identified the number of *Rab* genes formed in whole genome duplication (WGD) and tandem duplication in cotton. Among the 87 *GrRabs* in *G. raimondii*, we identified 60 pairs of *GrRabs* formed in whole genome duplication (WGD) (Additional file 2). We also analyzed the adjacent genes to investigate whether tandem duplication had taken place. Four of the 87 *GrRabs* showed tandem duplication. They were *GrRabA2a* vs *GrRabD1a* and *GrRabA2h* vs *GrRabD1b* respectively.

In 60 pairs of whole genome duplication gene pairs (Additional file 2), we obtained the *Ks* values of 58 pairs of whole genome duplication gene pairs. Of them, the *Ks* values of eight pairs of genes are greater than 1. They were *GrRabB1b* vs *GrRabB1a*, *GrRabC2c* vs *GrRabC2d*, *GrRabE1a* vs *GrRabE1g*, *GrRabE1e* vs *GrRabE1b*, *GrRabE1b* vs *GrRabE1f*, *GrRabE1b* vs *GrRabE1g*, *GrRabE1e* vs *GrRabE1f* and *GrRabE1g* vs *GrRabE1h*, respectively. It was found that the majority of the gene pairs come from *GrRabE* group. These gene pairs may be derived from the ancient event. For the remaining 50 gene pairs, the *Ks* values ranged from 0.331 to 0.8772, implying that these gene pairs originated from the *Gossypium* lineage WGD events. Among all the gene pairs, the *Ka*/*Ks* values were less than 1, which indicated the results of the purification selection, and help maintain the conservation of the duplicated genes in structure and function.

Exon–intron structural analysis and subcellular localization prediction

In order to better understand the genetic structure of 87 *Rab* gene family members in *G. raimondii*, we obtained the exon and intron structure distribution of each gene by combining their genome sequence with CDS sequence (Fig. 3) using the online website (http://wheat.pw.usda.gov/piece/GSDraw.php). The exon and intron gene structure analysis showed that all 87 *Rabs* had different introns. It is worth mentioning that the number of introns is different, but the genes clustered together have a very similar distribution of exons and introns. For 87 *Rab* protein gene family members, we compared the numbers of introns in each class. In all of 34 *RabA* family members, we found that there is only one intron among the 33 *Rabs* with one exception. Among 6 *RabB* family members, there are 5 introns. Then, we found that there are 5 introns among the most of *RabC* family members. All *Rabs* in *RabD* and *RabE* family contain 7 introns. Among 9 *RabF* family members, there are 8 *Rabs* containing 6 introns. Finally, all *Rabs* contained 6 introns in *RabG* family and 5 introns in *RabH* family. It can be concluded that intron numbers of most *Rabs* in the same group are basically the same, only a few genes exception.

Using the Softberry online software for subcellular localization prediction, we found that 87 *Rabs* mainly concentrated in organelles with membrane structure and cell membrane (Table 1). We further found that all members of *RabA*, *RabB* and *RabH* are located in the Golgi, *RabGs* are located in the vacuole, *RabCs* are distributed in the cell membrane or Golgi, *RabDs* and *RabFs* are distributed in the cytoplasmic or Golgi, *RabEs* are distributed in the cell membrane or cytoplasmic.

Taken together, the genes clustered together have similar gene structure and subcellular localization,

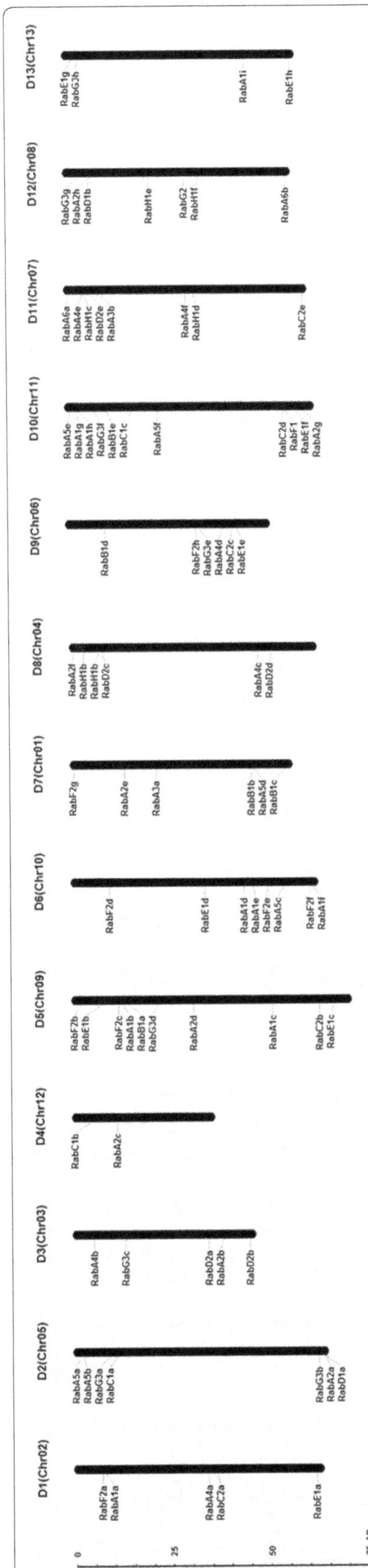

Fig. 2 Chromosomal distribution of *Rab* genes in *G. raimondii*. The chromosome numbers were consistent with the interspecific genetic map (D1 to D13) in allotetraploid cultivated cotton species (Wang et al. 2015) and the scaffolds (Chr.1 to Chr.13) in the genomic data of *G. raimondii* (Paterson et al. 2012). The nomenclature of *Rabs* was based on *Arabidopsis* nomenclature and the order of the chromosomes in *G. raimondii*

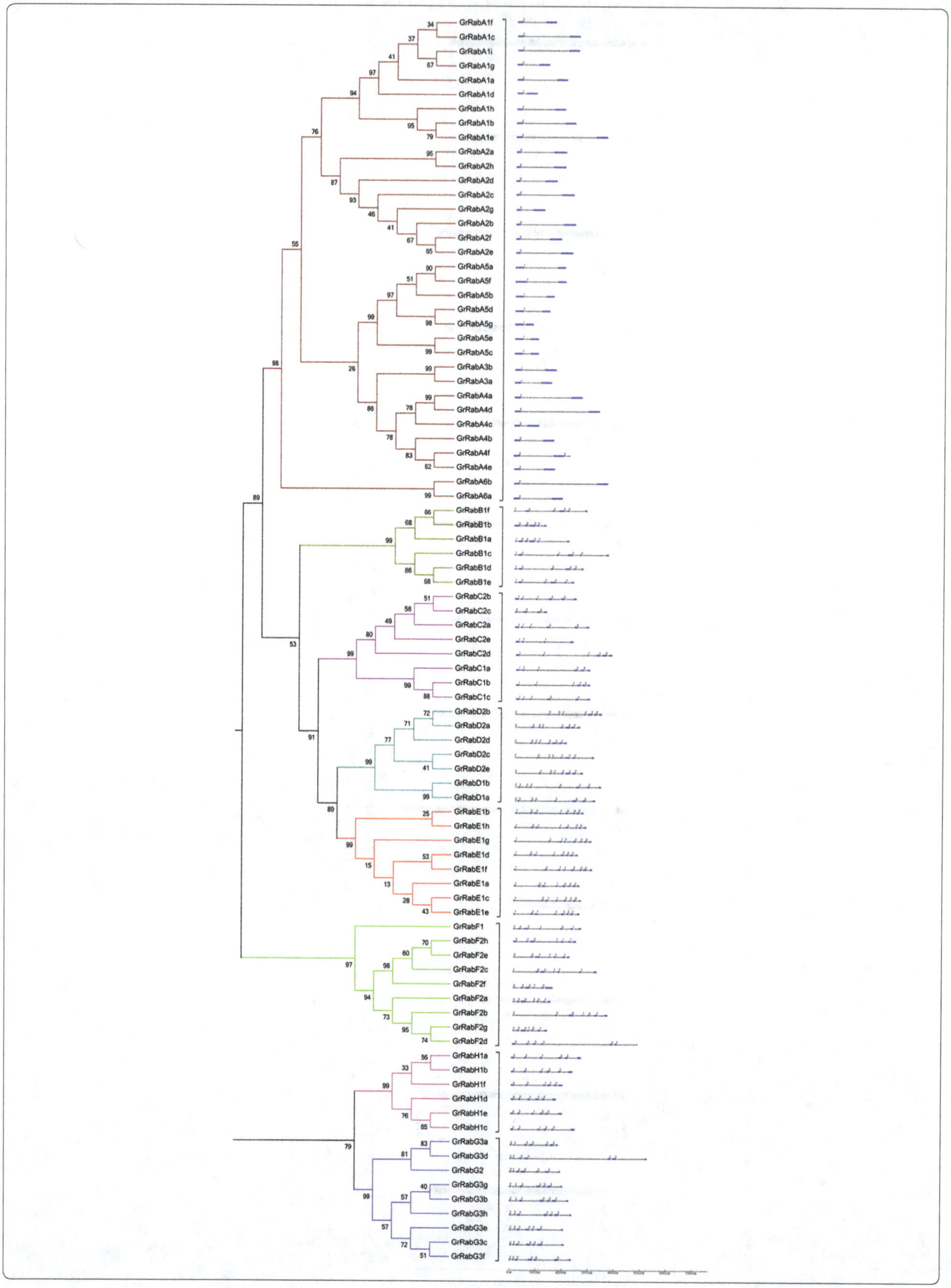
GrRabA1f
GrRabA1c
GrRabA1i
GrRabA1g
GrRabA1a
GrRabA1d
GrRabA1h
GrRabA1b
GrRabA1e
GrRabA2a
GrRabA2h
GrRabA2d
GrRabA2c
GrRabA2g
GrRabA2b
GrRabA2f
GrRabA2e
GrRabA5a
GrRabA5f
GrRabA5b
GrRabA5d
GrRabA5g
GrRabA5e
GrRabA5c
GrRabA3b
GrRabA3a
GrRabA4a
GrRabA4d
GrRabA4c
GrRabA4b
GrRabA4f
GrRabA4e
GrRabA6b
GrRabA6a
GrRabB1f
GrRabB1b
GrRabB1a
GrRabB1c
GrRabB1d
GrRabB1e
GrRabC2b
GrRabC2c
GrRabC2a
GrRabC2e
GrRabC2d
GrRabC1a
GrRabC1b
GrRabC1c
GrRabD2b
GrRabD2a
GrRabD2d
GrRabD2c
GrRabD2e
GrRabD1b
GrRabD1a
GrRabE1b
GrRabE1h
GrRabE1g
GrRabE1d
GrRabE1f
GrRabE1a
GrRabE1c
GrRabE1e
GrRabF1
GrRabF2h
GrRabF2e
GrRabF2c
GrRabF2f
GrRabF2a
GrRabF2b
GrRabF2g
GrRabF2d
GrRabH1a
GrRabH1b
GrRabH1f
GrRabH1d
GrRabH1e
GrRabH1c
GrRabG3a
GrRabG3d
GrRabG2
GrRabG3g
GrRabG3b
GrRabG3h
GrRabG3e
GrRabG3c
GrRabG3f

(See figure on previous page.)
Fig. 3 Phylogenetic relationships and gene structures of *Rabs* in cotton. The multiple alignment of the conserved *Rab* family domain were constructed with Clustal X (version 2.0). Phylogenetic tree was generated using the maximum likelihood method under WAG model in MEGA v5.2, and the reliability of interior branches was assessed with 1000 bootstrap resamplings. The gene structures were drawn using the online tool GSDS. Introns and exons were represented by *black lines* and *blue boxes*, respectively, and numbers at the exon–intron joints were intron phases

implying *Rab* genes in the same class generally have similar functions.

Expression pattern analysis

The analysis of gene expression patterns can provide powerful clues and help for prediction of gene function. We found that cotton *Rabs* are expressed in all tissues of plant and are regulated temporally and spatially depending on developmental stage and environmental conditions (Additional files 3, 4). Among the 169 *GhRabs* (Fig. 4), there were 161 genes with $\log_2$ FPKM > 1 in at least one of the 16 investigated organs and developmental stages, and these 161 genes were used to gauge the relative expression of each *Rab* gene. It indicates that most *Rab* family members play a certain role in different growth stages of cotton. The remaining 8 *GhRabs* may be pseudogenes or only expressed under special environmental conditions. About half of the genes (86 *GhRabs*) were expressed widely in vegetative organs, flower organs and reproductive organs. It shows that they are constitutively expressed and participate in the whole growth and development process of cotton. Approximately 10.65% (18 *GhRabs*) of the genes were expressed preferentially in cotton fibers, and most of them were highly expressed at the elongation stage of fiber development, indicating that these genes may be involved in regulating the growth and development of cotton fiber. A few genes are dominantly expressed in floral organs, suggesting that they may play a role in pollen formation and development. A small number of genes were preferentially expressed in the root, stem and leaf, suggesting that they might be involved in the process of plant stress resistance. We also observed the significant different expression patterns among each group of *Rab* family.

Discussion

It is reported that *Rab* genes played the various roles in all kinds of plants. Through cotton genome sequence information, we tried to analyze the basic biological information, gene structure, genome distribution, and expression characteristics of *Rab* gene family members. We hope to find some important genes related to stress response and fiber development in the whole genomic level, and help to utilize them to improve the quality and yield of cotton.

Consistency on classification and structure of 87 *GrRabs*

In *Arabidopsis*, a total of 57 *Rab* members are divided into eight groups. Similarly, 87 *Rabs* in *G. raimondii* are also divided into eight groups. We found that although the length of 87 *Rabs* and the numbers of introns are different, it is worth mentioning that the genes clustering together have a very similar distribution of exon and intron structure and subcellular localization. The intron numbers of 87 *GrRabs* in the same group are basically the same. It is believed that the genes clustered together have similar gene structure and the genes in the same class have similar functions.

Previous studies on many *Rab* proteins in *Arabidopsis* have been reported in detail. However, the study of Rab proteins in cotton is few. According to the function of *Rab* proteins in *Arabidopsis* (Rutherford and Moore 2002), we can infer the possible function roles of *Rab* proteins in cotton by clustering *Rab* members in *Arabidopsis* and *G. raimondii*. It has a guiding role for further study of gene function by genomic-wide comparative analysis among different plant species.

The functional diversity of *Rab* proteins

We used the transcriptome sequencing results of upland cotton genetic standard line TM-1 and carried out expression pattern analysis on 169 *GhRabs*. We found that 161 of 169 *GhRabs* have a certain amount of expression level in at least one of the 16 investigated organs and developmental stages. We speculate that the vast majority of *Rab* genes have certain function in cotton growth and development. About half of the genes (86 *GhRabs*) were expressed in all kinds of tissues and organs. In addition, some genes specifically expressed in certain tissues and organs, implying that *Rab* genes have diversity on the function. Referring the reports of *Rab*s function and their expression profile in cotton, cotton *Rabs* may involve in the biotic and abiotic stress response, pollen germination and pollen tube elongation, seed germination and fiber development.

In the previous study, it is reported that many *Rab* proteins are involved in important functions in a lot of plants. Among them, the hottest spot of current research is that *Rab* protein was involved in cell polarity growth. In the plant growth process, polar growth process is very important. The polarity growth process in plant includes elongation of pollen tubes (Peng et al. 2011) and root hairs (Preuss et al. 2006; Blanco et al. 2009). This process requires vesicle trafficking and microtubule actin proteins. *Rab* protein is an important molecular switch in the cell, which controls the transport of vesicles in

Fig. 4 Expression profiles of *Rab* genes in cotton. The RNA-seq relative expression data of 16 tissues was used to re-construct expression patterns of 169 *Rab* genes in *G. hirsutum* acc. TM-1. Analysis method of data model expression as follows: if the value of $\log_2$ FPKM is more than 1, we consider it to be expressed and if the value of $\log_2$ FPKM is less than or equal to 1, we consider it not to be expressed

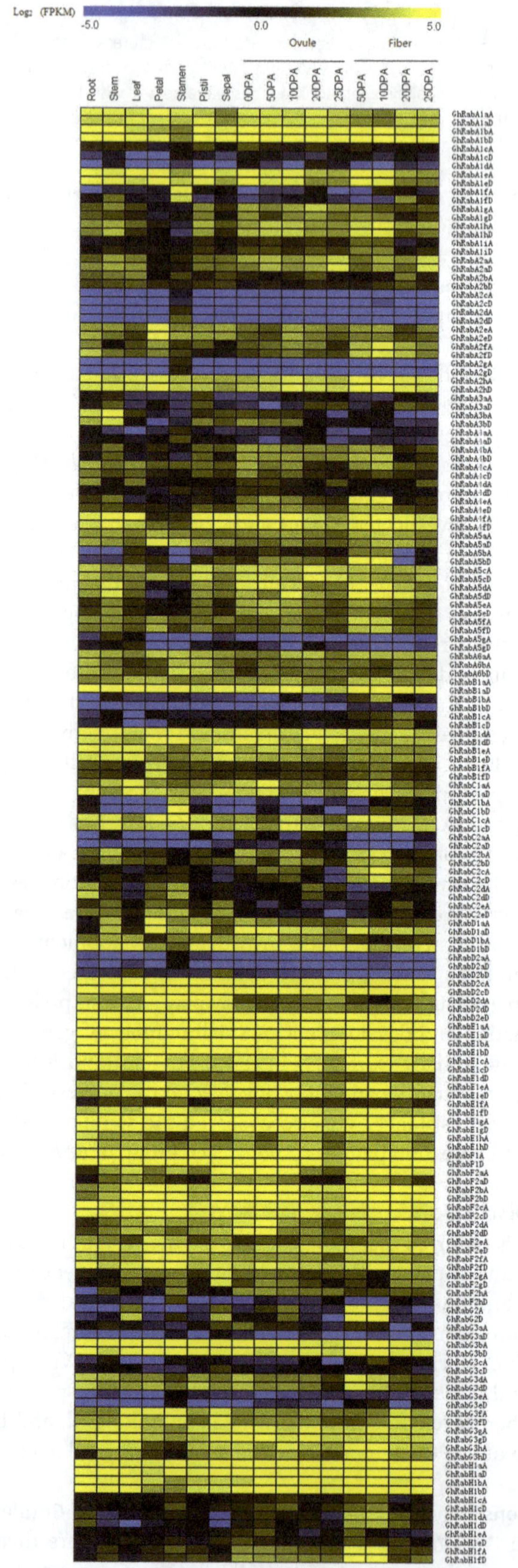

various ways, so as to promote the establishment and maintenance of apical polarity (Cole and Fowler 2006; Samaj et al. 2006).

In *Arabidopsis*, root hairs and pollen tubes elongate through the polar growth (Campanoni and Blatt 2007). Although they have different origins and functions, these cells promote cell elongation by Ca^{2+} enrichment at the cell tip and actin cytoskeleton regulation (Li and Yang 1999; Hepler et al. 2001; Samaj et al. 2005; Cole and Fowler 2006). The overexpression of *Rab11b* inhibits the elongation of pollen tube and influence the direction of pollen tube growth (de Graaf et al. 2005). *NtRab2*, which is overexpressed in tobacco, also inhibits the growth of pollen tube (Cheung et al. 2002). In *Arabidopsis*, *RabA4d* is necessary for the proper regulation of pollen tube growth. The loss of *RabA4d* leads to the destruction of polar growth and changes in the structure of the cell wall. These results indicate that *RabA4d* plays an important role in regulating the growth of pollen tube (Szumlanski and Nielsen 2009). Enhanced yellow fluorescent protein *RabA4b*-EYFP specifically located in the tip of the root hair cells of *Arabidopsis*. In root hair defective mutants, yellow fluorescence signal localization of *RabA4b*-EYFP in hair cells no longer exist. As a result, *RabA4b* potentially regulates membrane transport in plant cells through the involvement of cell wall components in the secretory process (Preuss and Nielsen 2004). *AtRab2* was specifically expressed in newly germinated seedlings and pollen tubes. The GUS connected to the promoter of the gene was clearly detected in mitosis of pollen tubes. After that, its expression level has been increased, indicating that *AtRab2* plays an important role in the growth and development of pollen tubes (Moore et al. 1997). *Rgp1* isolated from rice is a gene homologous to yeast *Ypt3*. After overexpression of this gene in tobacco plants, the plants exhibited dwarf and the abnormal development of floral organs. Of them, the transgenic plants were more than 6 times higher in the content of endogenous hormones than that in the wild plants. These results suggest that *Rgp1* may be involved in cytokinin signaling pathways (Kamada et al. 1992; Sano et al. 1994). Thus, the *Rab* protein is essential for root hairs and pollen tube elongation of plant. Cotton fiber cells, root hairs and pollen tubes are typical tissues for studying polar growth of plants. We hypothesized that *Rab* proteins would be involved in the fiber polarity elongation process and plant growth process in cotton.

Rab proteins not only play an important role in plant growth and development, but also play an important role in biotic and abiotic stress responses. When the relative water content of leaves decreased, the expression of *Rab2* increased. Further, the expression level of *Rab2* was increased when exogenous ABA was applied to the restored grass plants. The results show that *Rab2* plays an important role in drought stress (O'Mahony and Oliver 1999). Under a variety of environments (chilling, salinity, drying and ABA) induction, the expression level of *OsRab7* significantly altered. Through the transformation of *Arabidopsis* protoplasts, the *OsRab7* fusion protein of GFP was localized in the vacuole. It is suggested that *OsRab7* is transported to the vacuole through the vesicle transport in plant cells. The inducible expression of *OsRab7* suggests that it is involved in the response of these stresses, which suggests that *Rab* is related to abiotic stresses (Nahm et al. 2003). Overexpression of *AtRabG3e* in *Arabidopsis* could enhance the resistance of plants to salt and osmotic stress. The enhancement of plant resistance was related to the acceleration of endocytosis and the increase of sodium ion in vacuole (Mazel et al. 2004). In addition, when *AtRabG3b* was silenced, there was no difference in the silenced plants compared with the no transgenic plants. However, when *AtRabG3b* was overexpressed, it showed hypersensitive cell death on pathogenic fungi and mycotoxins. In addition, it also accelerated leaf senescence. These results suggest that *AtRabG3b* is involved in the regulation of cell apoptosis and the regulation of pathogen response during plant senescence (Kwon et al. 2009). Transgenic tobacco plants overexpressing *PgRab7* also showed resistance to drought and salt stress, which indicated that *PgRab7* was involved in plant stress response (Agarwal et al. 2008). *OsRab11* is widely expressed in various tissues and organs of plants, which is induced by jasmonic acid (JA). Overexpression of *OsRab11* in transgenic plants enhanced the resistance to pathogens by affecting the expression of genes involved in the metabolic pathway of jasmonic acid. Therefore, *OsRab11* is necessary in the jasmonic acid mediated signaling pathway (Hong et al. 2013). Based on above, we hypothesized that *Rab* genes in cotton play an important role in biotic and abiotic stress responses.

Conclusions

Rab proteins play an important role in plant growth and development, as well as in biotic and abiotic stress responses. In this study, we individually identified 87, 169, 136, 80 *Rabs* in the four sequenced cotton species. These *Rabs* are divided into eight groups. In each group, their intron numbers and subcellular localization are basically the same. Further, 60 pairs of segmental duplication due to whole genome duplication and two pairs of tandem duplication were detected, respectively. Expression patterns analysis indicated that most *Rab* family members play a certain role in different tissues/organs and different growth stages of cotton, implying their potential function in the polar growth and stress tolerance.

Abbreviations

Rab: ras-like in rat brain; PGDD: plant genome duplication database; WGD: whole genome duplication; FPKM: fragments per kilobase of exon model per million mapped reads; *Ka*: nonsynonymous substitutions per nonsynonymous site; *Ks*: synonymous substitutions per synonymous site.

Authors' contributions

GWZ conceived and designed the experiments. LP performed the experiments, analyzed the data and wrote the paper. Both authors read and approved the final manuscript.

Competing interests

The authors declare that they have no competing interests.

Funding

This program was financially supported in part by the JCIC-MCP project (No. 10).

References

Agarwal PK, Agarwal P, Jain P, Jha B, Reddy MK, Sopory SK (2008) Constitutive overexpression of a stress-inducible small GTP-binding protein *Pgrab7* from *Pennisetum glaucum* enhances abiotic stress tolerance in transgenic tobacco. Plant Cell Rep 27:105–115

Agarwal P, Reddy MK, Sopory SK, Agarwal PK (2009) Plant *Rabs*: characterization, functional diversity, and role in stress tolerance. Plant Mol Biol Rep 27:417–430

Barr FA (2009) *Rab* GTPase function in Golgi trafficking. Semin Cell Dev Biol 20:780–783

Blanco FA, Peltzer Meschini E, Zanetti ME, Aguilar OM (2009) A small GTPase of the *Rab* family is required for root hair formation and preinfection stages of the common bean-Rhizobium symbiotic association. Plant Cell 21:2797–2810

Brennwald P (2000) Reversal of fortune: do *Rab* GTPases act on the target membrane? J Cell Biol 149:1–4

Brighouse A, Dacks JB, Field MC (2010) *Rab* protein evolution and the history of the eukaryotic endomembrane system. Cell Mol Life Sci 67:3449–3465

Campanoni P, Blatt MR (2007) Membrane trafficking and polar growth in root hairs and pollen tubes. J Exp Bot 58:65–74

Cheung AY, Christine Y-HC, Glaven RH, Graaf BHJD, Luis V, Hepler PK, Wu HM (2002) *Rab2* GTPase regulates vesicle trafficking between the endoplasmic reticulum and the Golgi bodies and is important to pollen tube growth. Plant Cell 14:945–962

Cole RA, Fowler JE (2006) Polarized growth: maintaining focus on the tip. Curr Opin Plant Biol 9:579–588

de Graaf BH, Cheung AY, Andreyeva T, Levasseur K, Kieliszewski M, Wu HM (2005) *Rab11* GTPase-regulated membrane trafficking is crucial for tip-focused pollen tube growth in tobacco. Plant Cell 17:2564–2579

Finn RD, Clements J, Eddy SR (2011) HMMER web server: interactive sequence similarity searching. Nucleic Acids Res 39:29–37

Gallwitz D, Donath C, Sande C (1983) A yeast gene encoding a protein homologous to the human c-has/bas proto-oncogene product. Nature 306:704–707

Gurkan C, Lapp H, Alory C, Su AI, Hogenesch JB, Balch WE (2005) Large-scale profiling of *Rab* GTPase trafficking networks: the membrome. Mol Biol Cell 16:3847

Hepler PK, Vidali L, Cheung AY (2001) Polarized cell growth in higher plants. Annu Rev Cell Dev Biol 17:159–187

Hill D, Sylvester A (2007) Diversification of the *Rab* guanosine triphosphatase family in dicots and monocots. J Integr Plant Biol 49:1129–1141

Hong MJ, Yun ML, Son YS, Im CH, Yi YB, Rim YG, Bahk JD, Heo JB (2013) Rice *Rab11* is required for JA-mediated defense signaling. Biochem Biophys Res Commun 434:797–802

Kamada I, Yamauchi S, Youssefian S, Sano H (1992) Transgenic tobacco plants expressing rgpl, a gene encoding ras-related GTPbinding protein from rice, show distinct morphological characterization. Plant J 2:799–807

Kwon SI, Hong JC, Bae K, Jung JH, Jin HC, Park OK (2009) Role of an *Arabidopsis*, *Rab* GTPase *RabG3b* in pathogen response and leaf senescence. J Plant Biol 52:79–87

Li H, Yang Z (1999) Control of pollen tube tip growth by a *Rop* GTPase-dependent pathway that leads to tip-localized calcium influx. Plant Cell 11:1731–1742

Li F, Fan G, Wang K, Sun F, Yuan Y, Song G, Li Q, Ma Z, Lu C, Zou C, Chen W, Liang X, Shang H, Liu W, Shi C, Xiao G, Gou C, Ye W, Xu X, Zhang X, Wei H, Li Z, Zhang G, Wang J, Liu K, Kohel RJ, Percy RG, Yu JZ, Zhu YX, Wang J, Yu S (2014) Genome sequence of the cultivated cotton *Gossypium arboretum*. Nat Genet 46:567–572

Loytynoja A, Goldman N (2005) An algorithm for progressive multiple alignment of sequences with insertions. Proc Natl Acad Sci USA 102:10557–10562

Martinez O, Goud B (1998) *Rab* proteins. Biochim Biophys Acta 1404:101–112

Mazel A, Leshem Y, Tiwari BS, Levine A (2004) Induction of salt and osmotic stress tolerance by overexpression of an intracellular vesicle trafficking protein *AtRab7* (*AtRabG3e*). Plant Physiol 134:118–128

Moore I, Diefenthal T, Zarsky V, Schell J, Palme K (1997) A homolog of the mammalian GTPase *Rab2* is present in *Arabidopsis* and is expressed predominantly in pollen grains and seedlings. Proc Natl Acad Sci USA 94:762–767

Nahm MY, Kim SW, Yun D, Lee SY, Cho MJ, Bahk JD (2003) Molecular and biochemical analyses of *OsRab7*, a rice *Rab7* homolog. Plant Cell Physiol 44:1341–1349

Novick P, Zerial M (1997) The diversity of *Rab* proteins in vesicle transport. Curr Opin Cell Biol 9:496–504

O'Mahony PJ, Oliver MJ (1999) Characterization of adesiccation-responsive small GTP-binding protein (*Rab2*) from thedesiccation-tolerant grass *Sporobolus stapfianus*. Plant Mol Biol 39:809–821

Paterson AH, Wendel JF, Gundlach H, Guo H, Jenkins J, Jin D, Llewellyn D, Showmaker KC, Shu S, Udall J, Yoo MJ, Byers R, Chen W, Doron-Faigenboim A, Duke MV, Gong L, Grimwood J, Grover C, Grupp K, Hu G, Lee TH, Li J, Lin L, Liu T, Marler BS, Page JT, Roberts AW, Romanel E, Sanders WS, Szadkowski E, Tan X, Tang H, Xu C, Wang J, Wang Z, Zhang D, Zhang L, Ashrafi H, Bedon F, Bowers JE, Brubaker CL, Chee PW, Das S, Gingle AR, Haigler CH, Harker D, Hoffmann LV, Hovav R, Jones DC, Lemke C, Mansoor S, ur Rahman M, Rainville LN, Rambani A, Reddy UK, Rong JK, Saranga Y, Scheffler BE, Scheffler JA, Stelly DM, Triplett BA, Van Deynze A, Vaslin MF, Waghmare VN, Walford SA, Wright RJ, Zaki EA, Zhang T, Dennis ES, Mayer KF, Peterson DG, Rokhsar DS, Wang X, Schmutz J (2012) Repeated polyploidization of *Gossypium* genomes and the evolution of spinnable cotton fibres. Nature 492:423–427

Peng J, Ilarslan H, Wurtele ES, Bassham DC (2011) *AtRabD2b* and *AtRabD2c* have overlapping functions in pollen development and pollen tube growth. BMC Plant Biol 11:1–16

Pereira-Leal JB, Seabra MC (2001) Evolution of the *Rab* family of small GTP-binding proteins. J Mol Biol 313:889–901

Preuss ML, Nielsen E (2004) The *Arabidopsis Rab* GTPase *RabA4b* localizes to the tips of growing root hair cells. Plant Cell 16:1589–1603

Preuss ML, Schmitz AJ, Thole JM, Bonner HK, Otegui MS, Nielsen E (2006) A role for the *RabA4b* effector protein PI-4K1 in polarized expansion of root hair cells in *Arabidopsis thaliana*. J Cell Biol 172:991–998

Rutherford S, Moore I (2002) The *Arabidopsis*, Rab GTPase family: another enigma variation. Curr Opin Plant Biol 5:518–528

Saitou N, Nei M (1987) The neighbor-joining method: a new method for reconstructing phylogenetic trees. Mol Biol Evol 4:406–425

Salminen A, Novick PJ (1987) A ras-like protein is required for a post-Golgi event in yeast secretion. Cell 49:527–538

Samaj J, Read ND, Volkmann D, Menzel D, Baluska F (2005) The endocytic network in plants. Trends Cell Biol 15:425–433

Samaj J, Muller J, Beck M, Böhm N, Menzel D (2006) Vesicular trafficking, cytoskeleton and signalling in root hairs and pollen tubes. Trends Plant Sci 11:594–600

Sano H, Seo S, Orudgev E, Youssefian S, Ishizuka K (1994) Expression of the gene for a small GTP binding protein in transgenic tobacco elevates endogenous cytokinin levels, abnormally induces salicylic acid in response towounding, and increases resistance to tobacco mosaic virus infection. Proc Natl Acad Sci USA 91:10556–10560

Stenmark H, Olkkonen VM (2001) The *Rab* GTPase family. Genome Biol 2(reviews3007):1

Szumlanski AL, Nielsen E (2009) The *Rab* GTPase *RabA4d* regulates pollen tube tip growth in *Arabidopsis thaliana*. Plant Cell 21:526–544

Takai Y, Sasaki T, Matozaki T (2001) Small GTP-binding proteins. Physiol Rev 81:153–208

Thompson JD, Higgins DG, Gibson TJ (1994) CLUSTAL W, (improving the sensitivity of progressive multiple sequence alignment through sequence weighting, position-specific gap penalties and weight matrix choice). Nucleic Acids Res 22:4673–4680

Tuvim MJ, Adachi R, Hoffenberg S, Dickey BF (2001) Traffic control: *Rab* GTPases and the regulation of interorganellar transport. News Physiol Sci 16(2):56–61

Wang L, Guo K, Li Y, Tu Y, Hu H, Wang B, Cui X, Peng L (2010) Expression profiling and integrative analysis of the CESA/CSL, superfamily in rice. BMC Plant Biol 10(1):282

Wang K, Wang Z, Li F, Ye W, Wang J, Song G, Yue Z, Cong L, Shang H, Zhu S, Zou C, Li Q, Yuan Y, Lu C, Wei H, Gou C, Zheng Z, Yin Y, Zhang X, Liu K, Wang B, Song C, Shi N, Kohel RJ, Percy RG, Yu JZ, Zhu YX, Wang J, Yu S (2012) The draft genome of a diploid cotton *Gossypium raimondii*. Nat Genet 44(10):1098–1103

Wang S, Chen JD, Zhang WP, Hu Y, Chang LJ, Fang L, Wang Q, Lv FN, Wu HT, Si ZF, Chen SQ, Cai CP, Zhu XF, Zhou BL, Guo WZ, Zhang TZ (2015) Sequence-based ultra-dense genetic and physical maps reveal structural variations of allopolyploid cotton genomes. Genome Biol 16:108

Xu J, Xu X, Tian L, Wang G, Zhang X, Wang X, Guo W (2016a) Discovery and identification of candidate genes from the chitinase gene family for *Verticillium dahliae* resistance in cotton. Sci Rep 6:29022

Xu X, Feng Y, Fang S, Xu J, Wang X, Guo W (2016b) Genome-wide characterization of the β-1,3-glucanase gene family in *Gossypium* by comparative analysis. Sci Rep 6:29044

Yang Z (2002) Small GTPases: versatile signaling switches in plants. Plant Cell 14:S375–S388

Yang Z (2007) PAML 4: phylogenetic analysis by maximum likelihood. Mol Biol Evol 24:1586–1591

Yuan D, Tang Z, Wang M, Gao W, Tu L, Jin X, Chen L, He Y, Zhang L, Zhu L, Li Y, Liang Q, Lin Z, Yang X, Liu N, Jin S, Lei Y, Ding Y, Li G, Ruan X, Ruan Y, Zhang X (2015) The genome sequence of Sea-Island cotton (*Gossypium barbadense*) provides insights into the allopolyploidization and development of superior spinnable fibres. Sci Rep 5:17662

Zhang T, Hu Y, Jiang W, Fang L, Guan X, Chen J, Zhang J, Saski CA, Scheffler BE, Stelly DM, Hulse-Kemp AM, Wan Q, Liu B, Liu C, Wang S, Pan M, Wang Y, Wang D, Ye W, Chang L, Zhang W, Song Q, Kirkbride RC, Chen X, Dennis E, Llewellyn DJ, Peterson DG, Thaxton P, Jones DC, Wang Q, Xu X, Zhang H, Wu H, Zhou L, Mei G, Chen S, Tian Y, Xiang D, Li X, Ding J, Zuo Q, Tao L, Liu Y, Li J, Lin Y, Hui Y, Cao Z, Cai C, Zhu X, Jiang Z, Zhou B, Guo W, Li R, Chen ZJ (2015) Sequencing of allotetraploid cotton (*Gossypium hirsutum* L. acc. TM-1) provides a resource for fiber improvement. Nat Biotechnol 33:531–537

Zhang X, Xu X, Yu Y, Chen C, Wang J, Cai C, Guo W (2016) Integration analysis of MKK and MAPK family members highlights potential MAPK signaling modules in cotton. Sci Rep 6:29781

4

Studies of life history of *Gagea graeca* (*Liliaceae*) based on morphological and molecular methods

Martin Schnittler[1*], Akmaral Nursafina[4], Angela Peterson[2], Jens Peterson[3], Carl Barnick[1] and Anja Klahr[1]

Abstract

Background: We studied the life history of *Gagea graeca* (L.) A. Terracc. (sect. *Anthericoides*) by field surveys on the Greek island of Crete, including quantitative analyses of 405 individuals, estimation of resource allocation by measuring the nitrogen content of different plant organs, assessing seed set and recording genetic diversity via amplified fragment length polymorphism (AFLP) analyses. In contrast to most species of the genus *G. graeca* seems to be a short-lived perennial, developing several characters that are rather typical for annual plants.

Results: Although seed set varies largely, flowering plants produce many (68 ± 79) small, flattened seeds (mean weight 73 ± 22 µg) in comparison to a single bulbil. If measured as nitrogen content of the respective plant parts, investment in seeds (25%) is much higher than that in bulbils (4%). In addition, the threshold for flower formation (expressed as bulb size where 50% of the plants form the respective structure) is with 2.17 ± 0.05 mm lower than that for bulbils with 2.80 ± 0.16 mm. This is in accordance with AFLP analyses revealing predominantly sexual reproduction (only 9.1% of 110 investigated plants belonged to clones).

Conclusion: In the genus *Gagea* early, predominantly sexual reproduction seems to be characteristic for species from arid habitats, coupled with a low proportion of clonal plants.

Keywords: Amplified fragment length polymorphism (ALFP), Drought adaptation, Reproductive biology, Resource allocation, Seed set

Background

Gagea graeca (L.) A. Terracc. is one of the few white-flowering species in the genus *Gagea* Salisb. (*Liliaceae*). Together with the closely related *G. trinervia* (Viv.) Greuter (Peruzzi et al. 2008) it is assigned to section *Anthericoides* A. Terracc. This monophyletic section (e.g., Peruzzi et al. 2008; Peterson et al. 2008, 2011, 2016; Zarrei et al. 2009) within the genus *Gagea* (incl. *Lloydia* Rchb.; see Peterson et al. 2004, 2008; Peruzzi et al. 2008; Zarrei et al. 2011) is currently accepted in all infra-generic classifications (Levichev in Peterson et al. 2008; Zarrei et al. 2011; Peruzzi 2012a). Both species are endemic to the Mediterranean region and are diploid ($2n = 24$; Peruzzi 2003, 2008, 2012a). In several phylogenetic studies (e.g., Peruzzi et al. 2008; Peterson et al. 2011) the section *Anthericoides* was found to be in a sister position to other studied sections of the genus.

Plants of *G. graeca* germinate with a single leaf, but in subsequent years the number of basal leaves grows to 2, (sometimes to 3 or 4; Levichev in Peterson et al. 2008). According to Peruzzi et al. (2008) plants can produce a first flower already in their second year. In adult plants usually several flowers were observed. Compared with the size of the plant, flowers are showy, with white, elongated tepals with reddish veins, which are recurved at apex. This habit, resembling that of an annual plant, is unusual among the genus and points to prevailing reproduction via seeds, whereas most other species of the genus reproduce predominantly via bulbils (e.g., Levichev 1999, 2013; Schnittler et al. 2013). However, in addition to the renewal

*Correspondence: martin.schnittler@uni-greifswald.de
[1] Institute of Botany and Landscape Ecology, Ernst Moritz Arndt University Greifswald, Soldmannstr. 15, 17487 Greifswald, Germany
Full list of author information is available at the end of the article

bulb produced in the axil of the first leaf (obligatory for the genus, furthermore named parent bulb) plants with two leaves start to produce a first bulbil on a very short stolon (hypopodium according to Peruzzi et al. 2008). As such, vegetative reproduction is regularly available as a second option. The other species of section *Anthericoides*, *Gagea trinervia* differs from *G. graeca* in "having bulbils with stolon-like hypopodium, a regular sister bulbil at the immature stage, an unifacial, juvenile-like basal leaf to adult stage, few cauline leaves, few flowers and a low level of sexual reproduction" (Peruzzi et al. 2008).

Gagea graeca occurs in the eastern Mediterranean region, with an apparent center of occurrence in Greece and the Aegean islands (Fig. 1a; Jahn et al. 1995; Strid 2016). (Tison 2016) provide a preliminary map of the distribution of *G. graeca* together with a threat assessment of based on the IUCN categories. According to these authors, the species is found in southeastern Sterea Ellas and extends north to the southern Thessalian seashores, in most of the Peloponnese and many Aegean islands, including Crete and Rhodos (but seems to be absent from some northeastern islands and the Karpathos group). It is part of the flora of Turkey (Tekşen and Erkul 2015) and reported from southwestern Anatolia (provinces of Izmir, Antalya and Muğla) and Cyprus. Records from Israel are likely to be erroneous (the species is not mentioned in Danin and Danin 2015). On the contrary, *G. trinervia* occupies a small and highly disjunct range, occurring in on the Italian Island of Sicily and in the northeastern part of Libya (Peruzzi et al. 2008).

Gagea graeca can be found in various, usually slightly disturbed, open habitats like dry grasslands, rocky slopes and fallow fields especially on soils that are provided with some moisture to prevent the relatively small bulb from desiccation over the summer. In Crete, where the species was investigated, most of the ancient woodlands are now replaced by low, thorny shrubs (a vegetation called "phrygana") and open grasslands (Zimowski et al. 2014). Here *G. graeca* usually inhabits small gullies in where the vegetation cover does not exceed 50%. At such places, the plant can form large populations comprising thousands of individuals.

This investigation was undertaken to study adaptations of *Gagea graeca* on its Mediterranean environment, and compare these adaptations with those of another species of the genus adapted to arid environments, *G. bulbifera* (Pall). Salisb. (Beisenova et al. 2015). A quantitative morphology approach (Schnittler et al. 2009, 2013; Beisenova et al. 2015), including measurements on living plants of all ontogenetic stages, is used to quantify possible traits enabling *G. graeca* to persist in the only temporarily wet vegetation Mediterranean shrub land. In addition, amplified fragment length polymorphism (AFLP), (Vos et al. 1995) was employed to estimate the proportions of vegetative (leading to clones) and generative reproduction (producing new genotypes) in the species.

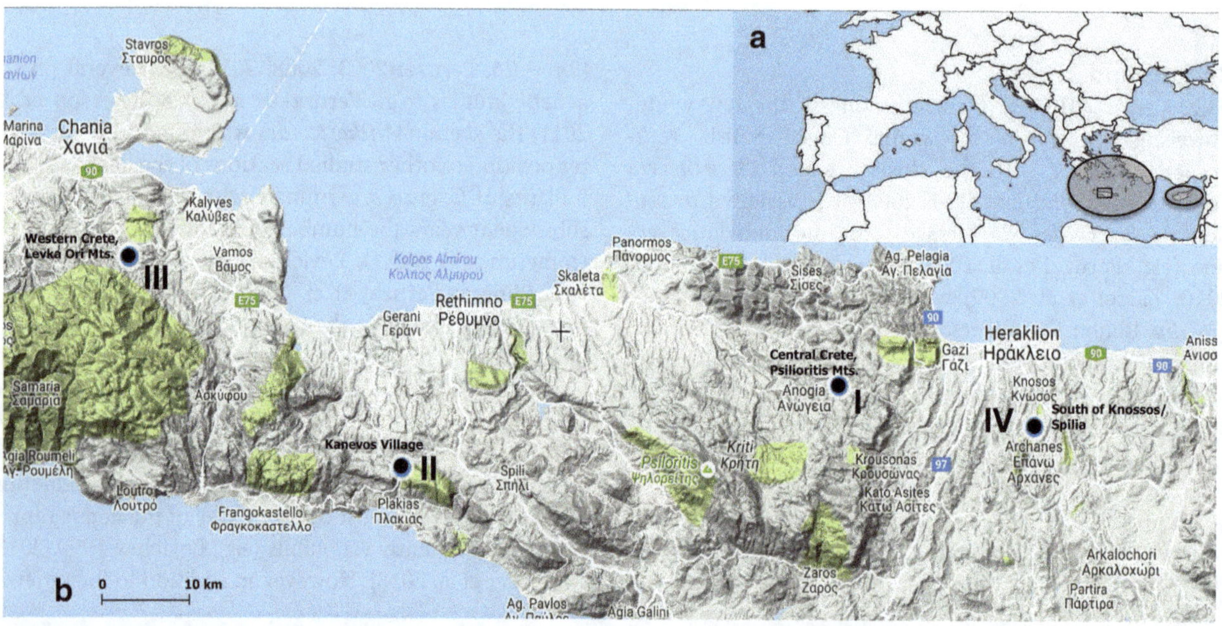

Fig. 1 **a** Distribution of *Gagea graeca* in the Mediterranean region (encircled regions). **b** Collecting sites for the four studied populations from the central part of the island of Crete (marked with an rectangle in **a**); map compiled using the geospatial conservation assessment tool GeoCAT (http://geocat.kew.org/)

Methods

Study region

Gagea graeca was investigated during spring of 2013 and 2014 at the island of Crete, where the species is common (Jahn et al. 1995). In these habitats that are often disturbed by grazing livestock, *G. graeca* inhabits slightly disturbed places especially where at least some soil moisture will be preserved during the summer, to prevent the bulbs from desiccation (Fig. 2). Table 1 gives numbers of plants investigated for morphological measurements, C/N data, seed set and AFLP analyses.

Four populations of Crete (Fig. 1) were investigated (numbered I to IV; Fig. 1b, Table 1):

I. Central Crete, Psilioritis Mts., Anogia, along a little brook in phrygana ca. 1 km SE Gonies, road fom Gonies to Anogia, at a branch that leads to a windmill, slopes with fine soil between rocks (flysch), elev. 600 ± 25 m a.s.l., 35°17′58 "N, 24°55′36″ E ± 25 m,

II. Central Crete, southern coast near Plakias, temporarily wet fallow field ca. 800 m SE Kanevos village, along the gravel road to a small chapel, ascent to Kouroupu hills, elev. 476 ± 25 m a.s.l., 35°13 ′37″ N, 24°24′10″ E ± 25 m,

III. Western Crete, Levka Ori Mts., Chania, rocks and fallow Olive orchards in *Quercus ilex*/*Pistacia lentiscus* woodlands: about 1 km N Tsakistra village, road between Tsakistra and Kampoi, elev. 581 ± 30 m a.s.l., 35°24′33″ N, 24°04′16″ E ± 25 m,

IV. Central Crete, south of Knossos/Spilia near Iraklio, phrygana in succession to *Quercus* shrubland: ca. 1450 m SW of Kera Eleousa village, elev. 222 ± 20 m a.s.l., 35°15′52″ N, 25°09′38″ E ± 25 m.

Morphological investigations

All morphological studies were carried out on living plants towards the end of the flowering season (March 15–30, 2014) from populations I and II (Table 1). At this time, most of the plants already developed fruits (capsules); a few plants were still in bloom. Of these 405 plants, 114 were in non-flowering condition.

Using a ruler and/or a digital caliper (precision ± 0.02 mm) we measured the diameter of the replacement bulbs (including the thin tunic of the former parent bulb; see Schnittler et al. 2013) and the diameter/length of the usually single basal bulbil and its short to extremely short stolon. In addition, we recorded number of flowers, shoot height above ground, and numbers of basal and stem leaves. For all measures, the standard deviation (SD) is given. To determine a threshold for the diameter of the replacement bulb needed to develop bulbils and/or flowers, these graphs were fitted against the equation $y = 1/[1 + e^{-(x-xo)/b}]$ which describes a sigmoid function with xo as the threshold (the turning point of its slope). For the graphs, flowers with only the male function (aborted and much smaller capsules) were counted as 0.5, perfect flowers as 1 (capsules are of normal size and fertile).

Seed production

About a month later (between May 1 and May 20), the number of seeds per plant was counted for 177 plants from populations I, II and IV (Table 1). To record variation in seed production, all formerly flowering plants (those with a developed shoot) for a certain area (1–2 m^2) were collected. At this time, capsules were fully developed and all vegetative plant parts were dried up. In a few plants capsules started to dehisce; such plants were not collected.

Resource allocation

Material from populations I, II and IV (Table 1) was used to determine absolute contents of carbon and nitrogen. Plants were dried for about 4 days in phosphorus pentoxide, and dry mass of the respective plant parts (seeds, replacement bulbs and vegetative structures) was determined with a precision of ± 1 µg (for seeds, 100 seeds were pooled for weighing). Several items of these plant structures were cut or pooled in a way that dry mass ranges between 6 and 15 mg. This weight corresponds with one average-sized replacement bulb, vegetative remains from a single plant, or ca. 100 seeds. Using replicate samples, a total of five replicates were measured except for basal bulbils. This procedure yielded 42 samples in total which were wrapped into tin foil and analyzed for carbon (C) and nitrogen (N) content with an automatic analyzer Vario EL III. Using the means of 3–10 samples, N-contents were calculated and used as a resource estimator (compare Ashman 1994). Figures for N content per mg dry mass were used to estimate nitrogen content of the respective plant parts, and calculate their relative investments in terms of nitrogen incorporated.

AFLP analyses

For a total of 110 sampled plants from populations I and II (Table 1) AFLP profiles were generated, using the methods described in Pfeiffer et al. (2011). For these analyses two transects (populations I and II with 27 and 28 pairs of plants, respectively, see Table 1) were sampled. We employed the sampling scheme of Pfeiffer et al. (2011), where pairs of plants (maximum distance 10 cm) were collected along a transect line (minimum distance between pairs 1 m).

We did five/six full replicates (including DNA extraction) for the first/second transect to determine the error rate. Due to the high DNA content of many *Liliaceae*

Fig. 2 Growth sites for *Gagea graeca*. **a**, **b** Population I along a small gully in the phrygana of the Psilioritis Mts. **c**, d Population II at a temporarily wet fallow field. **e**, **f** Population III at former olive orchards, now invaded by holm oak, at a slope

Table 1 Numbers of plants investigated for morphological measurements, C/N data, seed production and seed weight, and AFLP analyses from the four investigated populations

Population	I	II	III	IV	Total
Morphology	184	221	–	–	405
Seed production	59	64	–	54	177
Seed weight	8	10	–	6	24
C/N data	14	–	16	12	42
AFLP analyses	54	56	–	–	110

(Veselý et al. 2012; Zonneveld et al. 2015), we used a primer combination (EcoRI-AGCG plus Vsp-GCAT) with four selective bases. The first selective base was added to the pre-amplification primers, and all four were added to the primers used in the main amplification. The electroferograms obtained with a sequencer ABI 310 (applied biosystems) were binned and the alleles, defined as peak presence/absence, were read automatically to a 0/1 matrix using GenMapper 4.0 software (applied biosystems). To identify clones, an algorithm was programmed in Microsoft Excel which does not only check for identity, but allows to define a threshold to account for genotyping (biological, experimental, and scoring) errors (Douhovnikoff and Dodd 2003; Schnittler and Eusemann 2010). This threshold was derived from histograms of pairwise distances between samples. The histograms showed a bimodal distribution with two peaks, one for non-identical plants, and one for putative clones, with a local minimum (the threshold) in between. This threshold was below the pairwise distances between replicates. All samples with AFLP profiles deviating from each other in more alleles than defined by the threshold were regarded as independent genotypes.

Results

Morphology

Although *Gagea graeca* was always seen in populations of several hundred plants or more, the slender growth makes the plants quite inconspicuous. Only at flowering time, the plants are rather conspicuous (Fig. 3a), developing 1–3, rarely more, white flowers.

All measurements of morphological characters are summarized in Table 2. Large non-flowering plants develop two, more rarely three basal leaves. In flowering plants, which may have up to four basal leaves of 3–8 cm length, an upright shoot (82 ± 60 mm tall, range 25–414 mm) develops. In large flowering plants, the axils of the basal leaves can develop additional shorter shoots with usually one flower. The 2–6 scape leaves are much shorter and rarely exceed 4 cm in length, in the axils of the upper in addition to the single terminal flower further

Fig. 3 Morphology of *Gagea graeca*. **a** Strong flowering plant with two scapes and four basal leaves. **b** Closeup of a flower showing reddish veins (the color is even more pronounced at the outer side). **c** Longitudinal section through a flower, showing the blunt capsule. **d** Basal part of a flowering plant with one bulbil, showing the reddish tunics formed by the sheaths of basal leaves. **e** Three flowering plants showing variation in stolon length; the plant on the right comes close to the maximum. **f** Longitudinal section through a replacement bulb. Bar length indicates in **a** 1 cm, **b**–**e** 3 mm, **f** 2.5 mm

Table 2 Life history characteristics of two species of *Gagea* from arid environments (*Gagea graeca, G. bulbifera*) compared with a species from temperate deciduous woodlands (*G. lutea*)

Species	*G. graeca*	*n*	*G. bulbifera*	*n*	*G. lutea*	*n*
Reference	This study		Beisenova et al. (2015)		Schnittler et al. (2009), Pfeiffer et al. (2011)	
Morphology						
Bulb size, flowering plants	3.01 ± 0.63	*291*	2.55 ± 0.48	*291*	8.94 ± 1.62	*239*
Threshold (mm bulb diam.) for						
Flowers	2.17 ± 0.05	*405*	1.90 ± 0.13	*417*	6.94 ± 0.04	*505*
Bulbils	2.80 ± 0.16	*405*	4.96 ± 11.39	*417*	2.63 ± 0.21	*505*
Difference	−0.63	*405*	−3.06	*417*	4.31	*505*
Bulbils per plant showing bulbils (range)	1.01 ± 0.10 (1–2)	*194*	1.00 ± 0.00	*26*	8.65 ± 5.31 (1–23)	*226*
Average bulbil size	1.26 ± 0.39	*194*	1.11 ± 0.48	*26*	1.49 ± 0.42	*226*
Switch present?	No	*405*	No	*417*	Yes	*505*
Seed set						
Flowers per plant	1.9 ± 1.1 (1–7)	*177*	1.4 ± 0.9 (1–9)	*230*	3.2 ± 1.5 (1–8)	*217*
Capsules per fruiting plant	1.2 ± 0.6 (1–4)	*123*	1.4 ± 0.9 (0–9)	*210*	3.2 ± 1.5 (0–8)	*216*
Mean seed weight (μg)	73 ± 22	*24*	80 ± 20	*10*	1650 ± 340	*10*
Seeds per flowering plant	68 ± 79 (0–445)	*177*	119 ± 90 (0–827)	*230*	41 ± 30 (0–159)	*217*
Seeds per fruiting plant	98 ± 77 (0–445)	*123*	130 ± 86 (0–827)	*210*	41 ± 30 (0–159)	*216*
Seeds per capsule	80 ± 39 (0–184)	*123*	99 ± 40 (0–224)	*210*	12 ± 6 (0–33)	*216*
Resource allocation						
N-content (μg/mg dry weight)						
Replacement bulbs	27.9 ± 2.6	*9*	23.9 ± 1.8	*5*	8.3 ± 3.5	*10*
Seeds	41.2 ± 6.6	*24*	22.1 ± 1.8	*5*	25.7 ± 1.9	*10*
N investment (%) into						
Replacement bulbs	67.1	*9*	36.4	*5*	25	*10*
Vegetative structures	3.8	*9*	21.1	*5*	ca. 60	*10*
Basal bulbils	3.9	–	0.5	*1*	–	–
Axillary bulbils	–	–	4.5	*5*	–	–
Seeds	25.2	*24*	36.4	*5*	> 15	*10*
AFLP data						
Diversity R = (G−1)/(N−1)	0.95	*110*	1.00	*10*	0.56	*141*
Proportion of clonal plants	0.09	*110*	0.00	*10*	0.62	*141*

Mean values ± SD and ranges (in parentheses) are given as well as the number of plants or structures (n) investigated. For seed weight, 100 seeds were counted and weighed together. Due to the switch in *G. lutea* (flowering plants do not produce bulbils any more), investment in bulbils and flowers cannot be compared directly for this species. Data sets for morphology, seed set, resource allocation and AFLP analyses include different plants

flowers may develop. About one-third of all plants (98 of 291 flowering plants) developed only a single terminal flower; larger plants carry additional lateral flowers, but these are often much smaller and seem to be predominantly male, since ovaries are often underdeveloped. On average, flowering plants developed 1.9 ± 1.1 (1–7) flowers. The white tepals are obovate (Fig. 3b, c), broadest above the middle and show one larger central and two smaller lateral green veins, which turn conspicuously reddish towards the tip of the tepal. The six anthers are very pale yellow. Developing flowers are nodding and become upright orientated only during anthesis. Only a part of the flowers (mostly the terminal ones) develop into strictly upright capsules which can reach up to 12 mm length; the style remains usually on the tip of the capsule.

The replacement bulbs are always close (1–3 cm) to the surface, and are hidden by the reddish-brown sheaths of the basal leaves. The first basal leaf, which is as well the first to decay during flowering season, develops the replacement bulb; the second, very rarely the third as well, develops usually one basal bulbil which is on a saccate enlargement of the basal leaf, which forms a short stolon. Larger non-flowering plants (34 of 114, mean bulb diam. 2.20 ± 0.40) developed a bulbil (Fig. 3d–f). From 291 flowering plants measured, the weaker individuals (131 plants, mean bulb diam. 2.72 ± 0.56) did not yet develop a bulbil, the larger ones (160 plants, mean bulb diam. 3.24 ± 0.69 mm) supported a basal bulbil (diam. 1.26 ± 0.39 mm). Among these were two with a second, smaller basal bulbil (sister bulbil; diam. 1.17 ± 0.36 mm). Bulbils developed often on a short (0.42 ± 0.71 mm, range 0–4.01 mm) stolon. Compared with other species in the genus, the replacement bulbs of *G. graeca* stay small even in richly flowering plants; the maximum bulb size recorded by us was 5.12 mm.

Figure 4a, b shows flower and bulbil formation in dependency from the size of the replacement bulb. Only strong plants (exceeding 4 mm bulb diameter) develop always a bulbil. The threshold for flower formation is lower (2.17 ± 0.05 mm bulb diameter) than the threshold for the formation of the basal bulbil (2.80 ± 0.16 mm).

Seed production

A total of 177 plants from three populations (Table 1) with altogether 339 flowers and 165 seed capsules were counted. Seed production increased strongly with bulbil size class (Fig. 4c). Plants had on average 1.9 ± 1.1 (1–7) flowers, but the single flower of weak plants and the terminal flower(s) of larger plants did often fail to set seeds. Therefore, only 123 of 177 investigated plants were fertile; this cohort developed 1.2 ± 0.6 capsules with 98 ± 77 (0–445) seeds.

Apart from the 54 sterile plants, most individuals developed one capsule (n = 124, 69 ± 45 seeds), eight plants two capsules (112 ± 97 seeds), seven plants three capsules (333 ± 97 seeds), and one plant four capsules (263 seeds). Seed set in plants increased not proportionally with the number of flowers (Fig. 5a). Only about two-thirds, in plants with more flowers even less, of the flowers developed seed capsules (Fig. 5b). Seeds are very lightweight: average seed weight (measured for samples of 100 units) was 86 ± 22, 75 ± 14 and 48 ± 5 µg for populations I, II and IV (10, 8, and 6 samples are 100 seeds measured, respectively), and population IV differs significantly from the two other populations (Mann–Whitney U test, $p = 0.05$).

Resource allocation

Table 3 shows carbon and nitrogen content in different plant structures. For the three investigated plant structures replacement bulbs, seeds and vegetative parts C contents are comparable (Fig. 6a), whereas N content is much higher in seeds and bulbs (Fig. 6b). Except for population IV, where seeds had a much lower average weight, seeds store 1–1.5 times more N per unit dry mass than bulbs.

With known amounts of plant structures used for measurements, the relative investment of a plant into the respective structures can be estimated. The N content of an average-sized replacement bulb was measured as 470.40 ± 20.05 µg (n = 9). Bulbils were too small and occurred too rarely to measure N content. If the size of a bulb is approximated by the volume of a sphere with the diameter of the respective structure (replacement bulbs: 3.24 ± 0.69 mm; basal bulbils 1.26 ± 0.39 mm), the ratio between average volumes of replacement bulbs and bulbils is 17:1. If we assume that bulbils have the same N-content per unit volume as replacement bulbs, the N content of an average bulbil can be estimated as

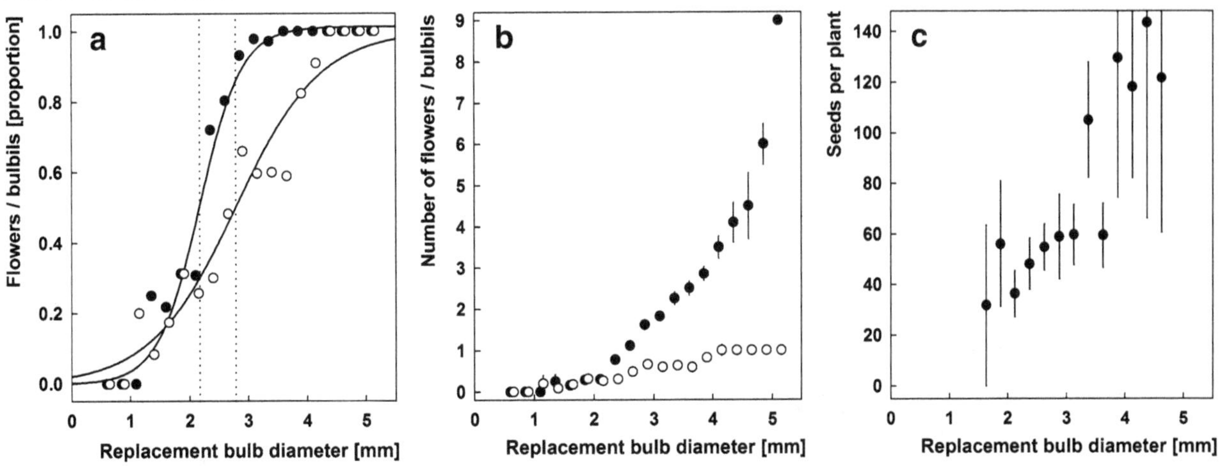

Fig. 4 Reproductive pattern of *Gagea graeca*, shown are cohorts of plants corresponding to classes of 0.25 mm diameter width of the replacement bulb (n = 405). **a** Proportions of plants forming flowers (black circles) and a basal bulbil (white circles, the latter are printed with a slight shift to the right to make overlaps with the black circles visible). Graphs were fitted against the equation $y = 1/[1 + e^{-(x-xo)/b}]$ with xo as the threshold for the formation of the respective structure (indicated by a vertical dotted line). **b** Average numbers of basal bulbils (white circles) and flowers (black) per plant in the respective diameter classes. **c** Seed set. Bars for the latter two plots denote the standard error of means

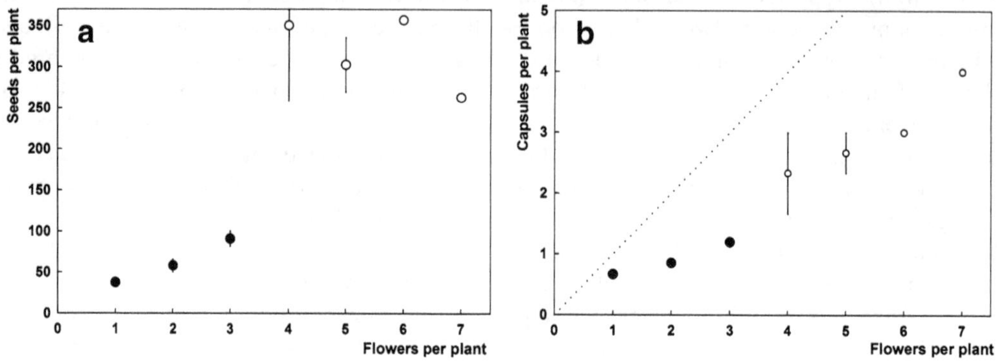

Fig. 5 Seed set for *Gagea graeca* in dependence form the numbers of flowers (**a**); and the relation of capsules per flower (**b**). Most plants hat 1–3 flowers (these classes are represented by at least 40 plants each, black dots), whereas only eight plants had 4–7 flowers (white dots). Bars indicate the standard error of means

Table 3 Mean N and C contents measured for bulbs, seeds and vegetative parts of *Gagea graeca*

Plant organ	*n*	N (μg/mg)	C (μg/mg)	C/N
Bulbs	9	27.9 ± 2.6	641.5 ± 62.5	22.9
Seeds	24	41.2 ± 6.6	486.0 ± 72.6	11.8
Veg. structures	9	5.6 ± 0.8	510.1 ± 62.6	90.3

Samples were pooled to reach a total weight between 6 and 15 mg (equaling one bulb, 100 seeds, and vegetative parts of 3 plants)

27.67 μg. The absolute N content for 100 seeds was measured as 259.97 ± 66.53 μg. Since a plant forms on average 68 ± 79 seeds (including plants with lacking seed set, see Table 2), the N content of 68 seeds would be 176.78 μg. Similarly, the vegetative parts of nine plants were measured (N content 79.10 ± 4.56 μg). Comparison of the absolute N contents of these structures gives the following picture: flowering plants of *G. graeca* allocate most of their resources (67.1%) to the replacement bulbs, 3.9% to the usually single basal bulbil, 25.2% to an average of 68 seeds, and 3.8% remain in vegetative plant structures.

Genetic diversity

The AFLP analysis with four selective bases resulted in a total of 299 (population I) and 259 (population II) readable alleles (peaks taller than 150 relative fluorescence units). With fully automated bin set and scoring we obtained on average 27.0 ± 4.5 (range 22–33, population I) and 25.9 ± 12.4 (range 10–43, population II) deviating peaks between replicates for populations, translating to error rates of 10.4 and 8.3%, respectively. Although this figures are rather high, the histogram for the pairwise comparisons of all samples (classes of five differing alleles) showed a clear gap between allele differences for replicates/clonal samples and all others, leading to upper thresholds for clonal identity of 40 and 28 deviating alleles, respectively. The non-clonal samples differed on average in 75.1 ± 8.1 (range 43–98) and 61.7 ± 10.9

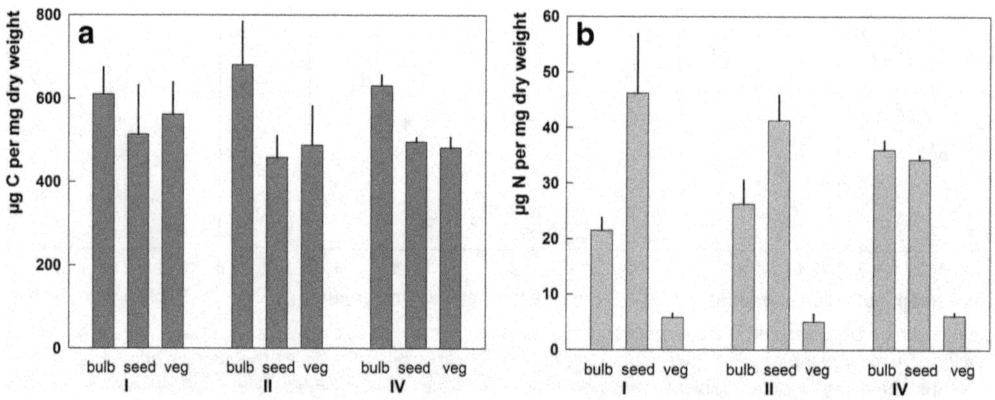

Fig. 6 Investment of carbon (**a**) and nitrogen (**b**) in different plant parts for three populations (I, II and IV) of *Gagea graeca*. Shown are mean ± SD of elementary contents (μg per mg dry weight) for bulbs (bulb, n = 3), seeds (seed, n = 6–10) and vegetative plant structures (veg, n = 3)

(range 29–103) alleles (populations I and II, respectively). Population I revealed five pairs of plants that were assigned to clones (Fig. 7), population II consisted of singletons only. This correlated with a lower genetic diversity for the first population, if expressed as the mean Jaccard distance (I: 0.48 ± 0.04; II: 0.57 ± 0.08). In total, 9.1% of all investigated individuals belonged to clones.

Discussion

Gagea graeca differs from other species of *Gagea* in the Mediterranean area (which are mostly members of the section *Didymobulbos*, Tison et al. 2013) by its slender habit and the solitary growth of plants. Typically, in species of *Gagea* relying on reproduction by subterranean bulbils the plants form clusters, whereas plants of *G. graeca* usually grow apart from each other. A high proportion of the plants in all populations collected by us were always seen in flowering condition. Therefore, sampling for the morphological measurements focused on small, non-flowering plants, to have all classes of parent bulbil size more evenly represented than it would be the case in a random sample from the population. We nevertheless could only achieve a proportion of 28% non-flowering plants. Although *G. graeca* is certainly a short-lived perennial indicated by the significant resource allocation to replacement bulbs (and true annuals are not known for *Liliaceae*, Peruzzi 2016) the habit of the plant with its comparatively large flowers resembles that of an annual. In addition, in agreement with Greuter (1970) in our quantitative study we rarely observed sister bulbils in *G. graeca* (two of 194 plants were seen with two bulbils). Peruzzi et al. (2008) did not found such a structure at any stage of the ontogeny (herbarium collections, cultivated plants).

Measuring or counting a number of morphological traits as demonstrated in Schnittler et al. (2009) quantifies these visible adaptations and allows to relate them to the resource status of the plant, here estimated by the size of the replacement bulb, which carries the plant through the summer dormancy period into the next year (Levichev 1999, 2013). These measurements confirm reports about plant ontogenesis (Peruzzi et al. 2008). Although the age of the plants cannot be determined, the measurements allow to determine threshold (in terms of bulbil size) for the development of seeds and bulbils. For the species of the genus *Gagea*, the relation between sexual (seeds) and asexual (bulbils) reproduction seems to be a crucial adaptive trait (Schnittler et al. 2013) and is determined by these thresholds (as the diameter of the replacement bulb where 50% of all plants develop the respective structure, Schnittler et al. 2009) and the number of seeds and bulbils. Therefore, resource allocation into seeds and bulbils can be assumed to be under high selective pressure. Several traits should influence the seed/bulbil ratio. First, seeds show a higher desiccation tolerance than bulbils, making them true dormant stages. A second factor, often seen in the genus *Gagea*, is partial or full sterility caused by high (and especially odd) ploidy levels. This constrains a higher investment into bulbils as the only remaining mode of reproduction. Third, large genome size, as often found in monocotyledonous plants, especially *Liliales*, (Leitch et al. 2010) tends to complicate meiosis and thus sexual reproduction. Taking these factors into account, species of *Gagea* in arid habitats should be diploid (or add least have a low even ploidy level) and have small genomes to allow predominant reproduction via seeds as the more drought resistant diaspore type. Indeed, a meta analysis of *Liliaceae* indicated counter selection against large genomes in arid climates (Carta and Peruzzi 2016).

If the environment allows, increased bulbil production can function as an exit strategy for missing seed set, as shown by a comparison between *G. lutea* (hexaploid) and *G. spathacea* (nonaploid, Schnittler et al. 2009): the latter species compensates missing seed set, most likely caused by incorrect pairing of chromosomes due to their odd number, with an increase in the number of basal bulbils and seems to be fully clonal (Pfeiffer et al. 2011). To compare the amount of resources invested into both types of diaspores, we used nitrogen content as a proxy (compare Ashman 1994). Shifts in the bulbil/seed ratio have a strong impact on clonal diversity of a species (since not

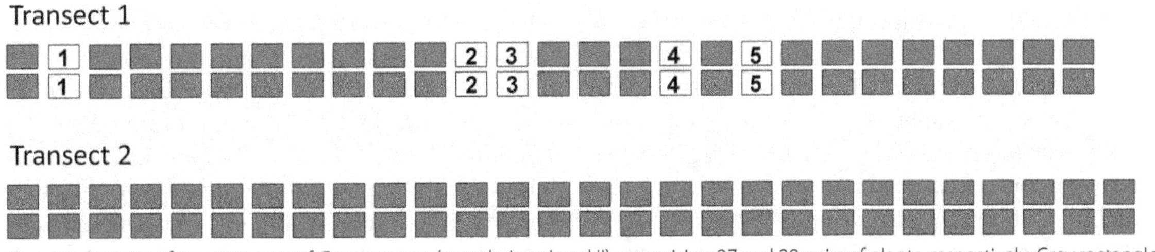

Fig. 7 Genetic diversity of two transects of *Gagea graeca* (populations I and II) comprising 27 and 28 pairs of plants, respectively. Grey rectangles without numbers represent plants with unique genotypes; white rectangles indicate plants belonging to clones which are numbered

only numbers, but also establishment probabilities of the respective diaspore types matter), but are as well recognizable by different resource allocation and changes in numbers of these propagules.

Table 2 shows a comparison of the life history traits of a species from humid environments (*G. lutea*, Schnittler et al. 2009) and two species from arid environments, (*Gagea graeca*, this study, *G. bulbifera*, Beisenova et al. 2015). *Gagea lutea* is a typical species of the species-rich sect. *Gagea* and inhabits deciduous woodlands and meadows of the humid temperate zone. *Gagea graeca* (sect. *Antericoides*) and *G. bulbifera* (sect. *Bulbiferae*) occur in arid habitats (Mediterranean shrub lands and Middle/Central Asian steppes, respectively). The latter two species assume different positions in phylogenies of the genus (Peterson et al. 2008), but share nevertheless several key traits that are different from *G. lutea*. First, both species from arid areas have high fecundity (as indicated by the number of seeds per produced per capsule, Table 2). Second, the threshold for flower formation is lower than for the formation of bulbils. Third, only a single basal bulbil is formed. Fourth, an ontogenetic switch between bulbil and seed production is absent: bulbils and seeds can be produced simultaneously (see discussion below). Fifth, the genetic diversity, calculated as the proportion of singleton genotypes, is rather high due to dominating reproduction via seeds.

Another difference between species from humid and arid regions regards seed morphology: Together with the species of the section *Bulbiferae*, the two species of sect. *Anthericoides* share one important character (which may be ancestral for the genus, since it is occurring as well in other *Liliaceae* genera, Peruzzi 2016): the platyspermous (flattened) seeds. Several other sections (*Platyspermum*, *Plecostigma*, *Stipitatae*) of the genus *Gagea* are platyspermous, and these assume different positions in phylogenetic trees (Peterson et al. 2008). Many of these species are adapted to arid conditions and inhabit steppes of the Irano-Turanian and Saharo-Arab regions. One possible explanation would be to assume an adaptation to wind dispersal of seeds (Peterson et al. 2008). The comparison with *G. lutea* (Table 2), a species from humid environments with terete seeds, shows that the low seed weight seems to allow platyspermous species to produce more seeds. This should be an important precondition for an ephemerous life style. Seeds of the humid-zone species are about 20 times heavier than these of the two arid-zone species. Not surprisingly, *Gagea lutea* produced on average 41 seeds per flowering plant, the average for the studied populations of *G. graeca* is 68 (but it should be noted that ca. one-third of the sampled plants failed to set seed); and for *G. bulbifera* 119 seeds per plant were counted (Table 2). The difference between the species from humid and arid environments becomes obvious if seed production is compared with the average size of replacement bulbs: if bulb volume is estimated as that of a sphere from the figures for diameter (Table 2), *G. lutea* produces 0.1 seeds/mm^3 bulb volume, but numbers for *G. graeca* and *G. bulbifera* are 4.8 and 13.7, respectively. Figures for bulbil production per mm^3 replacement bulb volume are comparable (0.13 vs. 0.07 and 0.11 bulbils/mm^3 bulb volume). Due to the switch for *G. lutea* the latter figure was calculated from the average dimeter of bulbil-producing, non-flowering plants (5.00 ± 1.51 mm). This indicates a clear shift in reproductive output from bulbils (*G. lutea*) towards seeds (*G. bulbifera, G. graeca*).

Gagea lutea differs from the two arid-zone species by another important character: a switch in ontogenesis between bulbil production and seed production (Schnittler et al. 2009). If plants gain resources (indicated by larger replacement bulbs), they produce first only bulbils, not flowers. Above a certain threshold in parent bulb size, only flowers, no bulbils, are developed. In such species, the resource threshold (expressed as diameter of the parent bulb) for bulbil production is always lower than that for flower/seed development (Schnittler et al. 2013). Species lacking a switch start to develop a single to several bulbils as soon as the replacement bulb has reached a certain diameter and then continue to form them indefinitely throughout the life of the plant. In such species, the threshold for flower/seed development may be lower than that for bulbil development, as it is the case in *G. bulbifera* and *G. graeca* (Table 2). From this reason, a direct comparison in terms of resources allocated to bulbils *vs.* seeds is impossible due to the switch in *G. lutea*: plants of this species never invest simultaneously into seeds and bulbils.

The results of AFLP genotyping reflect the consequences of a) seed and bulbil production and b) absence/presence of a switch for genetic diversity in a species. The AFLP data for *G. graeca* showed a small proportion of clonal plants (9%, Fig. 7), with only five pairs of clonal plants found in the first of the two transects. This is comparable with results from *G. bulbifera* (0% clonal plants, Beisenova et al. 2015) but not with the two remaining species of the genus with known data (*G. lutea*, 62%, Pfeiffer et al. 2011; *G. spathacea*, 100%, Pfeiffer et al. 2012). Only flowering plants of *G. graeca* were used for the AFLP analysis, which all can be expected to form one bulbil/year. In fact, in the neighborhood of a plant (the second member of a pair) we found in 50 cases a non-clonal plant (likely to be germinated from seeds, but we cannot rule out the possibility that they originate from the bulbil of a third plant which was not sampled), in five cases a clonal plant (derived from a bulbil of the first plant). If we compare these relationships (bulb to seed production 1:68 = 0.014; clonal to non-clonal descendants 5:50 = 0.100), we obtain roughly a sevenfold higher establishment rate of bulbils compared with seeds. The two structures are different in size: bulbils are nearly spherical and on average 0.48 mm in diam., equaling a volume of 0.46 mm^3; seeds are flattened and approximately 0.1 × 0.15 × 1.3 mm, equaling a volume (if estimated as that of an ellipsoid) of 0.08 mm^3. These differences should influence the amount of nutrients stored as well as the desiccation tolerance and explain why bulbils are likely to have a higher chance of establishment.

These considerations suggest that it is not the failure of bulbils to recruit new plants, but the comparatively high seed production of *Gagea graeca* that leads to a high genetic diversity of the populations. Seed production should nevertheless face a limit in arid climates: the more and the earlier in ontogenesis resources will be invested in seeds, the lesser resources remain for the replacement bulb which guarantees survival of the plant. But the relation volume to surface which we must assume to be an important parameter for desiccation tolerance, decreases with decreasing diameter of the bulb. In contrast to *G. bulbifera*, where bulbs can hold effectively water via sclerophyllous roots (see Levichev 1999), the bulbs of *G. graeca* lack this adaptation and may thus face a high risk of desiccation during the long Mediterranean summer. This limits the proportion of resources invested into seeds, since plants with large bulbs can be expected to have a higher chance of survival. Indeed, the average size of the bulbils is higher in *G. graeca* in comparison to *G. bulbifera* (3 vs. 2.5 mm), and so is the proportion of nitrogen invested into the replacement bulb (62% vs. 36%, Table 2). However, figures for relative investment into plant parts cannot be compared directly from two reasons: first, the measurements for *G. lutea* were obtained at the peak of flowering season, these for *G. bulbifera* near seed maturation, whereas the measured plants of *G. graeca* were already completely dried at the time of measurement. Therefore, storage processes may have been not yet completed in the first two species; which should lead to an underestimation of the relative investment into dormant parts (bulbs, bulbils, seeds) but overestimation for vegetative parts in the first two species. Second, *G. lutea* shows a switch – bulbils and seeds are never found at one plant, which does not allow a direct comparison.

Growth patterns in *Gagea* seem to be rather stable for a species (Levichev 2013). With a quantitative analysis of the morphology for a sampled population, thresholds for bulbil formation and flowering can be defined (Schnittler et al. 2009). For all hitherto investigated species from humid climates, thresholds for bulbil formation are lower than these for flower formation (Schnittler et al. 2013); the exception are the two species from arid climates, *Gagea bulbifera* and *G. graeca* (Table 2). Such shifts towards earlier generative reproductions may be a general trend for perennial herbs in arid habitats. Franks and Weis (2015) observed in a 5-year drought experiment with *Brassica rapa* that drought selected for plants that flowered at a smaller size and earlier ontogenetic stage. These changes seem to correspond with a lower level of neoteny and a much smaller genome in the arid-zone species (see discussion in Peruzzi et al. 2009). For *Gagea lutea*, direct measurements resulted in DNA contents (1C) of 19.75 pg (Greilhuber et al. 2000) and 21.35 pg (Zonneveld et al. 2015). Respective values for the species from arid environments, inferred from chromosome total haploid length are 4.00–6.96 (mean 5.48) pg for the species of sect. *Antericoides* and 8.65 pg for *G. bulbifera* (Peruzzi 2012a).

In contrast to the differences mentioned above, the trend towards andromonecy (weak plants produce male flowers only, and even in larger plants the last flower is often male) seems to be a general tendency in the genus. For *Gagea lutea*, this was reported by Nishikawa (1998), and male flowers were reported as well in the two species from arid zones. In the plants of *G. graeca* investigated for morphology we counted 95 flowering plants with one flower only, in 57 of these the flower was apparently hermaphroditic (mean bulb diam. 2.51 mm), in 38 it was apparently male (mean bulb diam. 2.31 mm). This is indirectly confirmed by the high proportion of weak sterile plants (30%) in the cohort sampled for seed set. An androdioecious breeding system was reported for several species in the genera *Gagea*, especially the species formerly assigned to *Lloydia* (Patterson and Givnish 2002; Manicacci and Despres 2001; Peruzzi et al. 2008). This fits into the pattern of frequent occurrence of female-sterile breeding systems observed in *Liliaceae* and was suggested to result from size-dependent sex allocation (Peruzzi 2012b). In addition to the finding that often the smallest plants of a cohort are male (*Lilium apertum*, Zhang et al. 2014; *Fritillaria montana*, Peruzzi et al. 2012), in *Gagea* spp. usually the last flowers, when the replacement bulb may run out of resources, are male. We did not investigate the pollination system for *G. graeca* but assume facultative autogamy, which was found as well in *G. lutea* (Pfeiffer et al. 2013). For *G. graeca*, Peruzzi et al. (2008) obtained viable seeds from isolated plants in cultivation.

In accordance with the general habit of the plants, our data suggest that the two species from arid environments (*Gagea graeca* and *G. bulbifera*), although belonging to divergent clades in phylogenetic trees, are more similar in life history than to *G. lutea* from a humid climate (Table 2). Not directly accessible is plant age, where the most remarkable differences may be expected. Since in

Gagea the plants replace all structures of the vegetation body completely every year, and only the replacement bulb survives dormancy periods, only a mark-revisit approach over several years would provide reliable age data. In the field we found for *G. graeca* only a low proportion (28%) of non-flowering plants. Looking for the distribution of plants over size classes of 0.25 bulb diameter, the highest number of plants (56) was found for 2.50–2.75 mm diam. We therefore assume that *G. graeca* is a short-lived perennial and may flower and produce seeds already in the second year of its life (see also Peruzzi et al. 2008).

Conclusions

Adaptations to arid environments in the genus *Gagea* (*Liliaceae*) are characterized by (i) a predominance of sexual reproduction and (ii) a lower reproductive age (lower threshold for flower development). This is connected with a reduction in size (especially bulb diameter). In contrast to most species of the genus, *Gagea graeca* seems to be a short-lived perennial, possessing several traits that are rather typical for annual plants.

Authors' contributions

MS designed the study, carried out most of the data analyses and wrote the manuscript, assisted by AP, JP, and AN. Field work and morphological analyses were carried out by AP, JP and MS, data for seed set were contributed by CB. AFLP analyses were performed by AN and AK; the C/N analyses by AN. All authors read and approved the final manuscript.

Author details

[1] Institute of Botany and Landscape Ecology, Ernst Moritz Arndt University Greifswald, Soldmannstr. 15, 17487 Greifswald, Germany. [2] Institute of Biology, Martin-Luther-University of Halle-Wittenberg, Weinbergweg 10, 06120 Halle/Saale, Germany. [3] State Office for Environmental Protection of Saxony-Anhalt, Reideburger Str. 47, 06116 Halle/Saale, Germany. [4] L.N. Gumilyov Eurasian National University, 5 Munaytpassov Str, 010008 Astana, Kazakhstan.

Acknowledgements

For generous help with accommodation and logistics we which to thank Andreas Traxler (Lentas, Crete), for help with plant measurements the student participants of the field trip to Crete 2013. Mohamed el Dadamouny, Greifswald, helped with compiling the map of occurrences.

Competing interests

The authors declare that they have no competing interests.

Funding

The project received no special funding. AN wants to thank as well the authorities of L.N. Gumilyev Natl. Univ., Astana, Kazakhstan, for supporting three research stays in Germany.

References

Ashman TL (1994) Reproductive allocation in hermaphrodite and female plants of *Sidalcea oregana* ssp. *spicata* (*Malvaceae*) using four currencies. Am J Bot 81:433–438

Beisenova S, Peterson A, Peterson J, Bersimbayev RI, Klahr A, Schnittler M (2015) On the limits of drought—life history of *Gagea bulbifera*. Flora 210:72–79

Carta A, Peruzzi L (2016) Testing the large genome constraint hypothesis: plant traits, habitat and climate seasonality in *Liliaceae*. New Phytol 210(2):709–716

Danin A, Danin B (2015) Flora of Israel online. Jerusalem. http://flora.org.il/en/plants/. Accessed 10 Nov 2016

Douhovnikoff V, Dodd RS (2003) Intra-clonal variation and a similarity threshold for identification of clones: application to *Salix exigua* using AFLP molecular markers. Theor Appl Genet 106:1307–1315

Franks SJ, Weis AE (2015) A change in climate causes rapid evolution of multiple life-history traits and their interactions in an annual plant. J Evol Biol 21:1321–1334

Greilhuber J, Ebert I, Lorenz A, Vyskot B (2000) Origin of facultative heterochromatin in the endosperm of *Gagea lutea* (*Liliaceae*). Protoplasma 212:217–226

Greuter W (1970) The taxonomic position of *Lloydia graeca* (*Liliaceae*) and related species. Israel J Bot 19:155–160

Jahn R, Schönfelder P, Mayer A, Scheuerer M (1995) Exkursionsflora für Kreta. Ulmer, p 446

Leitch IJ, Beaulieu JM, Chase MW, Leitch AR, Fay MF (2010) Genome size dynamics and evolution in monocots. J Bot 2010:e862516

Levichev IG (1999) Zur Morphologie in der Gattung *Gagea* Salisb. (*Liliaceae*) I Die unterirdischen Organe. Flora 194:379–392

Levichev IG (2013) Structural features of shoots in *Lloydia*, *Gagea*, *Kharkevichia* (*Liliaceae*) as evolutionary variability of the modules of mesome nature in monokotyledons. Bot Zhurn 98:409–453 **(In Russian)**

Manicacci D, Despres L (2001) Male and hermaphrodite flowers in the alpine lily *Lloydia serotina*. J Bot 79:1107–1114

Nishikawa Y (1998) The function of multiple flowers of a spring ephemeral, *Gagea lutea* (*Liliaceae*), with reference to blooming order. Canad J Bot 76:1404–1411

Patterson TB, Givnish TJ (2002) Phylogeny, concerted convergence, and phylogenetic niche conservatism in the core *Liliales*: insights from *rbcL* and *ndhF* sequence data. Evolution 56:233–252

Peruzzi L (2003) Contribution to the cytotaxonomical knowledge of *Gagea* Salisb. (*Liliaceae*) sect. *Foliatae* A. Terracc. and synthesis of karyological data. Caryologia 56:115–128

Peruzzi L (2008) Contribution to the cytotaxonomical knowledge of the genus *Gagea* Salisb. (*Liliaceae*). III New karyological data from the central Mediterranean area. Caryologica 61(1):92–106

Peruzzi L (2012a) Chromosome diversity and evolution in *Gagea* (*Liliaceae*). Bocconea 24:147–158

Peruzzi L (2012b) Male flowers in *Liliaceae* are more frequent than previously thought. Bocconea 24:301–304

Peruzzi L (2016) A new infrafamilial taxonomic setting for *Liliaceae*, with a key to genera and families. Plant Biosyst 150:1341–1347

Peruzzi L, Tison J-M, Peterson A, Peterson J (2008) On the phylogenetic position and taxonomic value of *Gagea trinervia* (Viv.) Greuter and *Gagea* sect. *Anthericoides* A. Terracc. (*Liliaceae*). Taxon 57(4):1201–1214

Peruzzi L, Leitch IJ, Caparelli KF (2009) Chromosome diversity and evolution in *Liliaceae*. Ann Bot 103:459–475

Peruzzi L, Mancuso E, Gargano D (2012) Males are cheaper, or the extreme consequence of size/age dependent sex allocation: sexist gender diphasy in *Fritillaria montana* (*Liliaceae*). Bot J Linn Soc 168(3):323–333

Peterson A, John H, Koch E, Peterson J (2004) A molecular phylogeny of the genus *Gagea* (*Liliaceae*) in Germany inferred from non-coding chloroplast and nuclear DNA sequences. Plant Syst Evol 245:145–162

Peterson A, Levichev IG, Peterson J (2008) Systematics of *Gagea* and *Lloydia* (*Liliaceae*) and infrageneric classification of *Gagea* based on molecular and morphological data. Mol Phyl Evol 46:446–465

Peterson A, Levichev IG, Peterson J, Harpke D, Schnittler M (2011) New insights into the phylogeny and taxonomy of Chinese species of *Gagea* (*Liliaceae*)—speciation through hybridization. Org Div Evol 11:387–407

Peterson A, Harpke D, Levichev IG, Beisenova S, Schnittler M, Peterson J (2016) Morphological and molecular investigations of *Gagea* (*Liliaceae*) in southeastern Kazakhstan with special reference to putative altitudinal hybrid zones. Plant Syst Evol 302:985–1007

Pfeiffer T, Klahr A, Heinrich A, Schnittler M (2011) Does sex make a difference? Genetic diversity and spatial genetic structure in two co-occurring species of *Gagea* (*Liliaceae*) with contrasting reproductive strategies. Plant Syst Evol 292:189–201

Pfeiffer T, Klahr A, Peterson A, Levichev IG, Schnittler M (2012) No sex at all? Extremely low genetic diversity in *Gagea spathacea* (*Liliaceae*) across Europe. Flora 207:372–378

Pfeiffer T, Harter DEV, Formella N, Schnittler M (2013) Reproductive isolation vs. inter-breeding between *Gagea lutea* (L.) Ker Gawl. and *G. pratensis* (Pers.) Dumort. (*Liliaceae*) and their putative hybrids in Mecklenburg-Western Pomerania. Plant Spec Biol 28:193–203

Schnittler M, Eusemann P (2010) Consequences of genotyping errors for estimation of clonality: a case study on *Populus euphratica* Oliv. (*Salicaceae*). Evol Ecol 24:1417–1432

Schnittler M, Pfeiffer T, Harter DEV, Hamann A (2009) Bulbils contra seeds: reproductive investment in two species of *Gagea* (*Liliaceae*). Plant Syst Evol 279:29–40

Schnittler M, Peterson A, Peterson J, Beisenova S, Bersimbaev RI, Pfeiffer T (2013) Minor differences with big consequences: reproductive patterns in the genus *Gagea* (*Liliaceae*). Flora 208:591–598

Strid A (2016) Atlas of the aegean flora. Part 2. Maps. Englera 33(2):1–878

Tekşen M, Erkul SK (2015) The synopsis of the genus *Gagea* (*Liliaceae*) in Turkey. Phytotaxa 230:101–129

Tison J-M (2016) *Gagea graeca*. The IUCN red list of threatened species 2016: eT13142235A18611875. http://dx.doi.org/102305. Accessed 10 Nov 2016

Tison J-M, Peterson A, Harpke D, Peruzzi L (2013) Reticulate evolution of the critical Mediterranean *Gagea* sect *Didymobulbos* (*Liliaceae*) and its taxonomic implications. Plant Syst Evol 299:413–438

Veselý P, Bureš P, Šmarda P, Pavlíček T (2012) Genome size and DNA base composition of geophytes: the mirror of phenology and ecology? Ann Bot 109:65–75

Vos P, Hogers R, Bleeker M et al (1995) AFLP: a new technique for DNA fingerprinting. Nucleic Acids Res 23:4407–4414

Zarrei M, Wilkin P, Fay MF, Ingrouille MJ, Zarre S, Chase MW (2009) Molecular systematics of *Gagea* and *Lloydia* (*Liliaceae*; *Liliales*): implications of analyses of nuclear ribosomal and plastid DNA sequences for infrageneric classification. Ann Bot 104:125–142

Zarrei M, Wilkin P, Ingrouille MJ, Chase MW (2011) A revised infrageneric classification for *Gagea* (*Tulipeae*; *Liliaceae*): insights from DNA sequence and morphological data. Phytotaxa 15:44–56

Zhang Z-Q, Zhu X-F, Sun H, Yang Y-P, Barrett SCH (2014) Size dependent gender modification in *Lilium apertum* (Liliaceae): does this exhibit gender diphasy? Ann Bot 114: 441–453

Zimowski M, Leuschner HH, Gärtner H, Bergmeier E (2014) Age and diversity of Mediterranean dwarf shrublands: a dendrochronological approach along an altitudinal gradient on Crete. J Veget Sci 25:122–134

Zonneveld BJM, te Linde B, van den Berg L-J (2015) Genome sizes of 227 accessions of *Gagea* (*Liliaceae*) discriminate between the species from the Netherlands and reveal new ploidies in *Gagea*. Springerplus 4:395

5

Aboveground biomass estimation at different scales for subtropical forests in China

Shunlei Peng[1,2*], Nianpeng He[1,3], Guirui Yu[1,3*] and Qiufeng Wang[1,3]

Abstract

Background: The accurate estimation of forest biomass at different scales is the critical step in the assessment of forest carbon stocks. We used three models at increasing scales: allometric model at ecoregional scale (model 1), dummy variable allometric model at both ecoregion and regional scales (model 2), and allometric model at regional scale (model 3) to estimate the aboveground biomass of six subtropical forests in China. Furthermore, we also tested whether wood density can improve the accuracy of the allometric model at regional scale.

Results: Aboveground biomass estimates for six subtropical forests were significantly affected by the ecoregions ($p < 0.05$). Model 1 and model 2 had good fitness with higher values of R^2, lower *RSE* (residual standard error) and *MPSE* (mean percent standard error) than model 3. The values of *MPSE* for model 1, model 2, and model 3 ranged from 2.79 to 30.40%, 5.15 to 40.94%, and 13.25 to 80.81% at ecoregion scale, respectively. At regional scale, *MPSE* of model 2 was very similar to that of model 1, and was less than model 3. New allometric models with wood density had greater R^2, lower *RSE* and *MPSE* than the traditional allometric models without wood density variable for six subtropical forests at regional scale.

Conclusion: The dummy variable allometric models have better performances to estimate aboveground biomass for six subtropical forests in China, which provided an effective approach to improve the compatibility of forest biomass estimations from different scales. New allometric models with wood density substantially improved accuracies of aboveground biomass estimation for subtropical forests at regional scale.

Keywords: Aboveground biomass, Dummy variable model, Wood density, Scale, Allometric equation

Background

Carbon (C) sequestration and accumulation in forests as aboveground biomass (AGB) is important for mitigating climate change. The estimation of forest biomass at a range of scales has been recognized as one of the most critical steps in the assessment of forest C stocks (Montagu et al. 2005; Tomppo et al. 2010). Tropical and subtropical forests have been reported to account for more than 40% of the global gross primary production (GPP) and net primary production (NPP) (Zhou et al. 2006; Beer et al. 2010; Pan et al. 2011). Long-term eddy covariance observations demonstrate that average net ecosystem production (NEP) of East Asian subtropical forests is 362 g C m^{-2} $year^{-1}$, greater than that of Asian tropical and temperate forests, and also higher than that of forests at the same latitude in North America, Europe and Africa (Yu et al. 2014). Subtropical forest biome in China covers 2.5×10^6 km^2, occupies about 25% of the total forest area in China (Wu 1995), and plays critical role in C sink and climate change regulating (Zhou et al. 2006; Tan et al. 2011; Yu et al. 2014). However, C budgets of these forests remain uncertain resulted in limited number of inventory plot biomass data and accuracies of allometric equations for estimating AGB of forests in subtropical region (Zhang et al. 2005; Xu et al. 2015; Xiang et al. 2016). Thus,

*Correspondence: pengshunlei@163.com; yugr@igsnrr.ac.cn
[1] Key Laboratory of Ecosystem Network Observation and Modeling, Institute of Geographic Sciences and Natural Resources Research, CAS, Beijing 100101, People's Republic of China
Full list of author information is available at the end of the article

developing allometric equations for subtropical forests is essential for accurate estimating C sequestration in subtropical region (Zaehle et al. 2006; Hudiburg et al. 2009).

Field inventory methods (e.g., harvest method, allomatric modeling, and biomass expansion factor methods) are often used to estimate forest biomass at local and regional scales (Brown et al. 1989; Fang et al. 2001; Wang 2006; Pajtik et al. 2008; Williams et al. 2012). Remote sensing methods can provide spatial information on AGB at large scales, but this method still linked with the relationship between remote sensing dataset and field inventory AGB dataset (Drake et al. 2003; Su et al. 2016). Thus, many scientists gave efforts to improve the tree allometric models at single tree, plot, regional, national, or even worldwide scales, using easily measured dimensional variables, such as diameter at breast high (*DBH*) and tree height (*H*) (Brown et al. 1989; Ter-Mikaelian and Korzukhin 1997; Chave et al. 2005; Návar 2009; Genet et al. 2011). However, different models may lead to greatly variation of biomass estimation because of difference in climatic conditions, site quality, and forest types (Muukkonen 2007; Fu et al. 2017). Therefore, sampling at different scales and creating general biomass model were very important to reduce the uncertainty of applying different models (Chave et al. 2014). Some studies have advanced the possibility of generalizing allometric equations across regional boundaries (Návar et al. 2013; Paul et al. 2013; Chave et al. 2014). Zeng et al. (2011) used dummy variable model to develop generalized biomass model of *Pinus massoniana* at regional scale in south China, and indicated dummy model had good performance. However, few study has compared the accuracy of these allometric equations for forest biomass estimations from site to regional scales (Návar et al. 2013). Moreover, allometric models at different scales for other main subtropical forests in china, such as evergreen broadleaf forest, deciduous broadleaf forest, and mixed forests were very lack (Xu et al. 2015; Xiang et al. 2016). Some studies indicate that wood density variable can greatly improve accuracies of biomass model for AGB estimates in tropical forests and subtropical evergreen broadleaved forest (Baker et al. 2004; Chave et al. 2005, 2014; Goodman et al. 2014; Xu et al. 2015). Chave et al. (2014) successfully developed the universal allometric model for tropical forests with wood density based on global database, which were widely used for AGB estimation in tropical forests. However, the performance of the allometric model with wood density was worth further testing in subtropical forests. Therefore, the development of generalized biomass allometric model at different scales was urgent to quantify the regional biomass and C storage of subtropical forests.

Subtropical region in China has varied ecological zones and forest types (Chinese Academy of Sciences 2001). It was very necessary to develop general biomass models based on ecological region to improve accuracies of AGB estimation. The main objectives of this study were (1) to develop the allometric models at different scales for aboveground biomass estimation of subtropical forests in China; (2) to assessment the accuracy of the allometric models at different scales for AGB estimates, and (3) to test the performance of the allometric model when wood density variable is available.

Materials and methods

The experimental site

The study region covered most of the subtropical regions of China (22°–34°N, 98°–123°E) extending across eight ecoregions (Fig. 1). The total forested area is approximately 2.5 × 10^6 km^2 in China (Wu 1995). The region was classified eight ecological zones (Table 1; Fig. 1) (Fu et al. 2013). The mean annual precipitation (MAP) ranges from 831 to 1342 mm and the mean annual temperature (MAT) varies from 12.5 to 19.1 °C (Table 1). The primary forests in this region were evergreen broadleaf forests (EBF). Due to long-term anthropogenic disturbances, the current forests were not primitive and were classified into six categories: *Cunninghamia lanceolata* (CL), coniferous mixed broadleaf forest (CMBF), subtropical deciduous broadleaf forest (DBF), evergreen broadleaf forest (EBF), *Eucalyptus* tree species forest (ETS), and *Pinus massoniana* (PM) based on China's vegetation classification system (Chinese Academy of Sciences 2001).

The dataset

The AGB data of the six subtropical forests (CL, CMBF, DBF, EBF, ETS, and PM) was collected from the large forest biomass dataset in China, which was compiled from published biomass studies and pre-existing datasets published between 1978 and 2008 (Luo et al. 2013).

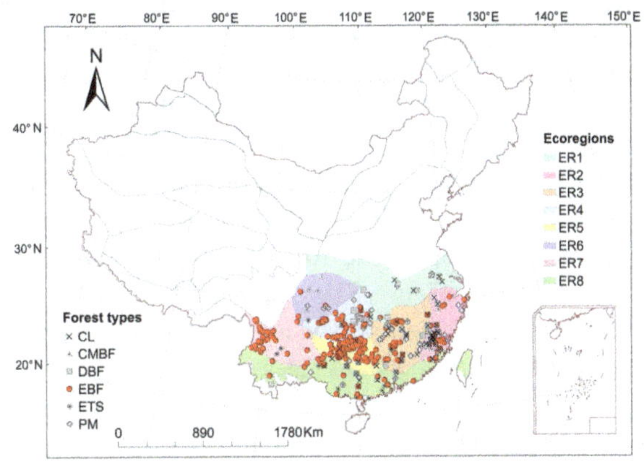

Fig. 1 The sampling plot locations for the six forest types in the subtropical region of China. See Table 1 for the abbreviations of the forest types and ecoregions

Table 1 Summary characteristics of six forests and eight ecoregions in the subtropical region of China

Forest types	N	n	D²H			AGB (kg)		
			Min	Max	Mean	Min	Max	Mean
Cunninghamia lanceolata forest (CL)	6	219	16.97	23,469.40	2878.59	1.25	358.95	49.86
Coniferous mixed broadleaf forest (CMBF)	6	77	20.39	9973.85	1587.48	1.35	156.32	42.84
Deciduous broadleaf forest (DBF)	5	40	50.69	9481.50	2157.55	2.76	240.26	59.28
Evergreen broadleaf forest (EBF)	6	277	80.69	42,360.52	4492.07	3.85	719.51	121.74
Eucalyptus tree species forest (ETS)	5	76	84.24	6149.86	1882.65	2.25	201.05	46.72
Pinus massoniana forest (PM)	7	283	16.25	19,822.5	2297.85	1.01	388.74	53.75
Ecoregions	M	n	MAT (°C)			MAP (mm)		
			Min	Max	Mean	Min	Max	Mean
Yangtze river delta ecological zone (ER1)	4	55	5.3	17.2	14.0	491	1381	907
Evergreen broadleaf forest ecological zone in the mountains of Zhejiang and Fujian provinces (ER2)	6	254	11.6	20.4	16.5	1022	1653	1342
Ecological zone in Jiangnan and Nanling mountains and hill (ER3)	6	185	13.0	20.7	16.9	1069	1626	1385
Evergreen broadleaf forest ecological zone in the mountains of the west Hunan, Guizhou and Hubei provinces (ER4)	6	161	7.4	17.7	14.8	798	1356	1073
Karst evergreen broadleaf forest and agricultural ecological zone in Guizhou and Guangxi provinces (ER5)	3	121	10.1	21.1	16.9	887	1493	1254
Ecological zone of Sichuan Basin (ER6)	2	39	7.3	18.1	15.4	677	1232	911
Ecological zone on Yunnan Plateau (ER7)	2	36	− 8.3	19.6	12.5	618	1129	831
South humid subtropical ecological zone (ER8)	6	121	6.9	22.9	19.1	833	1770	1252

Abbreviations in the brackets indicated each forest type and individual ecoregion, respectively in the subtropical region of China. Climate data, including mean annual temperature (MAT) and mean annual precipitation (MAP) was obtained from the National Climate Center (http://ncc.cma.gov.cn/cn/)

N number of forest distributed ecoregions, *n* number of sampling plots, *M* number of forest types

Moreover, we reviewed almost all related publications in China from 2008 to 2013 and recorded information of sampling plots on locations, forest types, stand age, stand density, DBH, tree height, wood density and AGB for the six subtropical forest types. The biomass components of sample trees (stems, branches, leaves, etc.) was measured using destructive harvesting and oven weighing method. Then, AGB of average tree (kg) for each plot was calculated from stand AGB (Mg ha^{-1}) and stand density. Our criteria for selected sampling tress was that the sample trees for each forest types were distributed as evenly as possible in the diameter classes in our dataset. Together, we collected 972 records of plot measured AGB data for the average trees of six subtropical forest types (Table 1; Additional file 1: Figure S1), 316 records of which had wood density information. The locations and forest types of the dataset were shown in Fig. 1.

Model description

Allometric model at different scales

We used general allometric equations to estimate AGB of six subtropical forests at individual ecoregional scale (model 1) and all subtropical regional scale (model 3), respectively. These allometric equations based on D^2H (D, diameter of the tree at breast height, cm, and H, tree height, m), which have been widely used to estimate AGB for forests (Jenkins et al. 2003; Muukkonen 2007; Návar 2009).

$$\ln(AGB) = a + b\ln(D^2H) + \varepsilon \tag{1}$$

where AGB is the aboveground biomass, a and b are parameters, and ε is the additive error. Then, the estimate of aboveground biomass is as follows:

$$AGB_{est} = \exp\left(a + RSE^2/2\right) \times \left(D^2H\right)^b \tag{2}$$

where RSE is the residual standard errors of the regressions.

We considered ecoregion as dummy variable and used dummy variable allometric model to estimate AGB for six subtropical forests at both ecoregion scale and regional scale (model 2). The general form of the dummy variable allometric model was as follows (Wang et al. 2008; Zeng et al. 2011).

$$\ln(AGB) = a_0 + \sum a_i z_i + b\ln(D^2H) + \varepsilon \quad (3)$$

$$AGB_{est} = \exp\left(a_0 + \sum a_i z_i + RSE^2/2\right) \times \left(D^2H\right)^b \quad (4)$$

where z_i is the dummy variable, a_i is the ecoregion-specific parameter, and other symbols are the same as Eqs. (1) and (2). The dummy variables are 0, 1. In model 2, we take each ecoregion as dummy variable, if forest has six distributed ecoregions, we used six dummy variables, z_1, z_2, z_3, z_4, z_5 and z_6, when $z_1 = 1$, the others = 0, when $z_2 = 1$, the others = 0, etc. (Zeng et al. 2011).

Testing the importance of wood density for accurate estimated AGB

At regional scale, we used 316 plot AGB data with wood density (*WD*) records to test whether *WD* improved accuracy of allometric model. $D^2H \times WD$ as variable was used to fit allometric model compared with the model without *WD* variable (Chave et al. 2014; Xu et al. 2015).

$$\ln(AGB) = a + b\ln(D^2H \times WD) + \varepsilon \quad (5)$$

then, the estimate of biomass is as follows:

$$AGB_{est} = \exp\left(a + RSE^2/2\right) \times \left(D^2H \times WD\right)^b \quad (6)$$

Accuracy assessment of models

The coefficient of determination (R^2), residual standard error of the regression (*RSE*), and mean percent standard error (*MPSE*) were used to assess the accuracies of the models (Zeng et al. 2011). *MPSE* were defined as follows:

$$MPSE = \frac{1}{n}\sum |(y_i - \hat{y})/\hat{y}| \times 100 \quad (7)$$

where n is number of plots; y_i and $\hat{y}$ are the observed and estimated values of AGB respectively.

Results

Allometric models for aboveground biomass estimation at different scales

The values of AGB and D^2H varied obviously in each forest type and each ecoregion, especially in DBF, ETS, CMBF and EBF (Table 1; Additional file 1: Figure S1). There was clear distinction of climate (e.g., MAT and MAP) among eight ecoregions (Table 1).

Model 1 had better performance for CL in ER4 ($R^2 = 0.987$) and ER5 ($R^2 = 0.986$), for CMBF in ER4 ($R^2 = 0.978$) and ER6 ($R^2 = 0.961$), for DBF in ER1 ($R^2 = 0.975$) and ER4 ($R^2 = 0.961$), for EBF in ER7 ($R^2 = 0.978$) and ER4 ($R^2 = 0.968$), for ETS in ER2 ($R^2 = 0.999$) and ER4 ($R^2 = 0.979$), and for PM in ER5 ($R^2 = 0.986$), ER6 ($R^2 = 0.979$), ER3 ($R^2 = 0.970$) and ER2 ($R^2 = 0.960$), and showed lower performance for CMBF in ER2 ($R^2 = 0.889$), and for EBF in ER8 ($R^2 = 0.885$) (Table 2; Additional file 1: Figure S2).

Model 2 considered the effects of the ecoregions on AGB estimation. The statistical results showed ER2 significantly influenced AGB estimations for CL ($p < 0.05$), and for PM ($p < 0.01$), respectively. ER8 significantly affected on AGB estimations for DBF ($p < 0.001$), EBF ($p < 0.001$), CMBF ($p < 0.05$) and ETS ($p < 0.05$). ER4, ER5, ER3, and ER8 significantly affected on AGB estimation for EBF, especially ER4 and ER5 ($p < 0.0001$) (Table 3; Additional file 1: Figure S3).

Model 2 had better performance to estimate AGB for PM ($R^2 = 0.964$), ETS ($R^2 = 0.954$), and CL ($R^2 = 0.947$), and showed lower performance to estimate AGB for EBF ($R^2 = 0.915$) and CMBF ($R^2 = 0.920$) (Table 3; Additional file 1: Figure S3).

Model 3 showed better performance to estimate AGB for PM ($R^2 = 0.959$) and CL ($R^2 = 0.944$), and had lower performance to estimate AGB for the other four forest types, especially for ETS ($R^2 = 0.759$) (Table 4; Additional file 1: Figure S4).

Assessment the accuracies of the allometric models at different scales

We compared measured AGB and predicted AGB estimated from three allometric models at different scales, and found that model 1 and model 2 had better accuracies for AGB estimations than model 3 (Fig. 2, Table 5). The *MPSE* values of three models varied obviously for AGB estimations in the same forest in different ecoregions, and showed increasing trend with increasing scales. At ecoregional scale, the values of *MPSE* from model 2 were similar to model 1, less than model 3 for CL, PM and DBF in the distributed ecoregions except for CL and PM in ER5, and for DBF in ER3 and ER4. For EBF and ETS, *MPSE* showed the same trend in the ecoregions except for EBF in ER1 and ETS in ER2. For all the forests, *MPSE* of Model 1, model 2, and model 3 ranged from 2.79 to 30.40%, 5.15 to 40.94%, and 13.25 to 80.81% at ecoregional scale, respectively. At regional scale, *MPSE* of model 2 was very similar to model 1, and was clearly less than model 3 in six subtropical forests (Table 5).

Testing importance of wood density for Aboveground biomass estimation

Figure 3 showed that WD variable used in allometric model greatly improved the estimate accuracies with

Table 2 Parameters of allometric models for estimating aboveground biomass of six forests at individual ecoregion scale in the subtropical region of China

Forests	Ecoregions	Allometric models at individual ecoregion scale (model 1)	*n*	R^2	*RSE*	*F* value
CL	ER1	$\exp(-3.006+0.008)\times(D^2H)^{0.857}$	14	0.918	0.123	134.9***
	ER2	$\exp(-2.065+0.029)\times(D^2H)^{0.762}$	92	0.946	0.241	1574.6***
	ER3	$\exp(-1.431+0.034)\times(D^2H)^{0.669}$	57	0.925	0.260	679.1***
	ER4	$\exp(-1.659+0.009)\times(D^2H)^{0.688}$	18	0.987	0.138	1203.8***
	ER5	$\exp(-3.611+0.017)\times(D^2H)^{0.936}$	12	0.986	0.184	690.6***
	ER8	$\exp(-2.067+0.033)\times(D^2H)^{0.749}$	26	0.954	0.256	499.8***
CMBF	ER1	$\exp(-4.280+0.011)\times(D^2H)^{1.169}$	7	0.955	0.148	104.8***
	ER2	$\exp(-1.746+0.057)\times(D^2H)^{0.744}$	34	0.889	0.338	255.6***
	ER3	$\exp(-0.527+0.044)\times(D^2H)^{0.552}$	7	0.912	0.295	51.5***
	ER4	$\exp(-0.615+0.001)\times(D^2H)^{0.590}$	7	0.978	0.053	222.3***
	ER6	$\exp(-1.953+0.043)\times(D^2H)^{0.777}$	14	0.961	0.294	293.6***
	ER8	$\exp(-1.391+0.048)\times(D^2H)^{0.72}$	8	0.932	0.102	82.2***
DBF	ER1	$\exp(-2.054+0.019)\times(D^2H)^{0.803}$	11	0.975	0.195	305.3***
	ER2	$\exp(-2.202+0.036)\times(D^2H)^{0.774}$	11	0.955	0.270	191.6***
	ER3	$\exp(-14.935+0.106)\times(D^2H)^{2.623}$	5	0.904	0.460	28.2*
	ER4	$\exp(-4.803+0.020)\times(D^2H)^{1.120}$	7	0.961	0.198	122.9***
	ER8	$\exp(-0.841+0.034)\times(D^2H)^{0.722}$	6	0.908	0.262	39.3**
EBF	ER2	$\exp(-2.909+0.054)\times(D^2H)^{0.920}$	55	0.932	0.328	729.7***
	ER3	$\exp(-0.482+0.022)\times(D^2H)^{0.592}$	29	0.901	0.209	245.2***
	ER4	$\exp(-1.699+0.011)\times(D^2H)^{0.820}$	48	0.968	0.149	1372.4***
	ER5	$\exp(-0.610+0.023)\times(D^2H)^{0.656}$	89	0.913	0.216	909.0***
	ER7	$\exp(-2.067+0.004)\times(D^2H)^{0.817}$	28	0.978	0.086	1168.7***
	ER8	$\exp(-2.761+0.068)\times(D^2H)^{0.911}$	28	0.885	0.368	200.8***
ETS	ER2	$\exp(-1.305+0.000)\times(D^2H)^{0.687}$	9	0.999	0.017	5074.8***
	ER3	$\exp(-5.646+0.029)\times(D^2H)^{1.245}$	12	0.900	0.240	90.0***
	ER4	$\exp(-3.615+0.001)\times(D^2H)^{0.859}$	6	0.979	0.044	184.0***
	ER7	$\exp(-1.352+0.021)\times(D^2H)^{0.631}$	8	0.924	0.204	73.3***
	ER8	$\exp(-3.062+0.026)\times(D^2H)^{0.905}$	41	0.904	0.230	368.8***
PM	ER1	$\exp(-2.515+0.020)\times(D^2H)^{0.843}$	23	0.948	0.201	379.7***
	ER2	$\exp(-2.071+0.040)\times(D^2H)^{0.804}$	53	0.960	0.283	1214.7***
	ER3	$\exp(-2.176+0.036)\times(D^2H)^{0.798}$	75	0.970	0.217	2374.4***
	ER4	$\exp(-2.589+0.037)\times(D^2H)^{0.839}$	75	0.942	0.271	1181.8***
	ER5	$\exp(-3.448+0.011)\times(D^2H)^{0.973}$	20	0.986	0.146	1301.2***
	ER6	$\exp(-2.366+0.018)\times(D^2H)^{0.831}$	25	0.979	0.187	1058.0***
	ER8	$\exp(-2.030+0.081)\times(D^2H)^{0.797}$	12	0.935	0.403	143.1***

*** Indicates significant at $p < 0.001$ level; ** indicates significant at $p < 0.01$ level, * indicates significant at $p < 0.05$ level. See Table 1 for the abbreviations of the forest types and ecoregions

higher R^2, lower *RSE* and *MPSE* than traditional allometric model without WD variable for six forests (Figs. 3, 4). The allometric model with WD variable developed by Chave et al. (2014) showed lower *MPSE* than traditional model for CL, DBF, EBF and PM, especially in EBF and DBF, and showed greater *MPSE* for AGB estimations in CL, CMBF, DBF, ETS and PM than allometric model with WD variable from our dataset (Fig. 4). In EBF, model created by Chave et al. (2014) showed similar lower *MPSE* to our model with WD variable (Fig. 4).

Discussion

Allometric models for aboveground biomass estimation at different scales

Many scientists gave efforts to improve the tree allometric models at single tree, plot, regional, national, or even worldwide scales (Brown et al. 1989; Chave et al. 2005; Návar 2009; Genet et al. 2011). In this study, we developed three allometric models from ecoregion to regional scales. Three allometric models using D^2H as the predictive variable offered good fitness of AGB allometric

Table 3 Parameters of dummy variable allometric model for estimating aboveground biomass of six forest types at both regional scale and ecoregion scale in the subtropical region of China

Forests	Ecoregions	Dummy variable allometric model (model 2)	R^2	*RSE*	*F* value
CL	General	$\exp(-2.064 + 0.166 + 0.142 + 0.083 - 0.009 + 0.071 + 0.031) \times (D^2H)^{0.739}$	0.947	0.249	631.5***
	ER1	$\exp(-2.064 + 0.031) \times (D^2H)^{0.739}$			
	ER2*	$\exp(-2.064 + 0.166 + 0.031) \times (D^2H)^{0.739}$			
	ER3	$\exp(-2.064 + 0.142 + 0.031) \times (D^2H)^{0.739}$			
	ER4	$\exp(-2.064 + 0.083 + 0.031) \times (D^2H)^{0.739}$			
	ER5	$\exp(-2.064 - 0.009 + 0.031) \times (D^2H)^{0.739}$			
	ER8	$\exp(-2.064 + 0.071 + 0.031) \times (D^2H)^{0.739}$			
CMBF	General	$\exp(-1.391 - 0.18 - 0.143 - 0.134 - 0.239 + 0.398 + 0.048) \times (D^2H)^{0.720}$	0.920	0.311	134.5***
	ER1	$\exp(-1.391 + 0.048) \times (D^2H)^{0.720}$			
	ER2	$\exp(-1.391 - 0.180 + 0.048) \times (D^2H)^{0.720}$			
	ER3	$\exp(-1.391 - 0.143 + 0.048) \times (D^2H)^{0.720}$			
	ER4	$\exp(-1.391 - 0.134 + 0.048) \times (D^2H)^{0.720}$			
	ER6	$\exp(-1.391 - 0.239 + 0.048) \times (D^2H)^{0.720}$			
	ER8*	$\exp(-1.391 + 0.398 + 0.048) \times (D^2H)^{0.720}$			
DBF	General	$\exp(-2.191 - 0.338 + 0.328 - 0.177 + 0.803 + 0.045) \times (D^2H)^{0.823}$	0.940	0.301	106.6***
	ER1*	$\exp(-2.191 + 0.045) \times (D^2H)^{0.823}$			
	ER2	$\exp(-2.191 - 0.338 + 0.045) \times (D^2H)^{0.823}$			
	ER3	$\exp(-2.191 + 0.328 + 0.045) \times (D^2H)^{0.823}$			
	ER4	$\exp(-2.191 - 0.177 + 0.045) \times (D^2H)^{0.823}$			
	ER8***	$\exp(-2.191 + 0.803 + 0.045) \times (D^2H)^{0.823}$			
EBF	General	$\exp(-1.896 - 0.141 + 0.451 + 0.321 + 0.103 + 0.184 + 0.038) \times (D^2H)^{0.785}$	0.915	0.277	481.8***
	ER2	$\exp(-1.896 + 0.038) \times (D^2H)^{0.785}$			
	ER3*	$\exp(-1.896 - 0.141 + 0.038) \times (D^2H)^{0.785}$			
	ER4***	$\exp(-1.896 + 0.451 + 0.038) \times (D^2H)^{0.785}$			
	ER5***	$\exp(-1.896 + 0.321 + 0.038) \times (D^2H)^{0.785}$			
	ER7	$\exp(-1.896 + 0.103 + 0.038) \times (D^2H)^{0.785}$			
	ER8**	$\exp(-1.896 + 0.184 + 0.038) \times (D^2H)^{0.785}$			
ETS	General	$\exp(-2.495 - 0.015 - 1.918 - 0.221 - 0.232 + 0.028) \times (D^2H)^{0.859}$	0.954	0.237	288.5***
	ER2	$\exp(-2.495 + 0.028) \times (D^2H)^{0.859}$			
	ER3	$\exp(-2.495 - 0.015 + 0.028) \times (D^2H)^{0.859}$			
	ER4***	$\exp(-2.495 - 1.918 + 0.028) \times (D^2H)^{0.859}$			
	ER7*	$\exp(-2.248 - 0.221 + 0.028) \times (D^2H)^{0.859}$			
	ER8*	$\exp(-2.248 - 0.232 + 0.028) \times (D^2H)^{0.859}$			
PM	General	$\exp(-2.369 + 0.168 + 0.053 - 0.102 + 0.048 + 0.072 + 0.156 + 0.032) \times (D^2H)^{0.821}$	0.964	0.252	1055.0***
	ER1	$\exp(-2.369 + 0.032) \times (D^2H)^{0.821}$			
	ER2**	$\exp(-2.369 + 0.168 + 0.032) \times (D^2H)^{0.821}$			
	ER3	$\exp(-2.369 + 0.053 + 0.032) \times (D^2H)^{0.821}$			
	ER4	$\exp(-2.369 - 0.102 + 0.032) \times (D^2H)^{0.821}$			
	ER5	$\exp(-2.369 + 0.048 + 0.032) \times (D^2H)^{0.821}$			
	ER6	$\exp(-2.369 + 0.072 + 0.032) \times (D^2H)^{0.821}$			
	ER8	$\exp(-2.369 + 0.156 + 0.032) \times (D^2H)^{0.821}$			

*** Indicates significant at $p < 0.001$ level; ** indicates significant at $p < 0.01$ level; * indicates significant at $p < 0.05$ level. See Table 1 for the abbreviations of the forest types and ecoregions

Table 4 Parameters of allometric models for estimating aboveground biomass of six forests at regional ecoregion scale in the subtropical region of China

Forests	Allometric model at regional scale (model 3)	*n*	R^2	*RSE*	*F* value
CL	$\exp(-1.920 + 0.032) \times (D^2H)^{0.736}$	219	0.944	0.252	3668.5***
CMBF	$\exp(-1.485 + 0.062) \times (D^2H)^{0.718}$	77	0.890	0.353	607.4***
DBF	$\exp(-1.536 + 0.110) \times (D^2H)^{0.733}$	40	0.839	0.466	198.3***
EBF	$\exp(-1.481 + 0.056) \times (D^2H)^{0.757}$	277	0.874	0.334	1905.5***
ETS	$\exp(-3.776 + 0.138) \times (D^2H)^{0.995}$	76	0.759	0.526	233.4***
PM	$\exp(-2.394 + 0.036) \times (D^2H)^{0.830}$	283	0.959	0.267	6547.2***

*** Indicates significant at $p < 0.001$ level. See Table 1 for the abbreviations of the forest types

models at different scales (R^2 ranged from 0.759 to 0.999) (Tables 2, 3, 4). This indicates D^2H as variable could improve accuracy of models (Muukkonen 2007; Návar 2009; Xu et al. 2015). Muukkonen (2007) found that allometric equations with only DBH as an independent variable provided lower overall estimations of tree biomass. Models at different scales may lead to variation of biomass estimation because of difference of climatic conditions, site quality, and forest structures (Muukkonen 2007; Fu et al. 2017). In this study, we found that model 1 and model 2 had better accuracies for AGB estimations than model 3 (Fig. 2; Table 3). The *MPSE* values of three models varied obviously for AGB estimations in the same forest in different ecoregions, and showed increasing trend with increasing scales. Case and Hall (2008) found that prediction error of generalized tree biomass equations for ten species in the boreal forest region of west-central Canada increased from regional to national scale. However, *MPSE* of model 2 was very similar to model 1, and obviously less than model 3 in six forests (Table 5), which indicated that dummy variable allometric model considered ecoregion factors could be proposed as general model to estimate AGB for subtropical forests, and provide a more effective new approach to improve the

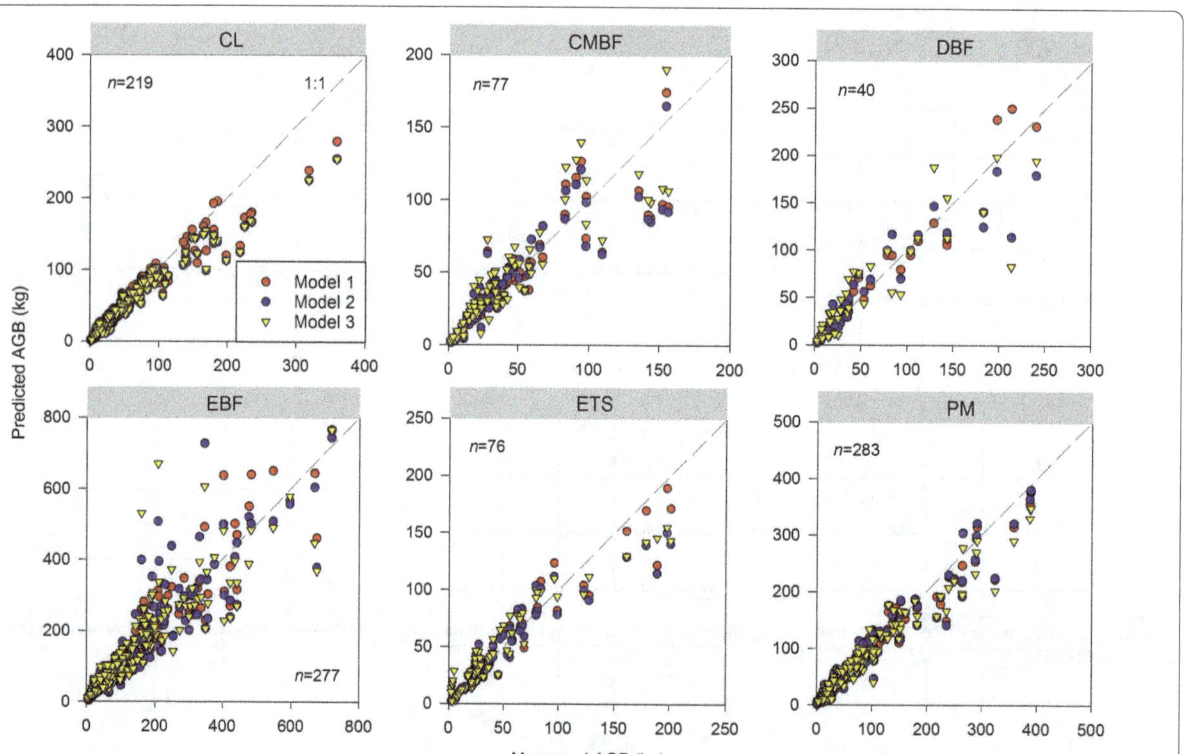

Fig. 2 Comparison measured AGB (aboveground tree biomass) and predicted AGB for six subtropical forests from three allometric models at different scales (model 1, model 2 and model 3) in China. Model 1: allometric model at ecoregion scale, model 2: dummy variable allometric model at both ecoregion scale and regional scale, and model 3: allometric model at regional scale. See Table 1 for the abbreviations of the six forest types

Table 5 ***MPSE*** **for three allometric models at different scales developed in the subtropical region of China**

Forests	Models	*MPSE* (%)								
		ER1	ER2	ER3	ER4	ER5	ER6	ER7	ER8	Region
CL	Model 1	9.51	19.38	20.26	10.06	11.72			20.41	17.92
	Model 2	10.83	19.27	21.49	11.71	29.08			21.07	19.44
	Model 3	26.04	21.66	23.99	21.99	27.36			24.02	23.01
CMBF	Model 1	9.66	26.40	17.89	3.64		22.08		7.75	19.31
	Model 2	20.97	26.16	30.69	7.45		25.55		23.19	23.98
	Model 3	20.68	28.28	28.79	13.25		31.29		53.38	29.42
DBF	Model 1	12.95	21.70	14.00	12.89				14.61	15.73
	Model 2	12.53	23.90	40.94	25.35				17.67	22.22
	Model 3	18.46	37.84	60.54	29.13				72.78	39.07
EBF	Model 1		25.24	13.26	9.82	16.41		5.61	30.40	17.02
	Model 2		30.74	17.25	9.87	18.32		5.29	30.65	19.14
	Model 3		32.27	29.85	24.61	21.70		11.01	31.03	25.02
ETS	Model 1		0.97	18.77	2.79			16.48	17.65	14.38
	Model 2		8.20	26.37	5.15			20.46	17.54	16.96
	Model 3		29.45	23.77	80.81			39.21	20.15	28.28
PM	Model 1	14.45	18.94	15.85	20.86	11.92	15.32		27.19	17.80
	Model 2	14.63	19.73	16.05	20.75	19.16	15.39		28.35	18.55
	Model 3	15.18	24.97	16.59	23.05	18.40	15.38		33.35	20.49

See Table 1 for the abbreviations of six forests and eight ecoregions

compatibility of forest biomass estimates at the ecoregional, and regional scales.

Assessment the accuracies of the allometric models at different scales

Regional climate data affected the precision of the regional model (Drake et al. 2003; Dewalt and Chave 2004; Chave et al. 2005; Wang 2006; Fu et al. 2017). In this study, MAT and MAP were clearly distinct among eight ecoregions (Table 1). Ecoregions including ER2, ER3, ER4, ER5, and ER8 significantly affected AGB estimations (Table 3). The aboveground biomass of PM, CL, and ETS was greater in the southern central regions with higher temperature and greater rainfall, than in the west regions with lower temperatures and less rainfall. The influences of climate were even more significant in EBF (Table 3). Therefore, forest regional climate data should be considered when the regional models were employed (Muukkonen 2007; Fu et al. 2017).

The number of plots applied to develop the allometric equations in ETS, DBF, and CBMF forest types ($N < 100$) may not be enough to represent the full range of species present at the study areas (Table 1). Návar (2009) reported that several hundred sampling plots were needed for fitting regional allometric equations. Moreover, three model at different scales that have been developed were more robust when there were not enough trees with diameters between 25 and 40 cm (Additional file 1: Figure S1), and the majority of samples had insufficient trees with a diameter of more than 25 cm, which would lead larger estimated error of AGB (Wang 2006; Zaehle et al. 2006; Hudiburg et al. 2009; Xiang et al. 2016).

Testing importance of wood density for aboveground biomass estimation

Wood density strongly varies among different geographical regions, climate gradients, and correlated to forest structure, tree architecture (Baker et al. 2004; Chave et al. 2005, 2014). Thus, wood density can improve the performance of allometric model. In this study, we compared the performance of the allometric model with wood density variable with traditional model without wood density variable at regional scale, and the model created by Chave et al. (2014), and found the model with wood density variable had better performance than other two models (Figs. 3, 4). It indicated that taking wood density as variables in the allometric model could greatly improve accuracy of biomass model (Chave et al. 2014; Xu et al. 2015). The

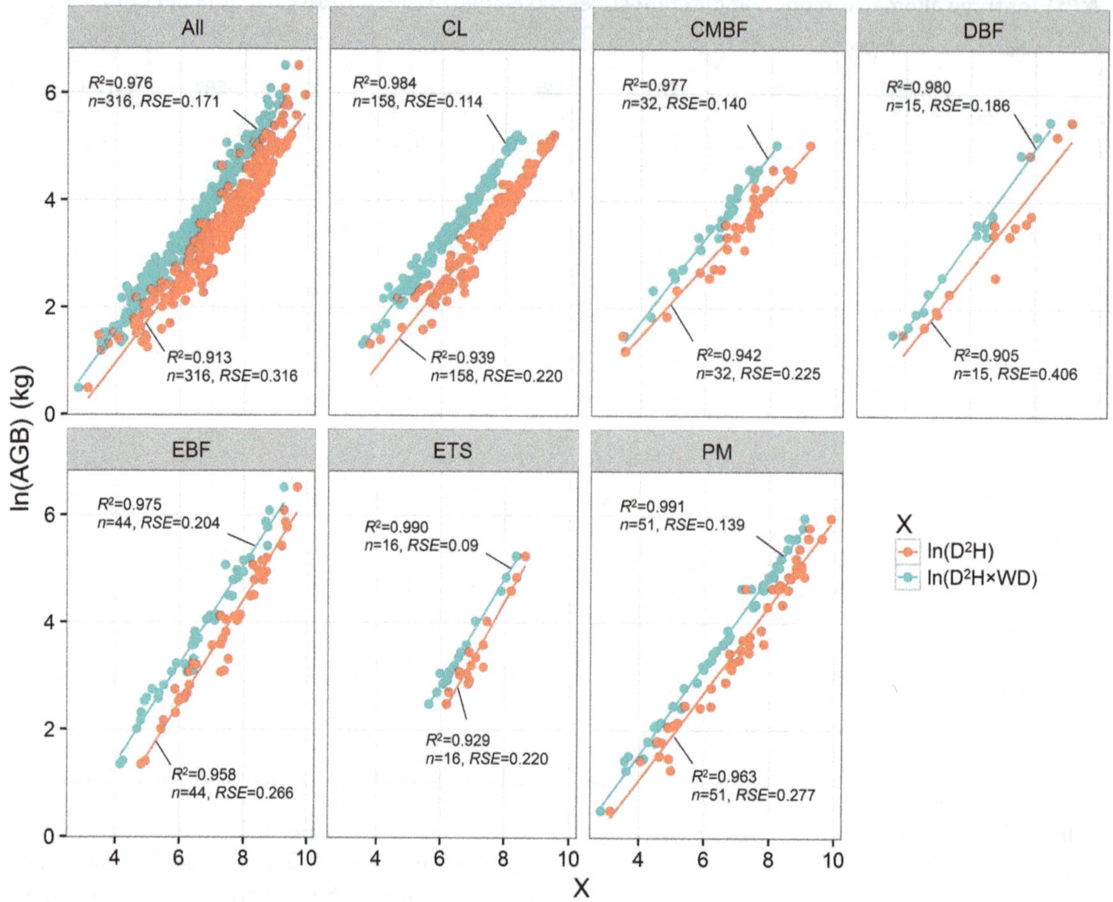

Fig. 3 Fitted curves for each forest type and all forest types at regional scale in subtropical region of China applied the allometric model with wood density variable, and the allometric model without wood density variable. See Table 1 for the abbreviations of the six forest types

reason was that wood density could reflect site climate, forest structure, and trees architecture, and reduce the effects of site climate and forest structure on AGB estimations. The model created by Chave et al. (2014) showed better performance for AGB estimation of EBF, similar to our model with wood density variable. This result suggests that the model created by Chave et al. (2014) can be used AGB estimation of EBF in subtropical region of China.

Conclusions

This study showed that ecoregions significantly affected AGB estimation for six subtropical forests in China. Dummy variable allometric model considered ecoregion as dummy variable had better performance similar to allometric model at both individual ecoregional scale and regional scale. Furthermore, we tested the performance of allometric model with wood density at regional scale and found wood density as an important variable in the allometric models greatly improved the accuracies of AGB estimations in six subtropical forests. Our findings showed that dummy variable allometric model considered ecoregion factors could be proposed as general model to estimate AGB for subtropical forests, and provide a more effective new approach to improve the compatibility of forest biomass estimates at the ecoregional, and regional scales. Moreover, the new allometric models with wood density, diameter, and tree height

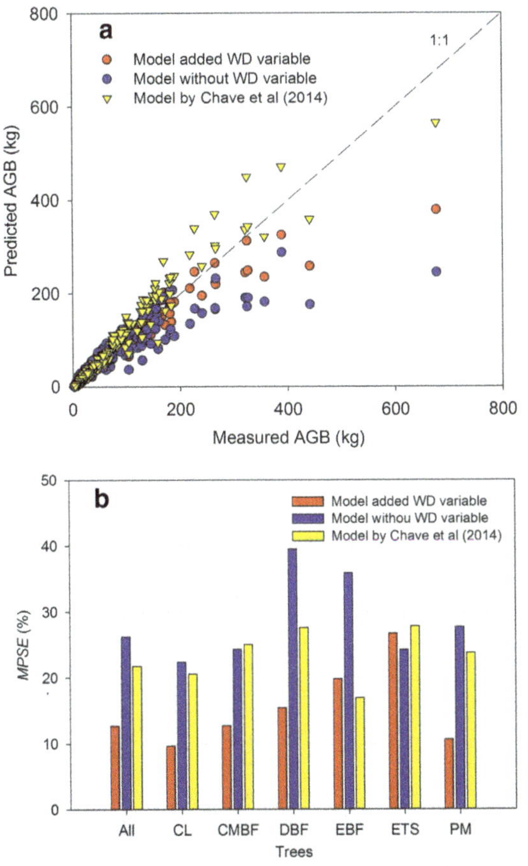

Fig. 4 Compared the measured AGB (aboveground tree biomass) and the estimated AGB in subtropical forests by the allometric model with wood density, the allometric model without wood density, and the model widely used in tropical trees created by Chave et al. (2014) $\left(AGB_{est} = 0.0673 \times (\rho D^2 H)^{0.976}\right)$, respectively (**a**), and compared *MPSE* among these three allometric models (**b**)

were more accurate than the traditional models without wood density in AGB estimations for subtropical forests at regional scales.

Authors' contributions

All authors contributed substantially to the work reported here. The four authors participated in the design of the study. SP and NH analyzed the data and wrote the manuscript. GY wrote and reviewed the manuscript. QW performed the data collection and reviewed the manuscript. All authors read and approved the final manuscript.

Author details

[1] Key Laboratory of Ecosystem Network Observation and Modeling, Institute of Geographic Sciences and Natural Resources Research, CAS, Beijing 100101, People's Republic of China. [2] Key Laboratory of Ecological Restoration in the Hilly Area, Pingdingshan University, Pingdingshan 467000, Henan, People's Republic of China. [3] College of Resources and Environment, University of Chinese Academy of Sciences, Beijing 100049, People's Republic of China.

Acknowledgements

We thank Professor Zeng Weisheng for the assistance with information on the dummy variable model, and thank Professor Martin Kent (Plymouth University, UK) for the improvement of English usage. We also greatly thank two anonymous reviewers for their insightful comments.

Competing interests

The authors declare that they have no competing interests.

Funding

This study is partially supported by the National Natural Science Foundation of China (No. 31770655, 31570471, 31290221).

References

Baker TR, Phillips OL, Malhi Y et al (2004) Variation in wood density determines spatial patterns in Amazonian forest biomass. Glob Change Biol 10:545–562

Beer C, Reichstein M, Tomelleri E et al (2010) Terrestrial gross carbon dioxide uptake: global distribution and covariation with climate. Science 329:834–838

Brown S, Gillespie AJ, Lugo AE (1989) Biomass estimation methods for tropical forests with applications to forest inventory data. For Sci 35:881–902

Case B, Hall RJ (2008) Assessing prediction errors of generalized tree biomass and volume equations for the boreal forest region of west-central Canada. Canada J For Res 38:878–889

Chave J, Andalo C, Brown S et al (2005) Tree allometry and improved estimation of carbon stocks and balance in tropical forests. Oecologia 145:87–99

Chave J, Rejou-Mechain M, Burquez A et al (2014) Improved allometric models to estimate the aboveground biomass of tropical trees. Glob Change Biol 20:3177–3190

Chinese Academy of Sciences (2001) Vegetation of China and geographic pattern-illustrition for vegetation map of the People's Republic of China (1:1000000). Geological Publishing House, Beijing

Dewalt SJ, Chave J (2004) Structure and biomass of four lowland neotropical forests. Biotropica 36:7–19

Drake JB, Knox RG, Dubayah RO et al (2003) Above-ground biomass estimation in closed canopy Neotropical forest using lidar remote sensing: factors affecting the generality of relationships. Glob Ecol Biogeogr 12:147–159

Fang JY, Chen AP, Peng CH et al (2001) Changes in forest biomass carbon storage in China between 1949 and 1998. Science 292:2320–2322

Fu BJ, Liu GH, Ouyang ZY et al (2013) Study on ecological regionalization in China. Science Press, Beijing

Fu LY, Lei XD, Hu HD et al (2017) Integrating regional climate change into allometric equations for estimating tree aboveground biomass of Masson pine in China. Ann For Sci 74:42–57

Genet A, Wernsdörfer H, Jonard M et al (2011) Ontogeny partly explains the apparent heterogeneity of published biomass equations for *Fagus sylvatica* in central Europe. For Eco Manag 261:1188–1202

Goodman RC, Phillips OL, Baker TR (2014) The importance of crown dimensions to improve tropical tree biomass estimates. Ecol Appl 24:680–698

Hudiburg T, Law B, Turner DP et al (2009) Carbon dynamics of Oregon and Northern California forests and potential land-based carbon storage. Ecol Appl 19:163–180

Jenkins JC, Chojnacky DC, Heath LS et al (2003) National- scale biomass estimators for United States tree species. For Sci 49:12–35

Luo YJ, Wang XK, Zhang XQ et al (2013) Biomass and its allocation of forest ecosystem in China. Chinese Forestry Press, Beijing

Montagu KD, Duttmer K, Barton CVM et al (2005) Developing general allometric relationship for regional estimates of carbon sequestration-an example using *Eucalyptus pilularis* from seven contrasting sites. For Ecol Manag 204:113–127

Muukkonen P (2007) Generalized allometric volume and biomass equations for some tree species in Europe. Eur J For Res 126:157–166

Návar J (2009) Allometric equations for tree species and carbon stocks for forests of northwestern Mexico. For Ecol Manag 257:427–434
Návar J, Ríos-Saucedo J, Pérez-Verdín G et al (2013) Regional aboveground biomass equations for North American arid and semi-arid forests. J Arid Environ 97:127–135
Pajtik J, Konopka B, Lukac M (2008) Biomass functions and expansion factors in young Norway spruce (*Picea abies*) trees. For Ecol Manag 256:1096–1103
Pan YD, Birdsey RA, Fang JY et al (2011) A large and persistent carbon sink in the world's forests. Science 333:988–993
Paul KI, Roxburgh SH, Jacquel RE et al (2013) Development and testing of allometric equations for estimating above-ground biomass of mixed-species environmental plantings. For Ecol Manag 10:483–494
Su YJ, Guo QH, Xue BL et al (2016) Spatial distribution of forest aboveground biomass in China: estimation through combination of spaceborne lidar, optical imagery, and forest inventory data. Remote Sens Environ 173:187–199
Tan ZH, Zhang YP, Douglas S et al (2011) An old-growth subtropical Asian evergreen forest as a large carbon sink. Atmos Environ 45:1548–1554
Ter-Mikaelian MT, Korzukhin MD (1997) Biomass equations for sixty-five North American tree species. For Ecol Manag 97:1–24
Tomppo E, Gschwantner T, Lawrence M et al (2010) National forest inventories: pathways for common reporting. Springer Press, New York
Wang CK (2006) Biomass allometric equations for 10 co-occurring tree species in Chinese temperate forests. For Ecol Manag 222:9–16
Wang ML, Bruce EB, Zhao DH (2008) An empirical comparison of two subject-specific approaches to dominant heights modeling: the dummy variable method and the mixed model method. For Ecol Manag 255:2659–2669
Williams RJ, Zerihum A, Montagu KD et al (2012) Allometry for estimating aboveground tree biomass in tropical and subtropical eucalypt woodlands: towards general predictive equations. Aust J Bot 53:607–619
Wu ZY (1995) Vegetation of China. Science Press, Beijing
Xiang WH, Zhou J, Ouyang S et al (2016) Species-specific and general allometric equations for estimating tree biomass components of subtropical forests in southern China. Eur J For Res 135:1–17
Xu YZ, Zhang JX, Franklin SB et al (2015) Improving allometry models to estimate the above and belowground biomass of subtropical forest, China. Ecosphere 6:1–15
Yu GR, Chen Z, Piao SL et al (2014) High carbon dioxide uptake by subtropical forest ecosystems in the East Asian monsoon region. Pro Natl Acad Sci 111:4910–4915
Zaehle S, Sitch S, Prentice C et al (2006) The importance of age-related decline in forest NPP for modeling regional carbon balances. Ecol Appl 16:1555–1574
Zeng WS, Zhang HR, Tang SZ (2011) Using the dummy variable model approach to construct compatible single-tree biomass equations at different scales- a case study for Masson pine (*Pinus massoniana*) in southern China. Can J For Res 41:1547–1554
Zhang L, Huang Y, Luo TX et al (2005) Age effects on stand biomass allocation to different components: a case study in forests of *Cunninghamia lanceolata* and *Pinus massoniana*. J Grad Sch Chin Acad Sci 22:170–178
Zhou GY, Liu SG, Li ZA et al (2006) Old-growth forests can accumulate carbon in soils. Science 314:1417

6

The effects of chilling-light stress on photosystems I and II in three *Paphiopedilum* species

Ying-Jie Yang[1,2,3], Wei Chang[1,2*], Wei Huang[1,2], Shi-Bao Zhang[1,2] and Hong Hu[1,2]

Abstract

Background: Low temperatures pose a critical limitation to the physiology and survival of chilling-sensitive plants. One example is the genus *Paphiopedilum* (Orchidaceae), which is mainly native to tropical and subtropical areas from Asia to the Pacific islands. However, little is known about the physiological mechanism(s) underlying its sensitivity to chilling temperature. We examined how chilling-light stress influences the activities of photosystem I (PSI) and photosystem II (PSII) in three species: *P. armeniacum*, *P. micranthum*, and *P. purpuratum*. All originate from different distribution zones that cover a range of temperatures.

Results: Photosystem II of three *Paphiopedilum* species was remarkable sensitivity to chilling stress. After 8 h chilling stress, the maximum quantum yield of PSII of three species of *Paphiopedilum* was significantly decreased, especially in *P. purpuratum*. The quantity of efficient PSI complex (P_m) value did not significantly differ after 8 h chilling treatment compared to the original value in three species. The stronger PSII photoinhibition and significantly less capacity for cyclic electron flow (CEF) were observed in *P. purpuratum*.

Conclusions: In conclusion, the three species of *Paphiopedilum* showed significant PSII photoinhibition when exposed to 4 °C chilling treatment. However, their PSI activities were not susceptible to chilling-light stress during 8 h. The CEF was important for the photoprotection of PSI and PSII in *P. armeniacum* and *P. micranthum* under chilling conditions. Our findings suggested that the photosynthetic characteristics of *Paphiopedilum* were well adapted to their habitat.

Keywords: Chilling temperature, *Paphiopedilum*, Photoinhibition, Photosystem I, Photosystem II, Cyclic electron flow

Background

The activities associated with energy capture and electron transfer are essential for photosynthesis. These reactions perform high-potential redox chemistry and lead to photodamage of the photosynthetic machinery. Photosynthesis can be very susceptible to many suboptimal environmental conditions, e.g., chilling temperatures, particularly in plants of tropical or sub-tropical origin (Havaux and Davaud 1994; Sonoike 1995; Zhang and Scheller 2004; Huang et al. 2010a, b). Members of *Paphiopedilum* are ornamental plants prized for their large, slipper-shaped floral labellums. Because of this, over-collection of the genus has become so extensive that many species are now sub-viable in their natural habitats. These plants usually occur in limestone or mountainous forests within tropical and subtropical zones ranging from Asia to the Pacific islands (Cribb 1998). Due to their tropical or subtropical origins, it is believed that these orchids cannot survive in regions with long-term natural chilling temperatures. However, the mechanism underlying their potential adaption to such conditions is unknown.

Generally, the decline in photosynthesis activity in response to chilling is attributed to a depression of Rubisco activity and RuBP regeneration, which consequently restricts the rate of CO_2 assimilation (Bernacchi et al. 2003; Sage and Kubien 2007). Such stress can

*Correspondence: changwei@mail.kib.ac.cn
[1] Key Laboratory of Economic Plants and Biotechnology, Kunming Institute of Botany, Chinese Academy of Sciences, 132# Lanhei Road, Heilongtan, Kunming 650201, Yunnan, China
Full list of author information is available at the end of the article

disrupt all of the key processes of photosynthesis, including thylakoid electron transport, the carbon reduction cycle, and control of stomatal conductance (Damian and Donald 2001). Because photochemical efficiency is suppressed by cold temperatures, the level of absorbed light energy exceeds that required for photosynthesis. If this excess energy cannot be dissipated as heat via non-photochemical quenching (NPQ) in a timely manner, a large amount of reactive oxygen species (ROS) is formed in the cells (Asada 1999).

The PSII reaction center is thought to be damaged when the structure of relevant protein components is destroyed (Barber and Andersson 1992; Aro et al. 1993; Havaux and Davaud 1994). However, in a recently introduced two-step scheme for PSII photoinhibition, it is now assumed that ROS does not directly induce oxidative damage to the PSII functional center but, instead, exacerbates this photoinhibition by suppressing the synthesis of proteins, notably the D1 protein, which has a vital role in the repair process for PSII (Nishiyama et al. 2001, 2004, 2011; Murata et al. 2007; Takahashi et al. 2009). Normally, plants have antioxidant systems, e.g., the water–water cycle, which eliminate ROS. However, under stress conditions such as chilling, excess ROS can accumulate and cause a cellular imbalance, ultimately leading to damage to photosynthetic apparatus (Asada 1999). In chilling-sensitive plants, such as cucumber (*Cucumis sativus*) and *Arabidopsis thaliana*, PSII photoinhibition is negligible during short-term treatment with chilling-light stress (Sonoike 1995; Zhang and Scheller 2004). In tropical tree species, however, PSII is selectively damaged when plants are exposed to chilling temperature associated with moderate light intensity (Huang et al. 2010a, b). Therefore, the response of PSII activity to cold temperatures differs among chilling-sensitive species. Furthermore, the effect of such stress on PSII activity in *Paphiopedilum* species is unclear.

Because low temperatures depress the rate of CO_2 assimilation, active electrons being transported through linear electron flow from PSII are accumulated at PSI, a phenomenon that contributes to the generation of hydroxyl radicals and over-reduction on the acceptor side in PSI (Sonoike 2006). As a result, the PSI acceptor side is attacked by oxidative hydroxyl radicals. Whereas PSI in sensitive cucumber and *Arabidopsis* is selectively damaged at chilling temperatures associated with moderate light intensity (Havaux and Davaud 1994; Terashima et al. 1994; Zhang and Scheller 2004), PSI activity in tropical trees is not susceptible to that stress combination (Huang et al. 2010a, b). Therefore, the performance of PSI shows obvious differences among species. Severe, irreversible photodamage to PSI activity can subsequently lead to PSII photoinhibition and even plant death (Suorsa et al. 2012). It remains unclear whether PSI activity in *Paphiopedilum* species is sensitive to chilling-light stress.

Here, we examined the influence of combined chilling and light stress on PSI and PSII activities in three species of *Paphiopedilum* that have different habitat zones of distribution. We postulated that, under stress conditions, species distributed in a tropical region would show stronger photoinhibition of PSI and/or PSII than species from a subtropical area.

Methods

Plant materials

Three species of *Paphiopedilum* were chosen, including *P. armeniacum*, which mainly occurs in the southwestern portion of Yunnan Province. This is one of the few species within the genus that can grow at elevations above 1500 m. Another species, *P. micranthum*, has a relatively wide-spread distribution, primarily in the subtropical regions of China, while *P. purpuratum* is found below 1000 m and only in the tropical regions of southernmost China and Vietnam. Before our trials began, all plant materials were cultivated for optimal growth in a greenhouse at the Kunming Institute of Botany (102°41′E, 25°01′N). Conditions included 20–25 °C and a photosynthetic photon flux density (PPFD) of approximately 15–20% of full sunlight.

Chilling treatment

Mature leaves collected from 12 healthy plants per species were exposed to 4 °C for 8 h in a cool room. For each species, six of the 12 detached leaves were placed on wet paper under a PPFD of 200 μmol photons m^{-2} s^{-1} while the other six were tested at 500 μmol photons m^{-2} s^{-1}. To determine the response of PSI and PSII to the chilling stress, detached leaves incubated in the presence or absence of lincomycin (1 mM) overnight at 25 °C in darkness were transferred to 4 °C and 500 μmol photons m^{-2} s^{-1} for 8 h. Relevant physiological measurements were conducted before and during the treatment period.

The 8 h chilling treatment was conducted in one phytotron at 4 ± 1 °C (chilling temperature). Twelve healthy plants per species were used. For leaves of both species, light intensity was controlled at 250–300 μmol photons m^{-2} s^{-1} for 8 h. Other experimental conditions included 60% relative humidity and a CO_2 concentration of 400 μmol mol^{-1}. Chlorophyll fluorescence and P700 parameters were recorded.

Chlorophyll fluorescence and P700 measurements

The chlorophyll fluorescence of PSII was measured with a Dual PAM-100 system (Heinz Walz, Effeltrich, Germany) connected to a computer with control software.

To develop light response curves, mature leaves were illuminated at 200 μmol photons m^{-2} s^{-1} for at least 20 min at 20 °C. Light-adapted photosynthetic parameters were then recorded after exposure to each light for 2 min. The following parameters were calculated: $F_v/F_m = (F_m - F_o)/F_m$, $Y(II) = (F_m' - F_s)/F_m'$ (Genty et al. 1989), and $NPQ = (F_m - F_m')/F_m'$ (Bilger and Björkman 1991). Here, F_v/F_m was the maximum quantum yield of PSII after 20 min dark adaptation; F_v represented variable chlorophyll fluorescence; F_o was the minimum fluorescence in the dark-adapted state (20 min); F_m and F_m' were the dark-adapted and light-adapted maximum fluorescence upon illumination with a pulse (300 ms) of saturating light (10,000 μmol photons m^{-2} s^{-1}), respectively; Y(II) was the effective quantum yield of PSII photochemistry; and F_s was the light-adapted steady-state fluorescence. The fraction of energy passively dissipated in the form of heat and fluorescence, $Y(NO) = F_s/F_m$ (Hendrickson et al. 2004; Kramer et al. 2004), F_s, light-adapted steady state fluorescence. The fraction of energy dissipated in the form of heat via the regulated non-photochemical quenching mechanism, $Y(NPQ) = F_s/F_m' - F_s/F_m$ (Kramer et al. 2004).

Simultaneously, the maximum photo-oxidizable P700 (P_m) were determined with saturation pulses using the Dual PAM-100 system. This variable represents the maximum alteration in signal upon quantitative transformation of P700 from the fully reduced to the fully oxidized state. At a defined optical property, the amplitude of P_m depends on the maximum amount of photo-oxidizable P700, which is a good parameter for reflecting the quantity of efficient PSI complex (Huang et al. 2010a, b, 2013). Leaves were dark-adapted for 20 min before P_m measurement (Huang et al. 2010a, b; Suorsa et al. 2012; Tikkanen et al. 2014). After far-red pre-illumination for 10 s, P_m was determined by applying a saturation pulse (Klüghammer and Schreiber 2008). P_m' was determined similarly, except that background actinic light was used instead of far-red illumination. The photochemical quantum yield of PSI, Y(I), is defined by the fraction of overall P700 that, in a given state, is reduced and not limited by the acceptor side (Pfundel et al. 2008). We used Dual-PAM-100 software to calculate $Y(I) = (P_m' - P)/P_m$. Parameter Y(ND) is estimated as P/Pm and represents the fraction of overall P700 that is oxidized in a given state that is enhanced by a trans-thylakoid proton gradient and photodamage to PSII. Y(NA) represents the fraction of overall saturation P700 that cannot be oxidized by a saturation pulse in a given state due to a lack of acceptors, which was calculated as $(P_m - P_m')/P_m$.

Photosynthetic electron flows through PSI and PSII were computed as: $ETRII = Y(II) \times PPFD \times abs\ I \times 0.5$ and $ETRI = Y(I) \times PPFD \times abs\ I \times 0.5$. Here, 0.5 was the proportion of absorbed light reaching PSI or PSII, and abs I was the absorbed irradiance set to 0.84 of incident irradiance. We estimated cyclic electron flow around PSI (CEF) as the difference in electron flow between PSI and PSII (Miyake et al. 2005; Huang et al. 2012).

Statistical analysis

Statistical analysis was performed with SPSS 16.0. Data were subjected to analysis of variance (ANOVA), and Tukey's multiple comparison tests were used to determine whether significant differences existed between treatments at $\alpha = 0.05$.

Results

Before the chilling-light treatment, the maximum quantum yield of PSII did not differ significantly among species. As the experiment proceeded, all of the tested leaves showed a decline in F_v/F_m, but the amplitudes differed (Fig. 1). For example, at a relatively low level of 200 μmol photons m^{-2} s^{-1}, photoinhibition of PSII was stronger in *P. purpuratum* than in *P. armeniacum* and *P. micranthum*. When exposed to the chilling temperature for 8 h, F_v/F_m from *P. armeniacum* and *P. micranthum* decreased slightly, from 0.75 to 0.70, while the value for that parameter dropped from 0.73 to 0.49 in *P. purpuratum*. When exposed to a strong light of 500 μmol photons m^{-2} s^{-1}, PSII photoinhibition was remarkably aggravated, with rapid F_v/F_m decreases of 36, 32, and 59% in *P. armeniacum*, *P. micranthum*, and *P. purpuratum*, respectively.

In order to determine whether the rate of PSII photodamage at chilling-light stress differs between the three species, detached leaves were pre-incubated with lincomycin and exposed to 500 μmol photons m^{-2} s^{-1}. Lincomycin, an inhibitor of chloroplast translation machinery, prevents the de novo synthesis of D1 protein and thus stops

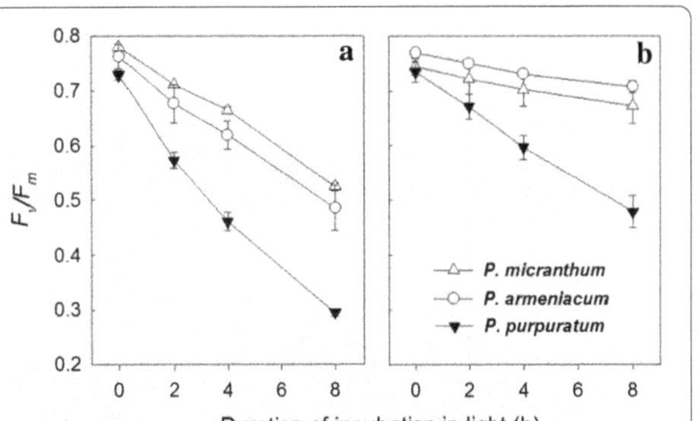

Fig. 1 Changes in F_v/F_m in three species of *Paphiopedilum* during chilling treatment at 4 °C under photosynthetic flux densities of 500 (**a**) or 200 (**b**) μmol photons m^{-2} s^{-1}. Mean ± SE was calculated from six independent plants per species

the turnover of PSII. The presence of Lincomycin made it possible to evaluate the rate of PSII photodamage. During chilling treatment at 4 °C and 500 μmol m^{-2} s^{-1} without lincomycin, both species showed the same extent of decrease in F_v/F_m (Fig. 2 −Lin). However, with pre-treatment by lincomycin, *P. purpuratum* showed significantly decrease in F_v/F_m than the other species (Fig. 2 +Lin), indicating that the rate of PSII photodamage at this chilling-light stress was greater in *P. purpuratum*. During 8 h treatment, no photoinhibition of PSI was observed in any of these species, with P_m values did not significantly differ compared to the original value in three species (Fig. 3), suggesting that PSI was insusceptible to chilling-light stress in three species.

Before chilling treatment, the light response curves of three *Paphiopedilum* species indicated that both the capacity for effective quantum yield of PSII [Y(II)] and non-photochemical quenching (NPQ) were significantly lower in *P. purpuratum* (Fig. 4a, d). The quantum yield of non-regulated energy dissipation [Y(NO)] was significantly higher in *P. purpuratum* than in the other species and gradually increased with PPFDs increased (Fig. 4b). Y(NPQ) and NPQ were increased at low PPFDs in three species (Fig. 4c, d), but significantly lower in *P. purpuratum* than in the other species. The effective quantum yield of PSI [Y(I)] and Y(NA) were significantly lower, but Y (ND) was significantly higher in *P. purpuratum* than in the other species (Fig. 5). These results showed that, compared with the other two species, *P. purpuratum* not only had a lower ability to utilize light energy, but also displayed less capacity to dissipate excess energy harmlessly as heat. Therefore, it was difficult to detect the light response curve of *P. purpuratum* after 8 h chilling treatment and the results did not provide the valid and believable data. The value of CEF at 200 μmol photons m^{-2} s^{-1} was detected to compare the CEF of three species after chilling treatment.

Compared with the plant untreated, values of Y(II) significantly decreased after chilling treatment and the values of *P. micranthum* was significantly lower than *P. armeniacum* (Fig. 6a, e). Y(NPQ) increased at low PPFDs in the two species (Fig. 6c, g) and significantly decreased with PPFDs increased. Values of NPQ of the two species were significantly lower than untreated plants, whereas Y(NO) of the two species were significantly higher than untreated plants (Fig. 6b, d, f, h).

The values of Y(I) of *P. armeniacum* and *P. micranthum* significantly decreased after chilling treatment (Fig. 7a, d). For the two species, Y(NA) was significantly decreased under low light intensity, however little

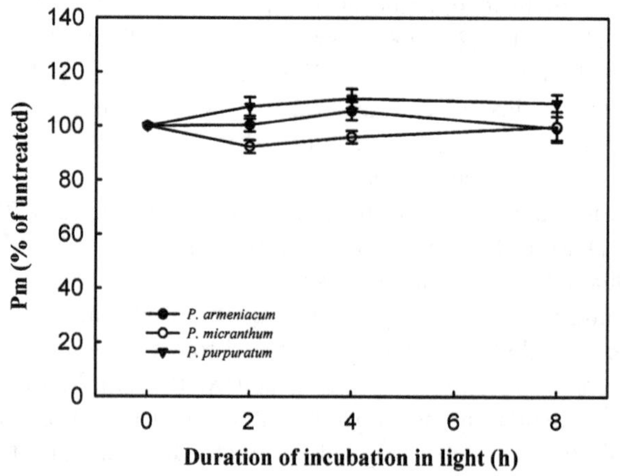

Fig. 3 Maximum photo-oxidizable P700 (P_m) during chilling treatment at 4 °C under photosynthetic flux densities of 500 μmol photons m^{-2} s^{-1} in three species of *Paphiopedilum*. Mean ± SE was calculated from six independent plants per species

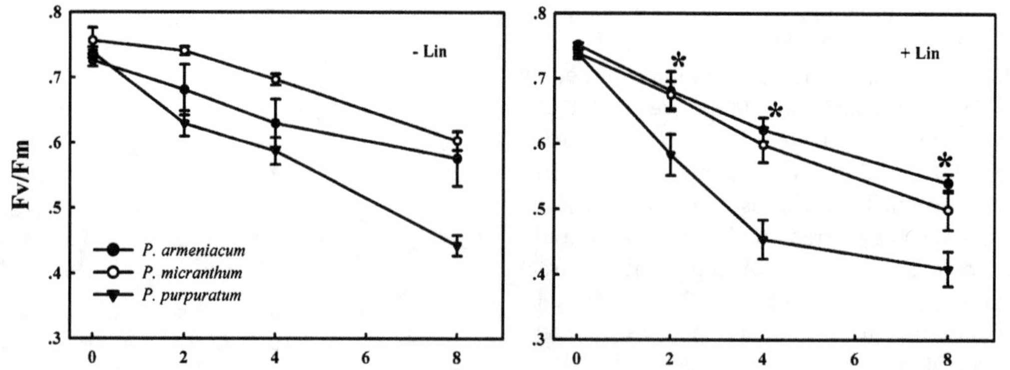

Fig. 2 Change in F_v/F_m in the three *Paphiopedilum* species during chilling treatment at 4 °C under photosynthetic flux densities of 500 μmol photons m^{-2} s^{-1} in the presence and absence of lincomycin. Mean ± SE was calculated from six independent plants per species

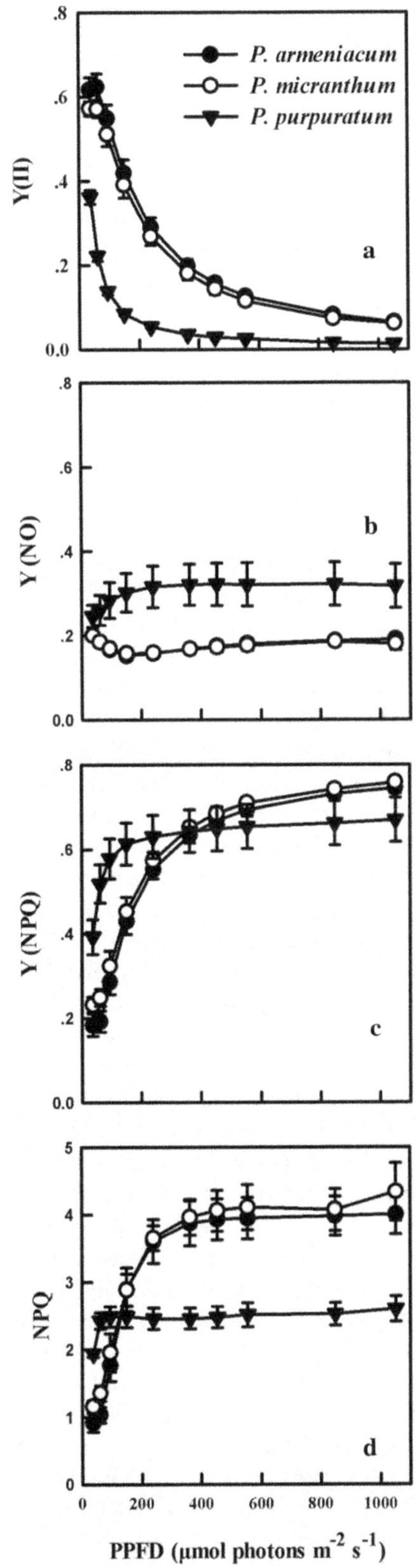

Fig. 4 Light response curves for Y(II) (**a**); Y(NO) (**b**); Y(NPQ) (**c**) and NPQ (**d**) from three species of *Paphiopedilum* at 25 °C. Mean ± SE was calculated from six independent plants per species

increased under higher light intensity (Fig. 7b, e). Compared with untreated status, the values of Y(ND) significantly increased after chilling treatment (Fig. 7c, f).

Two cyclic pathways around PSI have been identified in C3 plants; the first pathway is NDH pathway that is able to reduce plastoquinones from stromal NAD(P)H donors (Horvath et al. 2000) and the second pathway is PGR5 pathway, which is localized in the chloroplast and considered to be a factor for major cyclic electron transport activity in C3 plants (Munekage et al. 2002, 2004, 2008). The results indicated the NDH pathway in three species of *Paphiopedilum* (Fig. 8). Before chilling treatment, ETR(I) and ETR(II) of *P. purpuratum* was significantly lower than the other species. The ETRI/ETRII ratio can serve as an indicator of CEF activation (Yamori et al. 2011). Our data also showed that ETR(I)/ETR(II) ratio continued to increase when the light intensity changed from 0 to 400 μmol photon m^{-2} s^{-1}, with final values being 1.6 for *P. micranthum* and 1.4 for *P. armeniacum*. By contrast, the ETRI/ETRII ratio for *P. purpuratum* approximately remained stable at 1.6 under tested light intensity (Fig. 9). Furthermore, Y (ND) of three species was high under high light which indicated the presence of PGR5 pathway (Fig. 5). After chilling treatment, *P. armeniacum* and *P. micranthum* showed significantly lower ETR(I) and ETR(II) but significantly higher ETR(I)/ETR(II) ratio. *P. armeniacum* had higher values of the ETR(I)/ETR(II) ratio than *P. micranthum* (Fig. 10). After chilling treatment, the values of CEF at 200 μmol photons m^{-2} s^{-1} in *P. purpuratum* were significantly lower than the other species (Fig. 11). These results demonstrated that CEF activation had occurred in each species and albeit to different degrees.

Discussion

Paphiopedilum species are very well-known slipper orchids in horticulture and endangered. The genus *Paphiopedilum* occurs mainly in tropical and subtropical of Asia, and the most northerly distributed species occur in Yunnan and Guizhou province which must endure chilling-light stress in winter. However, the responses of PSI and PSII in *Paphiopedilum* to chilling-light stress are little known.

In our present study, the maximum quantum yield of PSII after dark adaptation was significantly decreased in three species of *Paphiopedilum* especially in *P. purpuratum* indicated that chilling-light stress caused significantly PSII photoinhibition in three species of *Paphiopedilum*, especially in *P. purpuratum*. This photoinhibition is a net result of photodamage and repair, stronger photoinhibition indicates that the rate of PSII photodamage exceeds that of repair. This repair of the photo-damaged PSII complex is mainly dependent on

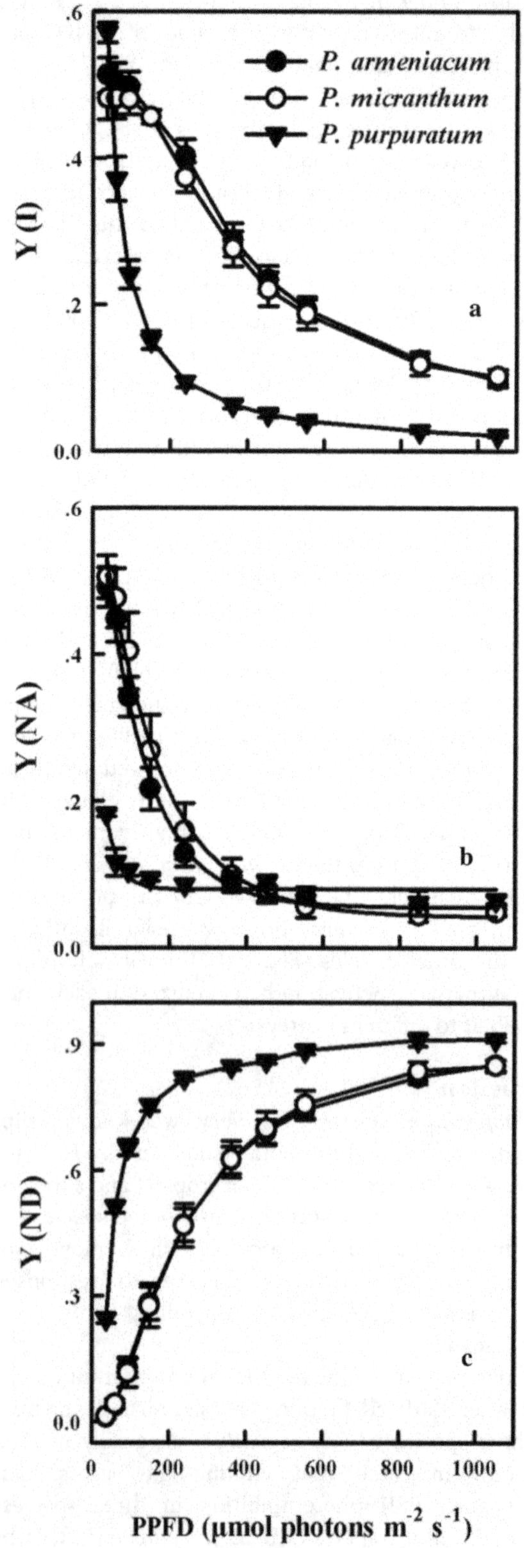

Fig. 5 Light response curves for Y(I) (**a**); Y(NA) (**b**) and Y(ND) (**c**) from three species of *Paphiopedilum* under 25 °C. Mean ± SE was calculated from six independent plants per species

ATP synthesis (Allakhverdiev et al. 2005). At chilling temperatures, photosynthetic electron transport from PSII to $NADP^+$ is blocked because of several mechanisms, including inhibition of the Calvin–Benson Cycle and depression of biochemical reactions (Murata et al. 2007). Consequently, the rate of PSII repair is very low under cold conditions (Allakhverdiev and Murata 2004). Furthermore, repair of that complex is based on new synthesis of D1 protein, which can be deterred by ROS (Nishiyama et al. 2001, 2004, 2011). At a chilling temperature, inhibition of CO_2 assimilation accelerates ROS production (Takahashi et al. 2007). Our data showed that the decrease in F_v/F_m was linearly over time in all three species, especially for leaves of *P. purpuratum* and in the presence of lincomycin, thereby suggesting that PSII repair was strongly inhibited by the reduced temperature. Before chilling treatment, *P. purpuratum* had significantly lower NPQ and higher Y(NO) than the other species also indicated the PSII photodamage (Takahashi et al. 2009). After chilling treatment, the values of Y(NO) significantly increased in *P. armeniacum* and *P. micranthum*, while the values of NPQ and Y(NPQ) of *P. micranthum* were strongly stimulated and significantly lower than *P. armeniacum*. These results suggested that excess light energy could not be consumed through photochemical quenching and NPQ. This indicated that the PSII reaction center activities were strongly down-regulated or reversibly damaged by the chilling stress. It is concluded that the photoinhibition of PSII in *Paphiopedilum* species was caused by a high level of non-regulated energy dissipation which was induced by 8 h chilling stress.

The rate of PSII photodamage is mainly correlated with CEF-dependent generation of ΔpH (Takahashi et al. 2009; Huang et al. 2012) and the production of ROS (Oguchi et al. 2009, 2011). For example, as light levels increase, PRG5-dependent CEF generates strong ΔpH by blocking proton transport from the lumen back to the stroma via ATP synthase (Tikkanen et al. 2014). The CEF-dependent generation of ΔpH can drive a Ca^{2+}/H^+ antiport to sequester Ca^{2+} in the lumen, which favors the stabilization of the oxygen-evolving complex (OEC) of PSII, where most of the photodamage occurs. Inactivation of the OEC can enhance the level of $P680^+$, subsequently damaging the PSII reaction centers (Takahashi and Murata 2008). Thus, impairment of PRG5-dependent CEF activity accelerates photodamage to both the OEC and PSII activity (Takahashi et al. 2009). Our results indicated that *P. purpuratum* had significantly less CEF capacity than the other species at 200 μmol photos m^{-2} s^{-1}, which was consistent with the tendency of PSII in our study. We therefore concluded that CEF played an important role in protecting PSII from photoinhibition

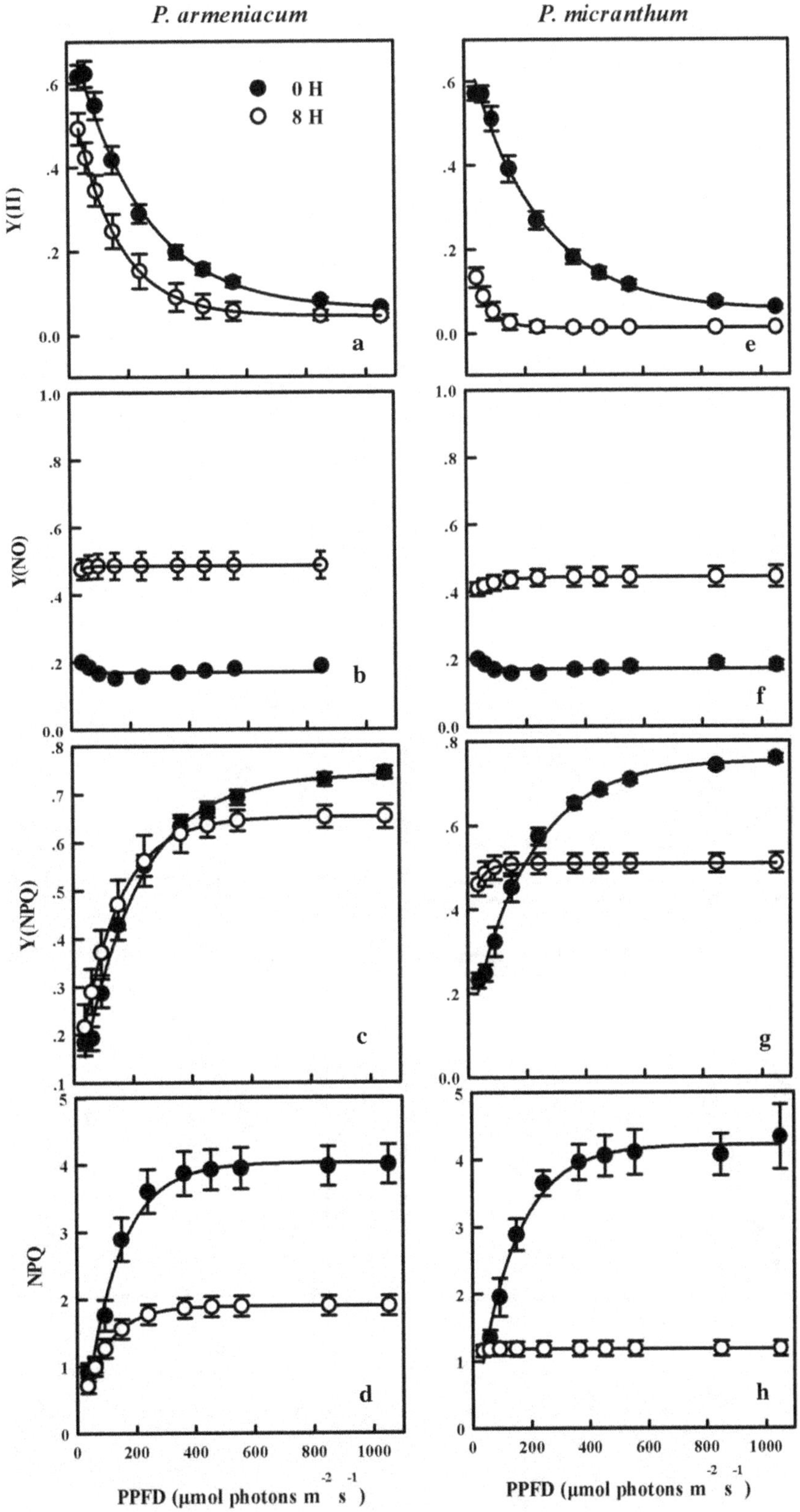

Fig. 6 Light response curves for Y(II) (**a**, **e**); Y(NO) (**b**, **f**); Y(NPQ) (**c**, **g**) and NPQ (**d**, **h**) from three species of *Paphiopedilum* after 8 h chilling stress. Mean ± SE was calculated from six independent plants per species

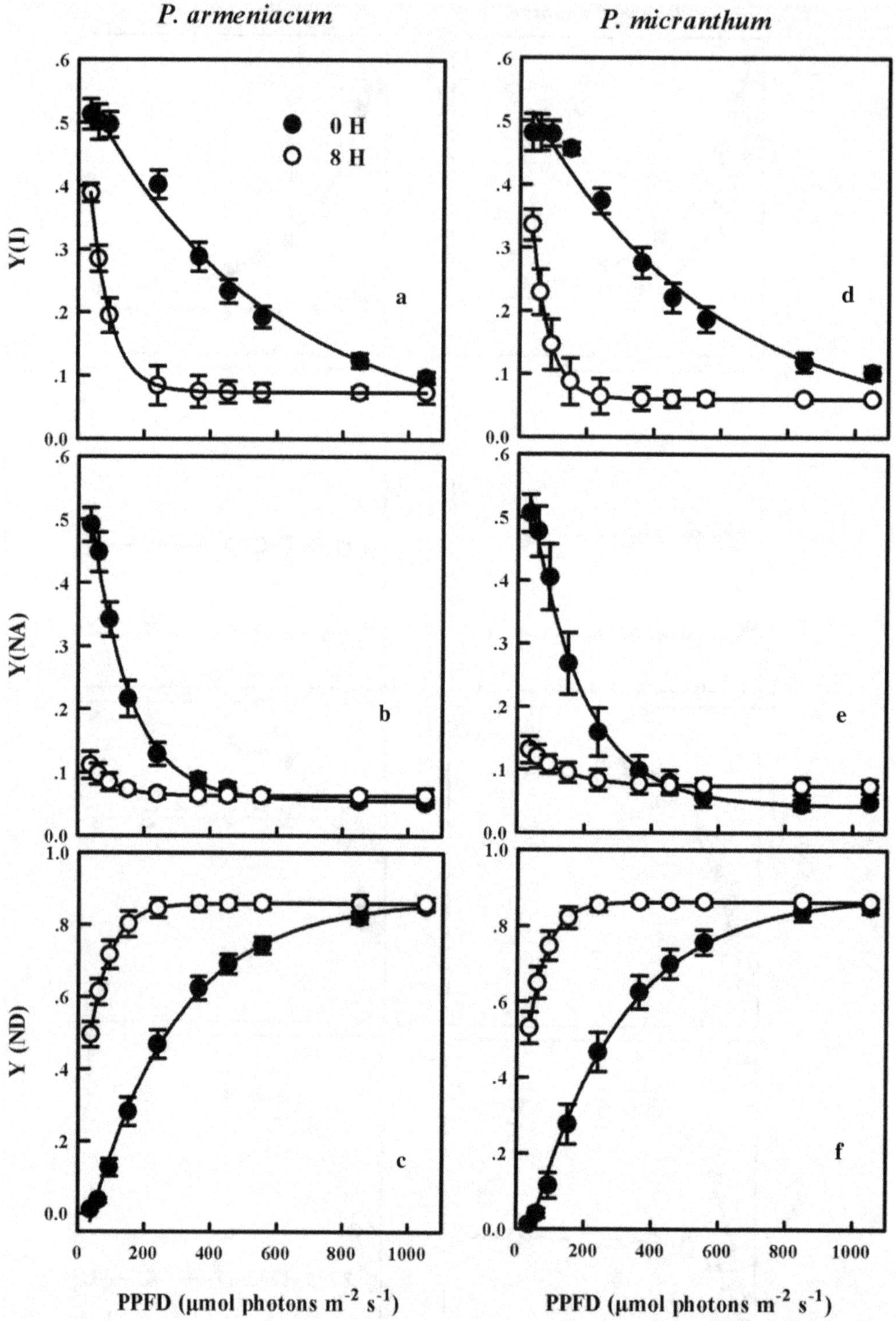

Fig. 7 Light response curves for Y(I) (**a**, **d**); Y(NA) (**b**, **e**) and Y(ND) (**c**, **f**) from three species of *Paphiopedilum* after 8 h chilling stress. Mean ± SE was calculated from six independent plants per species

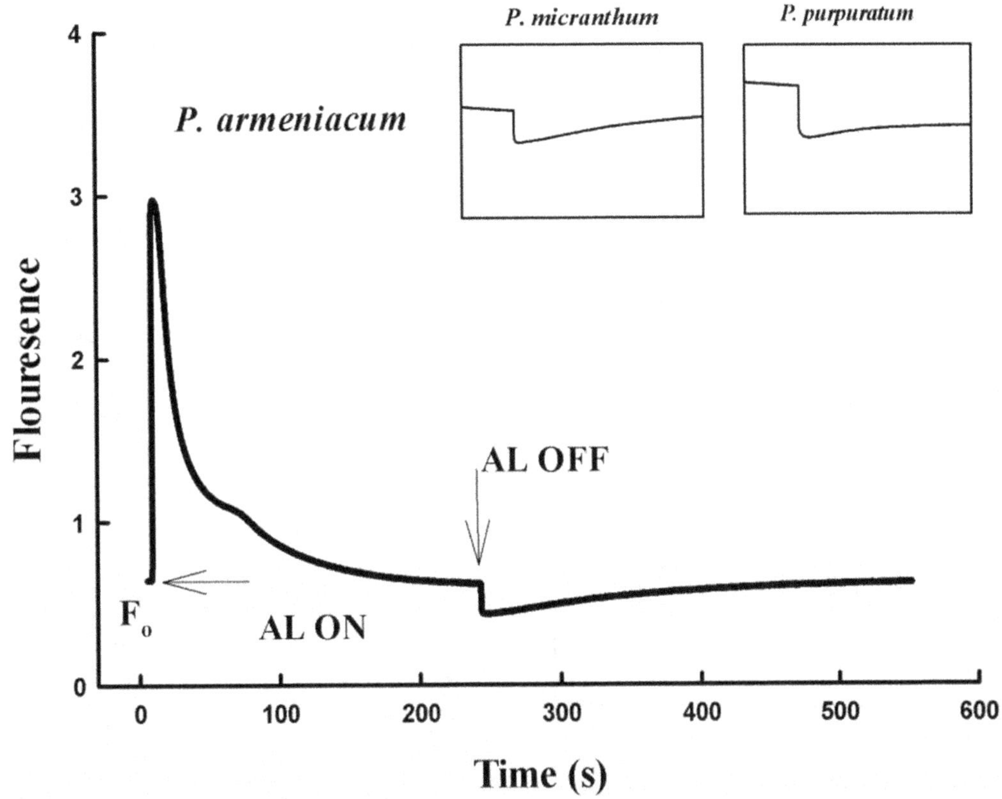

Fig. 8 Monitoring of NDH activity by chlorophyll fluorescence. Mature leaves of three *Paphiopedilum* species were exposed to actinic light (AL, 200 µmol photons m^{-2} s^{-1}) after the measuring light was turned on (Fo, the minimum level of chlorophyll fluorescence). The AL was turned off and the subsequent change in chlorophyll fluorescence was monitored as an indicator of NDH activity. The mature leaf was dark-adapted at 25 °C for 30 min, and the chlorophyll fluorescence was measured at 25 °C

under temporal chilling in the three species. In addition, the more chilling-sensitive of *P. purpuratum* showed lower NPQ capacity. Because the role of NPQ is to diminish ROS formation, we would expect the rate of production to be greater in *P. purpuratum* under chilling stress. Although the role of ROS in accelerating PSII photodamage remains controversial, Oguchi et al. (2009, 2011) have indicated that ROS can cause oxidative damage to PSII under high light. Therefore, the reduced NPQ capacity in *P. purpuratum* was probably another important mechanism that made its PSII more susceptible to chilling-light stress when compared with *P. micranthum* and *P. armeniacum*.

The sensitivity of PSI to chilling-light stress varies among species. Some chilling-sensitive species are more vulnerable to low temperature, including cucumber and *Arabidopsis*, display significant PSI photoinhibition after such combined stress treatments (Sonoike 1995, 1996; Sonoike et al. 1995; Zhang and Scheller 2004). However, PSI activity is not susceptible to chilling-light in other plants, such as tropical tree species and two *Cymbidium* species (Huang et al. 2010a, b, Li and Zhang 2016). We demonstrated that the quantity of efficient PSI complex, or P_m, remained relatively stable in all three species during the experimental period. After 8 h chilling treatment, the PSI activity of *P. armeniacum* and *P. micranthum* had the same varying trend.

Photoinhibition of PSI can occur under two scenarios: (1) over-reduction of the PSI acceptor side and (2) electron transport from PSII to PSI (Sonoike 2006, 2011). Strong illumination is, in principle, very harmful to PSI if the amount of electrons fed to the electron transfer chain by PSII exceeds the capacity of electron acceptors on the reducing side of PSI (Tikkanen et al. 2014). Indeed, PSI utilizes opposing strategies to cope with photo-oxidative stress. For example, the PGR5-dependent CEF pathway is essential for the photoprotection of PSI because it increases the fraction of oxidized P700 and prevents over-reduction of the acceptor side (Munekage et al. 2002, 2004). In *P. micranthum* and *P. armeniacum*, higher CEF activity prevented this over-reduction and minimized PSI photodamage whereas CEF activity was markedly lower in *P. purpuratum*. The higher values of Y(ND) and lower values of Y(NA) of *P. armeniacum* and

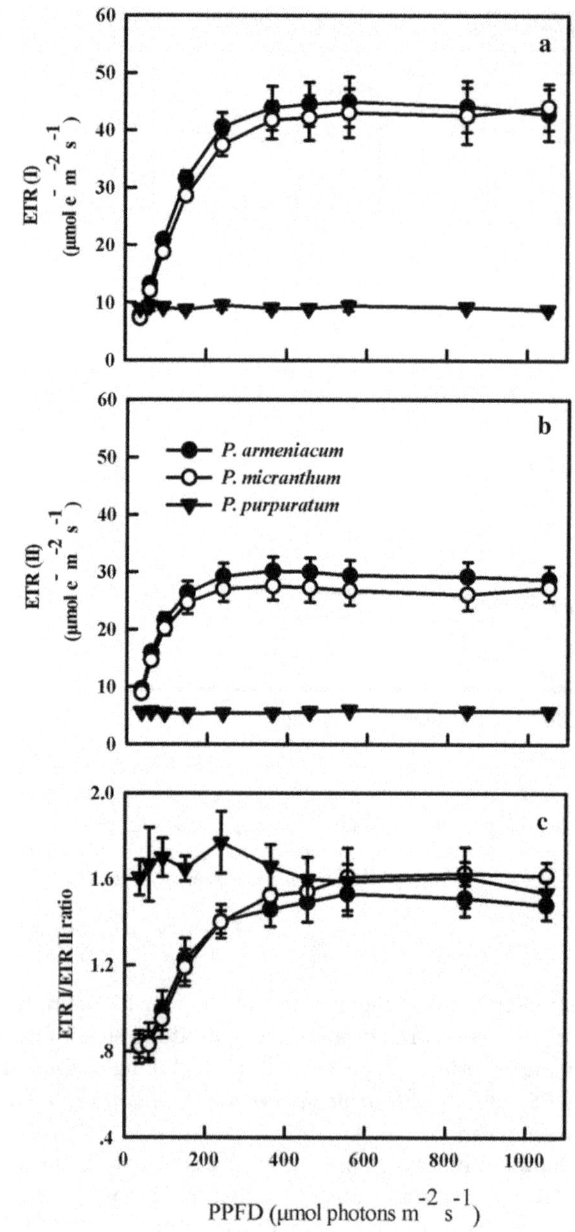

Fig. 9 Light-response changes in photosynthetic electron flow through PSI (ETRI) (**a**) and PSII (ETRII) (**b**), and ETRI/ETRII ratios (**c**) in leaves from three species of *Paphiopedilum* at 25 °C. Mean ± SE was calculated from 6 independent plants per species

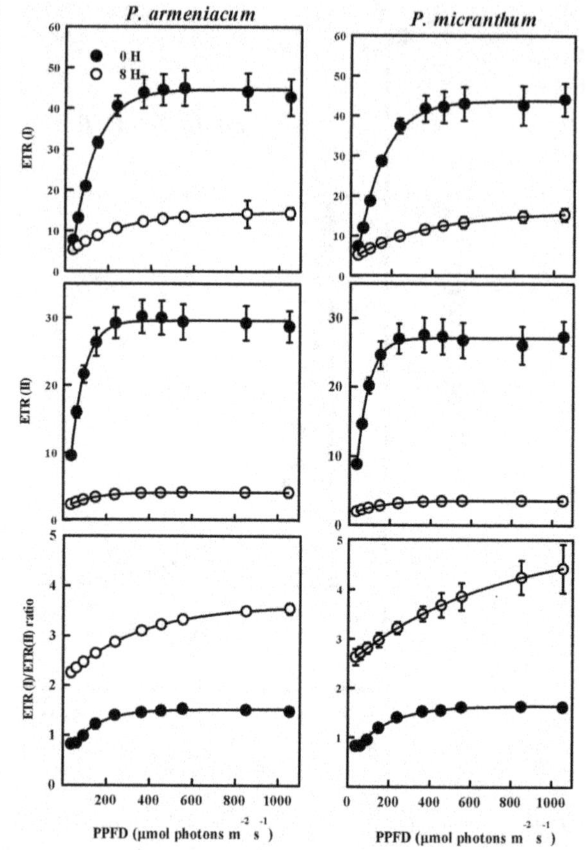

Fig. 10 Light-response changes in photosynthetic electron flow through PSI (ETRI) and PSII (ETRII), and ETRI/ETRII ratios in leaves from two species of *Paphiopedilum* after 8 h chilling stress. Mean ± SE was calculated from six independent plants per species

P. micranthum also indicated that PSI activity was being protected from chilling photodamage which was favored the survival of the two species during winter.

When electron transfer to PSI from PSII is strictly controlled, for example, in the presence of DCMU, PSI not only is protected against photo-oxidative stress (Sonoike 1996), but can also function as an extremely efficient quencher of excitation energy captured by the light harvesting machinery (Tikkanen et al. 2014). We showed that *P. purpuratum* had much lower values for ETR(II) under tested light intensities. This led to a high level of photo-oxidizable P700 and protected PSI from permanent photodamage. In addition, PSII photoinhibition is the ultimate regulator of photosynthetic electron flow and provides a photoprotective mechanism against damage to PSI (Tikkanen et al. 2014). Here, after 8 h of chilling at 500 μmol photons m^{-2} s^{-1}, F_v/F_m decreased by 59% in *P. purpuratum*. The severe PSII photoinhibition in that species was largely responsible for preventing excess electron flow to PSI, thereby allowing the amount of active PSII to be balanced with the capacity of the PSI electron acceptors. We assumed that the lack of susceptibility by PSI activity to chilling-light stress in *P. purpuratum* was probably not due to CEF activation but rather to the inhibition of electron transport from PSII to PSI.

In conclusion, the three species of *Paphiopedilum* showed significant PSII photoinhibition when exposed to 4 °C chilling treatment. However, their PSI activities

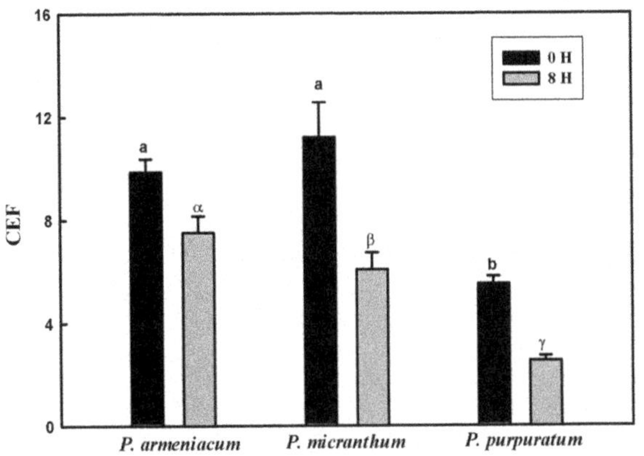

Fig. 11 The cyclic electron flow around PSI (CEF) at 200 μmol photons $m^{-2} s^{-1}$ in three *Paphiopedilum* species after 8 h chilling stress. Each vertical bar represents Mean ± SE for four measurements from individual plants. Different letters above bars indicate significant differences between treatment (P < 0.005, based on ANOVA, followed by Tukey's post hoc tests for comparison)

were not susceptible to chilling-light stress during 8 h. The relative higher CEF activity in *P. armeniacum* and *P. micranthum* alleviated PSII photoinhibition and protected PSI activity in stressed leaves. In the most sensitive species, *P. purpuratum*, lower CEF activity led to severe chilling-induced PSII photoinhibition. Compared with the other two species, greater CEF capacity in *P. armeniacum* alleviated chilling-induced PSII photoinhibition, which is correlated with its higher elevation distribution. Our findings suggested that the photosynthetic characteristics of *Paphiopedilum* were well adapted to their habitat.

Abbreviations

PSI: photosystem I; PSII: photosystem II; F_v/F_m: the maximum quantum yield of PSII; P_m: the maximum photo-oxidizable P700; CEF: cyclic electron flow; NPQ: non-photochemical quenching; PPFD: photosynthetic photon flux density; F_0: the minimum fluorescence in the dark-adapted state; F_m: the dark-adapted maximum fluorescence upon illumination with a pulse (300 ms) of saturating light (10,000 μmol photons $m^{-2} s^{-1}$); F_m': the light-adapted maximum fluorescence upon illumination with a pulse (300 ms) of saturating light (10,000 μmol photons $m^{-2} s^{-1}$); F_s: light-adapted steady state fluorescence; F_s: the light-adapted steady-state fluorescence; Y(II): the effective quantum yield of PSII photochemistry; Y(NO): the light-adapted steady state fluorescence; Y(NPQ): the fraction of energy dissipated in the form of heat via the regulated non-photochemical quenching mechanism; *Pm'*: the maximum change in P700 in a given light state; Y(I): quantum yield of PSI; Y(ND): the fraction of overall P700 that is oxidized in a given state that is enhanced by a trans-thylakoid proton gradient and photodamage to PSII; Y(NA): the fraction of overall saturation P700 that cannot be oxidized by a saturation pulse in a given state due to a lack of acceptors; ETRI: photosynthetic electron flow through PSI; ETRII: photosynthetic electron flow through PSII; abs I: the absorbed irradiance set to 0.84 of incident irradiance; OEC: the oxygen-evolving complex.

Authors' contributions

YJY and WC designed and performed the experiments, and wrote the manuscript. SBZ and WH supervised the research design and revised the manuscript. HH provides advice and guidance. All authors read and approved the final manuscript.

Author details

[1] Key Laboratory of Economic Plants and Biotechnology, Kunming Institute of Botany, Chinese Academy of Sciences, 132# Lanhei Road, Heilongtan, Kunming 650201, Yunnan, China. [2] Yunnan Key Laboratory for Wild Plant Resources, 132# Lanhei Road, Heilongtan, Kunming 650201, Yunnan, China. [3] University of Chinese Academy of Sciences, 19 A Yuquan Rd, Shijingshan District, Beijing 100049, People's Republic of China.

Acknowledgements

This work was supported by grants from the National Natural Science Foundation of China (31400289, 31670342, 31370362), National Science Foundation of Yunnan (2013FA044) and National Key Project of the Ministry of Science and Technology of China (2015BAD10B03).

Competing interests

The authors declare that they have no competing interests.

References

Allakhverdiev SI, Murata N (2004) Environmental stress inhibits the synthesis de novo of proteins involved in the photodamage-repair cycle of photosystem II in *Synechocystis* sp. PCC 6803. Biochim Biophys Acta 1657:23–32

Allakhverdiev SI, Nishiyama Y, Takahashi S, Miyairi S, Suzuki I, Murata N (2005) Systematic analysis of the relation of electron transport and ATP synthesis to the photodamage and repair of photosystem II in *Synechocystis*. Plant Physiol 137:263–273

Aro EM, McCaffery S, Anderson J (1993) Photoinhibition and D1 protein degradation in peas acclimated to different growth irradiances. Plant Physiol 103:835–843

Asada K (1999) The water-water cycle in chloroplasts: scavenging of active oxygens and dissipation of excess photons. Plant Physiol Plant Mol Biol 50:601–639

Barber J, Andersson B (1992) Too much of a good thing: light can be bad for photosynthesis. Trends Biochem Sci 17:61–66

Bernacchi CJ, Pimentel C, Long SP (2003) In vivo temperature response functions of parameters required to model RuBP-limited photosynthesis. Plant Cell Environ 26:1419–1430

Bilger W, Björkman O (1991) Temperature dependence of violaxanthin deepoxidation and non-photochemical fluorescence quenching in intact leaves of *Hedera canariensis* and *Malva parviflora* L. Planta 184:226–234

Cribb P (1998) The genus *Paphiopedilum*, 2nd edn. Natural History Publications, Borneo

Damian JA, Donald RO (2001) Impacts of chilling temperatures on photosynthesis in warm-climate plants. Trends Plant Sci 6:36–42

Genty B, Briantais JM, Baker NR (1989) The relationship between the quantum yield of photosynthetic electron transport and quenching of chlorophyll fluorescence. Biochim Biophys Acta 990:87–92

Havaux M, Davaud A (1994) Photoinhibition of photosynthesis in chilled potato leaves is not correlated with a loss of photosystem II activity-preferential inactivation of photosystem I. Photosynth Res 40:75–92

Hendrickson L, Furbank RT, Chow WS (2004) A simple alternative approach to assessing the fate of absorbed light energy using chlorophyll fluorescence. Photosynth Res 82:73–81

Horvath EM, Peter SO, Joet T, Rumeau D, Cournac L, Horvath GV et al (2000) Targeted inactivation of the plastid ndhBgene in tobacco results in an enhanced sensitivity of photosynthesis to moderate stomatal closure. Plant Physiol 123:1337–1350

Huang W, Zhang SB, Cao KF (2010a) The different effects of chilling stress under moderate illumination on photosystem II compared with photosystem I and subsequent recovery in tropical tree species. Photosynth Res 103:175–182

Huang W, Zhang SB, Cao KF (2010b) Stimulation of cyclic electron flow during recovery after chilling-induced photoinhibition of PSII. Plant Cell Physiol 51:1922–1928

Huang W, Yang SJ, Zhang SB, Zhang JL, Cao KF (2012) Cyclic electron flow plays an important role in photoprotection for the resurrection plant *Paraboea rufescens* under drought stress. Planta 235:819–828

Huang W, Fu PL, Jiang YJ, Zhang JL, Zhang SB, Hu H, Cao KF (2013) Differences in the responses of photosystem I and photosystem II of three tree species *Cleistanthus sumatranus*, *Celtis philippensis* and *Pistacia weinmannifolia* submitted to a prolonged drought in a tropical limestone forest. Tree Physiol 33:211–220

Klüghammer C, Schreiber U (2008) Saturation pulse method for assessment of energy conversion in PSI. PAM Appl Notes 1:11–14

Kramer DM, Johnson G, Kiirats O, Edwards GE (2004) New fluorescence parameters for the determination of Q(A) redox state and excitation energy fluxes. Photosynth Res 79:209–218

Li JW, Zhang SB (2016) Differences in the responses of photosystems I and II in *Cymbidium sinense* and *C. tracyanum* to long-term chilling stress. Front Plant Sci 6:1097

Miyake C, Horiguchi S, Makino A, Shinzaki Y, Yamamoto H, Tomizawa K (2005) Effects of light intensity on cyclic electron flow around PSI and its relationship to non-photochemical quenching of Chl fluorescence in tobacco leaves. Plant Cell Physiol 46:1819–1830

Munekage Y, Hojo M, Meurer J, Endo T, Tasaka M, Shikanai T (2002) *PGR5* is involved in cyclic electron flow around photosystem I and is essential for photoprotection in *Arabidopsis*. Cell 110:361–371

Munekage Y, Hashimoto M, Miyake C, Tomizawa K, Endo T, Tasaka M, Shikanai T (2004) Cyclic electron flow around photosystem I is essential for photosynthesis. Nature 429:579–582

Munekage Y, Genty B, Peltier G (2008) Effect of PGR5 impairment on photosynthesis and growth in *Arabidopsis thaliana*. Plant Cell Physiol 49:1688–1698

Murata N, Takahashi S, Nishiyama Y, Allakhverdiev SI (2007) Photoinhibition of photosystem II under environmental stress. Biochim Biophys Acta 1767:414–421

Nishiyama Y, Yamamoto H, Allakhverdiev SI, Inaba M, Yokota A, Murata N (2001) Oxidative stress inhibits the repair of photodamage to the photosynthetic machinery. EMBO J 20:5587–5594

Nishiyama Y, Allakhverdiev SI, Yamamoto H, Hayashi H, Murata N (2004) Singlet oxygen inhibits the repair of photosystem II by suppressing the translation elongation of the D1 protein in *Synechocystis* sp. PCC 6803. Biochemistry 43:11321–11330

Nishiyama Y, Allakhverdiev SI, Murata N (2011) Protein synthesis is the primary target of reactive oxygen species in the photoinhibition of photosystem II. Physiol Plant 142:35–46

Oguchi R, Terashima I, Chow WS (2009) The involvement of dual mechanisms of photoinactivation of photosystem II in *Capsicum annuum* L. plants. Plant Cell Physiol 50:1815–1825

Oguchi R, Douwstra P, Fujita T, Chow WS, Terashima I (2011) Intra-leaf gradients of photoinhibition induced by different color lights: implications for the dual mechanisms of photoinhibition and for the application of conventional chlorophyll fluorometers. New Phytol 191:146–159

Pfundel E, Klughammer C, Schreiber U (2008) Monitoring the effects of reduced PS II antenna size on quantum yields of photosystems I and II using the Dual-PAM-100 measuring system. PAM Appl Notes 1:21–24

Sage RF, Kubien D (2007) The temperature response of C_3 and C_4 photosynthesis. Plant Cell Environ 30:1086–1106

Sonoike K (1995) Selective photoinhibition of photosystem I in isolated thylakoid membranes from cucumber and spinach. Plant Cell Physiol 36:825–830

Sonoike K (1996) Degradation of psa B gene product, the reaction center subunit of photosystem I, is caused during photoinhibition of photosystem I: possible involvement of active oxygen species. Plant Sci 115:157–164

Sonoike K (2006) Photoinhibition and protection of photosystem I. In: Golbeck JH (ed) Photosystem I: the light-driven plastocyanin: ferredoxin oxidoreductase. Advances in photosynthesis and respiration. Springer, Dordrecht, pp 657–668

Sonoike K (2011) Photoinhibition of photosystem I. Physiol Plant 142:56–64

Sonoike K, Terashima I, Iwaki M, Itoh S (1995) Destruction of photosystem I iron–sulfur centers in leaves of *Cucumis sativus* L. by weak illumination at chilling temperatures. FEBS Lett 362:235–238

Suorsa M, Järvi S, Grieco M, Nurmi M, Pietrzykowska M, Rantala M, Kangasjärvi S, Paakkarinen V, Tikkanen M, Jansson S, Aro EM (2012) Proton gradient regulation5 is essential for proper acclimation of *Arabidopsis* photosystem I to naturally and artificially fluctuating light conditions. Plant Cell 24:2934–2948

Takahashi S, Murata N (2008) How do environmental stresses accelerate photoinhibition? Trend Plant Sci 13:178–182

Takahashi S, Bauwe H, Badger M (2007) Impairment of the photorespiratory pathway accelerates photoinhibition of photoinhibition of photosystem II by suppression of repair but not acceleration of damage processes in Arabidopsis. Plant Physiol 144:487–494

Takahashi S, Milward SE, Fan DY, Chow WS, Badger MR (2009) How does cyclic electron flow alleviate photoinhibition in *Arabidopsis*? Plant Physiol 149:1560–1567

Terashima I, Funayama S, Sonoike K (1994) The site of photoinhibition in leaves of *Cucumis sativus* L. at low temperatures is photosystem I, not photosystem II. Planta 193:300–306

Tikkanen M, Mekala NR, Aro EM (2014) Photosystem II photoinhibition-repair cycle protects photosystem I from irreversible damage. Biochim Biophys Acta 1837:210–215

Yamori W, Sakata N, Suzuki Y, Shikanai T, Maniko A (2011) Cyclic electron flow around photosystem I via chloroplast NAD(P)H dehydrogenase (NDH) complex performs a significant physiological role during photosynthesis and plant growth at low temperature in rice. Plant J 68:966–976

Zhang SP, Scheller HV (2004) Photoinhibition of photosystem I at chilling temperature and subsequent recovery in *Arabidopsis*. Plant Cell Physiol 45:1595–1602

7

Adansonia digitata and *Adansonia gregorii* fruit shells serve as a protection against high temperatures experienced during wildfires

Andreas Kempe*, Christoph Neinhuis and Thea Lautenschläger

Abstract

The thick and woody shell of the fruit of *Adansonia* species cannot be explained solely by adaptation to zoochory or hydrochory. Since the trunks of *Adansonia* possess a thick and fire-resistant bark and wildfires occur regularly in its habitat (savannah), we examined with the African *Adanonia digitata* and the Australian *Adansonia gregorii* whether the fruit offers protection against high heat typically experienced in wildfires. Heat-resistance tests were conducted by applying a simple heat test based on known temperature and temperature residence times occurring in savannah fires and complemented by tests to reveal the impact of heat on germination since long-term seed dormancy is known for *Adansonia*. Germination tests with acid treated and heat treated seeds were performed to establish if heat also increased germination rate as effectively as acid treatments have been found to do. Heat was found to increase germination rate, but not as effectively as treatment with acid, therefore fruits exposed to high temperatures experienced in wildfires may have a better chance of germination than fruits that were not exposed to wildfires. The ability of the investigated fruits to protect seeds from high temperatures suggests that wildfires may have played a role in the evolution of the hard-shell structure typically found in *Adansonia*.

Keywords: Angola, Australia, Heat-resistance, Fruit shell, Functional morphology, Germination capacity, Seed dormancy

Introduction

Adansonia digitata, also known as the baobab, monkey-bread tree or upside-down tree, is widely distributed in and characteristic of sub-Saharan Africa (Wickens and Lowe 2008). The Australian species, *A. gregorii* F. Muell., commonly named boab, occurs in the Northwest Territories of Australia (Wickens and Lowe 2008). The other six *Adansonia* species are endemic to Madagascar (Wickens and Lowe 2008). The noteworthy range of available literature dealing with ecology, traditional and present uses as a medical plant, food and fodder, etc. reflects the extent of attention the genus receives in science.

All *Adansonia* species develop large, ovoid or spherical fruits with a woody pericarp, commonly known as capsules; the botanical term is more precisely amphisarcum: a simple, indehiscent fruit with a pericarp differentiated externally into a dry crust and internally into one or more fleshy layers (Stuppy 2004).

The shape of *A. digitata* can vary considerably depending on distribution area (Munthali et al. 2012; Gebauer and Luedeling 2013). In Angola, the ripe fruits are rather elongated, ranging in length between approx. 30 and 65 cm. The hard woody fruit shell has a thickness of approx. 6–10 mm. The fruits of *A. gregorii* are smaller than *A. digitata* and typically reach 15 cm in length and 10 cm in diameter, with a shell thickness of 3–4 mm. Fruits of both species contain a light beige fruit pulp, which appears quite tough, chalky and crumbles when dried (Fig. 1).

*Correspondence: andreas.kempe@tu-dresden.de
Department of Biology, Faculty of Science, Institute of Botany, Technische Universität Dresden, 01062 Dresden, Germany

Fig. 1 Bisected fruits and position of nine thermocouples in *A. digitata* and five thermocouples in *A. gregorii*. Scale bar 50 mm

In spite of the fact that the baobab is a useful plant with plenty of applications, the shell of *A. digitata* has not yet been the subject of investigations. Our own microscopic investigations and analysis of composition in terms of cellulose, hemicellulose and lignin content is provided in the "Results" section. Detailed anatomical investigations of *A. gregorii* have been conducted by Bleechmore (2002) and were briefly summarised in the "Discussion".

The study was to investigate if the size and shape of baobab fruit play a role in protecting the seed from the heat experienced in wildfires. Two hypotheses about its woody fruit shell were initially considered: zoochory and hydrochory as two modes of seed dispersal. A hard or thick fruit shell can restrict access to the fruit pulp and hence the seeds to a particular distributor. Further, a woody fruit is able to float and drift to new shores. In order to explain the fruit's form and function in terms beyond zoochory and hydrochory, we also investigated whether the shell serves as heat protection during wildfires for seeds, bearing in mind that in savannahs, species must cope with wildfires. Moreover, heat may even enhance seed germination.

Zoochory and hydrochory

Wickens and Lowe (2008) noted that elephants, baboons, ungulates, bush pigs, small mammals, birds and other animals all eat the fruit pulp of *A. digitata*, either when capsules fall and break apart or sometimes when the shell has already opened on the tree, when the content is accessible to bats and birds. In Angola, rural people have reported that rat-like animals, most likely rock rats, gnaw at fruits hanging on the tree. Marsupials such as kangaroos and wallabies are dispersers in Australia (Wickens and Lowe 2008). Kamatou et al. (2011) noted that seed dormancy is broken when the seeds pass through the digestive tract of animals that consume the pulp. The seed shell is probably partly digested and thinned, which helps the embryo to break through. Similarly, Esenowo (1991) found that the most effective method to break dormancy was scarification of seeds.

It has been reported that Malagasy baobab fruits may be dispersed by marine hydrochory (Cornu et al. 2014). These authors were able to show that fruits found on the beach could spend a long time in the sea with little effect on their germination potential. Local dispersal by floodwater of rivers has been described for both *A. digitata* and *A. gregorii* (Wickens and Lowe 2008). Tsy et al. (2009) showed successful germination of *A. digitata* after storing seeds in sea water for 6 months (> 97% germination rates for each of four fruits immersed in seawater and maintained in dry conditions). Our own experiments with *A. digitata* in which five fruits were stored in a tub of salty water (3.5% salt content) and maintained a shaking plate to simulate wave action revealed that three fruits decayed after approx. 4–7 weeks (unpublished data). Once the fruit shell cracks, seeds were released and sank. Also, seed shells and seeds of the two fruits that remained floating decayed inside the fruits. Hence, long distance marine dispersal in *A. digitata* remains to be verified. Nevertheless, Baum et al. (1998, 2004) and Tsy et al. (2009) concluded long distance marine dispersal on the basis of phylogeography, at least for the common ancestor of present baobab species.

Distribution and fire adaptation

The distributions of *A. digitata* in Africa and *A. gregorii* in Australia coincide with the tropical savannah vegetation zone (Beard 1967; Owen 1974; Walker and Gillison 1982; Wilson 1988; Fukarek et al. 1995; Wickens and Lowe 2008). At the same time, it is known that fire plays an important ecological role in savannahs (Trollope 1984; Schüle 1990; Pausas 2015). Trees from fire-prone habitats therefore possess well-adapted barks to protect the cambium and to withstand wildfires. The key features are bark thickness and bark water content (Pinard and Huffman 1997; Chapotin et al. 2006; Lawes et al. 2011a, b; Rosell et al. 2014). *A. digitata* is recognized as well-adapted to fire, having a thick and fire-resistant bark (Owen 1974; Sidibé and Williams 2002; Chapotin et al. 2006; Watson 2007). Regeneration capacity after fire has also been described for *A. digitata* by e.g., Owen (1974) and Patrut et al. (2010). Some authors have noted that juvenile plants prefer fire protected sites (Duvall 2007; Venter and Witkowski 2013), whereas other authors assessed the lethal threat of fire to baobabs as limited because of lack of fuel load, notably in Zimbabwe (Romero et al. 2001) and South Africa (Venter and Witkowski 2013). Especially when trees are older than approx. 15 years, trunks are large enough to withstand fire and thus mortality due to fire decreases drastically (Johansson 1999). Juvenile trees of both species are capable to resprout from tuberous roots after fire (Bowman 1997; Nano and Clarke 2011; own observations).

Wildfire characteristics

The severity of wildfires depends on several parameters: fuel load, fuel distribution, compaction, moisture, air temperature, relative atmospheric humidity, wind etc. Early or late dry season fires differ significantly due to fuel load and drought (Trollope 1984; Williams et al. 1998). Notably, fire intensity is strongly correlated to fire frequency: frequency reduces intensity (Schüle 1990). It is known that surface fires are most common in savannah, while crown fires occur only under extreme fire conditions (Trollope 1984). Savadogo et al. (2007) and Dayamba et al. (2010) documented fire temperatures in relation to height for savannah woodlands in Burkina Faso. Regardless of season, maximum temperatures appeared near ground level, whereas at 1.5 m temperatures decrease to approx. 175 °C and drop below 100 °C near 5 m (Savadogo et al. 2007; Dayamba et al. 2010). In the same studies, temperature residence time above 60 °C was less than 2 min (late season) and less than 1 min (early season) at 1.5 m. In comparison, Miranda et al. (1993) measured up to 260 °C with annual fire frequency but as much as 650 °C at 1.6 m in Brazilian Cerrado fires when fire had been absent for 15 years; air temperature decreased to approx. 100 °C within 4 min. These information provided the basis for our own heat tests.

Approach of a heat test

We developed heat tests and measured temperatures over time inside and outside of fruits. We based our temperatures and heating duration on the data cited above, bearing in mind that a broad range of these two parameters can occur. In general, any heat applied to the outside of an object will be transmitted to the inside after a specific period of time that depends on the insulating properties of the material. We hoped to determine how long the fruit shell would insulate the inner tissue from a given heat load, given that bush fires occur over a period of just a few minutes (Vines 1968; Miranda et al. 1993; Savadogo et al. 2007). If the fruit shell is capable of delaying heat transmission, seeds can survive a bush fire undamaged if not exposed to naked flames.

We wished to simulate the conditions of a real savannah fire for fruits hanging on a tree, since safety and financial factors precluded a large-scale fire. Thus, for our heat test we abstracted the heat source and used charcoal on a grill that delivered a constant and regular heat and which could be repeated. Hence, heat was supplied primarily as radiation, whereas flames and convection were neglected. Consequently, the results serve primarily as the basis for discussion.

Germination and impact of heat

To complete our study, germination rate (GR) was determined to evaluate the impact of the heat on seeds. In general, temperatures above 60 °C are considered lethal for living tissues (e.g., Hare 1965; Vines 1968), although frequently that does not apply to seeds. For example, Esenowo (1991), Auld and O'Connel (1991), Gashaw and Michelsen (2002), and Dayamba et al. (2008, 2010) all reported that seeds from African and Australian savannah plants can withstand temperatures ranging from 60 to 200 °C, for at least 1 min. Esenowo (1991) argued that heat seems to play an important role in germination of *A. digitata*. He demonstrated that seeds achieved higher GR after 40 min soaking in 60–70 °C tempered water than soaking in cold water or after 1–5 min soaking in 80 °C tempered water, respectively. In contrast, Johansson (1999) documented that GR of *A. digitata* were no higher after treatment in boiling water compared to soaking in cold water and reported no germination after 5 min dry heat treatment at 100 °C. Maghembe et al. (1994) found GR for *A. digitata* greater than 80% regardless of whether seeds were untreated or treated with cold or hot water. Turner and Dixon (2009) documented GR for *A. gregorii* below 25% after treatment with boiling water for various durations. In general, fire is believed to help break the dormancy of *Adansonia* seeds (Watson 2007).

In order to break dormancy, experiments simulating passage through an animal's digestive tract have been carried out by soaking seeds in acidic solutions; the results have shown positive GR for both *A. digitata* [60–90% GR using the acids HCl, HNO_3, H_2SO_4 and various durations of soaking (Esenowo 1991), 2% GR after soaking in HCl (Johansson 1999), 70–100% GR after various durations of soaking in concentrated H_2SO_4 (Razanameharizaka et al. 2006), 5–97% GR after various durations of soaking in concentrated H_2SO_4 (Niang et al. 2015)] and *A. gregorii* [85–100% GR after various durations of soaking in concentrated H_2SO_4 (Turner and Dixon 2009)].

Methods

Material

Six fruits of *A. digitata* (two fruits per individual) were collected in the province Uíge in northern Angola (S8°33′05.9″, E13°45′07.8″, 139 m) in March 2015, at the end of the rainy season. The fruits were between 34 and 52 cm long with a diameter of 11.4 up to 14.8 cm at the widest part. They had a volume between 2.1 and 6.0 dm^3 and a dry weight between 0.7 and 1.4 kg. The shell's thickness ranged around 7.6 (± 1.6) mm.

Northern Territory Herbarium (DNA) in Palmerston, NT, Australia provided 20 ripe *A. gregorii* fruits in May 2014. The fruits were collected along Parry Creek Lagoon Road and Flametree Nursery, Riverfarm Road Kununurra from two trees, 9 and 10 m high with trunks of more than 1 m diameter at breast height. Six average fruits were used which had a length between 9.1 and 15.0 cm with a diameter of 7.2 up to 8.1 cm at the thickest part. The volume ranged between 0.23 and 0.42 dm^3 and the dry weight between 62 and 142 g. The shell's thickness was 3.0 (± 0.4) mm.

Anatomy

Fruit size and shell thickness were measured with yardstick with an accuracy of 1 mm and calliper with an accuracy of 0.1 mm, fruit volume by water-displacement with an accuracy of 10 ml. Sections of the pericarp were cut by a microtome, stained with Astra-blue/Safranin (Morphisto), and analysed using a light microscope (Carl Zeiss Axioskop 2) and a reflected light microscope (Olympus SZX16). Cell size and cell wall thickness were assessed. The density of shell was determined with five different samples according to DIN EN ISO 1183-1 using a precision balance (Kern, EMB 100-3) and a pycnometer (50 ml, Boro 3.3, Brand) with an accuracy of 0.01 g and 0.05 ml, respectively. Cellulose, hemicellulose and lignin content were determined by the Institute of Plant and Wood Chemistry (TU Dresden) according to standardized methods by Klason (1920), Kuerschner and Hoffer (1931), and Poljak (1948). Bremer et al. (2013) provide details of the extraction process.

Heat test

To investigate the delay of an experienced outside temperature to the inside of the fruit, temperatures outside and inside the fruit were recorded via thermocouples (tc's) at several positions (Fig. 1) during heating. Therefore small holes with a diameter of 3 mm were drilled from the top into the shell to insert tc's. The depth of the holes were measured with a calliper in order to place the tc's at the right position. Inner tc's were placed directly at the inner side of the fruit shell and in the mid axis of the fruit. Additional tc's were placed outside and fixed with tape (Fig. 2). Thermocouples type K (Jumo) according to DIN43710/IEC584 with a range of – 200–1150 °C were used. Temperature and time were logged every second by NI 9213 thermocouple input module (NI USB-9162 USB Single Module Carrier and Software SignalExpress 2012, National Instruments). Temperature gradients and corresponding time were analysed using the spreadsheet of Apache OpenOffice. The fruits were attached with a steel wire (0.5 mm diameter) at a wooden frame over the heat source. All frame elements had a distance of more than 1 m to the heat source. The wires of tc's were fixed upwards to the frame and then led towards the instruments. Beech charcoal on a grill served as heat source and delivered constant heat for more than 3 h. Heating value is given with 25–35 MJ kg^{-1}. The amount of used charcoal determines the temperature during burning. We used always the same amount of approx. 4.0 kg which delivered 500–600 °C. Heat transfer by radiation depends on temperature and radiating area. The radiation area was approx. 0.25 m^2. Thus the rate of heat flow, calculated by Stefan–Boltzmann law, was at least 4–6.5 kW. Between 100 and 200 °C were measured at approx. 10–20 cm above glow, which was the distance between glow and the fruits.

Fruits of *A. gregorii* were placed hanging vertically above heat source, as fruits hang naturally, whereas *A. digitata* fruits were placed horizontally for optimising heat transfer between heat source and fruit. Otherwise significant heating would only have occurred at the tip due to small radiating area and less heat would have reached the top of the fruit. Gradient of temperature over time was recorded during heat exposure and during cooling down as well.

Germination test

Seeds of three and four heat tested fruits of *A. digitata* and *A. gregorii* were used for germination tests. *A. digitata* seeds were divided into two batches, one from the glow facing side or lower half and one from the upper half. Heat tested seeds were compared to following control sets: soaking of seeds in concentrated H_2SO_4 for 10 min with subsequent rinsing in running water for 3 h (Esenowo 1991), in 70 °C hot water for 40 min (Esenowo 1991), cold water for 3 days (Johansson 1999), and no treatment. Therefore further two fruits of *A. digitata* and three fruits of *A. gregorii* were used. 50 seeds of each experimental variant were extracted out of its fruit pulp by cutting with a knife. Thus, 10 variants with in total 500 seeds are used for analysis of *A. digitata* (Table 1) and eight variants and 400 seeds for *A. gregorii* (Table 2).

Cultivation took place in the Botanical Garden of the TU Dresden from the beginning of March to the end of September 2016, with greenhouse conditions of an

Fig. 2 **a** *A. digitata* above glow with cracked shell by heat. **b** Decomposed fire-facing side after test. **c** Same fruit, view inside after test less than 10 min above glow with partly unburned and browned pulp. **d** Another fruit opened with heavily smouldered parts after test more than 10 min above glow. **e** *A. gregorii* above glow. **f** Same fruit, fire facing side after test. **g** Another fruit, in one case, ongoing smouldering of fruit shell (heat source was already removed) with successive shell decomposition and complete loss of seeds. All scale bars 50 mm

Table 1 Germination rates (GR) of *A. digitata*

Heat treatment		GR (%) (upper, lower side)
Average of T_{max} of inner tc's (°C) (SD)	Heat exposure time (min:s)	
76 (25)	06:25	3 (0, 6)
198 (103)	12:00	15 (24, 6)
445 (173)[a]	10:00	25 (34, 16)
Control		
No treatment		0
15 min H_2SO_4		76
40 min hot water, 70 °C		16
3 days cold water		2

[a] Local smouldering in fruit 3 led to extreme divergent temperatures (maximum of 595 °C at central tc)

Table 2 Germination rates (GR) of *A. gregorii*

Heat treatment		GR (%)
Average of T_{max} of inner tc's (°C) (SD)	Heat exposure time (min:s)	
120 (70)	10:45	2
128 (60)	11:45	14
152 (125)	11:45	10
429 (116)	12:00	0
Control		
No treatment		10
15 min H_2SO_4		24
40 min hot water, 70 °C		6
3 days cold water		4

average temperature of 21 °C and an average humidity of 82%. Seeds were irrigated every day. Germination was recorded every week.

Number of germinated seeds were analysed using the spreadsheet of Apache OpenOffice. Significant differences between treatments were analysed by Fisher's exact test (McDonald 2014).

Results

Anatomy—*Adansonia digitata*

Microscopic investigations revealed three layers of shell composition, which can be classified as exo-, meso- and endocarp (Fig. 3). Different cell dimensions characterized these layers. The outermost exocarp had densely packed, globular, and thick-walled cells, followed by radially elongated and thick-walled cells. Meso- and endocarp consisted of undirected elongated and spherical cells. The inner layer had the smallest and most thin-walled cells.

Density of fruit shell and of pulp was 0.61 (± 0.09), and 0.16 (± 0.01) g cm^{-3}, respectively. Due to visible inhomogeneity of *A. digitata*'s fruit shell, lignin, cellulose and hemicellulose were determined for three layers, which, however, did not represent exo-, meso- and endocarp but combined meso- and endocarp and divided exocarp in the outermost and the inner exocarp (Fig. 3). The outermost layer of exocarp consisted of 45% lignin, 21% cellulose and 28% hemicellulose. The subjacent layers counted for approx. 30% lignin, 36–38% cellulose and 33–37% hemicellulose.

Anatomy—*Adansonia gregorii*

Density of fruit shell and of pulp was 0.55 (± 0.06), and 0.20 (± 0.03) g cm^{-3} respectively. Further anatomy of *A. gregorii* has been investigated by Bleechmore (2002) and is given in the "Discussion" section.

Heat test

In the first 3 min the temperature gradients were nearly identical for all tested fruits. Once the fruit was placed over the heat source, all tc's recorded increasing temperatures. The outer temperature on the glow facing side jumped up immediately within seconds to 100–150 °C. The inner temperatures increased slowly and continuously: direct behind the shell faster than in the mid axis.

After 3 min: During exposure to glow the outer lower tc increased further up to 300 °C, whereas the temperature directly behind the glow facing shell eventually remained below 100 °C except in one case up to 110 °C. The temperature of central tc's did not exceed 50 °C.

6–14 min: All fruits became charred on the glow facing side. The tc's recorded a local increase up to a maximum of 587 and 240 °C directly behind the shell for *A. digitata* and *A. gregorii* respectively. When approx. 90 °C were attained in the centre of the fruit it was removed from the heat source. One fruit of each species was already removed after reaching 60 °C in the centre. Two *A. digitata* fruits showed cracks of a maximum of 0.5 cm width and 25 cm length. The difference of outside and central temperature was between 188 and 366 °C for *A. digitata* and between 116 and 406 °C for *A. gregorii* at the end of heat exposure.

End: The heat source was removed. Temperature of outer tc's dropped immediately nearly to ambient temperature, whereas the temperature inside decreased slowly. Fruit shell and inner parts of the two cracked *A. digitata* fruits started to glow and smoulder within 5 min after removal from heat source. A hole appeared in the shell from which some smouldering seeds fell down. At the same time central temperature increased locally further up to 595 °C after approx. 15 min. At that time recording was terminated. However, half of the fruit remained more or less unburned. In contrast, fruit shell and inner parts of one *A. gregorii* fruit started to glow and smoulder after removal from heat source. This fruit shell smouldered off completely and degraded to ashes (Fig. 2). All seeds from this fruit were lost. Representative gradients are displayed in Fig. 4.

Correlation of shell thickness, fruit volume, heat exposure time and temperature

Heating of the fruit was influenced by fruit volume, shell thickness, and most relevant by heat exposure time (Fig. 5). A long heat exposure caused generally higher temperatures at tc's (Fig. 5). Importantly, *A.*

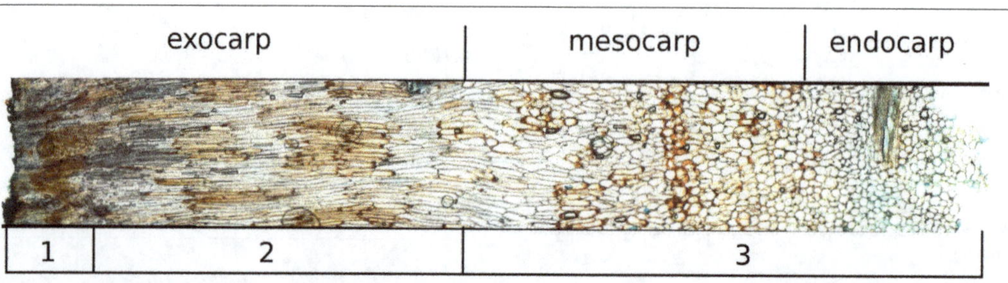

Fig. 3 Micrograph of *A. digitata* pericarp; shell thickness is approx. 7 mm

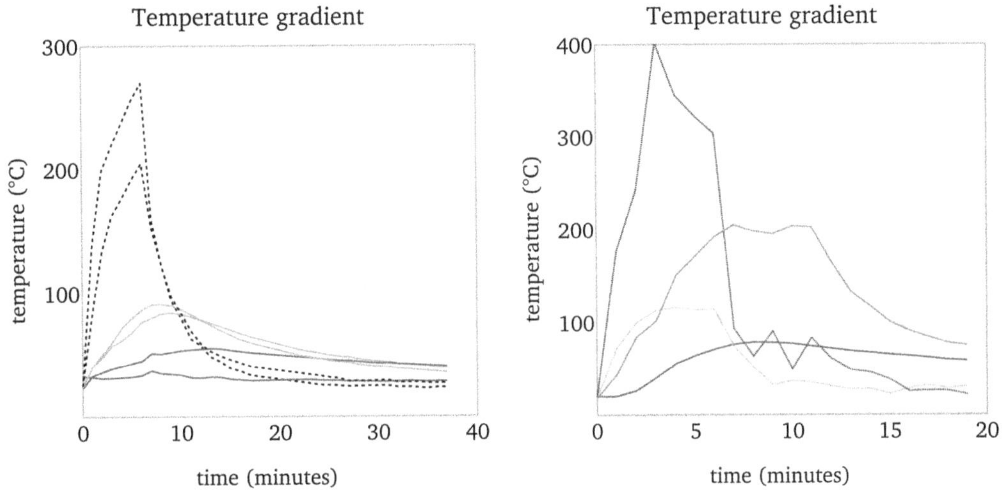

Fig. 4 Example of temperature gradients of *A. digitata*, left, and *A. gregorii*, right. Curves are smoothed, average temperature per 1 min. Left: Temperature gradient of 6 tc's. Each 2 black dotted, light grey and dark grey lines are temperatures outside downwards, inside the shell downwards and in the centre. Heat exposure 06:25. Record time 37 min. Right: temperature gradient of 4 tc's. Black dotted, light grey and dark grey lines are temperatures outside downwards, inside the shell downwards and in the centre. Grey dotted line is outside sidewards. Heat exposure 6 min. Record time 17 min

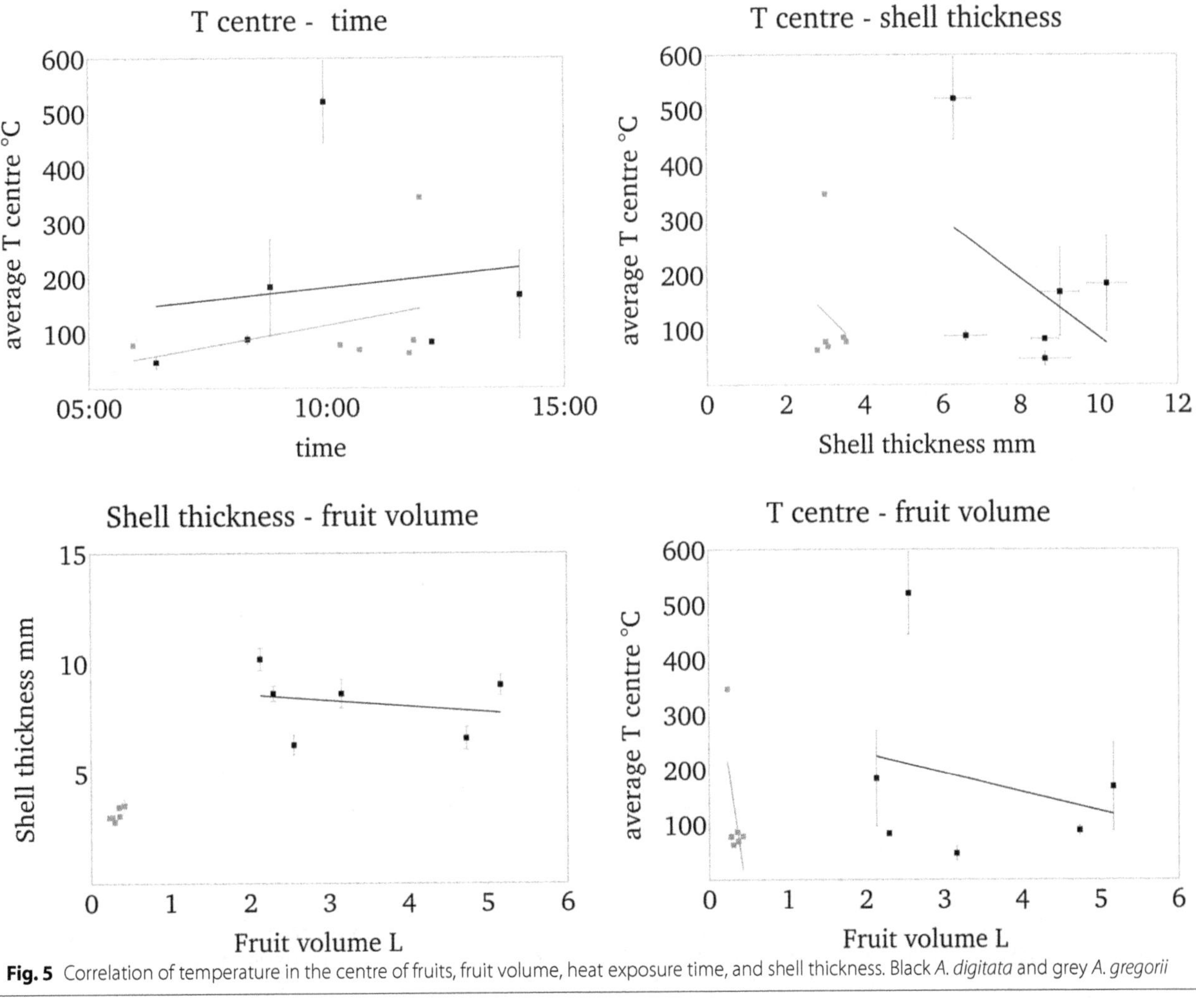

Fig. 5 Correlation of temperature in the centre of fruits, fruit volume, heat exposure time, and shell thickness. Black *A. digitata* and grey *A. gregorii*

digitata fruits with the thinnest shells reached highest measured inner temperatures (Fig. 5). Shell thickness and volume were not correlated (Fig. 5). Summing up, the maximum temperature outside the shell ranged between 210 and 490 °C within approx. 6 and 14 min for both species. The most important point was that in three of six *A. digitata* fruits the average central temperature remained below 90 °C within the test time and additionally as well in two further fruits at least at one of both central tc's. The *A. digitata* fruit with the highest central temperature had the thinnest shell and small volume. Shell thickness of *A. gregorii* was more homogeneous than in *A. digitata* and showed no relation to inner temperatures (Fig. 5). They depended rather on fruit volume. Smallest fruit had highest end temperatures (Fig. 5).

Germination test—*Adansonia digitata*

Treatment with sulphuric acid had the highest GR (Table 1). Seeds from heat tested fruits tended to germinate more when experienced temperatures above 100 °C. Hot water treatment had as well a positive effect on GR. Temperatures below 70 °C had little impact on GR (Table 1). There was a distinct difference between GR from glow facing side and upper side. Both fruits with higher experienced temperatures exhibited higher GR on upper side. The fruit with less heat exposure showed higher GR on the lower side (Table 1). A percentage of 90% of all germinated seeds sprouted within 6 weeks after sowing.

We have to emphasise that given temperatures were only measured very selective with five tc's inside the fruit. We observed maximum differences of more than 200 °C between two central tc's during heating, situated 10 cm away from each other. Thus, it was unclear how locally limited a measured maximum temperatures was. We assumed carefully a possible temperature distribution from maximum temperatures at the end of heat exposure of two fruits (Fig. 6). The values of fruit 3 must not mislead the reader that the higher a temperature the higher was the GR. Logically seeds cannot survive these high temperatures.

Germination test—*Adansonia gregorii*

Except from treatment with sulphuric acid, the GR's were up to 14% (Table 2). An amount of 75% of all germinated seeds did not sprout until the 12th week, but the earliest germination was observed for acid treatment after 4 weeks. The difference between highest GR after heat treatment (14%) and no treatment (10%) was not significant (P value = 0.48).

Discussion

In a preliminary work we asked if the shell of *A. digitata* would offer sufficient mechanical properties to restrict access by a frugivore. In particular we analysed two properties characterising rigidity and strength of the *A. digitata* shell employing mechanical stress tests. Young's modulus was determined by three-point bending test to be approx. 1.0 GPa and Charpy impact strength was measured to be 2.9 kJ m^{-2}. Both values were below one tenth of average angiosperm wood (Kretschmann 2010). Young's modulus for *A. gregorii* fruit is in the same range (approx. 1.5 GPa; Bleechmore 2002). Hence, we concluded that the fruit shell does not serve exclusively as seed protection against animals. In this study we investigated the fruit shell's capability of protecting seeds against high temperatures during wildfires.

Heat protection

When we talk about heat protection, we intended to evaluate heat protection for relevant conditions in savannah fires given by temperature-residence time and temperature. Wildfires in savannahs use up its fuel quickly and therefore usually last less than 10 min (Miranda et al. 1993; Savadogo et al. 2007; Dayamba et al. 2010). Heat impact during savannah fires were given with air temperatures ranging between 100 and 260 °C (rarely up to 650 °C) at 1.5 m height and still with approx. 100 °C at 5 m height (Miranda et al. 1993; Savadogo et al. 2007; Dayamba et al. 2010). Usually the lowest fruits of both investigated *Adansonia* species hang not below 1.5 m height. Maximum heights of mature trees of *A. digitata* and *A. gregorii* are up to 20 and 10 m respectively. Since fruits are randomly distributed over the entire tree

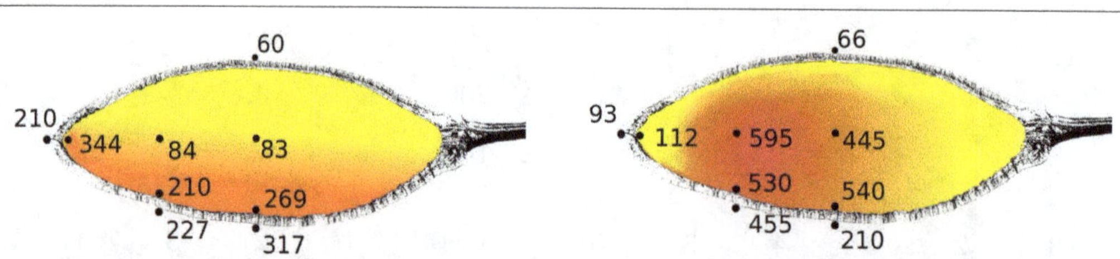

Fig. 6 Possible temperature distributions in *A. digitata* approximated from maximum temperatures (°C) at the end of heat exposure. See also charred region in fruit in Fig. 2d. Left: continual gradient. Right: smouldering seat. Yellow: low temperature; red: high temperature

crown, fruits are in threat of fire at least to a height of 5 m.

Based on that we arranged a fire test using beech charcoal, which had a temperature of 500–600 °C and a radiation area of approx. 0.25 m^2. Fruits were placed 10–15 cm above the glow. Both *A. digitata* and *A. gregorii* withstood at least 5 and 4 min, respectively, without reaching 60 °C in the centre while surface temperature of the fruits further increased to approx. 300 °C. A temperature of 90 °C in the centre was reached at the earliest after approx. 7 min in one *A. digitata* fruit and later than 9 min in four of six *A. digitata* fruits. This temperature was not exceeded in *A. gregorii* within 10 min. The lethal temperature for seeds is known to be higher than 90 °C (Esenowo 1991; Maghembe et al. 1994; Johansson 1999; Turner and Dixon 2009). The delay of temperature increase inside the fruits clearly demonstrates the shell's capability to protect seeds against heat. Regarding the fact that savannah fire duration is less than 10 min (Vines 1968; Miranda et al. 1993; Savadogo et al. 2007) the fruit shell delivers sufficient protection for the largest amount of fruits on a tree. Only in case of higher temperatures, close fire distance, and long heat exposure time, fruits may be damaged partially or totally leading to loss of seeds. Concerning our experimental set-up, decomposition conditions for fruit shells could be reached by more than 8 min heat exposure with 10 cm fruit—glow distance and heat source temperatures above 500 °C. When distance to glow were below 10 cm then shells of course decomposed faster.

Strong temperature gradients could be measured inside *A. digitata* fruits during heating. In one case a maximum difference of more than 200 °C occurred between two central tc's for at least 1 min in case of inner smouldering. Thus, the spongy fruit pulp definitely contributed to thermic insulation of seeds, too. In the case of inner smouldering peak temperatures of almost 600 °C occurred, the smouldering fire did not propagate into the whole fruit. It was striking that approximately the half of the fruit pulp remained unburned (Fig. 2d). Thus, in this case a considerable number of seeds would probably have survived.

We observed furthermore that sometimes shells were already split open on the tree or cracked during fire. Thereupon seeds could easily fall down into the surface fire, where they most probably were destroyed. Wickens and Lowe (2008) described frequent shell cracking for *A. gregorii* and argued that with enhanced seed dispersal (Wickens and Lowe 2008). If this has evolved actually as adaptation for Australian fauna, in our opinion, fire threat should have been limited.

The main differences between both *Adansonia* species investigated here were fruit size and pericarp thickness. The Angolan *A. digitata* fruit volume were approx. up to ten times larger than that of *A. gregorii* and its shell more than twice of the pericarp thickness of *A. gregorii*. But, *A. digitata* exhibits high fruit phenotype diversity in its African distribution range, and size and pericarp can also be similar, even identical to fruits of *A. gregorii*, examined for Sudanese and Malawian baobab populations (Munthali et al. 2012; Gebauer and Luedeling 2013). It is speculative if fruit size diversity has evolved depending on variations of fire intensity. We could not find any data about differing fire intensities and a correlation to fruit size in African *Adansonia*. Furthermore fire intensities in African or Australian savannah does not differ generally (Williams et al. 1998; Vigilante et al. 2004; Govender et al. 2006). Thus differences in fruit size between *A. digitata* and *A. gregorii* cannot be explained clearly by fire threat.

The expected time span until the temperature of 90 °C was reached in the centre of fruits should be shorter for small fruits than for large ones. In our experiments *A. digitata* exposed a greater surface to the glow, which led in consequence to its faster heating up as compared to *A. gregorii*. This was clearly a result of our experimental set-up.

Regarding heat-resistance of the baobab fruit shell, a comparison to fire-resistance of bark in savannah trees is obvious. The main parameter determining a tree's fire-resistance seems to be bark thickness (Hoffmann et al. 2003; Lawes et al. 2011b, 2013; Rosell et al. 2014; Schafer et al. 2015). Regarding studies on correlations of bark thickness, bark density and fire-resistance of trees, low bark density is not necessarily an indication for fire-resistance (Brando et al. 2012). High bark density due to high bark moisture content may be a good fire protection, as well (Lawes et al. 2011b). Conversely, Bauer et al. (2010) demonstrated that fire-resistance of the bark of seven tree species increased with decreasing bark density. They have argued that a less dense bark type with internal air spaces could be advantageous for fire-resistance by additional insulation effects due to the internal air spaces. Therefore the low density of fruit shell and pulp provide at least a good heat insulation of the seeds.

Anatomy

Our findings coincide evidently with Bleechmoore's (2002) results. Shells of both species consist of a very similar three-layered structure, characterized by the different tissue type and cell orientation: "...the exocarp, which is a thick outer wall, comprising radial elongated thick-walled cells; the mesocarp, consisting of thin walled parenchyma cells; and the endocarp, which comprises sclerenchyma cells of both fibres and sclereids." "The mesocarp comprises the bulk of the wall..., with slightly radially orientated cells until close to the endocarp, where the cell lumen increases and the cells become more randomly orientated." (Bleechmore 2002).

The mechanism of heat-resistance can be found in the chemical composition of the shell. Even though the shell material resembles wood, it is different from wood with regard to lignin and cellulose proportion. Reviewing 29 papers concerning ignition temperatures of wood, Babrauskas (2001) reported that softwoods, compared to hardwoods, have a smaller content of hemicellulose and a higher content of lignin [up to 34% (White 1987)], resulting in higher ignition temperatures. Hemicelluloses and lignin components are pyrolyzed in the ranges 200–300 °C and 225–450 °C, respectively (White and Dietenberger 2001). The higher lignin content in the outermost exocarp layer *of A. digitata* fruits (up to 45%) helps explain why the fruit may have the ability to protect against inflammability. This is substantiated by the occurrence of highly lignified cells in the exocarp of *A. gregorii*, as well (Bleechmore 2002).

Germination

Published literature reports differing results about successful germination of *A. digitata* after acid or heat treatment (Esenowo 1991; Maghembe et al. 1994; Johansson 1999; Razanameharizaka et al. 2006; Niang et al. 2015) but without unequivocal results, which treatment could explain seed distribution mechanisms. For *A. gregorii* Turner and Dixon (2009) showed higher germination rates (GR) for acid as well as for heat treatment. It is noteworthy that seeds of both species can withstand immersion in boiling water for up to 3 min without total loss of germination capability (Razanameharizaka et al. 2006; Turner and Dixon 2009) indicating an appropriate adaptation to wildfires.

Our data support a positive effect of heat on GR only for *A. digitata* reaching 25% while seeds without treatment did not germinate. Sulphuric acid treatment, however, resulted in 76% GR and 24% for *A. gregorii*, whereas heat treatment or no treatment did not differ significantly in the latter, GR was only up to 14 and 10% respectively.

Since acid treatment is assumed to simulate animal's digestion, our results substantiate the hypothesis of animals as seed propagators of *Adansonia*. Using only 2–5 tc's for inner temperature measurement provide only rough information about min./max. temperature distribution and gradients within fruit pulp. We can only carefully estimate from one fruit of *A. digitata* that heat treatment below 60 °C had no positive impact on germination. An upper temperature limit was not detectable due to a smouldering fire pocket close to the tc location in one case.

The seed shell is hard and up to 0.6 mm thick. Acid treatment thins the seed shell, facilitates water entering the seed, and helps the embryo to break through. Heat treatment may also crack and fissure the seed shell and contribute to an easier breaking through of the embryo as well but not as effective as acid. However, improved germination due to heat exposure does not contradict the argument of heat protecting fruit shell and can just be a side effect.

In our study the influence of the seed shell to thermic insulation was not investigated. It can be assumed that the seed shell contributes to heat protection. Which temperature eventually occurs in the inner of a seed is not known. Therefore lethal conditions for the seeds could not be derived from our experiments.

Determining characteristic values of flammability or thermal conductivity of standardised fruit shell samples were probably more meaningful, but we wanted to evaluate germinating capacity after realistic exposure to heat concerning the entire fruit. This led to the disadvantage that six fruits each of both *Adansonia* species did not deliver enough data to draw additional conclusions. This would require fruits of the same size, volume or shell thickness.

Furthermore using charcoal on a grill as heat source did not reflect a realistic savannah fire. Glow delivered radiation but only a small contribution of convection. Flames have large turbulent convection effects, influencing the speed at which the object is heated. Key sizes of radiation are area and temperature of heat source, i.e., a larger heat source area results in faster heating and higher temperatures. Forced convection appears due to up-current and wind. Consequently faster air movement causes greater heat exchange. Thus, our methodological approach clearly affects the heating time. As a result, the considered heat exposure times may be overestimated by a factor of two. On the other hand, the horizontal positioning of *A. digitata* fruits did not correspond to the natural exposure to fire. In fact, exposing the lateral surface to the heat source resulted in faster heating up of the fruit. Thus, our heat exposure times may be underestimated for *A. digitata*.

Conclusion

We investigated the role the shell structure in *Adansonia* ecology may play. Both, the African *A. digitata* and the Australian *A. gregorii* are species of fire-prone habitats possessing a well-adapted bark to withstand wildfires (Owen 1974; Bowman 1997; Sidibé and Williams 2002; Chapotin et al. 2006; Watson 2007; Nano and Clarke 2011). So far, hydrochory and zoochory gave us ambiguous explanation for the fruit shell's function.

Our data strongly indicate that the fruit shell of investigated *Adansonia* species serves as adequate protection for the seed from the heat generated in typical savannah wildfires. Even though heat treatment had a positive effect on seed germination compared to no treatment, acid treatment of seeds simulating the passage through digestive tracts of animals resulted in the highest germination rate. The structure of the fruit (size, volume, shell) was found in this study to protect the seed from high temperatures and the resultant increase of temperature was found to subsequently increase germination rates in *A. digitata*. Low density of shell and fruit pulp contributes to good heat insulation for the seeds. High lignin content in the outer exocarp inhibits most likely early ignition of the shell. Our investigation did not reveal the best heat-resistant material in nature; there are e.g., some *Banksia* species with fruits that can survive at least 500 °C for 2 min (Enright and Lamont 1989). Our results show that *A. digitata* and *A. gregorii* are well adapted to fire-prone savannahs.

Authors' contributions

All authors conceived the ideas and designed methodology; TL and AK collected and analysed the data; and led the writing of the manuscript. All authors contributed critically to the drafts and gave final approval for publication. All authors read and approved the final manuscript.

Acknowledgements

The fieldwork in Angola was supported by a travel fund from the German Academic Exchange Service (DAAD). We acknowledge support by the German Research Foundation and the Open Access Publication Funds of the TU Dresden. These published results were obtained in collaboration with the Instituto Nacional da Biodiversidade e Áreas de Conservação (INBAC) of the Ministério do Ambiente da República de Angola. We are much obliged to Northern Territory Herbarium (DNA) in Palmerston (Australia) for sending us fruits of *A. gregorii*. We further like to thank K. Bunk, D. Fietsch, M. Hielscher, M. Kallmeyer, C. Knaust, F. Leonhardt, M. Luft, N. Noack, G. Schäfer, S. Schroeder, K. Schwabe, and C. Senwitz for support.

Competing interests

The authors declare that they have no competing interests.

Funding

The fieldwork in Angola was supported by a travel fund from the German Academic Exchange Service (DAAD) and the program "Strategic Partnerships" of the TU Dresden.

References

Auld TD, O'Connel MA (1991) Predicting patterns of post-fire germination in 35 eastern Australian Fabaceae. Aust J Ecol 16:53–70

Babrauskas V (2001) Ignition of wood: a review of the state of the art. J Fire Prot Eng 12(3):163–189

Bauer G, Speck T, Blömer J, Bertling J, Speck O (2010) Insulation capability of the bark of trees with different fire adaptation. J Mater Sci 45:5950–5959

Baum DA, Small RL, Wendel JF (1998) Biogeography and floral evolution of baobabs *Adansonia*, Bombacaceae as inferred from multiple data sets. Syst Biol 47:181–207

Baum DA, Smith SD, Yen A, Alverson WS, Nyffeler R, Whitlock BA, Oldham RL (2004) Phylogenetic relationships of Malvatheca (Bombacoideae and Malvoideae; Malvaceae sensu lato) as inferred from plastid DNA sequences. Am J Bot 91:1863–1871

Beard JS (1967) Some vegetation types of tropical Australia in relation to those of Africa and America. J Ecol 55:271–290

Bleechmore H (2002) Australian boab nuts. An analysis of the microstructure and the potential of reshaping using solvent vapours. Dissertation, University of Canberra, Canberra

Bowman DMJS (1997) Observations on the demography of the Australian boab *Adansonia gibbosa* in the north-west of the Northern Territory, Australia. Aust J Bot 45:893–904

Brando PM, Nepstad DC, Balch JK, Bolker B, Christman MC, Coe M, Putz F (2012) Fire-induced tree mortality in a neotropical forest: the roles of bark traits, tree size, wood density and fire behavior. Glob Change Biol 18:630–641

Bremer M, Fischer S, Nguyen TC, Wagenfuehr A, Phuong LX, Dai VH (2013) Effects of thermal modification on the properties of two Vietnamese bamboo species. Part II: effects on chemical composition. BioRes 8:981–993

Chapotin SM, Razanameharizaka JH, Holbrook NM (2006) A biomechanical perspective on the role of large stem volume and high water content in baobab (*Adansonia* spp.; Bombacaceae). Am J Bot 93:1251–1264

Cornu C, Ramahafaly W, Danthu P (2014) *Adansonia madagascariensis*, a marine hydrochory hypothesis. Bois For Trop 320:7–14

Dayamba SD, Tigabu M, Sawadogo L, Oden PC (2008) Seed germination of herbaceous and woody species of the Sudanian savanna-woodland in response to heat shock and smoke. For Ecol Manag 256:462–470

Dayamba SD, Savadogo P, Zida D, Sawadogo L, Tiveau D, Oden PC (2010) Fire temperature and residence time during dry season burning in a Sudanian savanna-woodland of West Africa with implication for seed germination. J For Res 21:445–450

Duvall CS (2007) Human settlement and baobab distribution in south-western Mali. J Biogeogr 34:1947–1961

Enright NJ, Lamont BB (1989) Fire temperatures and follicle-opening requirements in 10 *Banksia* species. Aust J Ecol 14:107–113

Esenowo GJ (1991) Studies on germination of *Adansonia digitata* seeds. J Agric Sci 117:81–84

Fukarek F, Hübel H, König P, Müller GK, Schuster R, Succow M (1995) Die grosse farbige Enzyklopadie Urania-Pflanzenreich, vol 5. Vegetation, Urania-Verlag, Leipzig, pp 136

Gashaw M, Michelsen A (2002) Influence of heat shock on seed germination of plants from regularly burnt savanna woodlands and grasslands in Ethiopia. Plant Ecol 159:83–93

Gebauer J, Luedeling E (2013) A note on baobab (*Adansonia digitata* L.) in Kordofan, Sudan. Genet Resour Crop Evol 60:1587–1596

Govender N, Trollope WS, Van Wilgen BW (2006) The effect of fire season, fire frequency, rainfall and management on fire intensity in savanna vegetation in South Africa. J Appl Ecol 43:748–758

Hare RC (1965) Contribution of bark to fire resistance of southern trees. J For 63:248–251

Hoffmann WA, Orthen B, Nascimento PKV (2003) Comparative fire ecology of tropical savanna and fores trees. Funct Ecol 17:720–726

Johansson M (1999) The baobab tree in Kondoa Irangi Hills, Tanzania. Dissertation, Swedish University of Agricultural Sciences, Uppsala

Kamatou GPP, Vermaak I, Viljoen AM (2011) An updated review of *Adansonia digitata*: a commercially important African tree. S Afr J Bot 77:908–919

Klason P (1920) Ueber Lignin und Lignin-Reaktionen (About lignin and relevant chemical reactions, German). Ber Dtsch Chem Ges (A and B series) 53:706–711

Kretschmann DE (2010) Mechanical properties of wood. In: Ross RJ (ed) Wood handbook: wood as an engineering material. Forest Products Laboratory, Madison, pp 100–145

Kuerschner K, Hoffer A (1931) Eine neue quantitative Cellulosebestimmung (A new method of cellulose determination, German). Chem Ztg 17:161–168

Lawes MJ, Adie H, Russel-Smith J, Murphy B, Midgley JJ (2011a) How do small savanna trees avoid stem mortality by fire? The roles of stem diameter, height and bark thickness. Ecosphere 2:1–13

Lawes MJ, Richards A, Dathe J, Midgley JJ (2011b) Bark thickness determines fire resistance of selected tree species from fire-prone tropical savanna in north Australia. Plant Ecol 212:2057–2069

Lawes MJ, Midgley JJ, Clarke PJ (2013) Costs and benefits of relative bark thickness in relation to fire damage: a savanna/forest contrast. J Ecol 101:517–524

Maghembe JA, Kwesiga F, Ngulube M, Prins H, Malaya FM (1994) Domestication potential of indigenous fruit trees of the miombo woodlands of southern Africa. In: Leakey RRB, Newton AC (eds) ITE symposium, vol 29, Tropical trees: the potential for domestication and the rebuilding of forest resources. Institute of Terrestrial Ecology, Bush Estate, Penicuik, Midlothian, pp 220

McDonald JH (2014) Handbook of biological statistics, 3rd edn. Sparky House Publishing, Baltimore, pp 77–85

Miranda AC, Miranda HS, Dias IDFO, De Souza Dias BF (1993) Soil and air temperatures during prescribed cerated fires in Central Brazil. J Trop Ecol 9:313–320

Munthali CRY, Chirwa PW, Akinnifesi FK (2012) Phenotypic variation in fruit and seed morphology of *Adansonia digitata* L. (baobab) in five selected wild populations in Malawi. Agrofor Syst 85:279–290

Nano CE, Clarke PJ (2011) How do drought and fire influence the patterns of resprouting in Australian deserts? Plant Ecol 212:2095–2110

Niang M, Diouf M, Samba SAN, Ndoye O, Van Damme P (2015) Difference in germination rate of Baobab (*Adansonia digitata* L.) provenances contrasting in their seed morphometrics when pretreated with concentrated sulfuric acid. Afr J Agric Res 10:1412–1420

Owen J (1974) A contribution to the ecology of the African baobab *Adansonia digitata*. Savanna 3:1–12

Patrut A, Mayne DH, Von Reden KF, Lowy DA, Venter S, McNichol AP, Roberts ML, Margineanu D (2010) Age and growth rate dynamics of an old African baobab determined by radiocarbon dating. Radiocarbon 52:727–734

Pausas JG (2015) Bark thickness and fire regime. Funct Ecol 29:315–327

Pinard MA, Huffman J (1997) Fire resistance and bark properties of trees in a seasonally dry forest in eastern Bolivia. J Trop Ecol 13:727–740

Poljak A (1948) Pulping with peracetic acid. Angew Chem 60:45–46

Razanameharizaka J, Grouzis M, Ravelomanana D, Danthu P (2006) Seed storage behaviour and seed germination in African and Malagasy baobabs (*Adansonia* species). Seed Sci Res 16:83–88

Romero C, Dovie D, Gambiza J, Luoga E, Schmitt S, Grundy I (2001) Effects of commercial bark harvesting on *Adansonia digitata* (baobab) in the Save-Odzi valley, Zimbabwe, with considerations for its management. Hotspring Working Group, IES Working Paper 18, pp 28–50

Rosell JA, Gleason S, Mendez-Alonzo R, Chang Y, Westoby M (2014) Bark functional ecology: evidence for tradeoffs, functional coordination, and environment producing bark diversity. New Phytol 201:486–497

Savadogo P, Zida D, Sawadogo L, Tiveau D, Tigabu M, Oden PC (2007) Fuel and fire characteristics in savanna-woodland of West Africa in relation to grazing and dominant grass type. Int J Wildland Fire 16:531–539

Schafer JL, Breslow BP, Hohmann MG, Hoffmann WA (2015) Relative bark thickness is correlated with tree species distributions along a fire frequency gradient. Fire Ecol 11:74–87

Schüle W (1990) Landscapes and climate in prehistory: interaction of wildlife, man, and fire. In: Goldammer JG (ed) Fire in the tropical biota: ecosystem processes and global challenges. Springer, Berlin, pp 273–318

Sidibé M, Williams JT (2002) Baobab—*Adansonia digitata*. Southampton, International Centre for Underutilised Crops, p 66

Stuppy W (2004) Glossary of seed and fruit morphological terms. Seed Conservation Department, Royal Botanic Gardens, Kew

Trollope WSW (1984) Fire in savanna. In: Tainton NM, de Booysen V (eds) Ecological effects of fire in South African ecosystems. Springer, Berlin, pp 149–176

Tsy JMLP, Lumaret R, Mayne D, Vall AOM, Abutaba YIM, Sagna M, Raoseta SOR, Danthu P (2009) Chloroplast DNA phylogeography suggests a West African centre of origin for the baobab, *Adansonia digitata* L. (Bombacoideae, Malvaceae). Mol Ecol 18:1707–1715

Turner SR, Dixon KW (2009) Seed dormancy and germination in the Australian baobab, *Adansonia gregorii* F. Muell. Seed Sci Res 19:261–266

Venter SM, Witkowski ETF (2013) Where are the young baobabs? Factors affecting regeneration of *Adansonia digitata* L. in a communally managed region of southern Africa. J Arid Environ 92:1–13

Vigilante T, Bowman DM, Fisher R, Russel-Smith J, Yates C (2004) Contemporary landscape burning patterns in the far North Kimberley region of northwest Australia: human influences and environmental determinants. J Biogeogr 31:1317–1333

Vines RG (1968) Heat transfer through bark, and the resistance of trees to fire. Aust J Bot 16:499–514

Walker J, Gillison AN (1982) Australian savannas. In: Huntley BJ, Walker BH (eds) Ecology of tropical savannas. Springer, Berlin, pp 5–24

Watson R (2007) The African Baobab. Struik Publishers, Cape Town

White RH (1987) Effects of lignin content and extractives on the higher heating value of wood. Wood Fiber Sci 19:446–452

White RH, Dietenberger MA (2001) Wood products: thermal degradation and fire. In: Buschow KHJ, Cahn R, Flemings M, Ilschner B, Kramer E, Mahajan S, Veyssiere P (eds) Encyclopedia of materials: science and technology. Elsevier, Amsterdam, pp 9712–9716

Wickens GE, Lowe P (2008) The baobabs, pachycauls of Africa, Madagascar and Australia. Springer, Berlin

Williams RJ, Gill AM, Moore PHR (1998) Seasonal changes in fire behaviour in a tropical savanna in Northern Australia. Int J Wildland Fire 8:227–239

Wilson RT (1988) Vital statistics of the baobab (*Adansonia digitata*). Afr J Ecol 26:197–206

8

Orchid conservation: how can we meet the challenges in the twenty-first century?

Michael F. Fay[1,2*]

Abstract

With c. 28,000 species, orchids are one of the largest families of flowering plants, and they are also one of the most threatened, in part due to their complex life history strategies. Threats include habitat destruction and climate change, but many orchids are also threatened by unsustainable (often illegal and/or undocumented) harvest for horticulture, food or medicine. The level of these threats now outstrips our abilities to combat them at a species-by-species basis for all species in such a large group as Orchidaceae; if we are to be successful in conserving orchids for the future, we will need to develop approaches that allow us to address the threats on a broader scale to complement focused approaches for the species that are identified as being at the highest risk.

Keywords: Conservation priorities, Systematics, Phylogenetics, Population genetics, In situ conservation, Ex situ conservation, Integrated conservation, Mycorrhizas, Pollination, Illegal trade, CITES, Red List

Background

We live in exciting but challenging times for the study of biodiversity and its conservation. Novel technologies provide us with the opportunity to study species and their interactions in greater detail than ever before, but challenges associated with global change, habitat destruction, changing land use and unsustainable utilization of biodiversity make its conservation ever-more urgent, and the capacity to develop solutions at the species-level for more than a small proportion of species (e.g. those identified as being at the greatest risk) is likely be outstripped by the sheer scale and pace of change (Gale et al. 2018). In response to these challenges, scientists and conservation practitioners must make difficult choices in prioritizing their work. With c. 28,000 species, orchids are probably the second largest family of flowering plants after Asteraceae (Chase et al. 2003, 2015; Willis 2017). By the end of 2017, the IUCN Global Red List included assessments for 948 orchid species, of which 56.5% are threatened (IUCN 2017), but this leaves c. 27,000 species to be assessed for the Global Red List. Due to their complex biology, notably their interactions with mycorrhizal fungi, pollinators and host trees, orchids present particular challenges for conservation, and this is compounded by non-sustainable and often illegal collection for horticulture, medicine and food and by climate change (Fay 2015a; Gale et al. 2018).

> *The need for orchid conservation is paramount if we are to leave to future generations the rich and wildly fascinating orchid legacy we all enjoy today. Without effective conservation actions..., threatening process will continue to militate against the survival of rare orchids, resulting in their continued degradation and inevitable extinction.*

In this quote from their recent book, Swarts and Dixon (2017, p. 4) encapsulated the need for urgent and effective conservation action for orchids, a group that biologists including Linnaeus and Darwin have found fascinating due to their extreme specializations (Fay and Chase 2009). In this short review, I discuss the conservation status of orchids, the threats to their continued survival and approaches to addressing the challenges relating to orchid conservation in the twenty-first century.

Conservation status of orchids—what do we know?

With c. 28,000 species divided into five subfamilies, Orchidaceae are one of the largest and most widespread families of flowering plants, and they account for c. 8%

*Correspondence: m.fay@kew.org
[1] Royal Botanic Gardens, Kew, Richmond, Surrey TW9 3AB, UK
Full list of author information is available at the end of the article

of angiosperm species diversity (Chase et al. 2003, 2015; Willis 2017). However, only c. 1000 species have been assessed for the IUCN Global Red List to date (IUCN 2017), and an alarming 56.5% of those that have been assessed fall into one of the categories of threat (critically endangered, endangered and vulnerable). Major threats include habitat destruction and unsustainable (often illegal) harvesting, and because of their complex life histories orchids are thought to be particularly vulnerable to the effects of global environmental change (Fay and Chase 2009; Swarts and Dixon 2009a; Gale et al. 2018). All species of subfamily Cypripedioideae (the slipper orchids) were assessed for the Global Red List in a recent project, and due to a combination of habitat degradation and, in some cases, ruthless harvesting, c. 90% of species were assessed as threatened (Fay and Rankou 2016). Slipper orchids were chosen for that project because of their high profile and the expected high level of threat, but even the family-wide figure of 56.5% shows that many orchid species are threatened with extinction. Their conservation should be regarded as urgent if these iconic plants are not to decline further.

What are the threats?

Globally, habitats and the species that occur in them are under increasing pressure. Brooks et al. (2002), for example, reported that nearly 50% of vascular plant species are endemic to 25 "hotspots" of biodiversity, each with at least 1500 endemic plant species, but all of these hotspots had lost more than two-thirds of their pristine habitat. These authors predicted that, as a result of this habitat loss, many of the endemics in these hotspots are likely to become extinct or to be threatened with extinction in the near future. In a study of a Mediterranean island, Vogt-Schilb et al. (2016) presented evidence of high turnover in species composition of orchids in communities as result of change in land use. There is also increasing evidence that global change may also be influencing species distributions (e.g. Fay 2015b), and benefits of and problems associated with assisted migration and translocations to climatically suitable localities are increasingly being discussed in relation to orchids (e.g. Ramsay and Dixon 2003; Swarts and Dixon 2009a) and more generally (Pearman and Walker 2004; Ricciardi and Simberloff 2009). Like all plants, orchid species are affected by these pressures, but due to their often complex interactions with pollinators, mycorrhizal fungi and host trees, they are likely to be at greater risk as they are dependent on other organisms that are also being affected by habitat or climatic change. Thus, orchids present greater challenges than many other plant groups.

An additional area of threat relating to orchids is unsustainable harvesting, and the full impact of this is only slowly being understood. Indiscriminate collection for horticultural collections has been documented as having a major impact on some orchids, notably species of *Cattleya, Laelia, Renanthera* and some slipper orchids (*Cypripedium, Paphiopedilum, Phragmipedium*), and in some cases these have been systematically stripped from the wild to the point of (near) extinction. However, many orchids are not collected for horticulture or are collected in such small numbers that there is unlikely to be much impact (e.g. Cribb et al. 2003; Fay 2015a).

Largely as an attempt to control illegal smuggling of these desirable orchids and because of perceived problems with identification, all orchids were placed on the appendices of the Convention on International Trade in Endangered Species (CITES) (e.g. Cribb et al. 2003), and orchids account for >70% of the species listed on CITES. However, it is becoming increasingly clear that many orchid species are still being collected and transported across international borders, for use as medicine or food in addition to the horticultural trade, without the permits required under CITES (e.g. Fay 2015a; Hinsley et al. 2018). The extent of the illegal trade is difficult to assess, but attempts are being made to estimate the extent of non-compliance with CITES regulations (e.g. Ghorbani et al. 2014; Hinsley et al. 2017). Notable examples of poorly documented trade relate to orchids collected for traditional medicine in East Asia and for production of the foodstuffs salep in the eastern Mediterranean and the Middle East (e.g. Kreziou et al. 2016; de Boer et al. 2017) and chikanda in south-eastern Africa (e.g. Veldman et al. 2014); the development of novel DNA-based techniques is now providing the opportunity to identify the orchid species in these processed foodstuffs, and this will mean that documentation and policing of the trade will become increasingly feasible (e.g. de Boer et al. 2017). In addition to not halting illegal trade, an unintended consequence of the listing of all species of orchid on CITES has been a reduction in the collection of orchids for scientific purposes, including conservation research; (Roberts and Solow 2008).

Conserving orchids

Orchid conservation has been the subject of many reviews; to access the older literature, the reader is pointed to reviews written by Koopowitz (2001), Cribb et al. (2003), Koopowitz et al. (2003), Dixon and Phillips (2007) and Swarts and Dixon (2009b) and many of the chapters in the volume resulting from the First International Orchid Conservation Congress held in Western Australia in 2001 (Dixon et al. 2003). Most recently, Swarts and Dixon (2017) published a book focusing on conservation techniques for terrestrial orchids. Together, these publications represent a rich resource for conservation and should be consulted by anyone with an interest in orchid conservation. In their review, Swarts and Dixon (2009b) focused on the role of botanic gardens in supporting orchid conservation scientifically and horticulturally, and Pedersen et al. (2018) stressed the close link between collection-based research and conservation.

Documentation of orchid species (how many species should be recognized, where they occur and how they are related to each other) is improving, and this information provides the necessary background for establish conservation priorities, but effective conservation will also depend on the maintenance of the essential links with animals, fungi and other plants that allow them to survive. In recognition of the importance of these interactions, "Making the links" was chosen as the theme for the 5th International Orchid Conservation Congress, held on La Réunion in 2013 (Fay et al. 2015; Fay 2016; and references therein). Six of the papers from the 6th International Orchid Conservation Congress, held in Hong Kong in 2016, also focus on these links; see Gale et al. (2018) and references therein). With orchids, perhaps more than most other plant groups, integrated conservation involving in situ and ex situ approaches and studies of pollination, mycorrhizal associations and genetics (as summarized by Swarts and Dixon (2009a) for terrestrial orchids in Western Australia) will be necessary if we are to succeed.

Species-by-species approaches to conservation will continue to be the ideal for species identified as being the highest priorities or at the greatest risk, but for a group as large as Orchidaceae, in which so little is known about population genetics, etc. for many species, we will need to complement these with other approaches that allow us to conserve groups of species that have similar threats, that are closely related or that occur sympatrically. Approaches addressing conservation of process (rather than individual species) may be appropriate in groups which are undergoing relatively evolution due to hybridization and/or polyploidization (e.g. Ennos et al. 2012). Formal conservation planning has generally focused more on animals than plants, but, for example, the Conservation Breeding Specialist Group of IUCN has recently broadened the focus of its activities and changed its name to the Conservation Planning Specialist Group (http://www.cpsg.org/), and some publications now contain sections specifically focused on plants (e.g. IUCN 2017). The Orchid Specialist Group of IUCN looks forward to working with CPSG and similar organizations in developing effective conservation planning for orchids.

Systematics and genetics: helping to set conservation priorities

If we are to conserve biodiversity effectively, we first need to know what exists, and there is a clear role for systematic and taxonomic studies in circumscribing species and in identifying high priorities for conservation. This is particularly the case in taxonomically complex groups. In *Dactylorhiza*, Pillon et al. (2006) found greater phylogenetic and genetic diversity in the Caucasus and the Mediterranean Basin than in Western Europe (generally regarded as the centre of diversity for this genus), and they stated that conservation of lineages of *Dactylorhiza* in the Caucasus and Mediterranean should be given greater priority; because species number is correlated with taxonomic effort, it is not always an appropriate measure of biodiversity as it can be sensitive to "taxonomic inflation" in well studied areas. They also stressed the importance of conserving areas where allotetraploids are formed rather than conserving specific allotetraploids, thus conserving process rather than named taxa (see also Ennos et al. 2012). In *Ophrys*, there is little consensus regarding the number of species that should be recognised, with authors accepting < 20 species (Pedersen and Faurholdt 2007) or > 300 species (Delforge 2006); setting conservation priorities with so little consensus is impossible, and further studies aimed at sorting out the relationships and delimiting the units for conservation are badly needed.

Many groups of orchids are still poorly known, especially in tropical regions, and phylogenetic studies will be necessary to identify the number of species to be recognized and those that are phylogenetically isolated and consequently of high conservation value. Borba et al. (2014), for example, showed that the rare and poorly known monospecific genus *Cotylolabium* from Brazil falls in an isolated position as sister to the remainder of subtribe Spiranthinae, and it should thus be treated as a high priority for conservation as it represents the same amount of phylogenetic history as the other c. 40 genera in the subtribe. On a wider scale, Li et al. (2018) investigated the use of phylogenetic measures as a means of prioritizing members of Orchidaceae for conservation in the Indo-Burma Biodiversity Hotspot, revealing Thailand, South China and Vietnam as the areas harbouring the highest phylogenetic diversity and *Tropidia curculigoides*, *Thaia saprophytica* and *Risleya atropurpurea* as accounting for disproportionately great evolutionary distinctiveness.

At the population level, genetic studies can be used to identify regions or populations that should be treated as high priority for conservation. This is a rapidly developing area and reviews become outdated quickly and will not be dealt with in detail here. Previously, development of markers was time-consuming and expensive, but new technologies are speeding up marker development and allowing more loci to be studied than previously possible, and the quality of the information to be used in conservation planning will improve as a result (e.g. Gargiulo et al. 2018).

Conserving habitats

So long as climate change does not make conditions unsuitable for species, conserving the habitats where orchids and, for epiphytes, their host trees grow should be treated as the highest priority, and some countries have established reserves specifically for orchids (see, e.g., Cribb et al. 2003). Rasmussen and Rasmussen (2018) reviewed the relationship between epiphytic orchids and their host trees, calling for further research into the mechanisms controlling distribution of orchids on different species of trees.

Some orchid conservation organisations have the principle aim of establishing reserves, one being the Orchid Conservation Alliance (OCA), who state that "preservation of natural orchid habitat preserves the orchids, their pollinators, their genetic diversity, and other fauna, as well as the birds, frogs, insects, reptiles, and mammals in the forests where they live" (OCA 2017). However, this will not in itself be sufficient, given the pressures that orchids face from habitat destruction, unsustainable harvesting and climate change. Meeting these challenges will in many cases also involve a combination of creating new habitats, transplantation and ex situ conservation in seedbanks and living collections. Papers in this issue by Kendon et al. (2017), Zettler et al. (2017) and Higaki et al. (2017) focus on some aspects of ex situ conservation.

Conserving orchids in isolation from their pollinators, fungal associates and host plants means that the complexity of their biology is lost, even though the species still exists. For this reason, orchid conservationists are some of the keenest advocates of "integrated conservation", using ex situ techniques to support in situ conservation as appropriate.

Understanding pollinators and pollination

Orchids are renowned for the wide range of pollination mechanisms and syndromes (e.g. Darwin 1862; Micheneau et al. 2009) and the species diversity in the family has been attributed, in part, to the diversity of pollen mechanisms (e.g. Cozzolino and Widmer 2005). Because of the diversity of pollination mechanisms, Roberts (2003) stressed the importance of understanding pollination biology for effective orchid conservation, stating that "orchid conservation will require a case by case, functional ecosystem approach", and noted the need to conserve not only the orchid and the pollinator, but also in some cases the "pollinator food source, nesting site, larval host species, and in the case of parasitic pollinators, the larval host plant of its host species". If pollinators are not present in sufficient numbers (or at all), fruit production can be limited or absent, and this can have a major impact on the choice of sites for reintroduction programmes (Reiter et al. 2017). Hutchings et al. (2018) showed that climate change can decouple the phenology of pollinator and orchid species, potentially leading to reproductive failure of the orchid.

Despite the identified need for knowledge of the components necessary to ensure that pollinators are present and available for pollination, we are still far from understanding the pollination biology of many orchid species, and, despite the long history of the study of orchid pollination, new discoveries are still being made on a regular basis. Recent papers have investigated birds (e.g. Micheneau et al. 2006; van der Niet et al. 2015), crickets (Micheneau et al. 2010), fungus gnats (Phillips et al. 2014) and biting midges (Bogarín et al. 2018) as specialized pollinators. Many orchid species attract pollinators by deceipt, with the forms of deception including food deception, brood-site imitation, shelter imitation, rendezvous attraction and sexual deception (Jersáková et al. 2006), and recent discoveries of dual deceipt (pseudopollen lacking food value; Davies et al. 2013), carrion mimicry (van der Niet et al. 2011) and production of fruitfly aggregation pheromones (Karremans et al. 2015) demonstrate that we are far from understanding the full complexities of orchid pollination.

Understanding mycorrhizal associations

Seed and protocorm development is reviewed in this issue by Yeung (2017), including discussion of mycorrhizal associations and the survival of orchid seeds and plantlets in their natural habitats. Clearly, the role of mycorrhizal fungi is crucial to the survival of self-sustaining populations of orchids, but there is much research still to be conducted before we fully understand the mycorrhizal associations, especially with epiphytic orchid species. Even with temperate terrestrial species, the processes involved are not fully understood, but techniques including measurement of isotope enrichment of carbon, nitrogen and hydrogen now allow us to demonstrate the contribution that the fungi make to the nutrition of orchids, even when the plants are apparently capable of photosynthesis (e.g. Gebauer et al. 2016). These techniques were recently used, for example, by Fay et al. (2018) to demonstrate that reintroduced seedlings of *Cypripedium calceolus* had established mycorrhizal associations after planting out, despite being produced axenically.

Controlling trade: can harvesting of orchids from nature be sustainable?

With habitat destruction and global change, collection of orchids for horticulture, food or medicine represents one of the major threats to the survival of some groups of orchids, and Hinsley et al. (2018) highlighted four key priorities to address this problem. Much of the harvest of orchids and subsequent trade is unregulated and undocumented, and research into trade dynamics and the impacts of harvest will be of critical importance if we are to prevent orchids being driven into extinction. Strengthening the legal trade and tackling illegal trade were both identified as priorities—unless we provide the necessary support for the legal trade, the illegal trade will continue. Finally, raising the profile of orchid trade among policy makers, conservationists and the public was stressed by Hinsley et al. as being fundamental in underpinning the other priorities.

Conclusions

Due to their complex life histories, orchids are liable to be severely affected by habitat destruction and climate change, and unsustainable (often illegal and undocumented) harvest presents a major additional risk to some groups of orchids. To conserve orchids effectively, we will need to understand their biology, and this will require further research into areas including pollination, mycorrhizal associations, population genetics and demographics. However, due to the large number of species involved, species-by-species approaches will only be feasible for those species identified as the highest priorities (due, e.g., to phylogenetic distinction, extreme rarity or narrow distribution) or at the greatest risk, and we will need to complement these with broader-scale approaches. These should include habitat conservation, especially for orchid-rich environments, conservation planning for groups of species (e.g. those that are closely related, affected by similar threats or growing sympatrically) and increased monitoring and control of harvest to ensure that this is legal and sustainable. Without such combined approaches, we will not be able to ensure the survival of these charismatic species into the future.

Authors' contributions

The author read and approved the final manuscript.

Author details

[1] Royal Botanic Gardens, Kew, Richmond, Surrey TW9 3AB, UK. [2] School of Biological Sciences, University of Western Australia, Crawley, WA 6009, Australia.

Acknowledgements

The author thanks members of the Orchid Specialist Group of the Species Survival Commission (IUCN), especially Prof. Kingsley Dixon and the chairs of the newly formed trade and molecular identification groups (Drs Amy Hinsley, Jacob Phelps and Hugo de Boer) for their support and useful discussions.

Competing interests

The author declares that he has no competing interests.

Funding

Not applicable.

References

Bogarín D, Fernández M, Borkent A, Heemskerk A, Pupulin F, Ramírez S, Smets E, Gravendeel B (2018) Pollination of *Trichosalpinx* (Orchidaceae: Pleurothallidinae) by biting midges (Diptera: Ceratopogonidae). Bot J Linn Soc 186:510–543

Borba EL, Salazar GA, Mazzoni-Viveiros S, Batista JAN (2014) Phylogenetic position and floral morphology of the Brazilian endemic, monospecific genus *Cotylolabium*: a sister group for the remaining Spiranthinae (Orchidaceae). Bot J Linn Soc 175:29–46

Brooks TM, Mittermeier RA, Mittermeier CG, Da Fonseca GAB, Rylands AB, Konstant WR, Flick P, Pilgrim J, Oldfield S, Magin G, Hilton-Taylor C (2002) Habitat loss and extinction in the hotspots of biodiversity. Cons Biol 16:909–923

Chase MW, Cameron KM, Barrett RL, Freudenstein JV (2003) DNA data and Orchidaceae systematics: a new phylogenic classification. In: Dixon KW, Kell SP, Barrett RL, Cribb PJ (eds) Orchid conservation. Sabah Natural History Publications, Kota Kinabalu, pp 69–89

Chase MW, Cameron KM, Freudenstein JV, Pridgeon AM, Salazar G, van den Berg C, Schuiteman A (2015) An updated classification of Orchidaceae. Bot J Linn Soc 177:151–174

Cozzolino S, Widmer A (2005) Orchid diversity: an evolutionary consequence of deception. Trends Ecol Evol 20:487–494

Cribb PJ, Kell SP, Dixon KW, Barrett RL (2003) Orchid conservation: a global perspective. In: Dixon KW, Kell SP, Barrett RL, Cribb PJ (eds) Orchid conservation. Sabah Natural History Publications, Kota Kinabalu, pp 1–24

Darwin C (1862) On the various contrivances by which British and foreign orchids are fertilised by insects, and on the good effect of intercrossing. John Murray, London

Davies KL, Stpiczyńska M, Kamińska M (2013) Dual deceit in pseudopollen-producing *Maxillaria s.s.* (Orchidaceae: Maxillariinae). Bot J Linn Soc 173:744–763

de Boer HJ, Ghorbani A, Manzanilla V, Raclariu A-C, Kreziou A, Ounjai S, Osathanunkul M, Gravendeel B (2017) DNA metabarcoding of orchid-derived products reveals widespread illegal orchid trade. Proc R Soc B 284:20171182

Delforge P (2006) Orchids of Europe. North Africa and the Middle East. A&C Black, London

Dixon K, Phillips RD (2007) The orchid conservation challenge. Lankesteriana 7:11–12

Dixon KW, Kell SP, Barrett RL, Cribb PJ (eds) (2003) Orchid conservation. Sabah Natural History Publications, Kota Kinabalu

Ennos RA, Whitlock R, Fay MF, Jones B, Neaves LE, Payne R, Taylor I, de Vere N, Hollingsworth PM (2012) Process-based species action plans: an approach to conserve contemporary evolutionary processes that sustain diversity in taxonomically complex groups. Bot J Linn Soc 168:194–203

Fay MF (2015a) Undocumented trade in species of Orchidaceae. https://cites.org/sites/default/files/eng/com/pc/22/Inf/E-PC22-Inf-06.pdf. Accessed 11 Sept 2017

Fay MF (2015b) British and Irish orchids in a changing world. Curtis's Bot Mag 32:3–23

Fay MF (2016) Orchid conservation: further links. Ann Bot 118:89–91

Fay MF, Chase MW (2009) Orchid biology—from Linnaeus via Darwin to the 21st century. Ann Bot 104:359–364

Fay M, Rankou H (2016) Slipper orchids on the IUCN Red List. In: 2015 Annual Report to the Environment Agency—Abu Dhabi. Framework Support for Implementing the Strategic Plan of the IUCN Species Survival Commission. Pp 106–111

Fay MF, Pailler T, Dixon KW (2015) Orchid conservation: making the links. Ann Bot 116:377–379

Fay MF, Feustel M, Newlands C, Gebauer G (2018) Inferring the mycorrhizal status of introduced plants of *Cypripedium calceolus* (Orchidaceae) in northern England using stable isotope analysis. Bot J Linn Soc 186:587–590

Gale SW, Fischer GA, Cribb PJ, Fay MF (2018) Orchid conservation: bridging the gap between science and practice. Bot J Linn Soc 186:425–434

Gargiulo R, Ilves A, Kaart T, Fay MF, Kull T (2018) High genetic diversity in a threatened clonal species, *Cypripedium calceolus* (Orchidaceae), enables long-term stability of the species in different biogeographical regions in Estonia. Bot J Linn Soc 186:560–571

Gebauer G, Preiss K, Gebauer AC (2016) Partial mycoheterotrophy is more widespread among orchids than previously assumed. New Phytol 211:11–15

Ghorbani A, Gravendeel B, Zarre S, de Boer H (2014) Illegal wild collection and international trade of CITES-listed terrestrial orchid tubers in Iran. Traffic Bull 26:52–58

Higaki K, Rammitsu K, Yamashita Y, Yukawa T, Ogura-Tsujita Y (2017) A simple culture method promoting seed germination and seedling development in the mycoheterotrophic orchid *Gastrodia pubilabiata* symbiotic with a litter-decomposing fungus. Bot Stud 58:59

Hinsley A, Nuno A, Ridout M, St John FAV, Roberts DL (2017) Estimating the extent of CITES noncompliance among traders and end-consumers; lessons from the global orchid trade. Cons Lett 10:602–609

Hinsley A, de Boer HJ, Fay MF, Gale SW, Gardiner LM, Gunasekara RS, Kumar P, Masters S, Metusala D, Roberts DL, Veldman S, Wong S, Phelps J (2018) A review of the trade in orchids, and its implications for conservation. Bot J Linn Soc 186:435–455

Hutchings MJ, Robbirt KM, Roberts DL, Davy AJ (2018) Vulnerability of a specialized pollination mechanism to climate change revealed by a 356-year analysis. Bot J Linn Soc 186:498–509

IUCN (2017) The IUCN red list of threatened species. 2017-2. http://www.iucnredlist.org/. Accessed 25 Sep 2017

IUCN—SSC Species Conservation Planning Sub-Committee (2017) Guidelines for species conservation planning. Version 1.0. IUCN, Cambridge, Gland. https://doi.org/10.2305/IUCN.CH.2017.18.en. Accessed 23 May 2018

Jersáková J, Johnson SD, Kindlmann P (2006) Mechanisms and evolution of deceptive pollination in orchids. Biol Rev 81:219–235

Karremans AP, Pupulin F, Grimaldi D, Beentjes KK, Butôt R, Fazzi GE, Kaspers K, Kruizinga J, Roessingh P, Smets EF, Gravendeel B (2015) Pollination of *Specklinia* by nectar-feeding *Drosophila*: the first reported case of a deceptive syndrome employing aggregation pheromones in Orchidaceae. Ann Bot 116:437–455

Kendon JP, Rajaovelona L, Sandford H, Fang R, Bell J, Sarasan V (2017) Collecting near mature and immature orchid seeds for ex situ conservation: 'in vitro collecting' as a case study. Bot Stud 58:34

Koopowitz H (2001) Orchids and their conservation. B. T. Batsford, London

Koopowitz H, Lavarack PS, Dixon KW (2003) The nature of threats to orchid conservation. In: Dixon KW, Kell SP, Barrett RL, Cribb PJ (eds) orchid conservation. Sabah Natural History Publications, Kota Kinabalu, pp 25–42

Kreziou A, de Boer H, Gravendeel B (2016) Harvesting of salep orchids in north-western Greece continues to threaten natural populations. Oryx 50:393–396

Li J, Gale SW, Kumar P, Zhang J, Fischer G (2018) Prioritizing the orchids of a biodiversity hotspot for conservation based on phylogenetic history and extinction risk. Bot J Linn Soc 186:473–497

Micheneau C, Fournel J, Pailler T (2006) Bird pollination in an angraecoid orchid on Reunion Island (Mascarene Archipelago, Indian Ocean). Ann Bot 97:965–974

Micheneau C, Johnson SD, Fay MF (2009) Orchid pollination: from Darwin to the present day. Bot J Linn Soc 161:1–19

Micheneau C, Fournel J, Warren BH, Hugel S, Gauvin-Bialecki A, Pailler T, Strasberg D, Chase MW (2010) Orthoptera, a new order of pollinator. Ann Bot 105:355–364

OCA (2017) The orchid conservation alliance. https://orchidconservationalliance.org/. Accessed 27 Sep 2017

Pearman D, Walker K (2004) Rare plant introductions in the UK: creative conservation or wildflower gardening? Br Wildl 15:174–182

Pedersen HÆ, Faurholdt N (2007) Ophrys. The bee orchids of Europe. Kew Publishing, Kew

Pedersen HÆ, Find JI, Petersen G, Seberg O (2018) On the "Seidenfaden collection" and the multiple roles botanical gardens can play in orchid conservation. Lankesteriana 18:1–12

Phillips RD, Scaccabarozzi D, Retter BA, Hayes C, Brown GR, Dixon KW, Peakall R (2014) Caught in the act: pollination of sexually deceptive trap-flowers by fungus gnats in *Pterostylis* (Orchidaceae). Ann Bot 113:629–641

Pillon Y, Fay MF, Shipunov AB, Chase MW (2006) Species diversity versus phylogenetic diversity: a practical study in the taxonomically difficult genus *Dactylorhiza* (Orchidaceae). Biol Conserv 129:4–13

Ramsay MM, Dixon KW (2003) Propagation science, recovery and translocation of terrestrial orchids. In: Dixon KW, Kell SP, Barrett RL, Cribb PJ (eds) Orchid conservation. Sabah Natural History Publications, Kota Kinabalu, pp 259–288

Rasmussen HN, Rasmussen FN (2018) The epiphytic habitat on a living host: reflections on the orchid–tree relationship. Bot J Linn Soc 186:456–472

Reiter N, Vlcek K, O'Brien N, Gibson M, Pitts D, Brown GR, Bower CC, Phillips RD (2017) Pollinator rarity limits reintroduction sites in an endangered sexually deceptive orchid (*Caladenia hastata*): implications for plants with specialized pollination systems. Bot J Linn Soc 184:122–136

Ricciardi A, Simberloff D (2009) Assisted colonization is not a viable conservation strategy. Trends Ecol Evol 24:248–253

Roberts DL (2003) Pollination biology: the role of sexual reproduction in orchid conservation. In: Dixon KW, Kell SP, Barrett RL, Cribb PJ (eds) orchid conservation. Sabah Natural History Publications, Kota Kinabalu, pp 113–136

Roberts DL, Solow AR (2008) The effect of the convention on international trade in endangered species on scientific collections. Proc R Soc B 275:987–989

Swarts ND, Dixon KW (2009a) Terrestrial orchid conservation in the age of extinction. Ann Bot 104:543–556

Swarts ND, Dixon KW (2009b) Perspectives on orchid conservation in botanic gardens. Trends Plant Sci 14:590–598

Swarts ND, Dixon KW (2017) Conservation methods for terrestrial orchids. J. Ross Publishing, Plantation

van der Niet T, Hansen DM, Johnson SD (2011) Carrion mimicry in a South African orchid: flowers attract a narrow subset of the fly assemblage on animal carcasses. Ann Bot 107:981–992

van der Niet T, Cozien RJ, Johnson SD (2015) Experimental evidence for specialized bird pollination in the endangered South African orchid *Satyrium rhodanthum* and analysis of associated floral traits. Bot J Linn Soc 177:141–150

Veldman S, Otieno J, van Andel T, Gravendeel B, de Boer H (2014) Efforts urged to tackle thriving illegal orchid trade in Tanzania and Zambia for Chikanda production. Traffic Bull 26:47–50

Vogt-Schilb H, Pradel R, Geniez P, Hugot L, Delage A, Richard F, Schatz B (2016) Responses of orchids to habitat change in Corsica over 27 years. Ann Bot 118:115–123

Willis KJ (ed) (2017) State of the world's plants 2017. Report. Royal Botanic Gardens, Kew

Yeung EC (2017) A perspective on orchid seed and protocorm development. Bot Stud 58:33

Zettler LW, Rajaovelona L, Yokoya K, Kendon JP, Stice AL, Wood AE, Sarasan V (2017) Techniques for the collection, transportation, and isolation of orchid endophytes from afar: a case study from Madagascar. Bot Stud 58:54

9

Nuclear dynamics in the homothallic ascomycete *Sordaria fimicola*

Kulsumpun Krobanan and Wei-Chiang Shen*

Abstract

Background: The coprophilous ascomycete *Sordaria fimicola* usually reproduces sexually. Sexual differentiation in *S. fimicola* is accompanied by cellular and morphological changes, followed by multicellular tissue development to complete the sexual cycle. Although the morphological features of the sexual reproductive structure in *S. fimicola* have been well characterized, little is known about the nuclear dynamics and organization during these processes. Therefore, in this study, we successfully developed an *Agrobacterium*-mediated protoplast transformation protocol and generated histone H2B-mCherry-labeled *S. fimicola* strains. The life cycle of *S. fimicola* begins with germination of the ascospore and ends with the formation and discharge of new ascospores from the mature black sexual fruiting bodies, the so-called perithecia. The nuclear dynamics of the fluorescently labeled strains were examined during ascospore germination, hyphal elongation, and hyphal fusion using fluorescent microscopy.

Results: Live imaging revealed that the nuclei in the germlings and fusion hyphae during the pre-contact interaction are located adjacent to the tip.

Conclusions: This is the first report of the application of a fluorescence labeling technique in *S. fimicola*. This application will help researchers gain a better understanding of nuclear distribution and investigate the protein–protein interaction networks during fruiting body formation for advanced molecular genetic studies in *S. fimicola*.

Keywords: *Sordaria fimicola*, Histone H2B protein, mCherry

Background

Sordaria fimicola is a well-known homothallic ascomycete commonly found in fecal matter and has long been used to study genetic recombination, chromosome segregation, and the fungal life cycle as a whole (Ingold and Dring 1957; Olive 1956). *Sordaria fimicola* is easily cultivated in artificial cultures and can undergo sexual reproduction with a short life cycle under laboratory conditions (Olive 1956). The *S. fimicola* life cycle starts with mature sexual spores released from fruiting bodies embedded in animal dung, and spore germination is known to be induced by certain stimulants including sodium acetate (Bretzloff 1951). Once germinated, self-fertilization between hyphae in the same colony preferably initiates the sexual development. During hyphal fusion, pre-contact and contact interactions are observed between compatible hyphae, and the dissolution of cell walls allows nuclei to migrate to partner hyphae. The stages of coli formation followed by ascogenous hyphae differentiation have been reported (Piehl 1929). The multinucleate ascogenous hyphae then undergo septation to form binucleate penultimate cells where nuclear fusion occurs, followed by the subsequent development into asci. The young asci initially show lateral outgrowth and further elongation but maintain the primary nucleus before division occurs (Piehl 1929). Then, meiosis, one round of mitosis, and spore wall formation eventually lead to the production of eight haploid ascospores in a linear arrangement (Olive 1956; Piehl 1929). Coordination of these developmental events with tissue-specific differentiation leads to the final production of fruiting bodies, called perithecia, with mature asci and ascospores enclosed. Although the morphological and cytological features of the sexual structures in *S. fimicola* have been well described, the nuclear organization and dynamics during these processes remain unclear.

*Correspondence: wcshen@ntu.edu.tw
Department of Plant Pathology and Microbiology, National Taiwan University, No. 1 Roosevelt Road, Sec. 4, Taipei 10617, Taiwan

Fluorescently nuclear-tagged strains have been used to investigate the nuclear behavior and dynamics of live cells in many fungal systems such as the nuclear dynamics and their associated microtubules in *Neurospora crassa* (Freitag et al. 2004; Roca et al. 2010) and the nuclear dynamics during spore germination and hyphal fusion in *Fusarium oxysporum* (Ruiz-Roldan et al. 2010). Recently, the polar outgrowth and branching of young hyphae that emerge from the germination vesicles of ascospores were reported in *Sordaria macrospora* (Teichert 2017). Although *S. fimicola* has long been used for various aspects of fungal research (Ingold and Dring 1957; Ingold and Hadland 1959; Kitani and Whitehou 1974; Olive 1956), no strain generated via a genetic transformation method has yet been established for this organism to date. To expand the capabilities of *S. fimicola* in genetic analyses, in this study, we conducted the first successful transformation of *S. fimicola* and generated a histone H2B-mCherry-labeled strain, which was used to examine the nuclear dynamics during ascospore germination, sexual differentiation, and hyphal extension. In addition to providing a useful tool for genetic research, our time-lapse video of the nuclear dynamics should facilitate a better understanding of the detailed nuclear behaviors that occur during the life cycle and reproductive events in *S. fimicola*.

Methods

Strains, media, and culture conditions

Sordaria fimicola (Roberge ex Desm.) Ces. & De Not. (BCRC 33665) was obtained from the Bioresource Collection and Research Center (BCRC), Hsinchu, Taiwan and used as the experimental research strain in this study. Cultures of *S. fimicola* were grown on malt extract agar I [Blakeslee's Formula; 2% malt extract (Himedia, India), 2% glucose, 0.1% peptone and 1.5% (or 2%) agar], and malt extract agar III [2% malt extract and 1.5% (or 2%) agar] was used to induce the formation of sexual reproductive structures. The compositions of malt extract agar I and III media were provided by the BCRC website as previously reported (Lin et al. 2006). *Escherichia coli* DH5α and *Agrobacterium tumefaciens* EHA105 strains were routinely maintained on Luria–Bertani (LB) medium (Sambrook and Russell 2001). *E. coli* and *A. tumefaciens* transformants harboring targeted plasmids were selected on LB containing either 100 μg/ml ampicillin or 100 μg/ml kanamycin.

Nucleic acid manipulation and analysis

Genomic DNA (gDNA) of *S. fimicola* was extracted following slight modifications of a previously described protocol (Pitkin et al. 1996). The strains were grown on cellophane-covered malt extract agar I medium, and mycelia grown on the cellophane were peeled and dipped into liquid nitrogen. The mycelial pad was ground into powder using a mortar and pestle with liquid nitrogen. The fine powder was homogenized in 500 μl DNA extraction buffer [100 mM Tris/HCl, pH 7.5, 2% hexadecyltrimethylammonium bromide (CTAB), 1.4 M NaCl, 20 mM ethylenediaminetetraacetic acid (EDTA), 2% polyvinylpyrrolidone (PVP-40) and 1% mercaptoethanol] and then incubated for 1 h at 65 °C. The cell lysate was extracted with the same volume of phenol:chloroform (1:1) and centrifuged at 13,200 rpm, 4 °C for 15 min, and the supernatants were saved for subsequent chloroform extraction. DNA was precipitated by adding isopropanol and incubated for 15 min at 25 °C. DNA was harvested by centrifuging at 12,000 rpm, 4 °C for 10 min. The pellet was washed with 70% ethanol and collected by centrifugation. The DNA pellet was air-dried and re-suspended in sterile distilled water.

Sordaria fimicola transformation

Sordaria fimicola transformation was carried out using an *Agrobacterium*-mediated protoplast approach with slight modification (Idnurm et al. 2004; Minz and Sharon 2010; Walz and Kuck 1993). In brief, the mycelial plugs were grown in 100 ml malt extract broth I or complete liquid medium containing 1% glucose, 0.2% tryptone, 0.2% yeast extract, 0.15% KH_2PO_4, 0.05% KCl, 0.05% $MgSO_4$, 0.37% NH_4Cl, and 1 ml/l of a solution containing 1% of $ZnSO_4$, Fe(II)Cl_2, and $MnCl_2$ at 28 °C in 500-ml Erlenmeyer flasks and then further incubated at 28 °C for 3 days without shaking (Nowrousian et al. 1999). The mycelia homogenized in a blender were recovered for 1–2 h at 28 °C with shaking at 80–100 rpm. Recovered mycelia were harvested by centrifugation and subsequently washed with protoplast buffer (0.01 M Tris, 0.01 M $MgSO_4{\cdot}7H_2O$, 1 M KCl, pH 7.0). Protoplasts were generated by digesting with lysing buffer composed of 20 mg/ml of VinoTaste® Pro enzyme (Novozymes, Denmark), 2 mg/ml β-D-glucanase (InterSpex Product, USA), 5 mg/ml Driselase (InterSpex Product, USA), and 0.47 μl/ml of β-glucuronidase (Sigma, USA) in protoplast buffer at 28 °C for 12 h. To obtain protoplast cells, the lysate solution was filtered through Miracloth (Calbiochem, Germany), washed with protoplast buffer, collected by low-speed (1000–1200 rpm) centrifugation for 15 min at 4 °C, and adjusted to a final concentration of 10^7–10^8 cells/ml (Walz and Kuck 1993).

The *A. tumefaciens* EHA105 strain was used as the host strain to deliver the target gene to *S. fimicola* according to a slightly modified *Agrobacterium* mediated protoplast transformation protocol (Idnurm et al. 2004; Minz and Sharon 2010). First, *Agrobacterium* was subjected to plasmid electroporation. The *Agrobacterium* strain was pre-cultured in LB liquid medium supplemented with 100 μg/ml of kanamycin at 28 °C for 48 h. Bacterial cells were washed in minimal medium and then re-suspended in induction medium (minimal medium supplemented with 100 μM acetosyringone) until reaching an optical density of 0.15 at 660 nm. The culture was further incubated with shaking for 6 h at 28 °C, and then 200 μl of *A. tumefaciens* cells (in protoplast buffer) was added to 200 μl of the *S. fimicola* protoplasts obtained as described above (10^7–10^8 cells/ml). The mixtures were plated on induction medium agar (containing 5 mM glucose and 200 μM acetosyringone). After incubation at 28 °C for 48 h, the induction medium agar plates were washed with 5 ml of malt extract I or CM liquid medium, and 1 ml was plated onto corn meal agar medium supplemented with 200 μg/ml cefotaxime and 100 μg/ml hygromycin and incubated at 28 °C for 2 days. The colonies of the transformants were transferred to malt extract agar I meal agar medium containing 100 μg/ml hygromycin B.

Generation of fluorescence plasmids

To construct a nuclear fluorescence plasmid, the PR27p-H2B-mCherry plasmid was amplified by fusion polymerase chain reaction (PCR) using KAPA HiFi DNA Polymerase (KAPA Biosystems, USA) according to the manufacturer's protocol. The *PR27* promoter and *TrpC* terminator fragments were amplified with the primer pairs WC1218/WC2902 and WC2905/WC2906, respectively. *S. fimicola* histone H2B cDNA and mCherry fragments were then obtained by amplification with WC2903/WC2904 and WC601/WC602, respectively. The amplified fragments were cleaned with QIAquick Gel Extraction Kit and QIAquick PCR Purification Kit (QIAGEN, Germany). The fusion fragment of PR27p-H2B and mCherry-TrpCt was obtained by amplification with WC1218/WC2904 and WC601/WC2906, respectively. These two fragments were purified, mixed, and amplified with primers WC1218 and WC2906 to generate the PR27p-H2B-mCherry-TrpCt fragment. The overlap-extension PCR product was cloned into the pCR®-BluntII-TOPO® vector (Invitrogen, USA) and subsequently subcloned into pCAMBIA::hyg(S) by cutting with SpeI and SmaI to finally obtain pCAMBIA::hyg::RP27p-H2B-mCherry-TrpCt. The pCAMBIA::hyg(S) plasmid was generated by cutting with XhoI and SacI and ligating with a 1.4-kb *hph* cassette from pCR::hyg derived from the pCB1004 vector. For PCR screening, the reaction was carried out using BioTaq DNA Polymerase (Bioman Scientific, Taiwan). Primer sequences and strains are listed in Tables 1 and 2, respectively.

Table 1 Oligonucleotide primers used in this study

Primer	Sequence (5′–3′)	Purpose
WC33	TCCGTAGGTGAACCTGCGG	ITS1F
WC35	TCCTCCGCTTATTGATATGC	ITS4R
WC601	ATGGTGAGCAAGGGCGAGGAGGATAA	mCherryF
WC602	CTACTTGTACAGCTCGTCCATGCC	mCherryR
WC1218	CAGGTCTACAACCGTGATTA	PR27pF
WC2902	TTTGAAGATTGGGTTCCTAC	PR27pR
WC2903	GTAGGAACCCAATCTTCAAAATGCCTCCCAAGCCCG CCGAC	SfcH2BF
WC2904	TCCTCCTCGCCCTTGCTCACCATTTTCGTGCTGGAG GAGTACTTG	SfcH2BR
WC2905	CATGGACGAGCTGTACAAGTAG CACTTAACGTTA CTGAAATCAT	TrpCF
WC2906	GATCCTCTAGAAAGAAGGAT	TrpCR

Microscopic observations

The ascospores of the *S. fimicola* wild-type and fluorescently labeled strains were inoculated in the center of a malt extract agar III plate (6-cm Petri dish) and incubated at 28 °C under uniform constant white light. The cultures were observed based on the time course of the developmental structure, as shown in Table 3. For fluorescence microscopy, the images and time-lapse movies of live cells were captured with the Deltavision core system (Applied Precision, USA) equipped with an EMCCD (Electron-multiplying CCD) camera (Photometrics Evolve™ 512) on an Olympus IX71 inverted microscope, using UPlanSApo 100×/1.40 oil, UApo, 40×/1.35 oil, and UPlanSApo 20×/0.75 objective lenses. The mCherry fluorescence images were acquired using standard filter sets of 575/25 and 632/60 nm for excitation and emission, respectively. Images were analyzed using SoftWoRx-Acquire version 5.5.1 (Applied Precision, USA).

Results and discussion

Generation of the nuclear-labeled strains

To examine nuclear dynamics in *S. fimicola*, we aimed to generate fluorescent, nuclear-labeled strains to

Table 2 Strains used in this study

Strain	Genotypes	Source
A. tumefaciens EHA105	Wild type	Laboratory
S. fimicola BCRC 33665	Wild type	BCRC
S. fimicola WTH2B/2	WT:: RP27p-H2B-mCherry	This study
S. fimicola WTH2B/3	WT:: RP27p-H2B-mCherry	This study
S. fimicola WTH2B/4[a]	WT:: RP27p-H2B-mCherry	This study
S. fimicola WTH2B/5	WT:: RP27p-H2B-mCherry	This study
S. fimicola WTH2B/6	WT:: RP27p-H2B-mCherry	This study

BCRC Bioresource Collection and Research Center

[a] Used for fluorescent observation

Table 3 Time courses used in this study

Day post inoculation (days)	Hour after inoculation (h)	Developmental stage
	2, 3 and 6	Germination
	15	Hyphal elongation and fusion
1	27	Ascogonial coils formation
2	48	Protoperithecia formation
4	192	Mature perithecia with black ascospores

record live images of the nuclear distribution and behavior during different stages of the life cycle. Since histones are essential structural proteins for chromatin assembly and play functional roles in all living cell types, we used the fluorescent mCherry protein to tag H2B histone to monitor nuclear distribution in the ascospore germlings, vegetative hyphae, and initial fusing and other sexual structures. As shown in Fig. 1, we successfully generated the histone H2B-labeled strains using *Agrobacterium*-mediated protoplast transformation. Observation of the hygromycin B-resistant transformants under the fluorescent microscope revealed that five transformants clearly displayed strong mCherry signals. These transformants were further verified by PCR to detect the presence of the PR27p-H2B-mCherry fragment. Indeed, 974-bp and 1.1-kb fragments were amplified from these transformants with the primer pairs WC1218/WC2904 and WC2903/WC602, respectively, whereas no PCR product was obtained from the wild type (Fig. 1b, left and middle). Amplification of gDNA with the universal ITS1/ITS4 primer pair as a control resulted in a 584-bp product in both the wild-type and transformed strains (Fig. 1b right). Based on these results, the transformant WTH2B/4 was chosen for further investigation of nuclear dynamics.

Nuclear distribution during ascospore germination in *S. fimicola*

To examine the nuclear dynamics in different sexual structures, the ascospores of *S. fimicola* strains were first collected and inoculated onto malt extract agar III. Approximately 1–2 h post-inoculation, the ascospores of the H2B-mCherry-labeled strain formed a germination pore at one end of the spore, followed by the formation of the germination vesicle, which allowed the nuclei to enter the vesicle (Fig. 2). This differs from the germination behavior of *N. crassa*, whose ascospore was reported to germinate from one or both ends through the germ pore (Seale 1973), but is similar to the germination of *N. terricola*, which is mostly restricted to a single germ pore (Cai et al. 2006). The germ pore or emergence of the germination vesicle appeared to be located at the pointed terminus of the ascospore, while the opposite side of the germination pore seemed more rounded, indicating that *S. fimicola* has only one polar germ pore. Moreover, the ascospores in *Sordaria* spp. have been previously found to share common morphological characteristics, including smooth walls with a basal single germ pore and gelatinous sheath (Cai et al. 2006). Physically, the germination

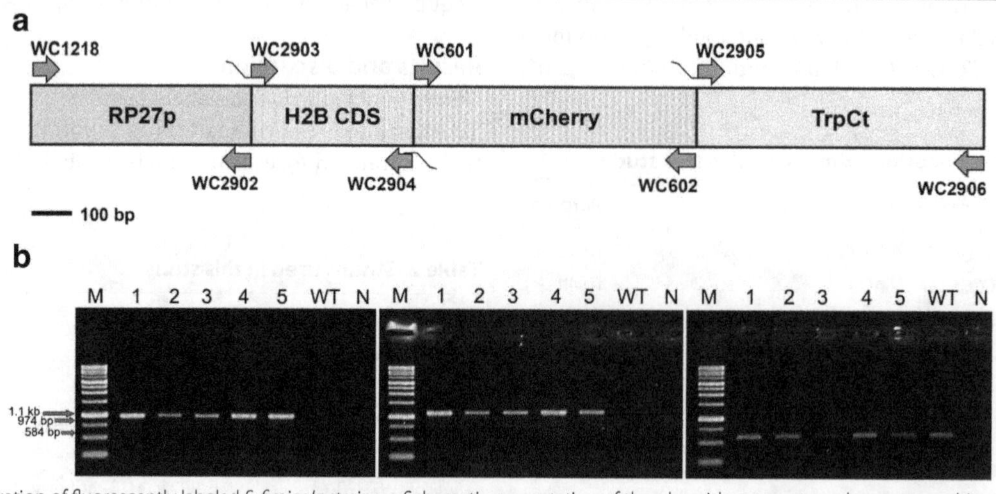

Fig. 1 Generation of fluorescently labeled *S. fimicola* strains. **a** Schematic presentation of the plasmid construct used to generate histone H2B-mCherry-labeled strains. **b** PCR was used to verify the successful generation of the H2B-mCherry-labeled strains. The plasmid construct was ectopically integrated into the *S. fimicola* wild type strain (WT: BCRC33665). The transformants and wild-type strain are shown in Lanes 1–5 and WT, respectively. Amplification with primer pairs WC1218/WC2904 (left), WC2903/WC602 (middle), and ITS1/ITS4 (right) generated fragments of 974 bp, 1.1 kb, and 584 bp, respectively

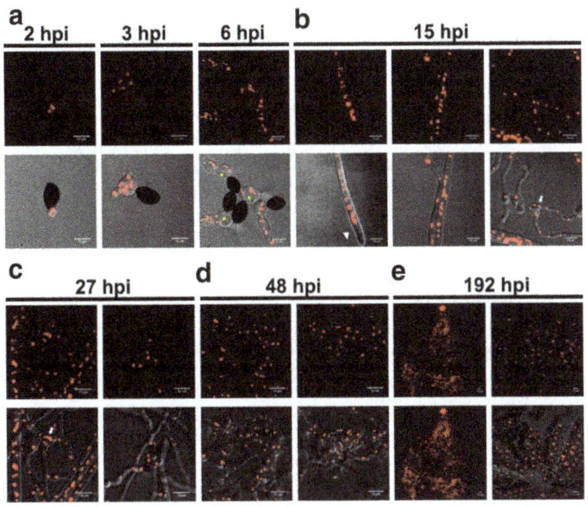

Fig. 2 Live-cell imaging of nuclear distribution during the *S. fimicola* life cycle. The H2B-mCherry-labeled strain was cultured on malt extract agar III under constant white light. The nuclear distribution in a single germination vesicle and germ tube from an ascospore was observed at 2, 3, and 6 h post-inoculation (hpi). At 15 hpi, nuclear fluorescence was distributed in the mature vegetative hyphae and hyphal fusion regions, whereas the nuclei were not detected in the hyphal tip regions of the mature vegetative hyphae. Nuclear distribution in the coil and ascogonia hyphae, including the early protoperithecia, was detected within 27 and 48 hpi, respectively. At approximately 192 hpi, mature perithecia with asci containing black mature and multinucleated ascospores were observed. Live images were captured with the Deltavision core system. Nuclear distribution in each developmental stages of *S. fimicola* including germination vesical formation (**a**), hyphal extension and contact (**b**), ascogonial coils formation (**c**), differentiated ascogenous hyphae development (protoperithecia) (**d**) and perithecia formation (**e**) are showed. White arrows and arrowheads indicate hyphal fusion and the extension tip of a vegetative hypha, respectively. Yellow asterisks indicate vacuoles. Scale bars represent 12 µm

of *S. fimicola* ascospores does not require any activation stimuli, which is similar to the situation for *S. macrospora* ascospores (Teichert 2017). In contrast, *N. crassa* germination requires either heat shock, a chemical agent, or both to activate the dormant ascospores (Emerson 1948, 1954). Moreover, the expansion of germination vacuoles was observed in the cytoplasm of the germination vesicle during the outgrowth of the germ tube, and the spore was eventually found to be empty (Hackett and Chen 1976). At 6 h post-inoculation, various inclusions were visible in this germling, including a single large or multiple medium-sized lipid droplet vacuoles in the center of the germination vesicle. These structures likely reflect vacuole inclusions that might help push nuclei toward the germination vesicle and the peripheral growth zone of a fungal colony and/or serve as nutrient reservoirs during establishment on solid agar (Figs. 2, 3; Additional file 1: Video S1). Interestingly, lipid droplet vacuoles in the germinating ascospores of *S. brevicollis* and *N. tetrasperma* have been similarly reported (Hackett and Chen 1976; Lowry 1968).

Nuclear behavior in the vegetative hyphae

After germination, the germ tube emerging from the vesicle continued elongating and branching to produce young vegetative hyphae. Initially, a uniform nuclear distribution was displayed along the length of the vegetative hyphae (Figs. 2, 3; Additional file 1: Video S1). However, at 15 h post-inoculation, the hyphal tips were devoid of nuclei (Figs. 2, 3; Additional file 2: Video S2). Such a nuclei-devoid region near the hyphal tip has also been described in other fungal systems, including *N. crassa* (Araujo-Palomares et al. 2007) and *S. macrospora* (Teichert 2017). This phenomenon is generally explained by the presence of a mature Spitzenkörper structure formed at the hyphal tip region, indicating the transition from a germling to vegetative hypha (Araujo-Palomares et al. 2007).

The pre-fusion interaction between vegetative hyphae was also observed at 15 h post-inoculation (Figs. 2, 3). The main morphology types of the hyphal fusions were tip-to-side and tip-to-tip. The hyphal tips approached other fusion-competent hyphae by growing toward their partners via a Spitzenkörper mechanism, whereby the approaching hyphal tips also made contact with other partner hyphae, and subsequently apical extension

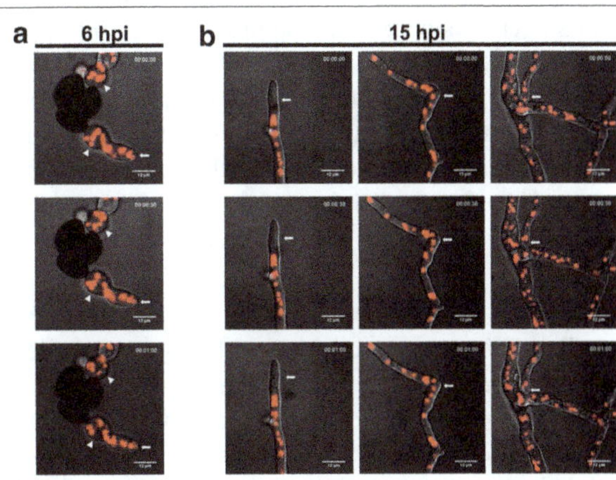

Fig. 3 Nuclear distribution of germ tube and mature vegetative hyphal tip and representative hyphal fusion events. The ascospores of the H2B-mCherry-labeled strain were grown on malt extract agar III under uniform constant white light. **a** Nuclear florescence was evenly distributed throughout the germ tube at 6 h post-inoculation (hpi). **b** At 15 hpi, the nuclei in the elongated vegetative hyphae (left) were absent in the tip regions, whereas nuclei were detected in the apical regions of pre-fusion hyphae [either tip-to-tip (middle) or tip-to-side (right)]. Live images were captured with the Deltavision core system. White arrows indicate differentiated sites of nuclear distribution among these fungal structures. White arrowheads indicate germination vesicles. Scale bars represent 12 µm

ceased to prevent protoplasmic flow through the fusion pore (Hickey et al. 2002). This observation indicated that the hyphal tip extension phase was replaced by hyphal orientation to form the mating structure (Brand and Gow 2009). Some of the approaching hyphae showed slight swelling of the apex but also displayed nuclei at the tip ends (Fig. 3), indicating that the end of Spitzenkörper function in the apical region resulted in a physiological change of the hyphal tip and allowed for the presence of nuclei.

Nuclear distribution during sexual development in *S. fimicola*

As sexual development further proceeded, ascogonial coils, likely formed as side branches of vegetative hyphae ca. 8–10 μm wide, were observed at 27 and 48 h post-inoculation (Fig. 2). At 48 h post-inoculation, in addition to ascogonial coil formation, multinucleate cellular compartments resulting from differentiated ascogenous hyphae were observed in this culture and defined as early protoperithecia (Fig. 2). These structures were ca. 30 μm in width and showed the packing of various hyphae. In fact, the protoperithecia developed from the ascogonium were formed by the aggregation of enveloping hyphae that emerged from either the ascogonial coil or neighboring hyphae, as previously described in detail (Lord and Read 2011). Through the growth of the protoperithecium accompanied by septation and tissue differentiation, the dark-pigmented mature fruiting body (perithecium) with its neck was observed at approximately 192 h post-inoculation, with a size of ca. 250–300 μm wide. Moreover, an ascus rosette containing asci of various stages of maturity derived from the perithecium of the mCherry-labeled strain was observed. The black, mature ascospore contained multiple nuclei derived from repetitive mitotic division, as shown in Fig. 2 (192 h post-inoculation).

Conclusions

Overall, in the present study, we established a gene delivery system for *S. fimicola* that allowed us to examine the nuclear dynamics that occurs during spore germination and the different stages of sexual development. Although much is known about nuclear distribution during hyphal extension and fusion in fungi such as *N. crassa* and *F. oxysporum*, little is known about hyphal fusion in self-fertilized fungi such as *S. fimicola*. Research on nuclear behavior using this tool could help unveil the complex pre- and post-fusion mechanisms, which would in turn provide valuable insights for understanding cell fusion in other eukaryotic species. Moreover, the link between the hyphal fusion event and fruiting body development is one of the most important questions in fungal biology, and the model genetic system of *S. fimicola* can be applied to further understand the fundamental processes in the sexual reproduction of eukaryotic cells.

Abbreviations

BCRC: Bioresource Collection and Research Center; LB: Luria–Bertani; CTAB: hexadecyltrimethylammonium bromide; EDTA: ethylenediaminetetraacetic acid; PVP-40: polyvinylpyrrolidone-40; EMCCD: electron-multiplying CCD; PCR: polymerase chain reaction; CM: complete medium; ITS1/ITS4: internal transcribed spacer 1/4 primers of nuclear ribosomal DNA.

Authors' contributions

KK and W-CS conceived the ideas and designed the experiments. KK performed the experiments and analyzed data. KK and W-CS contributed critically to the drafts and gave final approval for publication. Both authors read and approved the final manuscript.

Acknowledgements

We thank Ms. Sue-Ping Lee and the staffs of the Imaging Core Facility, Institute of Molecular Biology at Academia Sinica for the help of cell imaging manipulation.

Competing interests

The authors declare that they have no competing interests.

Funding

This research received no specific grant from any funding agency in the public, commercial, or not-for-profit sectors.

References

Araujo-Palomares CL, Castro-Longoria E, Riquelme M (2007) Ontogeny of the Spitzenkörper in germlings of *Neurospora crassa*. Fungal Genet Biol 44:492–503. https://doi.org/10.1016/j.fgb.2006.10.004

Brand A, Gow NA (2009) Mechanisms of hypha orientation of fungi. Curr Opin Microbiol 12:350–357. https://doi.org/10.1016/j.mib.2009.05.007

Bretzloff CW (1951) Fungus fruiting in submerged culture. Science 114:418–419

Cai L, Jeewon R, Hyde KD (2006) Phylogenetic investigations of Sordariaceae based on multiple gene sequences and morphology. Mycol Res 110:137–150. https://doi.org/10.1016/j.mycres.2005.09.014

Emerson MR (1948) Chemical activation of ascospore germination in *Neurospora crassa*. J Bacteriol 55:327–330

Emerson MR (1954) Some physiological characteristics of ascospore activation in *Neurospora crassa*. Plant Physiol 29:418–428

Freitag M, Hickey PC, Raju NB, Selker EU, Read ND (2004) GFP as a tool to analyze the organization, dynamics and function of nuclei and microtubules in *Neurospora crassa*. Fungal Genet Biol 41:897–910. https://doi.org/10.1016/j.fgb.2004.06.008

Hackett CJ, Chen KC (1976) Ultrastructure of developing ascospores in *Sordaria brevicollis*. J Bacteriol 126:883–894

Hickey PC, Jacobson D, Read ND, Glass NL (2002) Live-cell imaging of vegetative hyphal fusion in *Neurospora crassa*. Fungal Genet Biol 37:109–119

Idnurm A, Reedy JL, Nussbaum JC, Heitman J (2004) *Cryptococcus neoformans* virulence gene discovery through insertional mutagenesis. Eukaryot Cell 3:420–429. https://doi.org/10.1128/EC.3.2.420-429.2004

Ingold CT, Dring VJ (1957) An analysis of spore discharge in *Sordaria*. Ann Bot 21:465–477

Ingold CT, Hadland SA (1959) Phototropism and pigment production in *Sordaria* in relation to quality of light. Ann Bot 23:425–429

Kitani Y, Whitehou HLK (1974) Effect of proportion of parental nuclei in a heterokaryon on pattern of gene conversion in *Sordaria fimicola*. Mol Gen Genet 131:47–56

Lin JY, Wu TZ, Chou JC (2006) In vitro induction of fruiting body in *Antrodia cinnamomea*—a medicinally important fungus. Bot Stud 47:267–272

Lord KM, Read ND (2011) Perithecium morphogenesis in *Sordaria macrospora*. Fungal Genet Biol 48:388–399. https://doi.org/10.1016/j.fgb.2010.11.009

Lowry RJ (1968) Ultrastructural changes during germination of ascospores of

Neurospora tetrasperma. J Gen Microbiol 51:403–409

Minz A, Sharon A (2010) Electroporation and *Agrobacterium*-mediated spore transformation. Methods Mol Biol 638:21–32. https://doi.org/10.1007/978-1-60761-611-5_2

Nowrousian M, Masloff S, Poggeler S, Kuck U (1999) Cell differentiation during sexual development of the fungus *Sordaria macrospora* requires ATP citrate lyase activity. Mol Cell Biol 19:450–460

Olive LS (1956) Genetics of *Sordaria fimicola*. I. ascospore color mutants. Am J Bot 43:97–107. https://doi.org/10.2307/2438817

Piehl AE (1929) The cytology and morphology of *Sordaria fimicola* ces. and de not. Transact Wisconsin Acad Sci Arts Lett 24:323–341

Pitkin JW, Panaccione DG, Walton JD (1996) A putative cyclic peptide efflux pump encoded by the TOXA gene of the plant-pathogenic fungus *Cochliobolus carbonum*. Microbiolgy 142:1557–1565. https://doi.org/10.1099/13500872-142-6-1557

Roca MG, Kuo HC, Lichius A, Freitag M, Read ND (2010) Nuclear dynamics, mitosis, and the cytoskeleton during the early stages of colony initiation in *Neurospora crassa*. Eukaryot Cell 9:1171–1183. https://doi.org/10.1128/EC.00329-09

Ruiz-Roldan MC, Kohli M, Roncero MI, Philippsen P, Di Pietro A, Espeso EA (2010) Nuclear dynamics during germination, conidiation, and hyphal fusion of *Fusarium oxysporum*. Eukaryot Cell 9:1216–1224. https://doi.org/10.1128/ec.00040-10

Sambrook J, Russell DW (2001) Molecular cloning: a laboratory manual, 3rd edn. Cold Spring Harbor Laboratory Press, Cold Spring Harbor

Seale T (1973) Life cycle of *Neurospora crassa* viewed by scanning electron microscopy. J Bacteriol 113:1015–1025

Teichert I (2017) Nuclear dynamics during ascospore germination in *Sordaria macrospora*. Fungal Genet Biol 98:20–22. https://doi.org/10.1016/j.fgb.2016.11.004

Walz M, Kuck U (1993) Targeted integration into the *Acremonium chrysogenum* genome: disruption of the *pcbC* gene. Curr Genet 24:421–427

10

Diurnal regulation of the floral scent emission by light and circadian rhythm in the *Phalaenopsis* orchids

Yu-Chen Chuang[1†], Ming-Chi Lee[1†], Ya-Lan Chang[1], Wen-Huei Chen[2] and Hong-Hwa Chen[1,2*]

Abstract

Background: *Phalaenopsis bellina* and its closely related species, *P. violacea*, emit linalool, geraniol and their derivatives as the predominant monoterpenes at the full-bloom stages. Geranyl diphosphate synthase (PbGDPS) is the key enzyme that converts precursors for monoterpene biosynthesis. Besides the monoterpenes being synthesized in concert with floral development stages, we noticed that the scent emission of *P. bellina* and *P. violacea* was detected mainly in the daytime.

Results: The monoterpenes of *P. violacea* flowers displayed a diurnal emission pattern, which was regulated by an internal oscillator in the treatment of constant light. In contrast, constant dark diminished the scent emission levels, indicating that light also affects monoterpene emission in *P. violacea*. Further treating *P. violacea* with various light wavelengths showed that the monoterpene emission was greatest in white light condition. Other *Phalaenopsis* hybrids, including *P.* I-Hsin Venus 'KHM2212' and *P.* Meidarland Bellina Age 'LM128', responded differently to various light wavelengths but most of them still showed the highest scent emission under the whole spectra of light. A great number of light-responsive, HY5-interacting, and circadian-responsive elements was enriched on the promoters of both structural genes and transcription factors for monoterpene biosynthesis. Furthermore, several putative genes encoding components involved in light and circadian signaling pathways were also identified in the transcriptome of *P. bellina* flowers at consecutive stages (from the anthesis day to day 7 post anthesis).

Conclusions: Taken together, both circadian clock and light factors had positive effects on orchid floral scent emission, and the regulation resided on the control of both structural genes and transcription factors for monoterpene biosynthesis.

Keywords: Circadian clock, Floral scent, Light-regulated, Monoterpene, *Phalaenopsis*

Background

Floral scent is an important factor to attract pollinators for reproduction. To achieve this, the floral scent is often developmentally and rhythmically regulated to be associated with the activity of corresponding pollinators (Muhlemann et al. 2014). As insect pollinators exhibit rhythmic activities (Saunders 1997), the flowers emit scent on specific timing in a day (Hoballah et al. 2005; Fenske and Imaizumi 2016). This diurnal/nocturnal emission is proposed to be regulated by a circadian clock.

The circadian clock implicates an endogenous oscillator so as to control its downstream physiological process rhythmically even under constant condition. The oscillator is established by synchronizing to the environment stimuli including light and temperature; the two factors change extremely during day/night cycle (Devlin and Kay 2000). The light receptors are responsible for inputting the light signal into the circadian clock (McClung 2006).

Light directly affects floral scent emission as well. Plants treated with various light intensities and light wavelengths show fluctuation changes of their volatiles.

*Correspondence: hhchen@mail.ncku.edu.tw
†Yu-Chen Chuang and Ming-Chi Lee contributed equally to this work
[1] Department of Life Sciences, National Cheng Kung University, Tainan 701, Taiwan
Full list of author information is available at the end of the article

2-Phenylethanol from petunia flowers is increased under red and far-red light treatments (Colquhoun et al. 2013; Abe et al. 2003). Chiloglottone production in *Chiloglottis* orchids require UV-B light (Falara et al. 2012). The volatiles emitted from tea leaves are raised by blue light and red light (Fu et al. 2015).

Phalaenopsis bellina and its genetically as well as morphologically related species, *P. violacea* (Tsai 2003), are usually used in the breeding of scented cultivars. There are approximately 730 *Phalaenopsis* hybrids with *P. bellina* or *P. violacea* as parents registered in The Royal Horticultural Society. The floral scent profiles of *P. bellina* are dictated by monoterpenes, including linalool and geraniol and their derivatives (Hsiao et al. 2006). Geranyl diphosphate synthase of *P. bellina* (PbGDPS) is the key enzyme to supply geranyl diphosphate (GDP) for monoterpene biosynthesis. Its maximal expression is concomitant with the peak emission of monoterpenes at the full-bloom stage (day 5 post anthesis, D + 5) (Hsiao et al. 2008). Expression of *PbGDPS* and its downstream putative monoterpene synthases, TERPENE SYNTHASE 5 (*PbTPS5*) and *PbTPS10*, are regulated during flower development by several transcription factors (TFs), including PbbHLH4, PbbHLH6, PbbZIP4, PbERF1, PbERF9, and PbNAC1 (Chuang et al. unpublished).

We noticed that floral scent of *P. bellina* and *P. violacea* showed a pattern of diurnal emission in summer. The emission levels were reduced under cloudy and rainy days, even at full blooming stages, prompting us to investigate whether light or circadian rhythm plays an important role in the production of floral scent in orchids. In addition, the proposed components involved in the signal pathway were discussed as well.

Methods

Plant materials and growth condition

Two native species of *Phalaenopsis* orchids, *P. bellina* and *P. violacea* were obtained from Han-Lin Orchids (Tainan, Taiwan). Two commercial cultivars were also included in the analysis. *P.* I-Hsin Venus 'KHM221' (abbreviated as 'KHM221' in the following) is derived from a cross between '*P.* I-Hsin Viola Tris' and '*P.* Dragon's Gold' from I-Hsin Biotechnology Co. (Chiayi, Taiwan). The other cultivar, *P.* Meidarland Bellina Age 'LM128' (abbreviated as 'LM128' in the following) is derived from a cross between '*P.* K. S. Happy Eagle' and *P. bellina* from Meidarland Orchids (Tainan, Taiwan). The genetic constitution of both cultivars was downloaded from ORCHIDEYA.CA (http://www.orchideya.ca/) as follows: 'KHM221'—*P. amboinensis* (25%), *P. equestris* (25%), *P. venosa* (18.75%), *P. violacea* (15.63%), *P. rimestadiana* (5.71%), *P. amabilis* (4.69%), *P. lueddemanniana* (3.13%), and *P. aphrodite* (2.1%). For 'LM128', the parenthood is the following—*P. bellina* (50%), *P. violacea* (24.22%), *P. venosa* (15.63%), *P. amboinensis* (3.91%), *P. lueddemanniana* (3.13%), and *P. micholitzii* (3.13%). Among these native *Phalaenopsis* orchids, several are scented *Phalaenopsis*, including *P. amboinensis*, *P. bellina*, *P. lueddemanniana*, *P. venosa*, and *P. violacea*. All *Phalaenopsis* plants were grown in the greenhouse at National Cheng Kung University (NCKU) under natural light around 100 μmol m^{-2} s^{-1} (LI-COR photometer, United States) and controlled day/night temperature 30 °C/28 °C with 80% humidity in summer.

Light treatments

Plants were transferred to the growth chamber (F180 LED, Hi Point Corporation, Taiwan) on the day of anthesis (Dd). *P. violacea* was first analyzed by long day (16 h/8 h light/dark (L/D) or short day condition (8 h/16 h L/D). For analyzing whether floral scent emission is regulated by circadian rhythm or by light, six plants were treated with white light (16 h/8 h L/D) for 4 days (from Dd to D + 4), and then divided into two groups under the treatment of either constant light or constant dark for the following 4 days (from D + 5 to D + 8). Floral scent was collected from D + 4 to D + 8 during four periods of time (from 4 a.m. to 9 a.m., 10 a.m. to 4 p.m., 4 p.m. to 9 p.m., and 10 p.m. to 4 a.m.) each day.

To study the responses of floral scent emission to various light qualities, orchid plants were treated with various light sources for 16 h/8 h L/D at 27 °C with 60% humidity. Plants were domesticated for 3–5 days before harvest. LED of three-wavelengths were used, including blue light (central wavelength 450 nm) at 270 μmol m^{-2} s^{-1}, red light (central wavelength 650 nm) at 30 μmol m^{-2} s^{-1} and far-red light at 5 μmol m^{-2} s^{-1}. The scent release of *P. violacea* flowers in response to different light conditions was analyzed with triplicate biological replicates. Four individual seedling plants of 'LM128' were used and the replicate number for each condition was as the following: (a) No. 1 plant ($n = 3$ for white light (WL), $n = 4$ for blue light (BL), (b) No. 2 plant ($n = 1$ for WL, $n = 4$ for BL), (c) No. 3 plant ($n = 3$ for both WL and BL), and (d) No. 4 plant ($n = 3$ for WL, $n = 1$ for BL) (Fig. 3). For statistical analysis between two groups, assuming equal variance, one-tailed Student's t test was performed ($p < 0.1$). For 'KHM221' plants, 50 μmol m^{-2} s^{-1} white, blue, and red light conditions, and darkness were applied with duplicates.

Solid-phase extraction and GC–MS analysis

Floral scent was collected from plants by using solid-phase extraction system. The flower on the plants was placed in a plastic bag (8 × 8 × 5 cm^3) inserted with a silicone tube connecting the solid phase extraction columns (DSC-Si and DCS-18, Supelco, United States) using air pumps at 350 mmHg (Rocker, Taiwan). The volatiles were then eluted with the addition of hexane and identified by using (GC–MS; QP2010, SHIMADXU, Shimadzu Co, Tokyo, Japan) at the NCKU Instrument Center as previously described (Hsiao et al. 2006).

Isolation of upstream regulatory fragments and *cis*-element prediction

Upstream regulatory fragments of three monoterpene biosynthesis genes, including *PbGDPS*, *PbTPS5*, and *PbTPS10*, were isolated from *P. bellina* genomic DNA by use of the Universal GenomeWalker Kit (Clontech, USA) as previously described (Hsu et al. 2014). In addition, we also isolated 1.5-kb upstream regulatory fragments of the identified six TFs involved in the regulation of floral monoterpene biosynthesis in *P. bellina* (Chuang et al. unpublished) from *P. equestris* genomic sequences (Cai et al. 2015). The *cis*-elements related to light signaling and circadian clock on the upstream regulatory fragments of both monoterpene biosynthesis genes and TFs were predicted by using PlantPAN (Chow et al. 2015). The predicted results with 100% similar score were accepted.

Identification of genes involved in light and circadian clock signaling pathways

The *P. bellina* transcriptomic data of four floral development stages (Dd, D + 3, D + 5 and D + 7) were constructed, analyzed and annotated as described (Chuang et al. unpublished). The homologous genes of the *Arabidopsis* factors involved in light and circadian clock signaling pathways were identified in the *P. bellina* transcriptome according to Nr annotation. These included PHYTOCHROME A (PHYA), PHYTOCHROME B (PHYB), CRYPTOCHROME 1 (CRY1), CRYPTOCHROME 2 (CRY2), PHOTOTROPIN 1 (PHOT1), PHOT2, CONSTITUTIVE PHOTOMORPHOGENIC 1 (COP1), SUPPRESSOR OF PHYTOCHROME A1 (SPA1), PHYTOCHROME AND FLOWERING TIME 1 (PFT1), PHYTOCHROME-INTERACTING FACTOR 3 (PIF3), PIF7, HY5 (LONG HYPOCOTYL5), HY5-HOMOLOG (HYH), LATE ELONGATED HYPOCOTYL (LHY), CIRCADIAN CLOCK-ASSOCIATED 1 (CCA1), TIMING OF CAB EXPRESSION 1 (TOC1), and GIGANTEA (GI). The expression level of each individual gene in the transcriptomic data was determined by the log2 of fragments per kilobyte per million reads (FPKM) values. PHOT2, PFT1, PIF7, and TOC1 were excluded for final analysis since little or no expression for the four genes was detected.

Results

Effects of light on *P. violacea* scent emission

To determine how orchid floral scent emission varied throughout the day, *P. violacea* plants were first transferred to a 16 h/8 h L/D condition on Dd and grown under this photoperiod for 5 days. The floral scent of *P. violacea* was then collected at a 5–6-h interval on D + 4. Monoterpene emission increased at 4:00 a.m. to 10:00 a.m. internal, reached a peak at 10:00 a.m. to 16:00 p.m. interval, decreased thereafter, and was undetectable at midnight (Fig. 1a, b). These results indicate a diurnal monoterpene emission of *P. violacea* flowers.

To investigate whether the floral monoterpene emission is regulated by an internal oscillator, plants were further treated with constant light or constant dark conditions over 4 days (D + 5 to D + 8). Previously, monoterpene biosynthesis is found to be developmentally regulated (Hsiao et al. 2008). Indeed, the monoterpene emission levels increased and peaked on D + 5 during daytime (Fig. 1a). Under constant light, although the periodical peaks were gradually decreased from D + 6 to D + 8, they did exhibit a regular and similar pattern during the rest of the photoperiod, indicating the monoterpene emission is under rhythmic regulation (Fig. 1a). In contrast, monoterpene emission decreased significantly as the plants were placed in the constant dark (Fig. 1b). A minor peak of monoterpene emission (approximately one tenth of the level in the constant light condition) was still observed at the 10:00 a.m. to 16:00 p.m. internal on D + 5 (Fig. 1b), and the emission shifted to the 4:00 a.m. to 10:00 a.m. interval on D + 6, and even shifted to the midnight interval on D + 7 and D + 8 (Fig. 1b). The above results indicate a circadian regulation of floral monoterpene emission, and the effect of light as well.

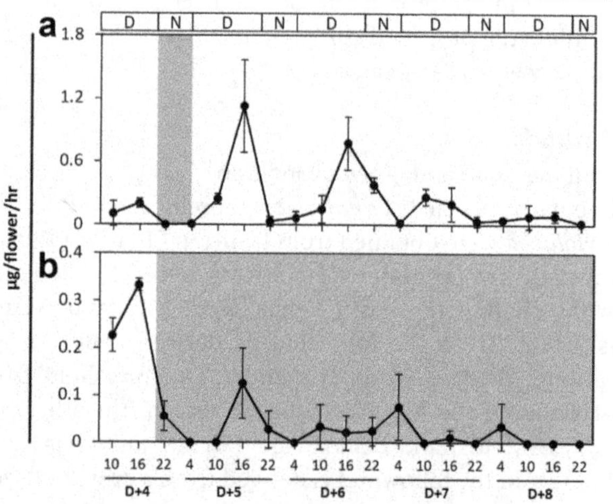

Fig. 1 Emission of monoterpenes from *P. violacea* flowers exposed to different light/dark conditions. **a** Flowers were exposed in 16 h/8 h L/D condition for 5 days and followed by constant light (~ 100 μmol m^{-2} s^{-1}) for 4 days. **b** Flowers were exposed in 16 h/8 h L/D condition for 5 days and followed by constant dark for 4 days. Floral scent was collected by applying solid-phase extraction at four time point in the interval of 5–6 h during 24-h. Shaded area was correspond to the dark condition. Experiments were based on triplicate biological replicates

Effects of various light qualities on monoterpene emission

To determine how the light quality could affect the orchid floral scent emission, we treated plants with various light wavelengths. Naturally, *P. violacea* is a native species in Malaysia and blooms in summer, and is considered as a long-day plant. Actually, the short-day condition (8 h/16 h L/D) reduced the floral monoterpene emission (Fig. 2). Therefore, we treated the orchids in the long-day condition (16 h/8 h L/D) with specified wavelengths, including the white light, blue light (270 μmol m^{-2} s^{-1}), red light (30 μmol m^{-2} s^{-1}), far-red light (5 μmol m^{-2} s^{-1}) and darkness. The emission of linalool and geraniol was higher under white light than those under the other light spectra for *P. violacea* (Fig. 2). Among them, blue light, far-red light and dark had more effects than red light on the monoterpene emission. In contrast to linalool, the emission of geraniol nearly diminished under far-red-light and dark treatment in *P. violacea*.

To see whether the offspring of *P. violacea* exhibits the similar response to different light treatment as their ancestor, we then analyzed the responses of two commercial *Phalaenopsis* cultivars containing *P. violacea* background to the light spectra. One is 'KHM2212' whose ancestry contains 15.63% of *P. violacea*. 'KHM2212' plants were basically with identical genetic background since they are micropropagated in tissue culture and their responses to different light were consistent (Fig. 3). 'KHM2212' emitted linalool and eucalyptol, and the emission of linalool was higher under white-light treatment than all the other light spectra (Fig. 3), similar to the observation in *P. violacea* (Fig. 2). The difference was that the red-light treatment reduced the emission of linalool more significantly than blue light and darkness in 'KHM2212' (Fig. 3). In addition, those various light spectra had little effect on the emissions of eucalyptol (Fig. 3),

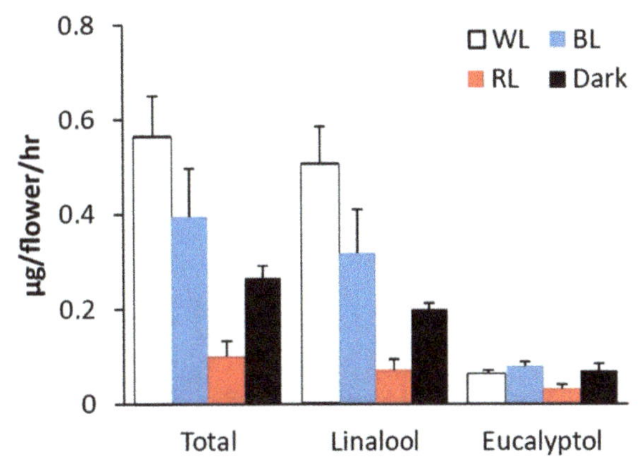

Fig. 3 Emission of monoterpenes from *P.* I-Hsin Venus 'KHM2212' in different light conditions. The plants were treated with 50 μmol m^{-2} s^{-1} white (WL), red (RL), and blue (BL) light conditions, and darkness (dark). Scent was collected from six flowers on D + 5. Experiments were based on duplicate biological replicates

a compound not detected in the volatile profile of *P. violacea*. We deduced that the differences might be resulted from the genetic constitution of 'KHM2212' other than *P. violacea*.

Another *Phalaenopsis* hybrid examined was 'LM128' containing the ancestry of *P. bellina* (50%) and *P. violacea* (24.22%). 'LM128' is an orchid hybrid generated from at least seven generations of breeding and six native species were involved, leading to its complicated genetic background. We selected four 'LM128' seedlings plants derived from independent lines and found that the floral scent emission for each individual plant responded differentially to white-light and blue-light conditions (Fig. 4). Among them, two plants showed similar responses to both light spectra as *P. violacea*, in which blue-light treatment reduced the scent emission (Fig. 4a, b). The other two responded differently, one seedling emitted higher monoterpene under blue light than white light (Fig. 4c) and the other showed the same level of emission (Fig. 4d). Although the responses were different among the four plants, the reactions of linalool and geraniol emission to both light spectra were consistent in each 'LM128' individual (Fig. 4).

Together, these results imply that whole spectra of light (i.e. white light) is conferred the optimal emission of scent compounds in most *Phalaenopsis* orchids with *P. violacea* background.

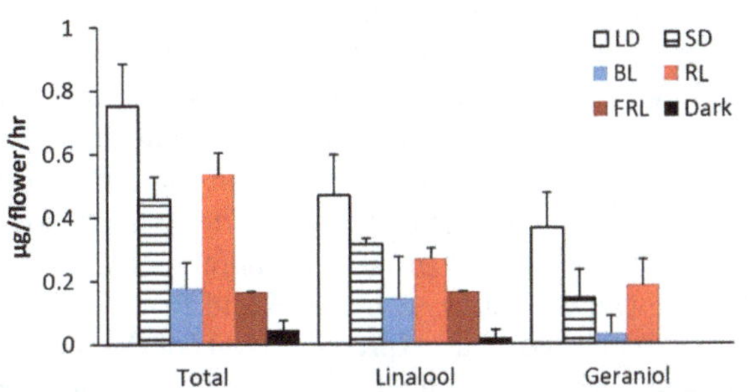

Fig. 2 Emission of monoterpenes from *P. violacea* flowers in different light condition. The plants were treated with 100 μmol m^{-2} s^{-1} white light in long day (18 h/6 h L/D, LD) and short day (6 h/18 h L/D, SD) conditions, 270 μmol m^{-2} s^{-1} blue light (BL), 30 μmol m^{-2} s^{-1} red light (RL), 5 μmol m^{-2} s^{-1} far-red light (FRL), and darkness (dark). Scent was collected on D + 5 of an individual *P. violacea* flower. Experiments were based on triplicate biological replicates

Analysis of the upstream regulatory fragments of structural genes and TFs for monoterpene biosynthesis

In the light signaling pathway, HY5 binds to the promoters of early light-responsive genes and TF genes to regulate plant development and photomorphogenesis (Lee

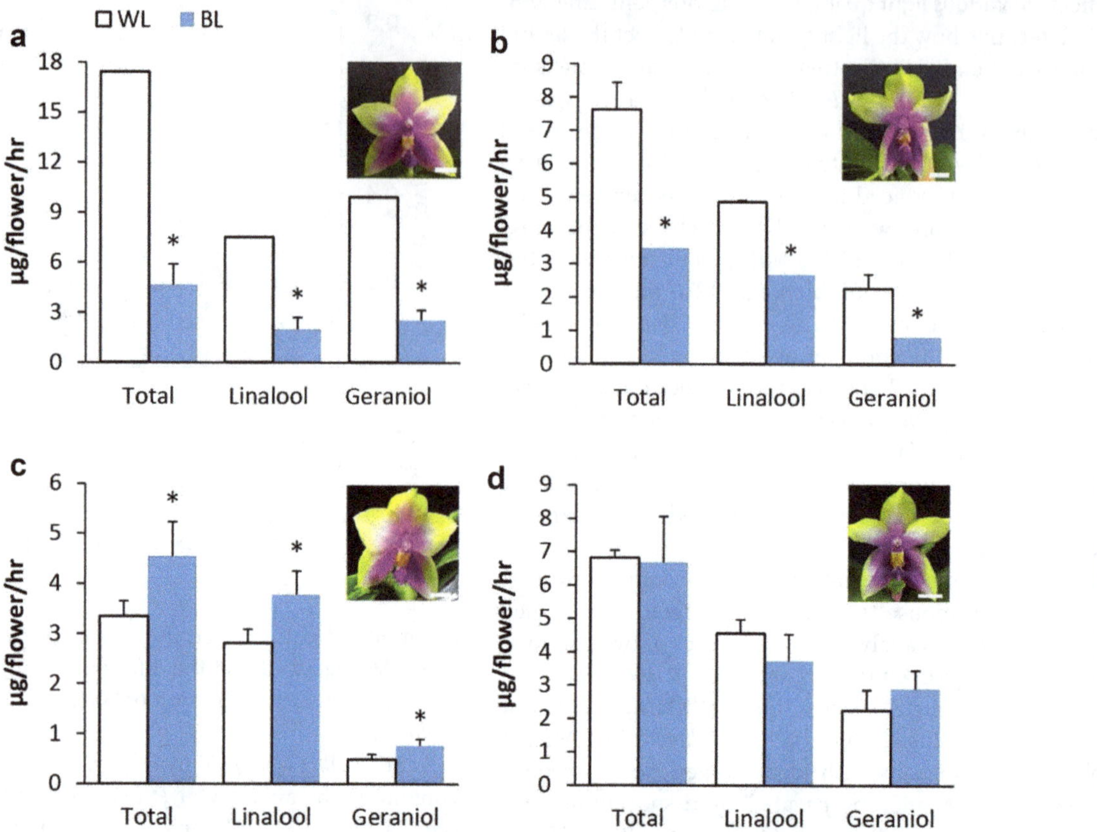

Fig. 4 Emission of monoterpenes of four individual plants of *P.* Meidarland Bellina Age 'LM128'. The plants were treated with 100 µmol m^{-2} s^{-1} white light (WL), and 270 µmol m^{-2} s^{-1} blue light (BL) conditions. The replicate number for each plants was listed as following: **a** No. 1 plant: $n = 3$ for white light (WL), and $n = 4$ for blue light (BL). **b** No. 2 plant: $n = 1$ for WL, and $n = 4$ for BL. **c** No. 3 plant: $n = 3$ for both WL and BL. **d** No. 4 plant: $n = 3$ for WL, and n = 1 for BL. Different letters on the column indicate significant difference by Student's *t* test (*p* value < 0.1). Scale bar = 1 cm

et al. 2007). We examined whether the upstream promoters of the structural genes and TFs for monoterpene biosynthesis in *Phalaenopsis* contain any HY5-interacting ACGT-containing elements (ACEs), such as C-box (GACGTC), Z-box (ATACGGT), CA-hybrid (GACGTA), CG-hybrid (GACGTG), as well as G-box (CACGTG) for PIF recognition (Lee et al. 2007; Zhang et al. 2011). Among the genes analyzed, *PbNAC1* contained a large number of ACE motifs ($n = 7$), implying that it might be a target of HY5 protein (Table 1, Fig. 5). In addition, several other *cis*-acting elements related to light and circadian clock signaling were also identified in the 1.5-kb promoter of *PbNAC1* by using PlantPAN (Chow et al. 2015). These included GATA-box (Luo et al. 2010), GT1-motif (Zhou 1999), and evening elements (EE) for CCA1 and LHY interaction (Alabadí et al. 2001) (Table 1, Fig. 5). These *cis*-elements were identified in the promoter of structural genes and TFs for monoterpene biosynthesis as well, although in a smaller number (Table 1, Fig. 5). *PbGDPS*, a key enzyme in the monoterpene biosynthesis (Hsiao et al. 2008), was the only one structural gene containing EEs, suggesting that it might be a downstream gene of circadian clock pathway.

Identification of factors involved in light and circadian clock signaling pathways

Both light and circadian signaling pathways were well studied in *Arabidopsis* (Dodd et al. 2015; Lau and Deng 2012). We identified their homologous genes in *P. bellina* floral transcriptome to explore the candidate factors responsible for the circadian- and light-regulating scent emission in *Phalaenopsis* flowers. The transcriptomic data were constructed from four floral developmental stages, including Dd, D + 3, D + 5 and D + 7 (Chuang et al. unpublished). The expression levels of the homologous genes were determined by log2 of FPKM values, and compared to that of *PbGDPS* (Fig. 6). The putative genes encoding the factors involved in light and circadian signaling pathways were successfully isolated in *P. bellina* transcriptome, and several ones showed higher expression, including putative genes encoding PHYTOCHROME A, PHYTOCHROME B, CRYPTOCHROME 2, COP1, LHY,

Table 1 *cis*-elements on the promoter fragments of scent-related structural genes and TFs

Gene name	Promoter length	HY5 interaction					Light-responsive		Circadian
		C-box[a]	Z-box	G-box[a]	CA Hybrid	CG Hybrid	GATA	GT1-motif	EE
PbGDPSp	1077	–[b]	–	–	–	–	3	1	2
PbTPS5p	1161	–	–	–	–	–	–	–	–
PbTPS10p	1506	1	–	–	–	–	3	–	–
PebHLH4p	1538	–	–	–	–	1	2	1	–
PebHLH6p	1673	–	–	1	–	–	2	1	–
PebZIP4p	1537	–	–	–	1	1	–	–	–
PeERF1p	1500	–	–	–	–	–	1	–	–
PeERF9p	1565	–	–	–	1	1	–	–	2
PeNAC1p	1501	1	2	1	–	3	3	2	1

[a] The number of palindromic sequences was counted by the sense strand

[b] The dash indicates that the *cis*-element is not detected

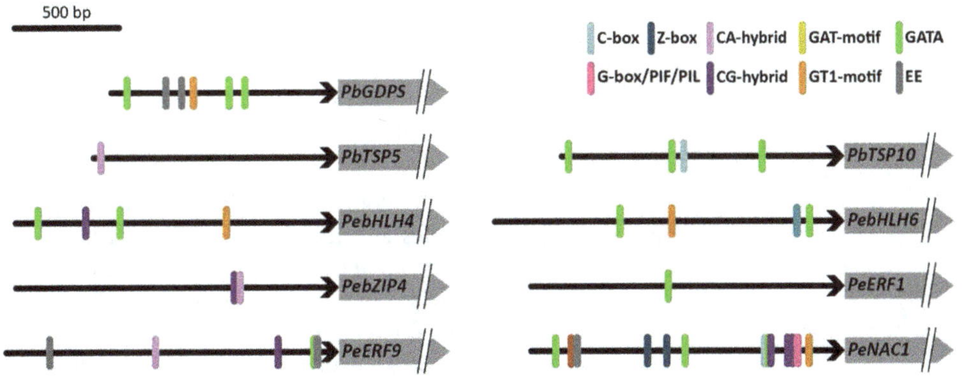

Fig. 5 Light- and circadian-responsive elements on the upstream regulatory fragments of three structural genes and six transcription factors for floral monoterpene biosynthesis in *P. bellina*. The distinct *cis*-elements on each fragments were represented by their location predicted by PlantPAN (Chow et al. 2015). The gene length was not shown in proportion to actual length. Scale bar of upstream regulatory fragments = 500 bp

CCA1, HY5 and GI (Fig. 6). As we identified the *cis*-elements related to the light and circadian responses in the structural genes and TFs for monoterpene biosynthesis, it is worth to study the function of these putative genes in *Phalaenopsis* orchids in the future.

Discussion

Light and/or circadian rhythm regulation on the emission of monoterpenes

Phalaenopsis bellina, or *P. violacea*, emits monoterpenes during daytime and is considered as day-pollinated flowers. Here, we present that this diurnal emission is affected mainly by circadian clock and to a lesser degree by the light factor. This is different from other plant species, whose diurnal rhythm is usually regulated by a single factor. *Rosa hybrida* L. cv. Honesty, *Antirrhinum majus*, and *Rosa damascena semperflorens* emit monoterpenes regulated by a circadian clock regardless of the light stimulus (Helsper et al. 1998; Dudareva et al. 2003; Picone et al. 2004). In contrast, in *Mahonia japonica* and *Rosa hybrida* cv. Fragrant Cloud, the diurnal emission of monoterpenes is determined primarily by the photoperiod (Picone et al. 2002; Hendel-Rahmanim et al. 2007).

The monoterpene emission patterns are likely to be parallel the expression of structural genes. In *A. majus* flowers, the terpene backbone biosynthesis pathway, *GDPS* and monoterpene synthase are all under circadian regulation (Dudareva et al. 2003, 2005; Tholl et al. 2004). The emission of monoterpenes in *P. bellina* matches the expression patterns of *PbGDPS* during flower development (Hsiao et al. 2008). Further study of whether these structural genes contribute to the diurnal emission will

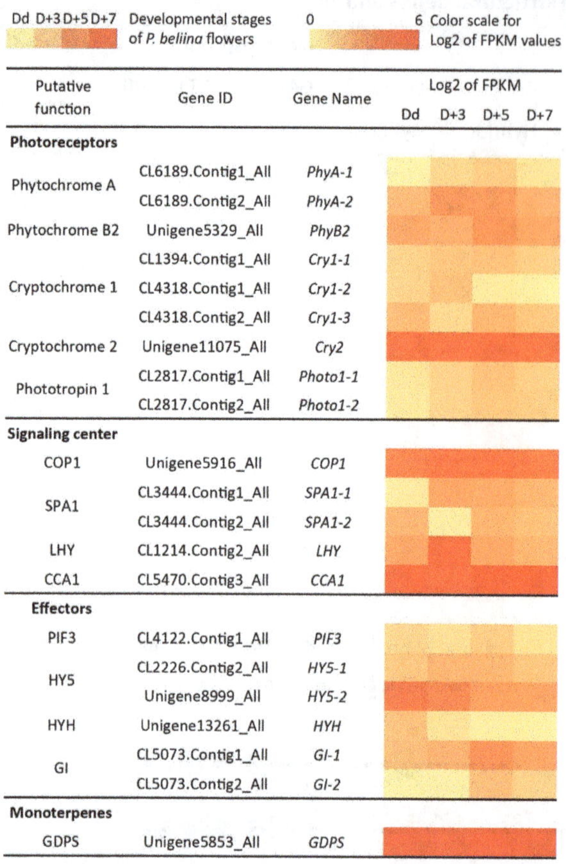

Fig. 6 Expression profiles of putative genes encoding components involved in light and circadian clock signaling pathway in *P. bellina* floral transcriptomes. The gene expression levels are represented by a color gradient from red to light-red

help to clarify how monoterpene biosynthesis is controlled in a day.

HY5 integrates the light and circadian signaling pathway for terpene biosynthesis

HY5, a specialized basic leucine zipper (bZIP) TF, plays an important role in mastering both light and circadian signaling pathway (Gangappa and Botto 2016). The role of HY5 on terpene biosynthesis is recently unraveled. *Arabidopsis* responds to illumination by integrating HY5 to repress the mevalonate (MVA) pathway and activate methylerythritol phosphate (MEP) pathway to increase plastidial terpenes for photosynthesis (Rodríguez-Concepción et al. 2004). As grape berries are shaded from light (sunlight exclusion), the down-regulation of the terpene biosynthesis leads to the reduction of terpene content and altered aroma (Zhang et al. 2014, 2017). HY5 might be involved in the light-induced terpene production in grapes (Costantini et al. 2017). Moreover, HY5-binding sites are identified in the promoter regions of terpene biosynthesis genes (Lee et al. 2007). In *Artemisia annua, a* circadian-regulated β-pinene synthase gene, *QH6*, contains G-box motif to interact with HY5 for retaining rhythmic expression under constant condition (Zhou et al. 2015). Other regulator in light signaling pathway, such as PIF5, is involved in terpene biosynthesis as well (Mannen et al. 2014). Here, we identified several HY5-interacting motifs on the upstream regulatory fragments of structural genes and TFs, especially PbNAC1, suggesting that the light and circadian clock signals might manage monoterpene biosynthesis in *P. bellina* by controlling these genes.

Naturally, the striking changes of light at dawn and dusk help to entrain the circadian clock. Phytochrome is responsible for the red-light input and cryptochrome for the blue-light (Devlin and Kay 2000). Our results show that *P. violacea* responded differently to various light qualities. It is intriguing to know how the signaling components isolated in this study work together to regulate monoterpene biosynthesis in *Phalaenopsis* orchids, including photoreceptors and COP1.

The floral scent of *P. bellina* is composed by monoterpenes and its biosynthesis is regulated by multiple factors. *PbGDPS* is expressed in the epidermal cells on the surface of perianth (Hsiao et al. 2008), suggesting a factor for its tissue specificity. Moreover, the development factor is involved to induce the scent biosynthesis at full-bloomed stages (Chuang et al. unpublished). In this study, we present that both circadian clock and light factor are required for the diurnal emission. These factors cooperate to establish the optimal spatial and temporal emission patterns of monoterpenes in *Phalaenopsis* flowers.

Conclusion

The present study indicates that the diurnal emission of monoterpenes from *P. violacea* flowers is controlled by a circadian clock and light factors. In addition, white light was the most optimal light condition to emit monoterpenes in four *Phalaenopsis* orchids examined. The light and circadian signal might be associate with floral monoterpene emission by regulating the transcription factors for monoterpene biosynthesis.

Authors' contributions

YCC, MCL, YLC, WHC and HHC conceived and designed the experiments. YCC, MCL, and YLC performed the experiments. YCC and MCL analyzed the data. YCC, MCL and HHC wrote the manuscript. All authors read and approved the final manuscript.

Author details
[1] Department of Life Sciences, National Cheng Kung University, Tainan 701, Taiwan. [2] Orchid Research and Development Center, National Cheng Kung University, Tainan 701, Taiwan.

Acknowledgements
We thank Dr. Shu-Hsing Wu (Institute of Plant and Microbial Biology, Academia Sinica, Taipei, Taiwan) for helpful discussions. We thank Dr. Chi-Kuang Wen (Shanghai Institutes for Biological Sciences, Chinese Academy of Sciences, China) for careful reading and suggestions of the manuscript. We thank Dr. Swee-Suak Ko (Academia Sinica Biotechnology Center in Southern Taiwan, Tainan, Taiwan) for assistance in performing experiments with their growth chamber. We thank Dr. Ming-Hsien Hsieh (Tainan District Agricultural Research and Extension Station, Council of Agriculture, Tainan, Taiwan) for kindly providing the photometer.

Competing interests
The authors declare that they have no competing interests.

Funding
This work was supported by Ministry of Science and Technology, Taiwan, Grant No.: 105-2313-B-006-002-MY3.

References

Abe H, Urao T, Ito T, Seki M, Shinozaki K, Yamaguchi-Shinozaki K (2003) *Arabidopsis* AtMYC2 (bHLH) and AtMYB2 (MYB) function as transcriptional activators in abscisic acid signaling. Plant Cell 15:63–78

Alabadí D, Oyama T, Yanovsky MJ, Harmon FG, Más P, Kay SA (2001) Reciprocal regulation between TOC1 and LHY/CCA1 within the *Arabidopsis* circadian clock. Science 293:880–883

Cai J, Liu X, Vanneste K, Proost S, Tsai WC, Liu KW, Chen LJ, He Y, Xu Q, Bian C (2015) The genome sequence of the orchid *Phalaenopsis equestris*. Nat Genet 47:65–72

Chow CN, Zheng HQ, Wu NY, Chien CH, Huang HD, Lee TY, Chiang-Hsieh YF, Hou PF, Yang TY, Chang WC (2015) PlantPAN 2.0: an update of plant promoter analysis navigator for reconstructing transcriptional regulatory networks in plants. Nucleic Acids Res 44:D1154–D1160

Colquhoun TA, Schwieterman ML, Gilbert JL, Jaworski EA, Langer KM, Jones CR, Rushing GV, Hunter TM, Olmstead J, Clark DG (2013) Light modulation of volatile organic compounds from petunia flowers and select fruits. Postharvest Biol Technol 86:37–44

Costantini L, Kappel CD, Trenti M, Battilana J, Emanuelli F, Sordo M, Moretto M, Camps C, Larcher R, Delrot S (2017) Drawing links from transcriptome to metabolites: the evolution of aroma in the ripening berry of Moscato Bianco (*Vitis vinifera* L.). Front Plant Sci 8:780

Devlin PF, Kay SA (2000) Cryptochromes are required for phytochrome signaling to the circadian clock but not for rhythmicity. Plant Cell 12:2499–2509

Dodd AN, Belbin FE, Frank A, Webb AA (2015) Interactions between circadian clocks and photosynthesis for the temporal and spatial coordination of metabolism. Front Plant Sci 6:245

Dudareva N, Martin D, Kish CM, Kolosova N, Gorenstein N, Fäldt J, Miller B, Bohlmann J (2003) (E)-β-ocimene and myrcene synthase genes of floral scent biosynthesis in snapdragon: function and expression of three terpene synthase genes of a new terpene synthase subfamily. Plant Cell 15:1227–1241

Dudareva N, Andersson S, Orlova I, Gatto N, Reichelt M, Rhodes D, Boland W, Gershenzon J (2005) The nonmevalonate pathway supports both monoterpene and sesquiterpene formation in snapdragon flowers. Proc Natl Acad Sci USA 102:933–938

Falara V, Amarasinghe R, Poldy J, Pichersky E, Barrow RA, Peakall R (2012) The production of a key floral volatile is dependent on UV light in a sexually deceptive orchid. Ann Bot 111:21–30

Fenske MP, Imaizumi T (2016) Circadian rhythms in floral scent emission. Front Plant Sci 7:462

Fu X, Chen Y, Mei X, Katsuno T, Kobayashi E, Dong F, Watanabe N, Yang Z (2015) Regulation of formation of volatile compounds of tea (*Camellia sinensis*) leaves by single light wavelength. Sci Rep 5:16858

Gangappa SN, Botto JF (2016) The multifaceted roles of HY5 in plant growth and development. Mol Plant 9:1353–1365

Helsper JP, Davies JA, Bouwmeester HJ, Krol AF, Van Kampen MH (1998) Circadian rhythmicity in emission of volatile compounds by flowers of *Rosa hybrida* L. cv. Honesty. Planta 207:88–95

Hendel-Rahmanim K, Masci T, Vainstein A, Weiss D (2007) Diurnal regulation of scent emission in rose flowers. Planta 226:1491–1499

Hoballah ME, Stuurman J, Turlings TC, Guerin PM, Connetable S, Kuhlemeier C (2005) The composition and timing of flower odour emission by wild *Petunia axillaris* coincide with the antennal perception and nocturnal activity of the pollinator *Manduca sexta*. Planta 222:141–150

Hsiao YY, Tsai WC, Kuoh CS, Huang TH, Wang HC, Wu TS, Leu YL, Chen WH, Chen HH (2006) Comparison of transcripts in *Phalaenopsis bellina* and *Phalaenopsis equestris* (Orchidaceae) flowers to deduce monoterpene biosynthesis pathway. BMC Plant Biol 6:14

Hsiao YY, Jeng MF, Tsai WC, Chuang YC, Li CY, Wu TS, Kuoh CS, Chen WH, Chen HH (2008) A novel homodimeric geranyl diphosphate synthase from the orchid *Phalaenopsis bellina* lacking a DD (X) 2–4D motif. Plant J 55:719–733

Hsu CC, Wu PS, Chen TC, Yu CW, Tsai WC, Wu K, Wu WL, Chen WH, Chen HH (2014) Histone acetylation accompanied with promoter sequences displaying differential expression profiles of B-class MADS-box genes for *Phalaenopsis* floral morphogenesis. PLoS ONE 9:e106033

Lau OS, Deng XW (2012) The photomorphogenic repressors COP1 and DET1: 20 years later. Trends Plant Sci 17:584–593

Lee J, He K, Stolc V, Lee H, Figueroa P, Gao Y, Tongprasit W, Zhao H, Lee I, Deng XW (2007) Analysis of transcription factor HY5 genomic binding sites revealed its hierarchical role in light regulation of development. Plant Cell 19:731–749

Luo XM, Lin WH, Zhu S, Zhu JY, Sun Y, Fan XY, Cheng M, Hao Y, Oh E, Tian M (2010) Integration of light-and brassinosteroid-signaling pathways by a GATA transcription factor in *Arabidopsis*. Dev Cell 19:872–883

Mannen K, Matsumoto T, Takahashi S, Yamaguchi Y, Tsukagoshi M, Sano R, Suzuki H, Sakurai N, Shibata D, Koyama T (2014) Coordinated transcriptional regulation of isopentenyl diphosphate biosynthetic pathway enzymes in plastids by phytochrome-interacting factor 5. Biochem Biophys Res Commun 443:768–774

Mcclung CR (2006) Plant circadian rhythms. Plant Cell 18:792–803

Muhlemann JK, Klempien A, Dudareva N (2014) Floral volatiles: from biosynthesis to function. Plant Cell Environ 37:1936–1949

Picone JM, Mactavish HS, Clery RA (2002) Emission of floral volatiles from *Mahonia japonica* (Berberidaceae). Phytochemistry 60:611–617

Picone JM, Clery RA, Watanabe N, Mactavish HS, Turnbull CG (2004) Rhythmic emission of floral volatiles from *Rosa damascena semperflorens* cv.'Quatre Saisons'. Planta 219:468–478

Rodríguez-Concepción M, Forés O, Martínez-García JF, González V, Phillips MA, Ferrer A, Boronat A (2004) Distinct light-mediated pathways regulate the biosynthesis and exchange of isoprenoid precursors during *Arabidopsis* seedling development. Plant Cell 16:144–156

Saunders D (1997) Insect circadian rhythms and photoperiodism. Invertebr Neurosci 3:155–164

Tholl D, Kish CM, Orlova I, Sherman D, Gershenzon J, Pichersky E, Dudareva N (2004) Formation of monoterpenes in *Antirrhinum majus* and *Clarkia breweri* flowers involves heterodimeric geranyl diphosphate synthases. Plant Cell 16:977–992

Tsai CC (2003) Molecular phylogeny, biogeography, and evolutionary trends of the genus *Phalaenopsis* (Orchidaceae). Dissertation, National Sun Yat-Sen University

Zhang H, He H, Wang X, Wang X, Yang X, Li L, Deng XW (2011) Genome-wide mapping of the HY5-mediated gene networks in *Arabidopsis* that involve both transcriptional and post-transcriptional regulation. Plant J 65:346–358

Zhang H, Fan P, Liu C, Wu B, Li S, Liang Z (2014) Sunlight exclusion from Muscat grape alters volatile profiles during berry development. Food Chem 164:242–250

Zhang E, Chai F, Zhang H, Li S, Liang Z, Fan P (2017) Effects of sunlight exclusion on the profiles of monoterpene biosynthesis and accumulation in grape exocarp and mesocarp. Food Chem 237:379–389

Zhou D-X (1999) Regulatory mechanism of plant gene transcription by GT-elements and GT-factors. Trends Plant Sci 4:210–214

Zhou F, Sun TH, Zhao L, Pan XW, Lu S (2015) The bZIP transcription factor HY5 interacts with the promoter of the monoterpene synthase gene *QH6* in modulating its rhythmic expression. Front Plant Sci 6:304

11

Soil respiration patterns and rates at three Taiwanese forest plantations: dependence on elevation, temperature, precipitation, and litterfall

Yu-Hsuan Huang[1], Chih-Yu Hung[1], I-Rhy Lin[1], Tomonori Kume[1], Oleg V. Menyailo[2] and Chih-Hsin Cheng[1*]

Abstract

Background: Soil respiration contributes to a large quantity of carbon emissions in the forest ecosystem. In this study, the soil respiration rates at three Taiwanese forest plantations (two lowland and one mid-elevation) were investigated. We aimed to determine how soil respiration varies between lowland and mid-elevation forest plantations and identify the relative importance of biotic and abiotic factors affecting soil respiration.

Results: The results showed that the temporal patterns of soil respiration rates were mainly influenced by soil temperature and soil water content, and a combined soil temperature and soil water content model explained 54–80% of the variation. However, these two factors affected soil respiration differently. Soil temperature positively contributed to soil respiration, but a bidirectional relationship between soil respiration and soil water content was revealed. Higher soil moisture content resulted in higher soil respiration rates at the lowland plantations but led to adverse effects at the mid-elevation plantation. The annual soil respiration rates were estimated as 14.3–20.0 Mg C ha^{-1} $year^{-1}$ at the lowland plantations and 7.0–12.2 Mg C ha^{-1} $year^{-1}$ at the mid-elevation plantation. When assembled with the findings of previous studies, the annual soil respiration rates increased with the mean annual temperature and litterfall but decreased with elevation and the mean annual precipitation. A conceptual model of the biotic and abiotic factors affecting the spatial and temporal patterns of the soil respiration rate was developed. Three determinant factors were proposed: (i) elevation, (ii) stand characteristics, and (iii) soil temperature and soil moisture.

Conclusion: The results indicated that changes in temperature and precipitation significantly affect soil respiration. Because of the high variability of soil respiration, more studies and data syntheses are required to accurately predict soil respiration in Taiwanese forests.

Keywords: Soil respiration, Elevation, Soil temperature, Soil water content, Litterfall, Temperature sensitivity

Introduction

Soil respiration, which refers to the efflux of carbon dioxide from the soil surface to the atmosphere, is a crucial component of carbon cycling. It is the primary pathway through which carbon moves from the ecosystem into the atmosphere, and it strongly affects the carbon balance and turnover (Raich and Schlesinger 1992; Davidson and Janssens 2006; Bond-Lamberty and Thomson 2010). In the forest ecosystem, the soil contributes to 60–80% of total ecosystem respiration (Law et al. 1999; Janssens et al. 2001). Understanding the rate of carbon release from the soil and its seasonal patterns will lead to a comprehensive understanding of how to tackle regional and global climate change.

Soil respiration is produced mainly by plant roots and soil organisms. The soil respiration rates are largely dependent on soil temperature and soil water content. Most related studies have shown that soil respiration is

*Correspondence: chengch@ntu.edu.tw
[1] School of Forestry and Resource Conservation, National Taiwan University, Taipei 106, Taiwan

positively correlated with soil temperature (Davidson et al. 1998), and that it can be modified or even impeded by soil water content under very wet or very dry conditions (Davidson et al. 2000; Raich and Schlesinger 1992). The effects of soil temperature and soil water content on soil respiration have been suggested to be interactive, rather than individual (Raich and Schlesinger 1992; Davidson et al. 1998; Reichstein et al. 2003; Saiz et al. 2006). Global meta-analysis data indicated that soil respiration is significantly correlated with mean annual temperature (MAT) and mean annual precipitation (MAP) (Raich and Schlesinger 1992) and possibly also with three variables—MAT, MAP, and vegetation productivity (Raich and Schlesinger 1992; Reichstein et al. 2003; Bond-Lamberty and Thomson 2010). However, the majority of these studies were conducted in temperate ecosystems; fewer measurements have been taken in the tropics and subtropics. Some measurements taken in the tropics and subtropics demonstrated a high deviation between measurements and estimations made using global regression models, with the differences ranging from − 73 to + 294% (Hashimoto et al. 2004; Chang et al. 2008; Katayama et al. 2009; Tan et al. 2013). These statistics exemplify the inherent variability of soil respiration in the tropics and subtropics; thus, soil resperation in these regions should be further investigated.

Because of the high cost of measurement instruments and the complex landscape, soil respiration has not been extensively investigated in Taiwan (Chang et al. 2008; Kao and Chang 2009; Hsieh et al. 2016). Therefore, soil respiration in different areas and its seasonal dynamics are not widely understood. In Taiwan, the temperature decreases with elevation, with a lapse rate of $-$ 5.3 °C km^{-1}. However, precipitation increases with elevation, with mountainous areas and lowlands receiving 2000–3500 and 1500–2000 mm of annual rainfall, respectively (Lee et al. 2004). Such climatic differences associated with their influences over primary production may have significant impacts on soil respiration on spatial and temporal scales. The mechanisms underlying the effects of these factors on soil respiration must be determined to enable the assessment of the interaction between climate and soil respiration and the processes involved in the interaction. Moreover, measuring the soil respiration rates in Taiwan would ensure the completeness of the global database, particularly for databases for tropical and subtropical ecosystems (Raich and Schlesinger 1992; Bond-Lamberty and Thomson 2010).

In this study, soil respiration at three forest plantations (two lowland and one mid-elevation) was investigated throughout an annual cycle. Soil temperature, soil water content, and litterfall production were also determined at each plantation. We investigated the soil respiration rate at each plantation and compared the rate among the plantations. Our objectives were as follows: (1) to determine how soil respiration varies between lowland and mid-elevation forest plantations, (2) to analyze the relative importance of soil temperature and soil moisture for temporal soil respiration patterns at the three studied plantations, and (3) to assess which factors are crucial for controlling the annual soil respiration rate across the plantations.

Materials and methods

Study sites

The first plantation in this study is located in the lowland plain in Rende District, Tainan (TN; 22°56′N, 120°16′E). This plantation was originally an orchard but had been abandoned for more than 15 years. It is now used as a representative plantation for predicting potential growth in current cropland afforestation areas because no such old plantation exists in Taiwan (Lin et al. 2011; Cheng et al. 2016a). Two stands, TN1 and TN2, were evaluated. TN1 is an abandoned lychee (*Litchi chinensis*) plantation with a stand age of 33 years. TN2 is an abandoned wax apple (*Eugenia javanica*) plantation with a stand age of 25 years. Because of the long-term abandonment, understory vegetation such as *Macaranga tanarius*, *Murraya paniculata*, *Celtis sinensis*, *Panicum maximum*, and *Alpinia speciose* grows rigorously at both stands.

The second plantation is located in the Datu tableland, Taichung (TC; 24°17′N, 120°35′E). This plantation is an *Acacia confusa* plantation situated at an elevation of 150 m above sea level. Two adjacent stands, TC1 and TC2, were studied. Both stands have the same *Acacia* overstory but differ in their understory vegetation because of the influence of fire disturbance and invasive grass. TC1 has not been disturbed by fire, and the dominant understory species are *Lantana camara*, *Zanthoxylum nitidum*, *Mallotus repandus*, and *Glochidion rubrum*. However, TC2 has been slightly influenced by ground fire, and the understory vegetation (more than 90% coverage) has been replaced by grass (*Panicum maximum*) (Cheng et al. 2013a).

The third plantation is located in mid-elevation montane areas in Xitou, Nantou (NT; 23°40′N, 120°46′E). The plantation is a Japanese cedar (*Cryptomeria japonica*) plantation situated at an elevation of approximately 1200 m above sea level. Three stands, NT1, NT2, and NT3, were studied. The stand ages of which are 90, 63, and 40 years, respectively, at the time of this study. In these three stands, the mean diameter at breast height (DBH), basal area, and living tree biomass carbon stocks increase with stand age. By contrast, tree density decreases with stand age (Cheng et al. 2013b) (Table 1).

Table 1 Stand characteristics and soil properties from surface level to a depth of 10 cm in the selected stands at TN, TC, and NT plantations in Taiwan

Site	Elevation (m)	MAT (°C)	MAP (mm)	Dominant vegetation	Trees (multi-trunks) density (No. ha^{-1})	Mean trunk DBH (cm)	Basal area (m^2 ha^{-1})	Canopy height (m)	LAI[a]	Litterfall (Mg C ha^{-1} $year^{-1}$)	Soil texture	pH	Soil organic carbon (%)
TN1	14	24.3	1698	*L. chinensis*	560 (1566) ± 134 (492)	16.7 ± 0.6	40.7 ± 3.2	11.8	4.4	5.4 ± 0.7	Loam	4.7 ± 0.1	2.2 ± 0.2
TN2	13			*E. javanica*	321 (2821) ± 91 (855)	12.8 ± 0.7	68.5 ± 12.1	10.6	3.9	5.8 ± 1.1	Sandy loam	4.8 ± 0.1	1.5 ± 0.1
TC1	110	23.0	1347	*A. confusa*	2360 ± 433	10.0 ± 0.9	21.7 ± 3.8	9.3	3.3	3.8 ± 0.2	Loam	3.9 ± 0.1	4.9 ± 0.4
TC2	110			*A. confusa*	2533 ± 985	9.9 ± 2.7	21.0 ± 3.6	8.4	3.3	3.8 ± 0.2	Loam	3.8 ± 0.0	4.5 ± 0.1
NT1	1150	16.6	2635	*C. japonica*	408 ± 22	48.6 ± 1.2	77.9 ± 5.9	29.0	4.0	2.0 ± 0.4	Loam	4.0 ± 0.3	17.2 ± 2.8
NT2	1250			*C. japonica*	617 ± 8	35.3 ± 1.0	63.7 ± 4.4	26.0	3.4	1.9 ± 0.2	Sandy loam	5.8 ± 0.1	3.4 ± 0.8
NT3	1370			*C. japonica*	1358 ± 162	23.1 ± 0.4	59.7 ± 4.0	20.8	4.5	3.6 ± 0.8	Loam	4.5 ± 0.1	4.3 ± 0.8

Climate data referred to Central Weather Bureau, Taiwan; Stand structure referred to Cheng et al. (2013a, b, 2016b) and Lin et al. (2011)

MAT mean annual temperature, *MAP* mean annual precipitation

[a] *LAI* leaf area index, was determined using a LI-COR 2200 (LI-COR Comp., Lincoln, Nebraska, USA)

Elevation, MAT, and MAP at the three study plantations are shown in Table 1. MAT is strongly affected by elevation and ranges from 16.6 °C at NT to 25.0 °C at TN. Elevation also exerts a significantly orographic influence on precipitation. A higher MAP was found at the mid-elevation NT plantation (2635 mm) than at the lowland plantations at TN (1698 mm) and TC (1347 mm). The climatic conditions at both TN and TC correspond to tropical and subtropical moist forests, and the climatic condition at NT is classified as a montane wet subtropical forest.

Previously studied stand characteristics and soil properties are shown in Table 1 (Lin et al. 2011; Cheng et al. 2013a, b, 2016b). Generally, all selected stands were closed canopy with leaf area indices between 3.3 and 4.5. The canopy height ranged between 8.4 and 29.0 m, and the mean stem DBH ranged between 9.9 and 48.6 cm. The soil textures were loam and sandy loam. Stand productivity was evaluated based on aboveground litterfall production, which has been correlated to the total net primary production and has been used to represent the labile input available to decomposition under a forest ecosystem in previous studies (Bond-Lamberty and Thomson 2010). Annual aboveground litterfall production over study stands ranged between 1.9 and 5.8 Mg C ha^{-1} (Table 1). Higher litterfall production was found at TC and TN than at NT.

In 2010, the mean air temperature was 24.6, 23.0, and 17.0 °C, and the total precipitation was 1778, 1182, and 2041 mm at TN, TC, and NT, respectively (Fig. 1). The annual air temperature and precipitation at TN and TC in 2010 were consistent with their long-term climate regimes, but lower precipitation was found at NT (2041 vs. 2500 mm) because of lower precipitation in summer. Generally, a hot and humid season from April to September and a cold and dry season from October to March can be observed at all three plantations.

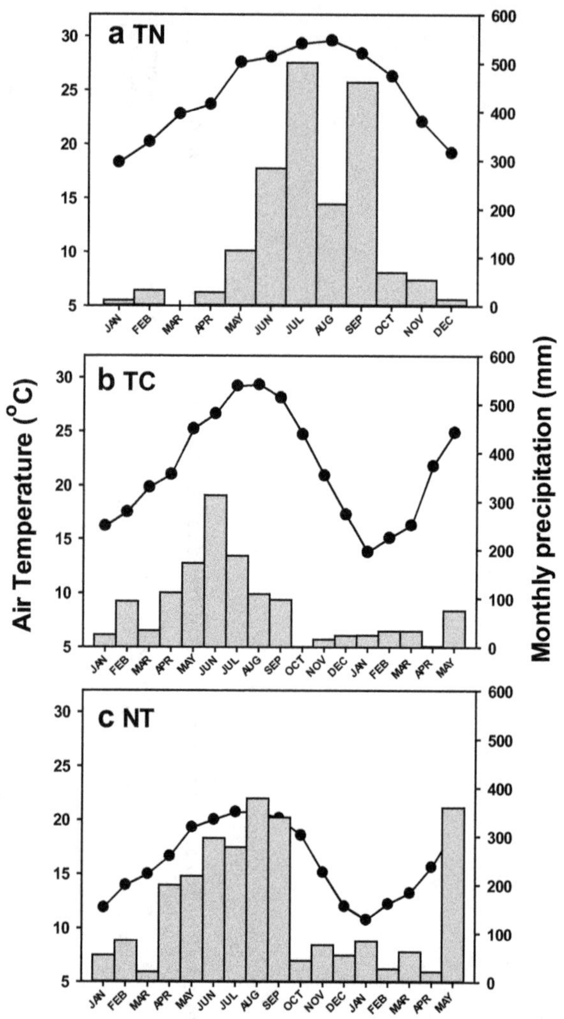

Fig. 1 Monthly mean air temperature (°C, black circle) and precipitation (mm, gray bar) from January 2010 to December 2010 at TN plantation and from January 2010 to May 2011 at TC and NT plantations

Soil respiration

Soil respiration was measured using a Li-6400 with a Li-6400-09 soil chamber (LI-COR Inc., Lincoln, Nebraska, USA). Measurements were taken using polyvinyl chloride (PVC) collars that were 11 cm in diameter and 7 cm in height. The collars were installed on the soil surface at a depth of 5 cm to prevent erosion from heavy rainfall and enable the growth of fine roots. Six PVC collars (duplicates from three 20 × 20-m^2 plots) were applied at each stand. The collars were installed 2 months before the first measurement and remained in place throughout the measurement period. Soil respiration at each collar was calculated as the average of three values generated from three continuous cycles. All chamber measurements were conducted between 08:00 and 14:00 local time to minimize sampling bias introduced by the time of day (Knowles et al. 2015; Hsieh et al. 2016), and measurements were taken monthly from January 2010 to May 2011 at TC and NT and from January 2010 to December 2010 at TN.

The temperature and water content of soil next to each collar were determined during the measurement period. Soil temperature at a depth of 10 cm was measured using the temperature sensor attached to the Li-6400. Soil water content was measured as volumetric soil moisture content by using time domain reflectometry probes (Hydrosense, Campbell Scientific Inc., Utah, USA). In addition, the full spectra of soil temperature and soil water content were recorded. A data logger (HOBO U30, Onset, Massachusetts, USA) connected to a temperature probe (S-TMB-M006, Onset) at a depth of 10 cm and a soil water content probe (S-SMx-M005, Onset) covering a depth of 0–20 cm were used at each stand. Both probes were programmed to measure values at 4-min intervals and store data hourly.

Modeling temporal soil respiration patterns

To determine whether temporal changes in soil respiration are related to soil temperature and soil water content, the relationships between soil respiration and soil temperature and soil water content were examined using three empirical regression models. The first model involved only soil temperature, and an exponential model was applied (Lloyd and Taylor 1994; Sheng et al. 2010).

$$\text{Soil respiration } (\mu \text{ mol } CO_2 \text{ m}^{-2} \text{ s}^{-1}) = a \times \exp(b \times T) \tag{1}$$

where T is soil temperature (°C) at a depth of 10 cm, and a and b are constants fitted using least squares techniques.

The second model used soil water content as the only indicator variable, and a quadratic equation was applied (Saiz et al. 2006).

$$\text{Soil respiration } = a + b \times \theta + c \times \theta^2 \tag{2}$$

where θ refers to soil water content (cm^3 cm^{-3}) over a surface at a depth of 20 cm, and a and b are constants fitted using least squares techniques.

In the third model, the soil respiration rate is the function of soil temperature and soil water content (Saiz et al. 2006; Sheng et al. 2010). The model is expressed as follows:

$$\text{Soil respiration} = a \times \exp(b \times T) \times \theta^c \tag{3}$$

where T is soil temperature, θ is soil water content, and a, b, and c are the model coefficients.

A Q_{10} function was used to determine the temperature–respiration relationship (Lloyd and Taylor 1994). The equation is expressed as follows:

$$Q_{10} = \exp(10 \times b) \tag{4}$$

where b is the model coefficient in Eqs. (1) and (3).

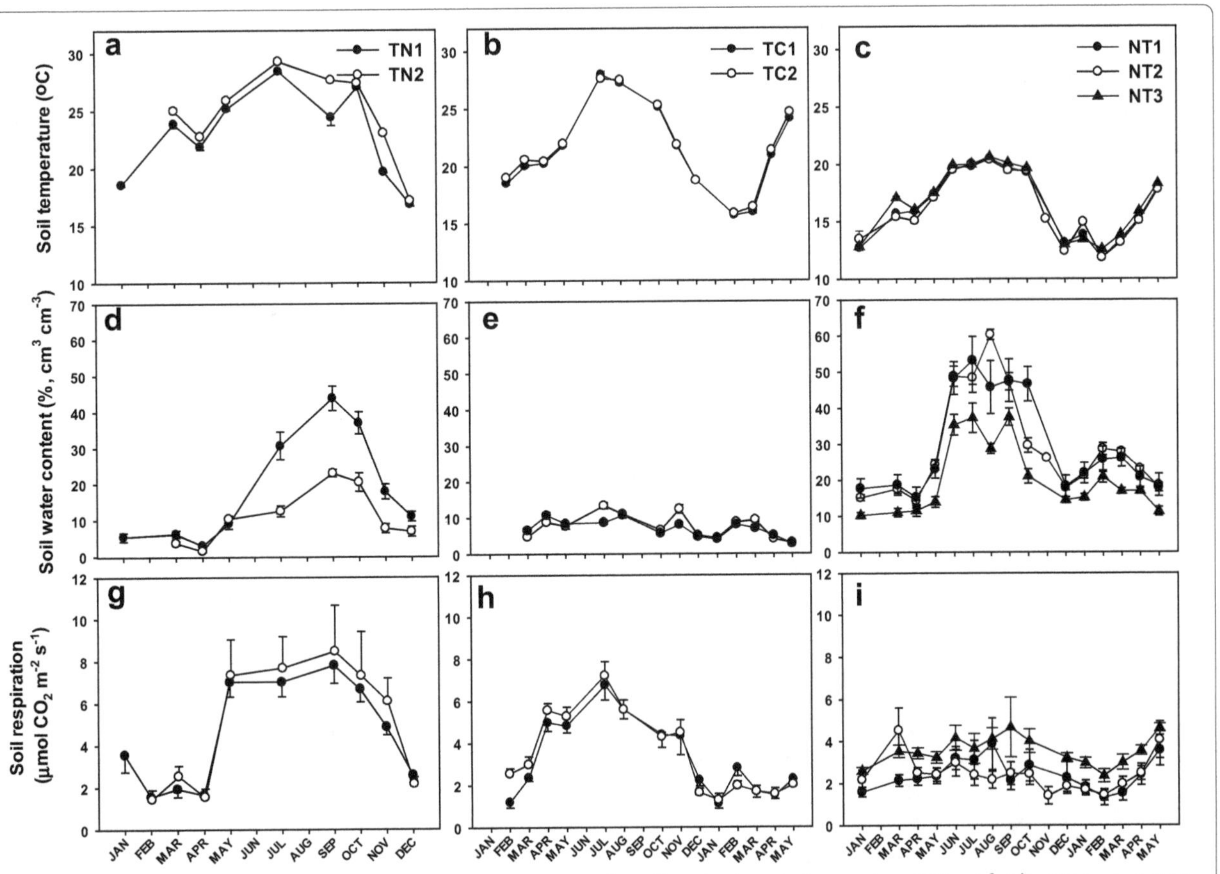

Fig. 2 Seasonal patterns of soil temperature (°C), soil water content (%, cm^3 cm^{-3}), and soil respiration rate (μmol CO_2 m^{-2} s^{-1}) at TN (**a**, **d**, **g**), TC (**b**, **e**, **h**), and NT (**c**, **f**, **i**) plantations in Taiwan. Each symbol is the mean of the measurements, and the standard deviation is plotted as error bars

Data analysis

The regression models were fitted using *R (2.15.2)* to estimate parameters a, b, and c; coefficients of determinations (R^2); and the root mean squared error (RMSE). The R^2 and RMSE were used to determine the significance and goodness-of-fit of the models. The soil respiration model was estimated based on the regression model with the highest R^2 and the lowest RMSE. We applied the continuous measurements of soil temperature and soil moisture from the data logger to the regressive equation to estimate the amount of carbon released through soil respiration. Although the values of the monthly field measurements and data logger were measured using different instruments and locations, their readings were closely correlated ($r > 0.98$). A linear equation that converted data logger data to field measurements was used before running the regressive equation. The daily soil respiration rate was firstly calculated. The annual soil respiration in 2010 was obtained by calculating the sum of daily estimations (Saiz et al. 2010).

To assess which factors are crucial for controlling the annual soil respiration rates across the plantations, the relationships among annual soil respiration rates, elevation, MAT, and litterfall production were examined. In addition to our study data, data from previous studies that used the same dynamic closed chamber method to measure soil respiration were assembled to generate a more robust relationship (Chang et al. 2008; Kao and Chang 2009; Hsieh et al. 2016).

Results

Temporal patterns of soil temperature, soil water content, and soil respiration

Soil temperature at a depth of 10 cm was low in winter and reached its peak in summer (Fig. 2), ranging between 15 and 30 °C at the lowland TN and TC plantations and between 12 and 20 °C at the mid-elevation NT plantation. The temporal patterns of soil water content were similar to those of soil temperature at all three plantations, namely high in summer and low in winter (Fig. 2). The highest soil water content was observed at NT, reaching > 50% $cm^3\ cm^{-3}$ from June to October because of higher precipitation. The lowest soil water content (< 10% $cm^3\ cm^{-3}$) was observed at TC and TN from December to March because of lower precipitation. According to monthly field measurements, soil water content at TC was below 15% $cm^3\ cm^{-3}$; however, data logger-recorded soil water content ranged between 3 and 34% $cm^3\ cm^{-3}$. The lower values of the monthly field measurements may be due to the placement of the probe at the same soil spot, which might have been desiccated after the measurements were taken. Nevertheless, a linear correlation was observed between the monthly field measurements and data logger data ($r = 0.98$, $p < 0.01$).

The seasonality of soil respiration rates at the lowland TC and TN plantations coincided strongly with soil temperature and soil water content (Fig. 2). The soil respiration rate in the hot and humid season was significantly

Table 2 Estimated parameters of a, b, and c; coefficients of determinations (R^2); root mean squared error (RMSE); and temperature sensitivity (Q_{10}) of regressions for modeling soil respiration at TN, TC, and NT plantations in Taiwan

Model	a	b	c	R^2	RMSE	Q_{10}	a	b	c	R^2	RMSE	Q_{10}
	TN1						TN2					
Temperature	0.592	0.089		0.430	1.965	2.44	0.269	0.120		0.605	1.920	3.31
Water content	1.709	0.246	− 0.003	0.654	1.648		0.053	0.870	− 0.022	0.824	1.400	
Temperature and water content	*0.862*	*0.033*	*0.349*	*0.705*	*1.529*	*1.39*	*0.657*	*0.053*	*0.371*	*0.847*	*1.310*	*1.70*
	TC1						TC2					
Temperature	0.333	0.106		0.611	1.182	2.88	0.374	0.103		0.573	1.359	2.80
Water content	1.073	0.095	0.032	0.565	1.290		0.971	0.214	0.013	0.523	1.506	
Temperature and water content	*0.181*	*0.070*	*0.738*	*0.843*	*0.783*	*2.01*	*0.051*	*0.099*	*1.133*	*0.669*	*1.255*	*2.69*
	NT1						NT2					
Temperature	0.573	0.086		0.676	0.444	2.37	1.347	0.036		0.100	0.826	1.44
Water content	3.864	− 0.138	0.002	0.336	0.663		4.057	− 0.104	0.001	0.085	0.865	
Temperature and water content	*0.744*	*0.128*	*− 0.289*	*0.755*	*0.402*	*3.60*	*3.030*	*0.116*	*− 0.652*	*0.453*	*0.668*	*3.19*
	NT3											
Temperature	1.394	0.055		0.721	0.376	1.74						
Water content	4.049	− 0.086	0.003	0.278	0.631							
Temperature and water content	*1.515*	*0.065*	*− 0.086*	*0.748*	*0.373*	*1.92*						

Italic text represents the best model with the highest R^2 and lowest RMSE, which was used for estimating annual soil respiration rates

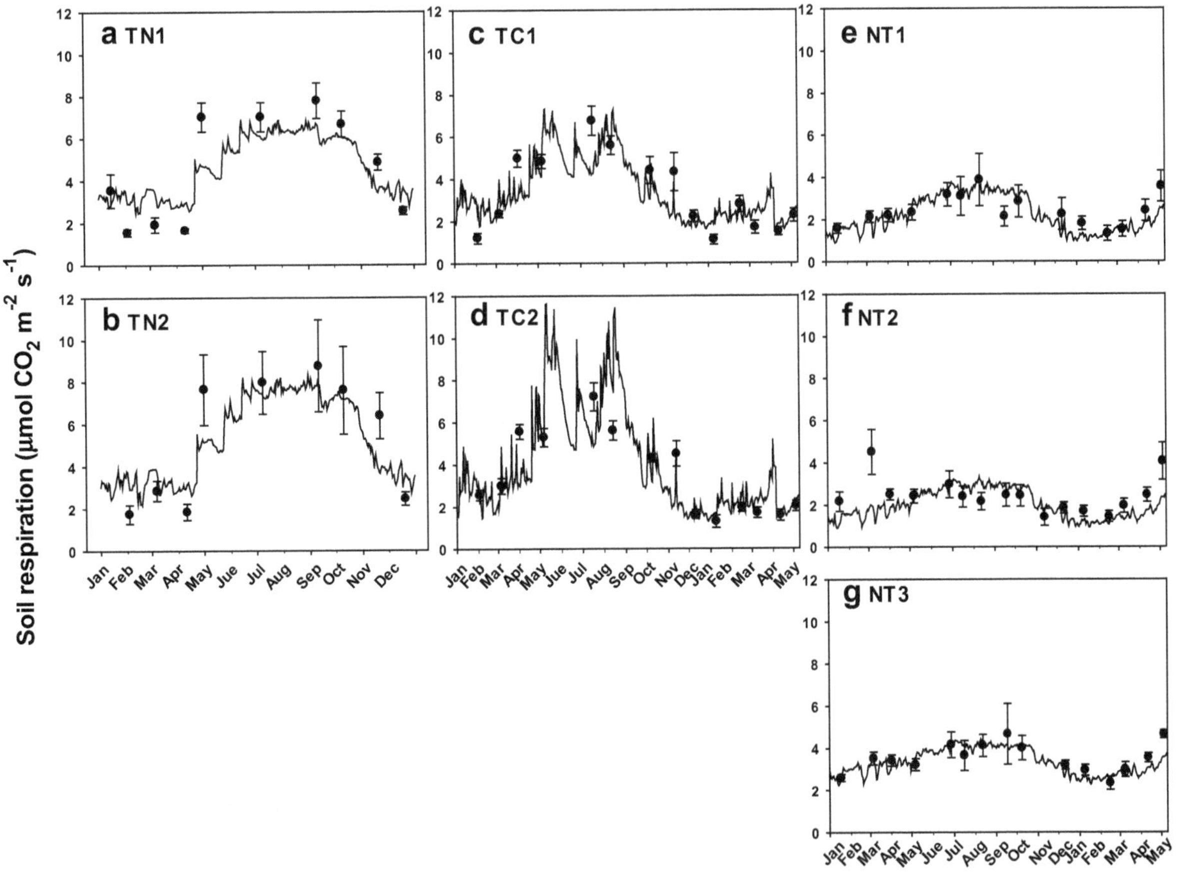

Fig. 3 Soil respiration rate (µmol CO_2 m^{-2} s^{-1}) measured in the field (black circles) and modeled by the regression equation (solid line) at TN (**a**, **b**), TC (**c**, **d**), and NT (**e**–**g**) plantations in Taiwan

Table 3 Estimated mean soil respiration rate and annual soil respiration rate in 2010 at TN, TC, and NT plantations in Taiwan

	Mean soil respiration rate (µmol CO_2 m^{-2} s^{-1})	Annual soil respiration rate (Mg C ha^{-1} $year^{-1}$)
TN1	4.67	17.6
TN2	5.30	20.0
TC1	3.78	14.3
TC2	4.58	17.3
NT1	2.02	7.6
NT2	1.86	7.0
NT3	3.22	12.2

higher than that in the cold and dry season. The soil respiration rates increased from 2 µmol CO_2 m^{-2} s^{-1} in winter to 8 µmol CO_2 m^{-2} s^{-1} in summer. However, the soil respiration rate at NT slightly increased from 2 µmol CO_2 m^{-2} s^{-1} in winter to 3.5 µmol CO_2 m^{-2} s^{-1} in summer.

Regression model of soil respiration

The soil temperature univariate model explained approximately 60% of the variations in soil respiration (Table 2). The soil water content model explained 52–82% of the variations at TN and TC but poorly explained those at NT (R^2 < 34%). The application of a combined soil temperature and soil water content model [Eq. (3)] yielded a higher R^2 (54–80%) and lower RMSE than did the univariate models. This finding suggested a better representation of the relationship by using both soil temperature and soil water content rather than a single factor. The combined models showed that soil temperature positively contributed to soil respiration at all three plantations. However, the models revealed bidirectional relationships between soil respiration and soil water content. Higher soil moisture content at TN and TC resulted in higher soil respiration rates, whereas higher soil water content at NT led to adverse effects and lower soil respiration rates.

The temperature-dependent Q_{10} values ranged between 2.44 and 3.31 at the lowland TN and TC plantations and

Table 4 Correlation coefficients at significant level p < 0.01 for annual soil respiration rate, elevation, mean annual temperature (MAT), mean annual precipitation (MAP), and litterfall production in 10 Taiwanese study stands

	Soil respiration (Mg C ha^{-1} $year^{-1}$)	Elevation (m)	MAT (°C)	MAP (mm)	Litterfall (Mg C ha^{-1} $year^{-1}$)
Soil respiration	1.00				
Elevation	– 0.93	1.00			
MAT	0.94	– 1.00	1.00		
MAP	– 0.92	0.94	– 0.94	1.00	
Litterfall	0.85	– 0.81	0.85	– 0.65	1.00

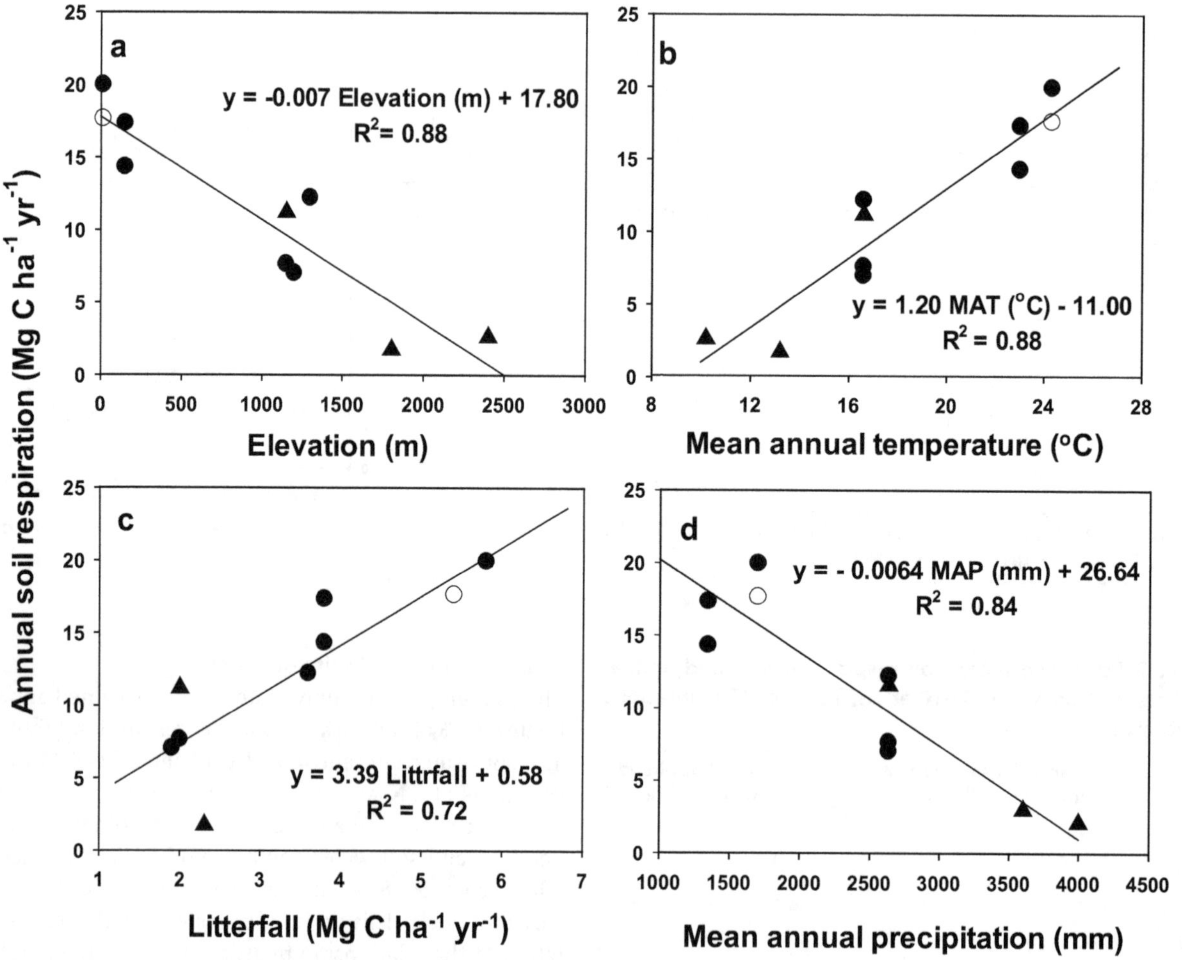

Fig. 4 Linear relationships of annual soil respiration with **a** elevation, **b** mean annual temperature, **c** litterfall, and **d** mean annual precipitation in Taiwan. Black circles refer to the current study and red circles refer to previous studies (Chang et al. 2008; Kao and Chang 2009; Hsieh et al. 2016)

between 1.44 and 2.37 at the mid-elevation NT plantation. However, when considering soil water content in Eq. (3), lower Q_{10} values from 1.39 to 2.69 at TN and TC plantations and a higher Q_{10} value from 1.92 to 3.60 at NT plantation were found. This finding suggested that the responses of soil respiration to soil temperature were confounded by soil water content.

Annual soil respiration rates at different plantations

Daily soil respiration rates calculated using the regression model [Eq. (3)] are shown in Fig. 3. The trend between measured and regression model are well corresponding, and the daily soil respiration variation could be found in the regression model. The annual soil respiration rates that summed by daily soil respiration rates were estimated as 17.6 and 20.0 Mg C ha^{-1} $year^{-1}$ at TN1 and TN2, 14.3 and 17.3 Mg C ha^{-1} $year^{-1}$ at TC1 and TC2, and 7.6, 7.0, and 12.2 Mg C ha^{-1} $year^{-1}$ at NT1, NT2, and NT3, respectively (Table 3). Although variations were observed on the stand scale, higher annual soil respiration rates were observed at the lowland TC and TN plantations than at the mid-elevation NT plantation.

When combined with the finding of previous studies (Chang et al. 2008; Kao and Chang 2009; Hsieh et al. 2016), the annual soil respiration rates were strongly correlated with elevation, MAT, MAP, and litterfall production ($p < 0.01$, Table 4), explaining 88, 88, 84, and 72% of the variation in annual soil respiration, respectively (Fig. 4). Notably, the annual soil respiration rates increased by 1.2 Mg C ha^{-1} $year^{-1}$ for every 1 °C increase in MAT and by 3.4 Mg C ha^{-1} $year^{-1}$ for every 1-Mg C ha^{-1} $year^{-1}$ increase in litterfall production. By contrast, the annual soil respiration decreased by 0.7 Mg C ha^{-1} $year^{-1}$ for every 100-m increase in elevation and by 0.64 Mg C ha^{-1} $year^{-1}$ for every 100-mm increase in MAP. Thus, higher annual soil respiration rates were found at the lowland plantations because of their higher MAT and litterfall production but lower MAP (< 1700 mm). By contrast, lower annual soil respiration rates were found at the plantations at higher elevations because of their lower MAT and litterfall production but higher MAP (> 2500 mm).

Discussion

Effects of soil temperature and soil water content on soil respiration

Soil temperature and soil water content are widely recognized as factors that control seasonal patterns of soil respiration (Raich and Schlesinger 1992; Saiz et al. 2006; Tang et al. 2006; Reichstein et al. 2003; Bond-Lamberty and Thomson 2010). In our study, we found that soil temperature or soil water content alone could somewhat account for the temporal pattern of soil respiration, and that the inclusion of both factors improved the predictive power of the model. The results suggested that the effects of the factors are interactive, rather than individual, for controlling the seasonal pattern of soil respiration. Generally, a higher soil respiration rate was observed in the hot and humid summer, whereas a lower soil respiration rate was observed in the cold and dry season.

Soil temperature and soil water content affected soil respiration differently. Soil temperature was positively correlated with the soil respiration rate at all three plantations. Soil water content was positively correlated with soil respiration at the lowland plantations but was negatively correlated at the mid-elevation plantation. We interpreted this finding as follows: soil water content at the lowland plantations was insufficiently high to reach the oxygen limitation for impeding soil respiration, whereas soil respiration at the mid-elevation plantation was inhibited by abundant precipitation (Davidson et al. 1998). Chang et al. (2008) and Hsieh et al. (2016) have reported the similar results of decreasing soil respiration with increasing soil water content in mountainous areas in Taiwan. Such a bidirectional relationship between soil water content and soil respiration has been previously observed on the watershed scale with a complex terrain or hydrological drainage (Riveros-Iregui et al. 2012; Knowles et al. 2015; Berryman et al. 2015). Our study provides the first evidence of a bidirectional relationship over an elevation gradient caused by different precipitation patterns.

The estimated Q_{10} values ranging between 1.44 and 3.31 were consistent with the global median value of 2.4 (Raich and Schlesinger 1992). The lower Q_{10} values at NT were possibly due to the oxygen limitation during summer, which rendered soil respiration less responsive to temperature (Davidson et al. 1998). When soil water content was added to the model, the Q_{10} values at NT increased from 1.44–2.37 to 1.92–3.60. The Q_{10} values at NT derived from the soil temperature and soil water content model [Eq. (3)] may be more accurate in representing the temperature dependence of soil respiration because the model considered the confounding effects of soil water content. However, the Q_{10} values at TN and TC derived from the combined soil temperature/soil water content model decreased from 2.44–3.31 to 1.39–2.69. The reduced Q_{10} values may have been masked by the correlation between soil temperature and soil water content (Tang et al. 2006).

Annual soil respiration rates

The average annual soil respiration rates at TN, TC, and NT were estimated as 19, 16, and 10 Mg C ha^{-1} $year^{-1}$, respectively. The annual soil respiration rates at TN and TC are consistent with those reported for other tropical forests such as the Amazon rainforest and those in Thailand, Malaysia, and Hawaii (Davidson et al. 2000; Hashimoto et al. 2004; Katayama et al. 2009; Litton et al. 2011). This finding indicated that both plantations are highly productive. In fact, the annual soil respiration rates at TN and TC are high compared with the rates reported by numerous studies worldwide (Raich and Schlesinger 1992; Bond-Lamberty and Thomson 2010). This result indicated that the rates at these plantations are significant to global soil respiration models. The annual soil respiration rate at NT is high compared with those recorded at Japanese cedar plantations in Japan (Shutou and Nakane 2004; Lee et al. 2006), possibly because of the higher annual temperature and active plant growth rates at NT (Cheng et al. 2013b).

In the present study, tight linkages were observed among annual soil respiration, elevation, MAT, MAP, and litterfall production at Taiwanese forest plantations (Table 4; Fig. 4). These results were expected because both abiotic (temperature and precipitation) and biotic (litterfall production) factors are critical drivers of soil respiration (Raich and Schlesinger 1992; Reichstein et al. 2003; Bond-Lamberty and Thomson 2010). The MAT, MAP, and litterfall production at all plantations were correlated with elevation (Fig. 4). Thus, elevation can be considered the main determinant factor influencing MAT, MAP, and primary production and defining the magnitude of the soil respiration rate. Previous studies have similarly reported that elevation is a crucial factor controlling the soil respiration rate (Kane et al. 2003; Litton et al. 2011; Berryman et al. 2014). Based on Fig. 4a, the annual soil respiration rate decreased by 0.7 Mg C ha^{-1} for every 100-m increase in elevation. However, such a relationship may be modified if the findings of new studies are added. For example, Raich and Tufekcioglu (2000) suggested that soil respiration rates in coniferous forests are 10% lower than those in broad-leaved forests. Broad-leaved forests in mountainous areas were not included in the present study.

In addition to climatic influence, differences in stand characteristics such as species composition, stand production, soil properties, and disturbances contribute to the variation in soil respiration (Raich and Tufekcioglu 2000; Tang et al. 2006; Bond-Lamberty and Thomson 2010; Sheng et al. 2010). In the present study, higher soil respiration rates at TN2 and NT3 might be because of their larger litterfall production and active plant growth rates (Table 1) (Lin et al. 2011; Cheng et al. 2013b). The higher soil respiration rate at TC2 was likely due to the invasive grass that accelerated the turnover rate of soil organic matter (Liao et al. 2008; Ward et al. 2009).

Conceptual model for soil respiration in Taiwan

A conceptual model of biotic and abiotic factors affecting soil respiration was developed to improve the understanding of soil respiration in Taiwan. We proposed three determinant factors influencing soil respiration patterns and rates: (i) elevation, (ii) stand characteristics, and (iii) soil temperature and soil moisture.

(i) *Elevation* On the interregional scale, elevation significantly affected MAT, MAP, and primary production and thus defined the magnitude of the soil respiration rate. Higher annual soil respiration rates were found at lowland plantations, whereas lower annual soil respiration rates were found at the plantations at higher elevations.

(ii) *Stand characteristics* On the stand scale, differences in stand characteristics exerted further influence on the soil respiration rate. In this study, larger litterfall production, active plant growth rates, and invasive grass tended to enhance soil respiration rates.

(iii) *Soil temperature and soil water content* On the temporal scale, soil respiration rates were found to be driven by the seasonality of soil temperature and soil water content. The soil respiration rates were higher in the hot and humid season than in the cold and dry season. The combined soil temperature and soil water content model indicated that soil temperature positively contributed to the soil respiration rate. In addition, a bidirectional relationship was observed between soil respiration and soil water content. At the lowland plantations, higher soil water content led to higher soil respiration, whereas excessive amounts of water at the mid-elevation plantation limited the oxygen supply, yielding lower soil respiration.

Based on the conceptual model, we anticipate that climate change will play a significant role in the responses of soil respiration to the determinant factors. In the scenario of global warming, rising temperatures can induce higher soil respiration. Changes in precipitation patterns may also affect soil respiration. Drying at mid-elevation might lead to the release of more soil organic carbon into the atmosphere; therefore, causing a shift from a carbon sink to a carbon source. Because of the high variability of soil respiration, more studies and data syntheses are required to accurately predict soil respiration in Taiwanese forests.

Conclusions

Soil temperature and soil water content are widely recognized as factors that control the seasonal patterns of soil respiration. The study results suggested that the effects of soil temperature and soil water content are interactive, rather than individual, for controlling the seasonal pattern of soil respiration. A combined soil temperature and soil water model explained 54–80% of the seasonal variations. However, these two factors affected soil respiration differently. Soil temperature positively contributed to soil respiration at all three plantations, and a bidirectional relationship between soil respiration and soil water content was observed among different plantations. Higher soil moisture content at TN and TC resulted in higher soil respiration rates; however, higher soil water content at NT led to adverse effects. The average annual soil respiration rates at TN, TC, and NT were estimated as 19, 16, and 10 Mg C ha^{-1} $year^{-1}$, respectively. When assembled with the findings of previous studies, tight linkages were observed among annual soil respiration, elevation, MAT, MAP, and litterfall production. Higher annual soil respiration rates were found at the lowland plantations because of their higher MAT and litterfall production but lower MAP. By contrast, lower annual soil respiration rates were found at the plantations at higher elevations because of their lower MAT and litterfall production and higher MAP. We proposed three determinant factors affecting soil respiration patterns and rates: (i) elevation, (ii) stand characteristics, and (iii) soil temperature and soil moisture. Based on the conceptual model, we anticipate that climate change will play a significant role in the future responses of soil respiration to the aforementioned factors. Because of the high variability of soil respiration, more studies and data syntheses are required to accurately predict soil respiration in Taiwanese forests.

Authors' contributions

YHH, CYH, and CHC planned the study, analyzed the data and developed the paper. IRL, KT, and OVM aided the field study and provided input to the manuscript. All authors read and approved the final manuscript.

Author details

[1] School of Forestry and Resource Conservation, National Taiwan University, Taipei 106, Taiwan. [2] Institute of Forest Research RAI SR, Krasnoyarsk 660036, Russia.

Acknowledgements

We gratefully acknowledge Chang-Ya Chen and Hong-Jai Ruan for their assistance with field and laboratory work.

Competing interests

The authors declare that they have no competing interests.

Funding

This study was supported by the Ministry of Science and Technology of Taiwan and a cooperative grant from the National Science Council of Taiwan and the Siberian Branch of the Russian Academy of Sciences.

References

Berryman EM, Marshall JD, Kavanagh K (2014) Decoupling litter respiration from whole-soil respiration along an elevation zone in a Rocky Mountain mixed-conifer. Can J For Res 44:432–440

Berryman EM, Barnard HR, Adams HR, Burns MA, Gallo E, Brooks PD (2015) Complex terrain alters temperature and moisture limitations of forest respiration across a semiarid to subalpine gradient. J Geophys Res-Biogeo 120:707–723

Bond-Lamberty B, Thomson A (2010) A global database of soil respiration data. Biogeosciences 7:1915–1926

Chang SC, Tseng KH, Hsia YJ, Wang CP, Wu JT (2008) Soil respiration in a subtropical montane cloud forest in Taiwan. Agric For Meteorol 148:788–798

Cheng CH, Chen YS, Huang YH, Chiou CR, Lin CC, Menyailo OV (2013a) Effects of repeated fires on ecosystem C and N stocks along a fire induced forest/grassland gradient. J Geophys Res-Biogeo 118:215–225

Cheng CH, Hung CY, Chen CP, Pei CW (2013b) Biomass carbon accumulation in aging Japanese cedar plantations in Xitou, central Taiwan. Bot Stud 54:60

Cheng CH, Huang YS, Menyailo OV, Chen CT (2016a) Stand development and aboveground biomass accumulation in cropland afforestation in Taiwan. Taiwan J For Sci 31:105–118

Cheng CH, Hsiao SC, Huang YS, Hung CY, Pai CW, Chen CP, Menyailo OV (2016b) Landslide-induced changes of soil physicochemical properties in Xitou, Central Taiwan. Geoderma 265:187–195

Davidson EA, Janssens IA (2006) Temperature sensitivity of soil carbon decomposition and feedbacks to climate change. Nature 440:165–173

Davidson EA, Belk E, Boone RD (1998) Soil water content and temperature as independent or confounded factors controlling soil respiration in a temperate mixed hardwood forest. Glob Change Biol 4:217–227

Davidson EA, Verchot LV, Henrique Cattânio J, Ackerman IL, Carvalho JEM (2000) Effects of soil water content on soil respiration in forests and cattle pastures of eastern Amazonia. Biogeochemistry 48:53–69

Hashimoto S, Tanaka N, Suzuki M, Inoue A, Takizawa H, Kosaka I, Tanaka K, Tantasirin C, Tangtham N (2004) Soil respiration and soil CO_2 concentration in a tropical forest, Thailand. J For Res 9:75–79

Hsieh IF, Kume T, Lin MY, Cheng CH, Miki T (2016) Characteristics of soil CO_2 efflux under an invasive species, Moso bamboo, in forests of central Taiwan. Tree 30:1749–1759

Janssens IA, Lankreijer H, Matteucci G, Kowalski AS, Buchmann N, Epron D, Pilegaard K, Kutsch W, Longdoz B, Grunwald T, Montagnani L, Dore S, Rebmann C, Moors EJ, Grelle A, Rannik U, Morgenstern K, Oltchev S, Clement R, Gudmundsso J, Minerbi NS, Berbigier P, Ibrom A, Moncrieff J, Aubinet M, Bernhofer C, Jensen NO, Vesala T, Granier A, Schulze ED, Lindroth A, Dolman AJ, Jarvis PG, Ceulemans R, Valentini R (2001) Productivity overshadows temperature in determining soil and ecosystem respiration across European forests. Glob Change Biol 7:269–278

Kane ES, Pregitzer KS, Burton AJ (2003) Soil respiration along environmental gradients in Olympic National Park. Ecosystems 6:326–335

Kao WY, Chang KW (2009) Soil CO2 efflux from a mountainous forest-grassland ecosystem in central Taiwan. Bot Stud 50:337–342

Katayama A, Kume T, Komatsu H, Ohashi M, Nakagawa M, Yamashita M, Otsuki K, Suzuki M, Kumagai T (2009) Effect of forest structure on the spatial variation in soil respiration in a Bornean tropical rainforest. Agric For Meteorol 149:1666–1673

Knowles JF, Blanken PD, Williams MW (2015) Soil respiration variability across a soil moisture and vegetation community gradient within a snow-scoured alpine meadow. Biogeochemistry 125:185–202

Law BE, Ryan MG, Anthoni PM (1999) Seasonal and annual respiration of a ponderosa pine ecosystem. Glob Change Biol 5:169–182

Lee PF, Ding TS, Hsu FH, Geng S (2004) Breeding bird species richness in Taiwan: distribution on gradients of elevation, primary productivity and urbanization. J Biogeogr 31:307–314

Lee MS, Mo WH, Koizumi H (2006) Soil respiration of forest ecosystems in Japan and global implications. Ecol Res 21:828–839

Liao C, Peng R, Luo Y, Zhou X, Wu X, Fang C, Cheng J, Li B (2008) Altered ecosystem carbon and nitrogen cycles by plant invasion: a meta-analysis. New Phytol 177:706–714

Lin YR, Cheng CH, Tseng TY (2011) Carbon sequestration potential of afforestation in the plain areas in Taiwan: examples from longterm abandoned orchards and afforested sites. Quart J Chin For 44:567–588 **(In Chinese)**

Litton CM, Giardina CP, Albano JK, Long MS, Asner GP (2011) The magnitude and variability of soil-surface CO2 efflux increase with mean annual temperature in Hawaiian tropical montane wet forests. Soil Biol Biochem 43:2315–2323

Lloyd J, Taylor JA (1994) On the temperature dependence of soil respiration. Funct Ecol 8:315–323

Raich JW, Schlesinger WH (1992) The global carbon dioxide flux in soil respiration and its relationship to vegetation and climate. Tellus 44:81–99

Raich JW, Tufekcioglu A (2000) Vegetation and soil respiration: correlations and controls. Biogeochemistry 48:71–90

Reichstein M, Rey A, Freibauer A, Tenhunen J, Valentini R, Banza J, Casals P, Cheng Y, Grunzweig JM, Irving J, Joffre R, Law BE, Loustau D, Miglietta F, Oechel W, Ourcival J, Pereira JS, Peressotti A, Ponti F, Qi Y, Rambal S, Rayment M, Romanya J, Rossi F, Tedschi V, Tirone G, Xu M, Yakir D (2003) Modeling temporal and large-scale spatial variability of soil respiration from soil water availability, temperature and vegetation productivity indices. Glob Biogeochem Cycles 17:1104

Riveros-Iregui DA, McGlynn BL, Emanuel RE, Epstein HE (2012) Complex terrain leads to bidirectional response of soil respiration to inter-annual water availability. Glob Change Biol 18:749–756

Saiz G, Byrne KA, Butterbach-Bahl K, Kiese R, Blujdea V, Farrell EP (2006) Stand age-related effects on soil respiration in a first rotation Sitka spruce chronosequence in central Ireland. Glob Change Biol 12:1007–1020

Sheng H, Yang Y, Yang Z, Chen G, Xie J, Guo J, Zou S (2010) The dynamic response of soil respiration to land-use changes in subtropical China. Glob Change Biol 16:1107–1121

Shutou K, Nakane K (2004) Change in soil carbon cycling for stand development of Japanese cedar (*Cryptomeria japonica*) plantations following clear-cuttting. Ecol Res 19:233–244

Tan ZH, Zhang YP, Liang N, Song QH, Liu YH, You GY, Li LH, Yu L, Wu CS, Lu ZY, Wen HD, Zhao JF, Gao F, Yang LY, Song L, Zhang YJ, Munemasa T, Sha LQ (2013) Soil respiration in an old-growth subtropical forest: patterns, components, and controls. J Geophys Res-Atmos 118:1–10

Tang XL, Zhou GY, Liu SG, Zhang DQ, Liu SZ, Li J, Zhou CY (2006) Dependence of soil respiration on soil temperature and soil moisture in successional forests in southern China. J Integr Plant Biol 48:654–663

Ward SE, Bardgett RD, McNamara NP, Ostle NJ (2009) Plant functional group identify influences short-term peatland ecosystem carbon flux: evidence from a plant removal experiment. Funct Ecol 23:454–462

12

Post genomics era for orchid research

Wen-Chieh Tsai[1,2,3], Anne Dievart[4,5], Chia-Chi Hsu[3], Yu-Yun Hsiao[2,3], Shang-Yi Chiou[3], Hsin Huang[3] and Hong-Hwa Chen[1,2,3*]

Abstract

Among 300,000 species in angiosperms, Orchidaceae containing 30,000 species is one of the largest families. Almost every habitats on earth have orchid plants successfully colonized, and it indicates that orchids are among the plants with significant ecological and evolutionary importance. So far, four orchid genomes have been sequenced, including *Phalaenopsis equestris*, *Dendrobium catenatum*, *Dendrobium officinale*, and *Apostaceae shengen*. Here, we review the current progress and the direction of orchid research in the post genomics era. These include the orchid genome evolution, genome mapping (genome-wide association analysis, genetic map, physical map), comparative genomics (especially receptor-like kinase and terpene synthase), secondary metabolomics, and genome editing.

Keywords: Comparative genomics, Genome editing, Genome evolution, GWAS, Orchidaceae, *Phalaenopsis*, Post genomics era, Receptor-like kinase, Secondary metabolomics, Terpene synthase

Background

Containing about 30,000 species, Orchidaceae plants account for one tenth of all angiosperms containing 300,000 species (Hsiao et al. 2011b). More than 70% of orchids are epiphytic growth with distinct physiological characteristics (Hsiao et al. 2011b). Orchids are highly adapted to unfavorable environment. They have colonized successfully every habitat on earth wherever the sun shines. According to the molecular clock revealed by whole genome sequencing of *Phalaenopsis equestris* (Cai et al. 2015), the emergence of orchids have occurred in the late Cretaceous (76 Mya), and allowed them to cross the border of Jurassic mass extinction (66 Mya). This is consistent to the observation of the date for the most ancient fossil record (76–84 Mya) (Ramirez et al. 2007). It has been speculated that the speciation rates of orchids are exceptionally high (Gill 1989), with the fact that, even now, new species of orchids are still recorded worldwide suggesting that the evolution of orchids has never ceased.

The enormous number of Orchidaceae species allows it divided into five subfamilies: Apostasioideae, Vanilloideae, Cypripedioideae, Orchidoideae, and Epidendroideae (Fig. 1). They are all extraordinary floral diversified, and this heterogeneity has been related to the specialized interaction between the pollinators and orchid flowers (Cozzolino and Widmer 2005). The unique features for orchids include the obligate interactions between orchids and mycorrhizal fungi (Otero and Flanagan 2006), with C3 and/or crassulacean acid metabolism photosynthesis (Mascher et al. 2017), and epiphytic growth forms (Silvera et al. 2009). Orchids have exclusive reproductive strategies contributing to their successful adaptation to their ecological exploitations (Yu and Goh 2000).

The orchid cultivation and hybridization in Taiwan is very popular to the worldwide orchid market. The elegant appearance, some even with charming fragrance, and prolonged long life for orchid flowers have promoted attractiveness of *Phalaenopsis* orchids among breeders, nurseries and customers. For the past 20 years, orchid researchers have devoted to establish the foundation of orchid genomics research. These include karyotype analysis (Kao et al. 2001; Lin et al. 2001), genome size analysis (Chen et al. 2013, 2014), establishment of expressed sequence tags (ESTs) (Hsiao et al. 2006), genomics and transcriptomics databases (Fu et al. 2011; Su et al. 2011, 2013; Tsai et al. 2013), bacterial artificial chromosome

*Correspondence: hhchen@mail.ncku.edu.tw
[3] Department of Life Sciences, National Cheng Kung University, Tainan 701, Taiwan
Full list of author information is available at the end of the article

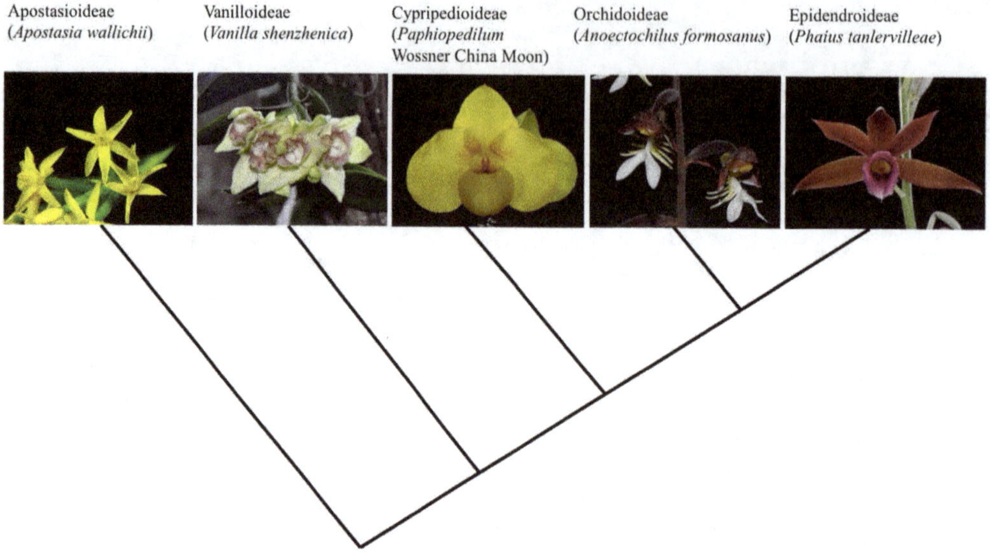

Fig. 1 The phylogenetic relationship among five subfamilies of Orchidaceae, and their example plants

(BAC) end sequencing (BES) (Hsu et al. 2011), chloroplast genome sequencing (Chang et al. 2006; Pan et al. 2012), miRNA (Lin et al. 2013a), and whole genome sequencing (Cai et al. 2015; Yan et al. 2015; Zhang et al. 2016a, 2017). In addition, orchid functional genomics studies has been available with virus-induced gene silencing using *Cymbidium* mosaic virus infectious clones for the assessing gene functions involved in flower color, floral morphogenesis and floral scent studies (Lu et al. 2007; Hsieh et al. 2013a, b). In micro-propagation, the induction of polyploidy has been developed to circumvent the hybrid incompatibility (Sattler et al. 2016).

In this article, we review the relevant orchid research progress and the future directions of the orchid investigation at the post genomics era. These include the orchid genome evolution, genome mapping, comparative genomics, secondary metabolomics, and genome editing.

Orchid genome evolution

Genome size variation

Angiosperms have variable genome sizes ranging nearly 2400-fold from *Genlisea margaretae* (Lentibulariaceae) with just 0.065 pg to the huge genome in *Paris japonica* (Melanthiaceae) with 152.23 pg. Many species with large genomes are observed in monocots, such as species in Alliaceae, Asparagaceae, Liliaceae, Melanthiaceae and Orchidaceae (Leitch et al. 2009). Among these, Orchidaceae with the genome sizes ranging 168-fold (1C = 0.33–55.4 pg) are perhaps the most diverse angiosperm families (Leitch et al. 2009).

As the species-richest subfamily, Epidendroideae with genome contents ranging over 60-fold (1C = 0.3–19.8 pg) harbor the most variable genome size in Orchidaceae. Orchidoideae, where the largest descending/offspring-ing from species in subtribe Orchidinae, are pictured by a more restricted range of genomes (1C = 2.9–16.4 pg) varying not more than sixfold. Cypripedioideae show genome sizes ranging only tenfold (1C = 4.1–43.1 pg). Cypripedioideae contain the largest mean genome size (1C = 25.8 pg) among all the subfamilies. Only genome size of few species in Vanilloideae was estimated, ranging from 1C = 7.3 to 55.4 pg. In this subfamily, *Pogonia ophioglossoides* presents the largest genome size (1C = 55.4 pg) (Leitch et al. 2009). Apostasioideae, the primitive subfamilies, contain calculated 1C-values ranging from 0.38 pg in *Apostasia nuda* to 5.96 pg in *Neuwiedia zollingeri* var. *javanica*, a close to 16-fold range (Jersáková et al. 2013). *P. equestris* and *P. aphrodite* subsp. *formosana*, the two native *Phalaenopsis* species usually be used as parents for breeding in Taiwan, respectively have a relative small genome size of 1.6 and 1.4 pg/1C (Chen et al. 2013, 2014).

Recent progress of orchid transcriptomic sequencing

High-throughput EST sequencing provides a gateway into the genome by reasons of the many data covered in the genomewide expression information. Before the next generation sequencing technologies (Zhang et al. 2013) were developed, the most popular sequencing method was the Sanger method, which was applied to the EST sequencing project. 1080 subtractive ESTs were obtained from an *Oncidium* Gower Ramsey pseudobulb subtractive EST library (Tan et al. 2005). Most ESTs were revealed as being related to carbohydrate metabolism and regulatory function, biosynthesis of mannose, pectin and starch, stress-related, and transportation (Tan et al. 2005). To illustrate expressed genes in reproductive organs of *Phalaenopsis*, *P. equestris* mature flower buds were collected and 5593 ESTs were sequenced and annotated (Tsai et al. 2006). In addition, 2359 ESTs were sequenced from scented *P. bellina* flower buds cDNA library to deduced ESTs involved in scent biosynthesis pathway (Hsiao et al. 2006).

The sudden rise of rapid and low-cost next-generation sequencing technologies is substantially promoting our competence to examine the sequences information comprehensively at unparalleled resolution and depth in a cell (Delseny et al. 2010). The technologies were adopted rapidly for orchid transcriptome analysis (Table 1). 454 technology was independently applied to generate 8233 contigs and 34,630 singletons sequenced from the mixed tissues of three *Phalaenopsis* species (Tsai et al. 2013), and 50,908 contig sequences released from six different tissues of *O.* Gower Ramsey (Chang et al. 2011). These data set expansively increased information of expressed genes in *Phalaenopsis* and *Oncidium* and speed identifying sets of ESTs associated with a broad range of biological processes (Chang et al. 2011; Hsiao et al. 2011a; Huang et al. 2015). A total of 121,917 unique ESTs were obtained from the *Ophrys* species by using 454 pyrosequencing and Illumina (Solexa) technologies to identify genes responding to pollinator attraction (Sedeek et al. 2013). A traditional Chinese herb together with 454 pyrosequencing and Illumina technology were able to generate plentiful ESTs for mining the genes participated in alkaloid biosynthetic pathway and polysaccharide biosynthesis in *Dendrobium officinale* (Guo et al. 2013; Zhang et al. 2016b). To provide a general resource for studying the pod development of *Vanilla planifolia*, one of the most valued flavour species for its flavour qualities, the combined 454 and Illumina RNA-seq technologies produced de novo transcriptome with high quality assembly for this important orchid cash crop (Rao et al. 2014). In addition, to improve the horticultural value of *Phalaenopsis* and *Cymbidium*, transcriptome derived from browning leaf of *Phalaenopsis* explant (sequencing by Illumina HiSeq 2000), and variable colour of *Cymbidium* leaf (sequencing by 454 pyrosequencing) were investigated (Xu et al. 2015; Zhu et al. 2015). To study the symbiotic orchid–fungus relationship and the molecular mechanism of orchid seed germination, 454 and Illumina were adopted to explore transcriptomes derived from *Serapias vomeracea* (Perotto et al. 2014), *C. hybridum* (Zhao et al. 2014), *Anoectochilus roxburghii* (Liu et al. 2015), and *Gastrodia elata* (Tsai et al. 2016).

In Orchidaceae, about 40% species adopt crassulacean acid metabolism (Mascher et al.) to fix carbon dioxide suggesting Orchidaceae to be the largest CAM clade in plants (Silvera et al. 2009). To illuminate the origin and evolution of CAM pathway, transcriptomes derived from leaves of CAM orchids *P. equestris*, *D. terminale* and *C. mannii* were sampled at different time interval and sequenced by Illumine HiSeq 2000 (Deng et al. 2016; Zhang et al. 2016c). Their results showed that key carbon fixation pathway genes might primarily evolve by changes at the transcription level in CAM plants. Several techniques were applied to study the development of spectacular orchid flower morphology. This includes the developing floral transcriptomes originating from *Phalaenopsis* (Hsiao et al. 2011a), *Cymbidium* (Zhang et al. 2013; Li et al. 2014; Yang and Zhu 2015), and *Orchis* (De Paolo et al. 2014). The molecular model of MADS-box genes associated with floral development was proposed and discussed.

Recently, root transcriptome from *Paphiopedilum concolor* was also produced to explore genes involved in orchid root development (Li et al. 2015b). Over 1195 unique genes participating in secondary metabolic pathways, and 609 ESTs involved in plant hormone biosynthesis, and plant signal transduction were revealed. The accumulated transcribed sequences could be directly used to develop microarray platform. It is also a resource for phylogenetic analysis. For example, an developed oligomicroarray harboring 14,732 unique expressed sequences based on the information of ESTs collected from *Phalaenopsis* orchids was applied to compare transcriptomes among sepal, petal, and labellum (Hsiao et al. 2013). Three hundred fifteen single-copy orthologs characterized in the transcriptomes of 10 species distributed in all five subfamilies of Orchidaceae were used to investigate the phylogenetic association of orchids (Deng et al. 2015). The results indicated this strategy appeared to be more reliable and efficient than using a few markers of genes for phylogenic analyses, particularly for those orchids whose DNA sequences are difficult to be amplified or the holomycotrophic species (Deng et al. 2015).

NGS technologies are not only applied to characterize orchid transcriptomes but also used to systematically analyze small RNAs in orchids (Table 1). The roles of small RNAs were studied on the regulation of flowering in *P. aphrodite* and *Erycina pusilla* (An et al. 2011; An and Chan 2012; Chou et al. 2013; Lin et al. 2013a), flower development in *Orchis italica* and *Cymbidium ensifolium* (Aceto et al. 2014; De Paolo et al. 2014; Li et al. 2015c), and interaction between the fungus *Piriformospora indica* with an *Oncidium* hybrid orchid (Ye et al. 2014). Later, comprehensive collection of small RNAs derived from *P. aphrodite* (Chao et al. 2014), and *D. officinale* (Meng et al. 2016) were performed. These efforts provided valuable messages about the expression,

Table 1 Characteristics of findings in the literature for the application of next generation sequencing (NGS) to orchid transcriptomes

Subfamily	Species	Sequencing platform	Tissue	Study aim	References
Apostasioideae	*Apostasia shenzhenica*	Illumina/Solexa	Mature flower buds	Study of floral development and evolutionary trends of orchid flowers	Tsai et al. (2013)
Apostasioideae	*Neuwiedia malipoensis*	Illumina/Solexa	Mature flower buds		
Vanilloideae	*Vanilla shenzhenica*	Illumina/Solexa	Mature flower buds		
Vanilloideae	*Galeola faberi*	Illumina/Solexa	Mature flower buds		
Cypripedioideae	*Paphiopedilum armeniacum*	Illumina/Solexa	Mature flower buds		
Cypripedioideae	*Cypripedium singchii*	Illumina/Solexa	Mature flower buds		
Orchidoideae	*Habenaria delavayi*	Illumina/Solexa	Mature flower buds		
Orchidoideae	*Hemipilia forrestii*	Illumina/Solexa	Mature flower buds		
Epidendroideae	*Phalaenopsis equestris*	Illumina/Solexa	Mature flower buds		
Epidendroideae	*Cymbidium sinense*	Illumina/Solexa	Mature flower buds		
Vanilloideae	*Vanilla planifolia*	Illumina/Solexa Roche/454	Pod tissues, seeds	Study of biosynthetic routes to flavor components	Rao et al. (2014)
Cypripedioideae	*Paphiopedilum concolor*	Illumina Hiseq 2000	Roots	Identify the genes that control root growth and development	Li et al. (2015b)
Orchidoideae	*Ophrys* species	Roche/454 Illumina/Solexa	Flowers, labellums, leaves, flower organ from open flowers and buds	Identify genes responding for pollinator attraction	Sedeek et al. (2013)
Orchidoideae	*Orchis italica*	Illumina/Solexa (MiSeq)	Inflorescences	The roles of small RNAs on the flower development	Aceto et al. (2014)
Orchidoideae	*Orchis italica*	Illumina Hiseq 2500	Florets of inflorescence before anthesis	Analyzing transcripts potentially involved in flower development	De Paolo et al. (2014)
Orchidoideae	*Serapias vomeracea*	Roche/454	Protocorms	Investigate the molecular bases of the orchid response to mycorrhizal invasion	Perotto et al. (2014)
Orchidoideae	*Anoectochilus roxburghii* (Wall.) Lindl.	Illumina HiSeq 4000	Dry seeds, seeds from asymbiotic or symbiotic germination	Study of seed germination process	Liu et al. (2015)
Orchidoideae	*Gastrodia elata* Blume	Illumina Hiseq 2000	Vegetative tissues, corms, juvenile tubers	Address the gene regulation mechanism in gastrodin biosynthesis	Tsai et al. (2016)
Epidendroideae	*Phalaenopsis aphrodite*	Sanger: EST	Protocorms	Gene discovery and genomic annotation	Fu et al. (2011) , Hsiao et al. (2011a)
Epidendroideae	*Phalaenopsis equestris*	Sanger: EST	Mature flower buds		
Epidendroideae	*Phalaenopsis bellina*	Sanger: EST	Mature flower buds without column		
Epidendroideae	*Phalaenopsis aphrodite* *Phalaenopsis equestris* *Phalaenopsis bellina*	Roche/454	Mixed tissues		
Epidendroideae	*Phalaenopsis equestris*	Illumina/Solexa	Leaves		
Epidendroideae	*Phalaenopsis aphrodite*	Roche/454 Illumina/Solexa	Leaves, stems, roots, young inflorescences, stalks, flower buds, flowers, germinating seeds	Investigate expressed genes involved in many biological processes of orchids	Su et al. (2011)

Table 1 continued

Subfamily	Species	Sequencing platform	Tissue	Study aim	References
Epidendroideae	*Phalaenopsis aphrodite*	Illumina/Solexa	Leaves, stalks, flower buds	Study the roles of small RNAs on the regulation of flowering	An et al. (2011), An and Chan (2012)
Epidendroideae	*Phalaenopsis aphrodite*	Illumina/Solexa	Leaves, roots, flowers, germinating seeds, young inflorescences	Identify species- and tissue-specific miRNAs	Chao et al. (2014)
Epidendroideae	*Phalaenopsis* Brother Spring Dancer 'KHM190'	Illumina Hiseq 2000	Petals, sepals or labellums from flower buds of wild-type and peloric petal mutant plants	Study regulation of floral- organ development	Huang et al. (2015)
Epidendroideae	*Phalaenopsis sp.*	Illumina Hiseq 2000	Explants	Examine *Phalaenopsis* leaf explant browning	Xu et al. (2015)
Epidendroideae	*Oncidium* 'Gower Ramsey'	Roche/454	Leaves, pseudobulbs, young inflorescences, inflorescences, flower buds, mature flowers	Identify genes associated with flowering time	Chang et al. (2011)
Epidendroideae	*Oncidium* 'Gower Ramsey'	Illumina/Solexa	Roots with or without fungus	Study the roles of small RNAs on the interaction between root and the fungus	Ye et al. (2014)
Epidendroideae	*Erycina pusilla*	Illumina/Solexa	Roots, leaves, peduncles, flowers, capsules	Investigate photoperiod-dependent flowering genes	Chou et al. (2013)
Epidendroideae	*Erycina pusilla*	Illumina/Solexa	Roots, leaves, peduncles, flowers, capsules	Study the roles of small RNAs on the regulation of flowering	Lin et al. (2013a)
Epidendroideae	*Cymbidium ensifolium* 'Tiegusu'	Illumina HiSeq 2000	Flower buds, mature flower	Identify genes associated with floral development	Li et al. (2014)
Epidendroideae	*Cymbidium sinense* 'Qi Jian Bai Mo'	Illumina HiSeq 2000	Plants in vegetative phase/floral differentiation phase/reproductive phase	Identify genes associated with floral development	Zhang et al. (2013)
Epidendroideae	*Cymbidium hybridum* 'Golden Boy'	Illumina Hiseq 2000	Roots with or without fungus	Study of orchid-mycorrhizal fungi interactions	Zhao et al. (2014)
Epidendroideae	*Cymbidium ensifolium* 'Tiegusu'	Illumina/Solexa	Flower bud	Identify miRNAs related to floral development	Li et al. (2015c)
Epidendroideae	*Cymbidium ensifolium* 'tianesu'	Roche/454	Sepals, petals, labellums, gynostemia from flower buds and mature flowers	Reveal genes associated with floral organ differentiation	Yang and Zhu (2015)
Epidendroideae	*Cymbidium sinense* 'Dharma'	Roche/454	Roots, leaves, pseudobulbs, flowers	Analyze molecular mechanism underlying leaf-color variations	Zhu et al. (2015)
Epidendroideae	*Cymbidium sinense* *Cymbidium atropurpureum* *Cymbidium mannii*	Illumina Hiseq 2000	Leaves	Explore the evolution and molecular regulation of CAM plants	Zhang et al. (2016c)
Epidendroideae	*Dendrobium officinale*	Roche/454	Stems	Study of alkaloid biosynthesis	Guo et al. (2013)
Epidendroideae	*Dendrobium officinale*	Illumina Hiseq 2000	Juvenile and adult plants	Identify genes associated with polysaccharide synthesis	Zhang et al. (2016b)
Epidendroideae	*Dendrobium officinale*	Illumina Hiseq 2500	Flower, roots, leaves, stems	Study of the regulatory networks of the production and accumulation of the medicinal constituents	Meng et al. (2016)

composition and function of small RNAs which help us to better understand functional genomics of orchids.

For the storage and manage the massive expressed gene sequences from orchid, OrchidBase was developed to collect the transcriptomic sequences from 11 various tissues/organs of *Phalaenopsis* spp. and flower tissues of 10 species distributed in five subfamilies of Orchidaceae (Fu et al. 2011; Tsai et al. 2013; Niu et al. 2016). Both deep sequencing with ABI 3730, Roche 454 and Illumina/Solexa were applied to generate EST sequences collected in OrchidBase. OrchidBase is generously accessible at http://orchidbase.itps.ncku.edu.tw/. The database delivers a prominent feature of genetic resource for both data mining and experimental researches of orchid biology and biotechnology. Orchidstra (http://orchidstra.abrc.sinica.edu.tw), another orchid transcriptomic database, was developed to collect 233,924 unique contigs of *P. aphrodite* transcriptomic sequences by use of a Illumina/Solexa and Roche 454 platform. Profiling analysis with RNA-Seq was applied to categorize the genes with tissue-tropism expression patterns (Su et al. 2011, 2013). In addition, 50,908 contigs of sequences generated by using Roche 454 platform from various organs of *Oncidium* were assembled into the OncidiumOrchidGenomeBase (http://predictor.nchu.edu.tw/oogb/) (Chang et al. 2011). These EST dataset are valuable for the identification of specificity of orchids, annotation of genes for genomic sequencing, and assistance in the organization of the orchid genome.

Current status of orchid genome sequencing

With the quick development and lower cost of NGS, whole genome sequencing of non-model species can be implemented. The first milestone was the sequencing of the *Phalaenopsis equestris*, a tropical epiphytic orchid and recurrently be used as parent species for orchid breeding (Cai et al. 2015). The *P. equestris* genome was whole-genome shotgun and sequenced by Illumina technology. The genome size was estimated to 1.16 Gb containing 29,431 predicted protein-coding genes (Cai et al. 2015). Analysis of the *P. equestris* genome showed that the majority of the genome (about 62%) was occupied by repetitive DNAs, mostly transposable elements (TEs). In addition, before the radiation of most of the orchid clades, an orchid-specific paleopolyploidy event was discovered. This species is also the first CAM plant that has been whole-genome sequenced, and the gene family (α carbonic anhydrase) involved in CAM pathway has an obvious expansion. The result suggested that the evolution of CAM photosynthesis in *P. equestris* might associate with gene duplication. In addition, genes located at the heterozygous regions might relate to self-incompatibility. Genes in type II MADS-box clades, including the E-class, C/D-class, B-class AP3 and AGL6 clades, contain extra orthologs than other plant species. These expanded clades are involved in orchid floral organ development that can provide the unique evolutionary paths of these floral organ identity genes accorded with the innovative development of lip and column in orchids. Furthermore, the *Phalaenopsis* genome sequence was practical to identify MYB genes controlling floral pigmentation patterning (Hsu et al. 2015) and TCP genes participated in the ovule development (Lin et al. 2016).

Dendrobium officinale, with both ornamental value and therapeutic effects, is the second sequenced orchid plant by joining both the NGS Illumina Hiseq 2000 and the third-generation PacBio machineries (Yan et al. 2015). The assembled genome of *D. officinale* had a predicted gene number of 35,567, and that was higher than that in *Phalaenopis*. For example, the number of B-class MADS-box genes presented in *D. officinale* was much higher than that in *Phalaenopsis* with its four members in B-class AP3-like subfamily, and one member in B-class PI-like subfamily. In contrast, 19 AP3-like genes and five PI-like genes were present in this *Dendrobium* genome. It is possible that the *D. officinale* plants used for the whole genome sequencing were hybrids rather than a native species. Later, another *Dendrobium* species (*Dendrobium catenatum*) was whole genome sequenced by Illumina HiSeq 2000 platform (Zhang et al. 2016a). The predicted 28,910 protein-coding genes were comparable with those of *Phalaenopsis*, and a whole genome duplication event could be shared with *Phalaenopsis*. Adaptation to a wide range of ecological niches of *Dendrobium* might relate to the expansion of many resistance-related genes. In addition, it seems that extensive duplication of genes encoding glucomannan synthase associated with the synthesis of medicinal polysaccharides. It also found that MADS-box gene clades ANR1, StMADS11, and MIKC* were expanded in the *Dendrobium* suggesting the function of these clades might relate to the astonishing diversity of plant architecture in *Dendrobium* (Zhang et al. 2016a).

Most recently, the primitive orchid *A. shenzhenica*, a representative of one of two genera that form a sister lineage to the rest of the Orchidaceae, was whole genome sequenced to provide a reference for improving our understanding of orchid origins and evolution (Zhang et al. 2017). The *A. shenzhenica* genome was sequenced by use of a combination of different approaches including Illumina, PacBio and 10× genomics technologies. The total length of the genome assembly was 349 Mb containing 21,841 protein-coding genes. *A. shenzhenica* shows changes within MADS-box gene classes, which control a diverse suite of developmental processes, during orchid evolution. This study provides new insights on the genetic evidence underlying key orchid innovations, including the development of the labellum and gynostemium, pollinia, and seeds without endosperm, as well as the evolution of epiphytism (Zhang et al. 2017).

Whole-genome duplication

A striking feature of plant genomes is the whole genome duplication (WGD) has occurred for many times (Cui et al. 2006; Van de Peer et al. 2009). WGD is a major evolutionary force in angiosperm genomes. The complete genome sequences of angiosperm trees have provided information on polyploidy and genome evolution. It is now generally recognized that in the predecessor of all seed plants, one WGD occurred, and in the ancestor of all angiosperms, an extra one happened (Jiao et al. 2011). Furthermore, a hexaploidy event preceded the radiation of core eudicots (Jiao et al. 2012). Whereas a WGD shared by most of the monocots was also suggested (Paterson et al. 2004). Based on the *Phalaenopsis* and *Dendrobium* genome sequences, two turns of WGDs in the *D. catenatum* lineage was inferred from the examination of both the allocation of synonymous substitutions per synonymous site (Ks) throughout all paralogous genes and for duplicated genes sitting in synteny blocks (Zhang et al. 2016a). This most recent WGD event is shared with *Dendrobium*, *Phalaenopsis*, and *Apostasia* and might occur near the Cretaceous–Paleogene (K/Pg) boundary. Putative peaks at older Ks age distribution might point to further ancient WGD events in the monocot lineage that could be shared by monocot ancestors (Cai et al. 2015; Zhang et al. 2016a, 2017).

Since it is suggested that species diversification might be facilitated by WGDs (Van de Peer et al. 2009), it is intriguing to perceive whether the WGD is common with the subfamily Orchidoideae containing 3630 species, and the subfamilies Cypripedioideae, and Vanilloideae, which include 180, and 185 species, respectively. Orchidoideae separated from the Epidendroideae (about 20,000 species including *P. equestris* and *D. catenatum*) about 59 million years ago (Gustafsson et al. 2010). Cypripedioideae and the ancestor of Orchidoideae and Epidendroideae subfamilies are predicted to have disconnected from each other around 68 million years ago (Gustafsson et al. 2010). Fortunately, extensive transcriptomic data sets and/or whole genome sequences from members of these other subfamilies are ongoing. It will be optimistic for uncovering the mystery of orchid genome evolution after more orchid genomic and transcriptomic data obtained.

Transposable elements

Complements of genomic elements and extraordinary variations in sizes were observed in the genomes of flowering plant containing approximately 300,000 species (Wendel et al. 2016). An obvious route to genome expansion is WGD events, but many species with great genomes are diploid. For example, *Oryza australiensis* and *O. sativa* are from the same genus, however, the former is more than twice the size of the latter due to the addition of ~ 400 Mbp of DNA majorly by three individual retrotransposable element families in the past few million years (Piegu et al. 2006). And thus, the dynamics of transposable element proliferation and clearance might play the important role in the majority of variation in plant genome size, which is in turn superimposed by the history of WGD (Bennetzen and Wang 2014).

The assembled *P. equstris* genome is composed of about 62% repetitive DNA. This is relatively higher than those found in rice (29%) and grape (41%), but similar in sorghum (61%). About 59% of the genome is occupied by interspersed repeats and TEs and 3% is accounted for tandem repeats. Among the TEs, ~ 46% of the genome is occupied by long terminal repeats (LTRs). Long interspersed nuclear elements (LINEs) account for ~ 8% of the genome (Cai et al. 2015). Most LTRs (71%) appeared in the *P. equstris* genome during 11.7–43 million years ago, far behind the origin of the last common ancestor of orchids (74–86 million years ago). *Copia* and *Gypsy* TEs might experience a recent burst (Cai et al. 2015).

About 78.1% of the *D. catenatum* genome is occupied by 789 Mb repetitive elements. Retrotransposable elements constitute a huge portion of the *D. catenatum* genome including LINE/RTE (5.68%), LINE/L1 (8.44%), LTR/Gypsy (18.49%), and LTR/Copia (27.36%). In addition, the quantity of de novo predicted repeats was predominantly more than that composed based on Repbase, demonstrating that *D. catenatum* has arose numerous unique repeats contrasted to other sequenced plant genomes (Zhang et al. 2016a). During the last 5 million years, a burst of LTR activity was observed, suggesting that these LTRs were integrated into the genome of *D. catenatum* after it was separated from *P. equestris* (22.6–59.6 million years ago) (Zhang et al. 2016a).

Angiosperms have a high frequency of hybridization and polyploidy, with resultant increased TE activity and genome shuffling, and profitable gene duplication (Oliver et al. 2013). TE activity enhanced genomic plasticity provides a more satisfactory and complete explanation to "abominable mystery" described by Darwin. Although explanations for orchid diversity are persuasive and valid, the extreme diversity of orchids still cannot fully understand. However, due to a few of genomic data of orchids, the role of TE involved in orchid diversity and evolution has not been accessed yet. Probably, artificial arena of *Phalaenopsis* orchid breeding might supply direct and relatively recent clues for the pivotal of TEs in the generation of selectable cultivars in *Phalaenopsis* orchids.

Genome mapping

Several fundamental studies are necessary to well-assemble the whole-genome sequences. These studies include high-density genetic map and bacteria artificial chromosomes (BACs) or yeast artificial chromosomes (YACs)-based physical map.

Genetic map

Even though the whole-genome sequence of *Phalaenopsis equestris* is published (Cai et al. 2015), the current scaffold number of *P. equestris* genome is still high (236,185 scaffolds), and the assembly of *P. equestris* scaffolds remains to be improved. A saturated genetic linkage map is regarded as an effective tool to facilitate scaffolds assembly. Besides, genetic mapping is referred to a step-in genome mapping, which can integrate with a physical map and a cytogenetic map to access genes in genome. For many important crops, genome mapping using genetic linkage maps, physical maps and cytogenetic maps are well developed, such as rice, wheat, barley, etc. (Kurata et al. 1994; Ramsay et al. 2000; Hearnden et al. 2007; Li et al. 2015a). In Orchidaceae, the genetic linkage maps have been constructed for the medicinal orchids, *Dendrobium* (Xue et al. 2010; Lu et al. 2012a, b; Feng et al. 2013), but not for the ornamental orchids, *Phalaenopsis*. Hence, it is vital needed to construct a saturated genetic map for *Phalaenopsis*.

Usually, a genetic linkage map is constructed based on the crossing over events during meiosis. By estimating the recombination frequency between molecular markers, linkage groups can be built while several markers are linked together. A genetic linkage map is comprised of the numbers of linkage groups corresponding to the chromosome numbers.

First, to identify the crossing over events, one of the significant factors is a segregated mapping population. For constructing genetic maps in rice and wheat, F_2 population or recombinant inbred lines (RILs) are often developed as suitable mapping populations (Kurata et al. 1994; Li et al. 2015a). However, unlike other crops, *Phalaenopsis* orchids have long juvenile periods. It usually takes 2–3 years to generate a new generation. Consequently, it is inefficient and time-consuming to generate *Phalaenosis* F_2 population or RILs. As an alternative approach, an outcrossing F_1 population is used as a mapping population. Yet, the segregation pattern may be more complex due to the segregated alleles that may vary in number. Currently, a F_1 population with 116 progenies from the cross between *P. aphrodite* subsp. *formosana* and *P. equestris*, the two native species in Taiwan and important parents in breeding, has been adopted as the mapping population for constructing a genetic map for *Phalaenopsis*.

After determination of the mapping population, various types of molecular markers are applied for linkage analysis. For instance, restriction fragment length polymorphisms (RFLPs), amplified fragment length polymorphisms (AFLPs), expressed sequence tags (ESTs), and simple sequence repeats (SSRs) are widely utilized in the constructing genetic maps for several crops. Among them, SSR markers particularly have benefits for classification and mapping of genes due to their extraordinary reproducibility, co-dominant inheritance, relative large quantity, extreme polymorphism, simplicity of genotyping and easily amplified through PCR (Varshney et al. 2005). Previously, 950 SSRs from *P. equestris* have been identified, and among them 206 SSRs primer sequences were developed for *Phalaenopsis* genetic mapping (Hsu et al. 2011). *P. equestris* contains 38 chromosomes ($2N = 2X = 38$) and its genome size is 1600 Mb per haploid genome. It is a relatively large genome compared to other crops, *ex.:* rice 400–430 Mb, and sorghum 750–770 Mb (Eckardt 2000). Hence, development of abundant molecular markers for *Phalaenopsis* genome mapping is necessary. Nowadays, as NGS technologies are rapidly developed and spread, huge amount of binary single nucleotide polymorphism (SNP) markers is easily generated. Due to this characteristic of SNP markers generation, constructing a saturated genetic map in a relative short time is feasible.

After a saturated, high-density *Phalaenopsis* genetic map is plotted, it is capable of sequence alignment and boosting *P. equestris* scaffolds assembly. In addition, a genetic linkage map can be adopted for useful trait analysis and applied for quantitative trait loci (QTL) mapping. Combining the phenotypic data of the mapping population, QTLs may be able to be mapped on the saturated genetic linkage map. Moreover, after QTL mapping, the molecular markers corresponding to QTLs will be adopted for marker-assisted selection (MAS) for interested agronomic traits. For orchid marketing, *Phalaenopsis* breeders usually constant cross or backcross for about 5–10 generations to select potentially commercial cultivars. Considering 2–3 years for each generation, a period up to 20–30 years is required to breed valuable and popular cultivars, such as *P.* Sogo Yukidian "V3" and *P.* I-Hsin Sesame, for markets. By applying the saturated genetic map to MAS, the time-consuming breeding process will be shortened. Thus, a saturated genetic map is considered as the first step for breeding by design and to reveal interested traits controlled by QTLs and genes.

Physical map

The construction of physical map is using restriction enzymes (RE) digestion of large insert clone libraries, such as BACs and YACs (Chaney et al. 2016). For example, the BAC libraries are treated with one RE, and the digested fragment patterns are analyzed by using electrophoresis. Different BAC clones with similar fragment patterns mean that they contain overlapping regions and can be assembled together to form fingerprinting contigs (FPCs). Therefore, physical map provides a framework for accelerated position cloning of genes for important agronomic traits, and assembly of genome sequence. With the integration of genetic map, the molecular markers identified on the genetic map can be aligned with physical map, and show the physical location of these markers.

Alternatively, optical mapping is another approach to construct a physical map. Similar to FPCs, optical mapping uses RE to digest DNA and stained with an intercalating dye. Then DNA fragments are imaged, separated according to their sizes, and then ordered with restriction map. Although optical mapping is effective, it is too complex and slow in lengthening the DNA, data imaging and processing. Hence, the optical mapping has mainly applied for small genome organisms, ex.: bacteria (Zhou et al. 2004; Haas et al. 2009). Further improvement of automatic technology for imaging and data processing is now available, so that optical mapping is granted for mapping of large genomes, such as human genome (Jo et al. 2009; Teague et al. 2010).

As an improved high throughput layout, the BioNano Genomics Irys system is much efficient than the traditional optical mapping. This system uses a modified RE to introduce DNA breakage on a single strand or incisions in DNA at sequences-specific site. The breaks are then labeled with fluorescence, and the DNA molecular is passed through automated electrophoresis system to form linearized DNA in nanochannels. For each molecular, the length and distances between labeled sites are measured and analyzed to create a molecular map. Two DNA molecules containing similar distance patterns are assembled into a large contig. Therefore, the BioNano Genomics Irys system is automated and more efficient for mapping complex plant genomes, including wheat and barley (Xiao et al. 2015; Stankova et al. 2016; Mascher et al. 2017). The physical mapping of *Phalaenopsis* using the upmost technology is currently underway.

Genome-wide association studies (GWAS)

Plant breeding is the practice to produce novel varieties with better phenotypes which are then applied in farming and husbandry, and for human utility (Pawełkowicz et al. 2016). The basic strategy for plant breeding is to select the best individuals and their progenies containing the desired agricultural traits. The association mapping strategy commonly evaluates the statistical analyses to calculate the significance of the association between various phenotypes and the genetic polymorphism in a set of defined persons with genetic variations (Ogura and Busch 2015).

In the past, a number of molecular markers have already been applied for genome mapping or association analysis, such as SSRs, AFLPs, and various kinds of SNPs. With the emergence of NGS technologies, genome-wide association studies (GWAS) is developed for the application of genetic variations that are largely and compactly distributed all over the genome. Therefore, GWAS has replaced the traditional mapping strategy with more opportunities to identify large collections of molecular markers and find new genes and regulatory sequences responsible for specific traits (Pawełkowicz et al. 2016).

With the great progresses in the methods of detecting trait differences and genetic variations, GWAS has emerged as a promising approach to inspect the relationships between complex genetic polymorphisms and distinctive phenotypes for most species (Slate et al. 2009). Over last decades, the employment of the automated image system encourages high-throughput phenotyping of a lots of plants by using large-scale image snapping and phenotypic trait quantification from the images (Ogura and Busch 2015). GWAS can combine the quantitative traits to their genetic variations in distinct individuals, and inspire the quantitative regulation of the growth and development.

As NGS technologies emerge, the genotyping-by-sequencing (GBS) approach rapidly gets hundreds of thousands to millions of markers to construct the high-resolution genetic map, and generates sufficient information and coverage of plant genomes (Edwards and Batley 2010). GWAS through the GBS approach provides major assistant for traditional crop improvement with the identification of great amount of SNP molecular markers and the production of genetic linkage maps with high-density markers (Poland and Rife 2012).

GBS method is a useful and powerful approach to rapidly get lots of markers to construct genetic maps with high resolution. This tactics generates sufficient information and coverage to the plant genomes (Edwards and Batley 2010). The first step of GBS is to digest genome DNAs from different plants in a population of genetically distinct individuals with an appropriate RE (Elshire et al. 2011). Each digested genome DNA is ligated with unique, short DNA sequences (barcodes adaptors) to make different plant DNAs be assembled and sequence determined together in a sole sequencing lane. Then the genome DNAs were recovered and prepared for NGS to characterize high amount of SNPs between each individual in this population, and these SNPs are applied as molecular markers for the construction of a genetic map with high density (Elshire et al. 2011).

There are three advantages by using GBS for genome-wide genetic analysis. First, the usage of unique short barcode for each plant sample makes all the DNAs be pooled together for sequencing. Second, an appropriate RE is necessary for getting highly reproducible results and important genome regions, as well as reducing the repetitive regions and genome complexity. Third, with the NGS technology, a high amount of SNP molecular markers will be identified from the sequencing data, and be applied to construct a genetic map with high density.

For example, by using the GBS approach, 41,371 SNP markers were identified in 254 advanced wheat breeding lines from the wheat breeding program of International Wheat and Maize Improvement Center (Poland et al. 2012b). In addition, 12.4 Gb of high-quality genome sequence and 129,156 SNP markers have been identified for potato. These SNPs were mapped to 2.1 Mb of potato reference genome, and the average read depth was 63× coverage per cultivar (Uitdewilligen et al. 2013). In durum wheat, 9983 putative SNP markers were characterized between their two parents and used for genotyping 91 RILs (van Poecke et al. 2013).

Moreover, GBS approach is a powerful method to develop high-density markers even lack of the sequenced reference genome or previous identification of DNA polymorphism. The high-density markers identified from GBS can be used to anchor the orders of physical maps as well as the whole genome shotgun sequence to improve the genome assembly (Poland et al. 2012a). The application of genomics-assisted breeding to valuable cash crops especially with complex genomes is a very important process. Similar transitions are found in rapeseed, lupin (Fabacea), lettuce, switchgrass (perennial warm season bunchgrass in north America), soybean, and maize (Bus et al. 2012; Poland and Rife 2012; Truong et al. 2012; Yang et al. 2012; Lu et al. 2013; Sonah et al. 2015). For example, in *Manihot esculenta* Crantz (ICGMC 2015), the complex 2412-cM map is incorporated into 10 biparental maps (containing 3480 meioses), and 22,403 genetic markers are anchored on 18 chromosomes (ICGMC 2015). This map was arranged 71.9% of the draft genome assembly and 90.7% of the predicted protein-coding genes (ICGMC 2015). It is beneficial with the chromosome-anchored genome sequence for breeding amendment by providing the prominent characterization of markers linked to chief traits.

In *Phalaenopsis* orchids, the GBS analysis has been carried out in the first generation offsprings (F_1) from the cross between *P. aphrodite* subsp. *formosana* (white-flower) and *P. equestris* (red-flower). This is beneficial to the assembly of whole-genome sequences of *P. equestris* into 19 linkage groups (unpublished data). The regulatory regions for various phenotypes in the F_1 plants, including plant size, flower color and size, have been investigated (unpublished data).

Most economically important traits are usually inherited in a quantitative manner, such as the height, weight, and stress resistance. The genetic basis of the quantitative traits are regulated by many genes, the synergistic effects of genes, and the mutual interactions between genes and environment. The major challenge to identify the QTL is to precisely locate the gene via high-density linkage map. This is an extremely costly and time-consuming process. GWAS with NGS technologies offer an opportunity for a powerful strategy to identify large numbers of SNP markers, and construct high-density genetic association map that is suitable for underling these important and complex QTLs. For example, complete genotyping of 2815 maize inbred accessions have characterized 681,257 SNP markers distributed across all over the maize genome. Among them, SNPs associated to the plausible candidate genes are identified for kernel color, sweetness, and flowering time (Romay et al. 2013).

The orchid floral scent trait could be regulated by QTL. Recently, GBS has been applied to study the deceptive orchid *Orphis*, and found several SNPs linked to odour related genes (Sedeek et al. 2014). The study of the ecological speciation was surveyed to understand the reproductive hurdles and variations in floral phenotypes in four closely related species, sexually deceptive *Ophrys* species with various flower morphology and distinct labellum coloration. Regardless of the flower odour chemistry may fundamentally cause the reproductive hurdles, GBS showed common polymorphism all over the *Ophrys* genome but highly distinguished polymorphisms in genes involved in floral odour biosynthesis (Sedeek et al. 2014). This finding suggests that these species are marked mostly by genic divergence rather than genome-wide variations. This result may be applied to other orchids with sexual deception, where floral odour genes may be amongst the first to differentiate (Sedeek et al. 2014).

Recently, the molecular omics, such as epigenome, transcriptome, proteome, and metabolome, have been used as complementary manners to improve the SNP-trait association studies. Epigenomic changes, that are responded to environment stimulate and reflect in phenotypic changes, provide another meaning of the "molecular phenotype" definition. Recent advances in population epigenomics studied the association between genetic SNPs and epigenomic variations in a global accession collection of *Arabidopsis thaliana* (Kawakatsu et al. 2016). Moreover, the transcriptomic variations, which reflect the variation in both genetic and epigenetic regulatory and other omics studies of the proteome and metabolome, have proved to be great resources as molecular phenotype (He et al. 2013).

Alternatively, it would also be valuable if the epigenome, transcriptome, proteome, and metabolome are applied as molecular markers, the "genotypes", to calculate their associations to the downstream phenotypic traits. For example, the differential transcriptomes from 368 maize diverse inbred lines was used to identify the expression presence/absence variation (ePAV; genes were expressed, or not) and served as as "phenotype" to perform association analysis with 15 agronomic phenotypes and 526 metabolic traits (Jin et al. 2016). As for the quantitative traits of QTLs, the genomic variations from these molecular omics are also quantitative and range mutable. However, they are not only binary or with limited numbers of alleles. Therefore, the quantitative GWAS (qGWAS) has recently been proposed to solve the continuous genotype issue. It has also been used to explore the regulatory network by treating expression level of genes as both "genotypes" and "phenotypes" (Wen et al. 2016).

The major goal for plant breeding is to select the best progenies with the desired traits. GWAS can guild the genetic variations with the high-density SNP markers for mapping the regulatory locus of these traits. In *Phalaenopsis* orchids, several agricultural traits have been emphasized, including flower sizes, length of floral inflorescence, flower colors and pigmentation patterning, and floral fragrances. With the GWAS analysis, it will be plausible to construct the high-density genetic map, and the regulatory regions responsive for the important traits can be identified. Moreover, this will be beneficial for the *Phalaenopsis* breeding for new varieties with the traits of interests.

Comparative genomics

Receptor-like kinases in orchid genomes

Phosphorylation is a reversible addition of phosphate to proteins. This post-translational modification is involved in all signaling pathways and every cellular activity in living organisms. In eukaryotes, this process is conducted by a superfamily of protein kinases (ePKs for eukaryotic protein kinases) which represent ~ 1.5 to 2.5% of all genes in average, making ePK one of the largest protein families in eukaryotes (Manning et al. 2002). In *Arabidopsis thaliana* and rice (*Oryza sativa*), the two model plant species, ~ 1000 and ~ 1500 Ser/Thr protein kinases have been identified, corresponding to 2.9 and 2.3% of the proteomes in respective organisms (http://plantsp.genomics.purdue.edu/index.html) (Shiu et al. 2004; Dardick et al. 2007).

According to the taxonomy of Hanks and Hunter and the studies based on kinase domain similarity and phylogeny of eukaryotes, the kinomes of many eukaryotic genomes have been divided into subfamilies (Hanks and Hunter 1995; Martin et al. 2009). In the plant kingdom, the largest subfamily is composed of plant-specific receptors named receptor-like kinase (RLK) (Shiu et al. 2004; Dardick et al. 2007). These receptors play important roles in signal transduction to relay external signals across cell membranes. They are involved in developmental processes and/or represent guard molecules which are able to recognize pathogen attacks in a process called pathogen-associated molecular patterns (PAMP)-triggered immunity (PTI) (Boller and Felix 2009; Gish and Clark 2011; Wu and Zhou 2013; Antolín-Llovera et al. 2014; Haruta and Sussman 2017). Many of the RLKs are also activated under a whole range of abiotic stress responses (Ye et al. 2017). These receptors typically contain an amino-terminal extracellular domain (ECD), a transmembrane (TM) domain, and an intracellular domain composed of the kinase domain (KD). Several phylogenetic studies of the RLK subfamily have been conducted. They focus mainly on *Arabidopsis* but also on other plant species (Shiu and Bleecker 2001, 2003; Shiu et al. 2004; Lehti-Shiu et al. 2009; Liu et al. 2009; Sakamoto et al. 2012; Zan et al. 2013). *Arabidopsis* RLK phylogenetic analyses inferred from KD alignments have led to the classification into 12 subgroups. These subgroups contain similar motifs in their ECD, including CRINKLY 4 like (CR4L), CrRLK1L (named after the first member identified in *Catharanthus roseus*), cystein-rich (CRK), extensin, lectin (Pawełkowicz et al.), leucine-rich repeats (LRR), lysin motif (LysM), proline-rich extensin-like (PERK), RKF3, wall-associated (Kawakatsu et al.), LRK10L-2 and receptor-like cytoplasmic kinase (RLCK) (Shiu and Bleecker 2001, 2003; Shiu et al. 2004).

To analyze the RLK subfamily in *Dendrobium catenatum* and *Phalaenopsis equestris*, first the hmmsearch program was used to seek the kinase hidden Markov model (HMM) profile (PF00069.16) within the Arabidopsis, rice, and the two orchid proteomes (Sonnhammer et al. 1998; Eddy 2009). The KD sequences of these proteins were extracted and aligned using the MAFFT program (Katoh and Standley 2013). A phylogenetic tree was then built by the maximum likelihood method FastTree (Price et al. 2009). The tree leaves were annotated according to previous Arabidopsis and rice annotations to classify orchid sequences within one of the RLK subgroups (Fig. 2) (Shiu and Bleecker 2003; Shiu et al. 2004). This analysis confirms previous observations that with ~ 650 and ~ 1100 RLK genes in *Arabidopsis* and rice respectively. The expansion rate in rice is approximately twice that of *Arabidopsis*. The subgroup expansions observed in rice have been shown to mainly involve receptors with roles in PTI, like the WAK or LEC subfamilies (Lehti-Shiu et al. 2009; Vaid et al. 2013; Delteil et al. 2016). In the orchid genomes, ~ 400 and ~ 300 RLKs have been detected in

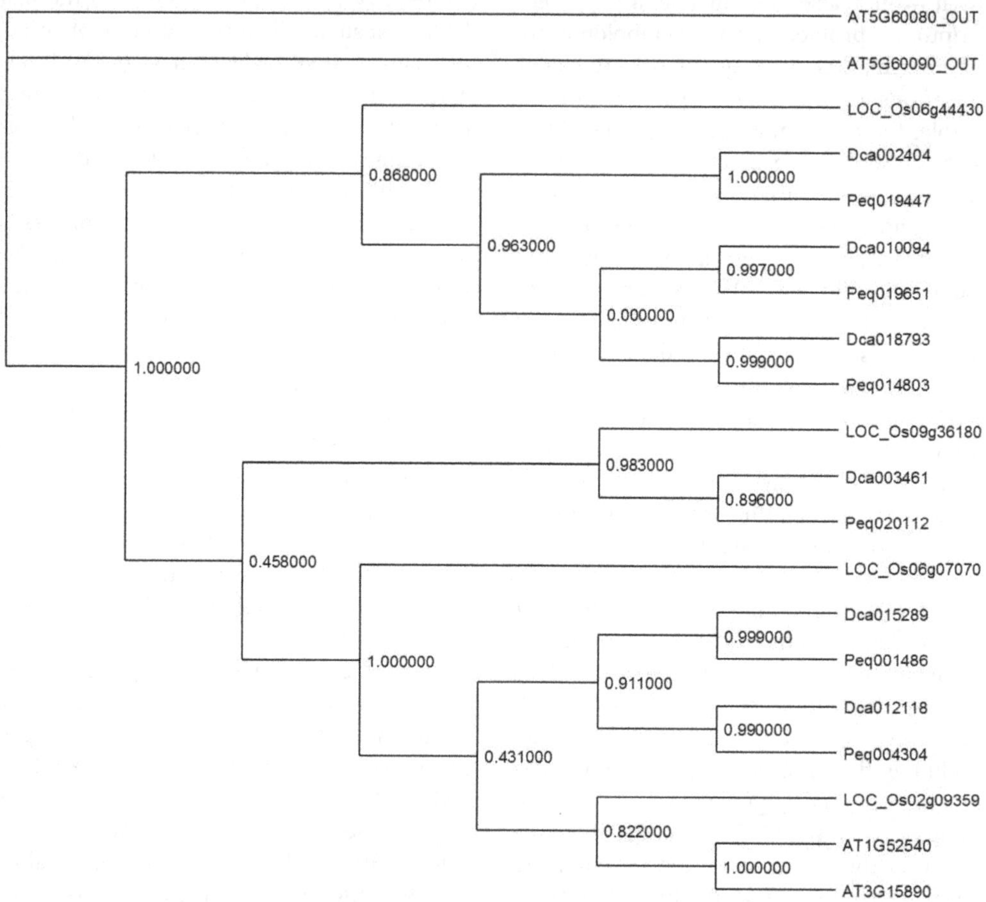

Fig. 2 Phylogenetic tree of the RLCK-XV SG. The phylogenetic tree of the RLCK-XV classified sequences has been built with PHYML (default parameters,) (Guindon et al. 2009) based on the MAFFT (Katoh and Standley 2014) alignment of full-length amino-acid sequences. Two Arabidopsis sequences (noted "OUT") have been added as outgroup to build the tree

Dendrobium and *Phalaenopsis* respectively, accounting for 1.4 and 1% of the respective proteomes (containing 29,400 and 29,679 annotated proteins) (see Table 2 for the complete list of accessions and their classification). Compared to the Arabidopsis and rice proportion of RLKs in genomes (~ 1.8 and ~ 1.7% respectively), the number of RLK in orchid genomes is lower. Considering that the *Phalaenopsis* proteome has been established on ~ 85% of the estimated total genome, ~ 15% more RLK sequences could be present in the *Phalaenopsis* genome, increasing the proportion of RLK in this genome to ~ 1.1%. Nevertheless, these proportions are below the range of what has been observed previously in other plant genomes. These results suggest that the *Dendrobium* and *Phalaenopsis* RLK subgroups did not experience large-scale expansions such as the ones observed in the rice genome. Looking at small-scale expansions, it has to be noted that one RLCK clade seems to have expanded specifically in the orchid genomes. Indeed, in the Arabidopsis and rice genomes, 2 and 4 genes are classified into the RLCK-XV subset, while 6 genes belonging to this SG have been retrieved in each orchid genome. The RLCK SG has a particularity among the RLKs since these receptors are lacking the ECD and TM domains. Some of them have been shown to be membrane-anchored and physically associated with membrane-spanning RLK. These protein kinases are then involved in signal relays (not signal perception) via transphosphorylations with other receptors to modulate signaling related to development and stress responses (Lin et al. 2013b). In Arabidopsis, the function of the two genes belonging to the RLCK-XV SG has not been characterized yet. In silico expression data show that these genes are ubiquitously expressed along all developmental stages (Hruz et al. 2008).

Terpene synthases (TPSs) in orchid genomes

Terpenoids are the largest class of metabolites found to date, including over 40,000 structures (Croteau et al. 2000; Chappell 2002; Gershenzon and Dudareva 2007), responsible to the interaction between plants and

Table 2 Proportion of *Arabidopsis, Oryza* and orchid sequences belonging to RLK subgroups

RLK subgroups	*Arabidopsis thaliana*	*Oryza sativa*	*Dendrobium catenatum*	*Phalaenopsis equestris*
LEC	8.8	19.1	8.9	6.3
CR4L	1.3	1.9	1.7	1.7
CRK	11.6	7.8	8.6	7.3
CrRLK1L	2.4	1.6	3.2	3.3
Extensin	0.5	0.4	1.5	0.3
LRK10L-2	2.1	4.6	2.7	0.7
LysM	1.0	1.0	1.5	2.0
PERK	2.9	1.8	2.2	4.0
RKF3	0.3	0.3	0.5	0.7
LRR	39.3	33.7	38.7	37.2
WAK	4.5	11.9	2.5	2.7
RLCK	24.4	15.4	26.8	32.2

environment, including defense to insects, pathogens, diseases and other stresses, also work to attract pollinators. These compounds are now utilized as pharmaceuticals (e.g., taxol), fragrance (e.g., limonene from orange or lemon oil), industrial materials (e.g., diterpene resin acid), and even as biofuel ingredient (Chen et al. 2011). With the great use of the terpene, understanding the genes that related to the biosynthesis of terpenes is now under study. Terpene synthases (TPSs), the key enzymes that generate structure diversity of terpene are usually followed by cytochromes P450 (CYPs), which further modify the products from TPSs and cause much more diverse molecules of terpenes.

The *TPS* gene family number are different among species, *Physcomitrella patens* (moss) has only one functional *TPS* gene, while in *Eucalyptus grandis* (flooded gum) 113 putative functional genes are found (Külheim et al. 2015). However, through the analysis of plant genome, researchers have shown most of the plant *TPS* gene family have their gene numbers ranging from 20 to 150, belonging to mid-size family (Chen et al. 2011). The functions of TPSs can be mono- or multi-functional, and the enzymes can be highly identical to each other. For instance, the diterpene synthases levopimaradiene/abietadiene synthase and isopimaradiene synthase showed 91% identity in Norway spruce, moreover, the functional bifurcation of these two enzymes was proved to cause by four amino acid residues only (Keeling et al. 2008). This suggests that *TPS* genes have gone through gene duplication, neofunctionalization, and/or subfunctionalization, so that lead to specialized metabolites of large family of *TPS* genes.

The large *TPS* family is divided into seven clades, from TPS-a to TPS-g, according to their protein sequence (Bohlmann et al. 1998; Dudareva et al. 2003; Martin et al. 2004). *Phalaenopsis* equestris genome has 23 TPSs belonging to TPS-a, -b, -c, e/f, -g (Figs. 3, 4). We investigated the TPS evolutionary relationship among orchids, to see whether duplication and then sub- or neo-functionalization of the *TPSs* have occurred during evolution. Cao et al. (2010) has proposed that diterpene synthases are the origin of mono- and sesqui-terpene synthases. Interestingly, the evolution of *TPSs* with other related genes may also create unexpectable mechanism. For instance, the evolution pathways of *TPS/CYP* pairs are different in monocot and dicot: *TPS/CYP* pairs duplicated with ancestral *TPS/CYP* pairs as template to evolve in dicot, but evolutionary mechanism of monocot showed genome rearrangement of *TPS* and *CYP* individually (Boutanaev et al. 2015). Moreover, *TPS* cluster density showed that *TPSs* occupied 0.008 gene/Mb in *P. patens* genome, 0.07 gene/Mb in Sorghum, and 0.3 gene/Mb in Arabidopsis and *Vitis vinifera* (Chen et al. 2011). *Phalaenopsis* tends to have dense TPS cluster compared to others, with 1.4 genes/Mb found in *P. equestris* and 2.09 genes/Mb in *Dendrobium.* It is possible that the higher density of TPSs genes in both *Phalaenopsis* and *Dendrobium* are related to that they both are CAM plant, and requires the TPS cluster to synthesize terpenoid compounds quickly for them to circumvent the adverse environment.

The *TPS* number, expression pattern, gene structure and phylogeny relationship among orchids may reveal the initial architecture of the *TPS* genes role in the orchids. Based on functions of these TPSs, mono-, di- and sesquiterpene synthases are found in orchids, Arabidopsis, and *Oryza sativa.* Total numbers of TPSs found in *P. equestris* are lesser than in *Arabidopsis*, rice, and *Dendrobium*, but the annotated monoterpene synthases are more than the others. *Dendrobium* seems to have significant larger

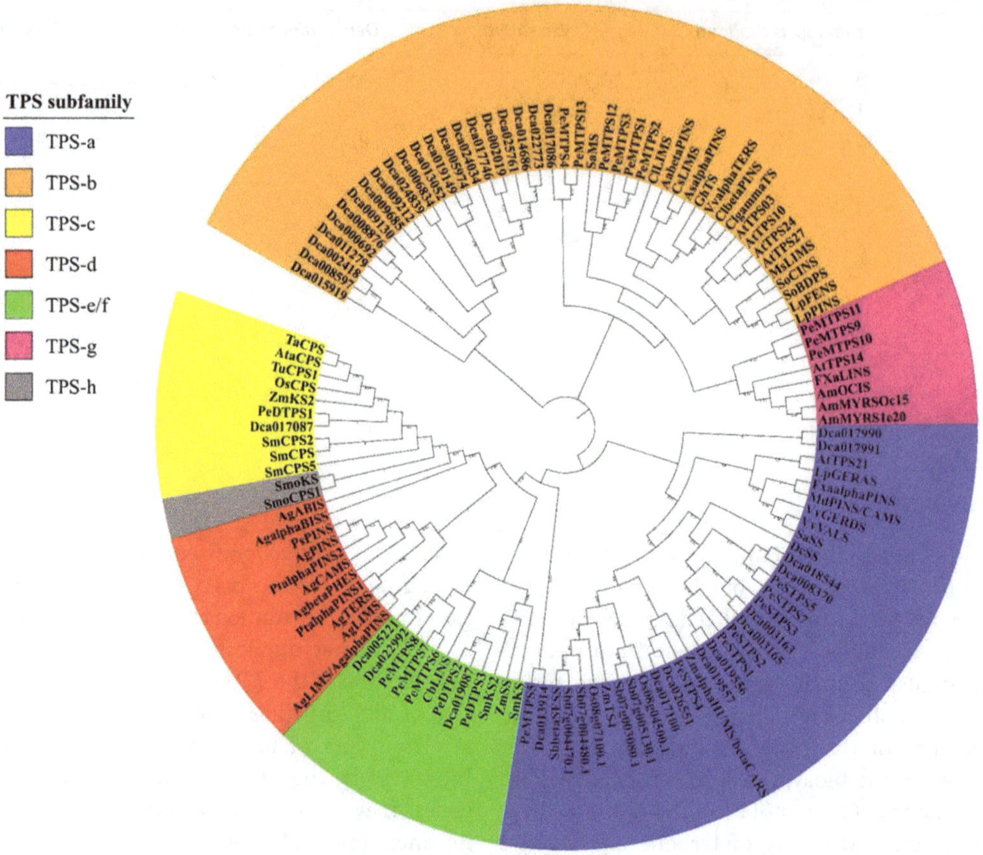

Fig. 3 Phylogeny of *P. equestris* and *Dendrobium catenatum's TPS* subfamily. The evolutionary history was inferred using the Neighbor-Joining method (Saitou and Nei 1987). The bootstrap consensus tree inferred from 100 replicates (Felsenstein 1985) is taken to represent the evolutionary history of the taxa analyzed (Felsenstein 1985). Branches corresponding to partitions reproduced in less than 50% bootstrap replicates are collapsed. The evolutionary distances were computed using the JTT matrix-based method (Jones et al. 1992) and are in the units of the number of amino acid substitutions per site. The analysis involved 123 amino acid sequences. All ambiguous positions were removed for each sequence pair. There were a total of 1411 positions in the final dataset. Evolutionary analyses were conducted in MEGA7 (Kumar et al. 2016)

group of annotated diterpene synthases than *Arabidopsis*, rice, and even the closest related *P. equestris* (Table 3).

The genes related in terpene synthesis are usually found to be lined together, forming functional clusters in plants (Matsuba et al. 2013). The functional clusters of *TPS* genes are already found in several species, such as *Eucalpyus, solanum, Vitis vinifera, Arabidopsis,* rice, etc. (Shimura et al. 2007). Based on functions of these TPSs, mono-, di- and sesquiterpene synthases are found in orchids, Arabidopsis, and *Oryza sativa* (Table 3). In some plant, such as *Eucalyptus*, only the *TPSs* in same subfamily would form clusters (Külheim et al. 2015). Moreover, the TPS can form cluster with genes that are related within the same biosynthesis pathway. In *solanum*, TPS form functional cluster with *cis*-prenyl transferase (Matsuba et al. 2013). In our analysis of *P. equestris* genome, total 23 TPSs were found and predicted as mono-, di-, and sesqui-terpene synthase. In addition, those TPSs form clusters on four different scaffolds, and each scaffold only has the terpene synthase with the same function (Fig. 4).

Secondary metabolomics

Plants produce assorted specified metabolites, but the biosynthesis genes responsible for the production and regulation of these metabolites stay mostly unknown, hampering attempts to monitor plant pharmacopeia. Assumed that genes encompassing particular metabolites pathways display environmentally reliant co-regulation, we supposed that genes participated in a specified metabolites biosynthesis pathway are arranged in strong relations (modules) in gene coexpression networks, accelerating their discovery (Wisecaver et al. 2017). Many geneticists have developed an effective method

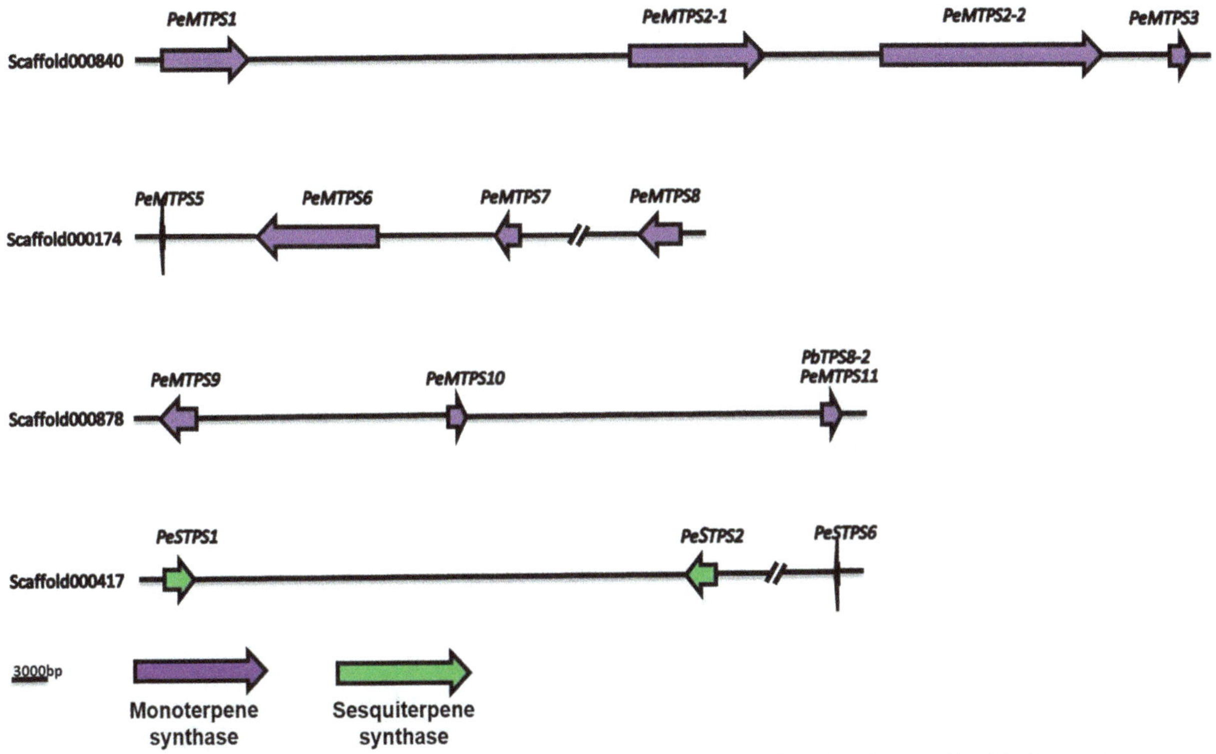

Fig. 4 TPS of *P. equestris* form several clusters. The genes are located on the scaffolds which is form from the assembly of the *P. equestris* genome. The arrow direction shows the translation direction of genes, yellow arrow indicated the monoterpene synthase and green arrow represented sesquiterpene synthase

Table 3 The annotated function of *TPS* of *Arabidopsis*, *Oryza* and orchids

Annotated function	*Arabidopsis thaliana*	*Oryza sativa*	*Dendrobium catenatum*	*Phalaenopsis equestris*
Monorpene synthase	7	2	1	13
Sesquiterpene synthase	22	18	14	7
Diterpene synthase	3	12	21	3

for identifying the plant genes that produce the chemical compounds to protect plants from predation, and it is a natural source of many important drugs.

Floral scent

Floral volatile compounds (VOCs) have been characterized in several orchids, such as lilac aldehydes in *Platanthera bifolia*, phenylacetaldehyde in *Gymnadenia odoratissima*, the green-leaf volatiles in *Epipactis helleborine*, chiloglottones in *Chiloglottis*, 1-octen-3-ol in *Dracula lafleurii*, and monoterpenes in *P. bellina*.

Identification of the fragrant vandaceous orchids that are endemic to Malaysia is performed by using similar strategy to excavate probable fragrance-related EST-SSRs as the molecular markers in the *Vanda* Mimi Palmer (Teh et al. 2011). The unique transcripts were obtained from the *Ophrys* species by using 454 pyrosequencing and Illumina (Solexa) technologies to identify genes responding for pollinator attraction (Sedeek et al. 2013).

Medicinal orchids

To study the genes involved in alkaloid biosynthetic pathway and polysaccharide biosynthesis in *Dendrobium officinale*, an important traditional Chinese herb, 454 pyrosequencing and Illumina technology was respectively applied to generate plentiful ESTs (Guo et al. 2013; Zhang et al. 2016b). Among them, 69 sequences denoting 25 genes in the biosynthesis of alkaloid backbone are detected, and 170 and 37 genes encoding to glycosyltransferase and cellulose synthase, respectively, showed differential expression patterns (Guo et al. 2013; Zhang et al. 2016b). *Dendrobium catenatum* whole genome sequence predicted 28,910 protein-coding genes are comparable with those of *Phalaenopsis*. In *Dendrobium*, the expansion of several resistance-related genes indicate a potent immune system accountable for adaptation to a broad extent of ecological niches. In addition, the genes participating glucomannan synthase actions are generally linked to the biosynthesis of medicinal polysaccharides (Zhang et al. 2016b).

Mycoheterotrophic metabolomics

Orchidaceae contain typical mycorrhizal plants. Their seeds possess no endosperm therefore they are devoided of nutrient supply. For seed germination and growth during seedling stage, the majority of orchid species are reliant on mycorrhizal fungi. Orchids are absolutely mycoheterotrophic (reliant on symbiotic fungi for the supply of carbon and nitrogen) during the achlorophyllous protocorm stage that follows seed germination in nature (Rasmussen and Rasmussen 2009; Fochi et al. 2017; Suetsugu et al. 2017).

Orchids are unique among plants in that during their life history stages, they require mycorrhizal symbioses with soil fungi from seed germination to adulthood. To identify the molecular process of orchid seed germination and the symbiotic orchid–fungus correlation, 454 and Illumina have been applied to explore transcriptomes derived from *Serapias vomeracea* (Perotto et al. 2014), *Cymbidium hybridium* (Zhao et al. 2014), *Anoectochilus roxburghii* (Liu et al. 2015), and *Gastrodia elata* (Tsai et al. 2016). The genes related to mycorrhizal symbiosis in autotrophic orchids and arbuscular mycorrhizal plants exist with common molecular mechanisms among various mycorrhizal types.

Proteome analysis of 2D-LC–MS/MS different nutrient changed stages of symbiotic germination has been performed in *Oncidium sphacelatum* (Valadares et al. 2014). A bidirectional carbon flow even in the mycoheterotrophic symbiosis in *Epipactis helleborine* has been indicated (Suetsugu et al. 2017). This is in contrast to most fully mycoheterotrophic and partially mycoheterotrophic species, which interact mainly with ectomycorrhiza-forming fungi, such as the *Sebacinales*, *Russulaceae* and *Thelephoraceae*, suggesting that their ultimate source of carbon is the photosynthate produced by nearby trees (Lee et al. 2015; Gebauer et al. 2016).

Genome editing in orchids

In the past decade, new technologies for genetic modification have emerged and known as genome-editing technologies. These technologies depend on engineered endonucleases that show sequence-specific manner of DNA cleavage due to the presence of a sequence-specific DNA-binding domain or RNA sequence (Gaj et al. 2013; Carroll 2014). Via the binding of the specific DNA sequence, the engineered nucleases can proficiently cleave the targeted genes. The cleaved DNA with double-strand breaks (DSB) subsequently result in DNA repair and lead to gene modification at the target sites. The DNA repair mechanisms include homology-directed repair and non-homologous end joining breaks (Wyman and Kanaar 2006). Recently, the CRISPR (clustered regularly interspaced short palindromic repeats)/Cas (CRISPR-associated) system was derived from the adaptive immune system (type II) of prokaryotic organism (Jinek et al. 2012). CRISPRs were identified as an unusual sequence element containing a series of 29-nucleotide repeats separated with 32-nucleotide "spacer" sequences in *Escherichia coli* genome (Ishino et al. 1987; Wiedenheft et al. 2012), and function as RNA interference mechanism to bind and cleave target DNA. A short CRISPR RNA (crRNA), the type II CRISPR/Cas from *Streptococcus pyogenes*, can recognize a complementary strand in foreign DNA with sequence specificity. Furthermore, formation of a ribonucleoprotein complex with Cas9 nuclease to generate site-specific DSBs requires a trans-activating crRNA (tracrRNA) (Walsh and Hochedlinger 2013). The modules of crRNA and tracrRNA are joined into a single RNA molecule, and termed as guide RNA (Mali et al. 2013). Effective cleavage includes the presence of the protospacer adjacent motif (PAM) in the complementary strand of foreign DNA succeeding the recognition sequence (Jinek et al. 2012).

Genome editing in other crops

The crucial requirement of genome editing is the accessibility of accurate genomic data as well as gene functions which are mostly available in model plants. The deficiency for the genomic information on horticultural crops has significantly limited breeding efficiency. So far, several cashcrops for horticulture have been whole-genome sequenced (Bolger et al. 2014). These comprise grapevine, papaya, strawberry, sweet orange, etc. In addition, a substantial increased number of transcriptomes of many horticultural crops are also feasible. The genomics information include reference genome data, transcriptomic sequences and genomic resequencing data for several horticultural crops, may provide unrestricted targets for genome editing for depicting the gene functions, which sequentially can assist engaging the CRISPR/Cas technology to design better crops.

Plant genome editing alleviates governing worries associated to genetically modified (GM) plants. Up to now, plant protoplasts of *Arabidopsis thaliana*, tobacco, lettuce and rice have been transfected using CRISPR/Cas9 system and achieved targeted mutagenesis in regenerated plants at frequencies up to 46% (Woo et al. 2015). The genome targeted sites enclosed small insertions or deletions cannot be distinguished from the naturally occurring genetic variations, and that they are germline-transmissible (Woo et al. 2015).

Genome editing in orchids

The two criteria for a crop to perform genome editing are the accessibility of genomic data and confirmation of gene functions as well as the availability for the platform of genetic transformation and regeneration process. So far, the two sequenced orchid genomes are *Phalaenopsis equestris* and *Dendrobium officinale*. The transformation of *D. officinale* is well established and has a shorter regeneration time. For *P. equestris*, the transformation system has been established, but the regeneration time is long, about 2–3 years. With these advantages, also being a popular medicinal plant for multiple pharmaceutical effectives such as immunomodulation, anti-oxidation, anti-fatigue, genome editing has been recently reported in *D. officinale* to knockout the expression of several lignocellulose biosynthesis genes, including C3H, C4H, 4CL, CCR and IRX (Kui et al. 2016). In addition, Kui et al., has adopted Agrobacterium to deliver the CRISPR/Cas9 construct and compare five promoters with Cauliflower mosaic virus 35S promoter, and identified their compatible promoter activities. Even though Kui et al. showed 100% success rate for the genome editing, they did not show the reduction of the content of lignocellulose in the knockout plants.

In *Phalaenopsis* orchids, the whole genome sequence is available, and the gene functions for floral morphogenesis, flower color, and floral scent are well studied. These have laid the groundwork for genome editing in *Phalaenopsis* orchids with the genetic transformation system. It is well expected that breeding of many more cultivars can be feasible via genome editing of these trait genes.

Conclusions and perspectives

With the available whole genome sequences of *P. equestris, D. catenatum, D. officinale*, and *A. shenzhenica*, the genetic blueprint of orchids provides a fundamental knowledge of the genetic basis of orchids. Furthermore, the whole genome sequences of one of the most popular aromatic orchids, *Vanilla*, will be available soon. The efforts by many scientists to use a plethora of genome information and genomics tools will lead to a promising understanding of the biological, physiological, molecular and genetic mechanisms of orchids in years to come. Comparative genomics analysis revealed expanded RLCK-XV clade was detected in *Phalaenopsis* genome, and the investigation of their expression patterns and putative interactors should give new insights into their functions.

In addition, the genome sequences will also be an important resource for genetic transformation for molecular breeding, including molecular marker-assisted breeding (MAB), or genome-assisted breeding (Hruz et al. 2008), and the production of transgenic or genome-edited plants. These are necessary to aid orchid horticultural research and speed up the orchid breeding. The major goal for plant breeding is to select the best progenies with the desired traits, and GWAS can guild the genetic variations with the high-density SNP markers for mapping the regulatory locus of these traits. In *Phalaenopsis* orchids, several agricultural traits have been focused, the flower sizes, length of floral inflorescence, flower colors and pigmentation patterning, and floral fragrances. With the GWAS analysis, the high-density genetic map can be constructed, and the regulatory regions responsive for the important traits can be identified. Moreover, this will be benefit for the *Phalaenopsis* breeding for new varieties with the traits of interests.

Abbreviations

BACs: bacteria artificial chromosomes; CAM: crassulacean acid metabolism; Cas: CRISPR-associated system; CRISPR: clustered regularly interspaced short palindromic repeats; ECD: extracellular domain; ePAV: expression presence/absence variation; ESTs: expressed sequence tags; FPCs: fingerprint contigs; GBS: genotyping-by-sequencing; GDPS: geranyl diphosphate synthase; GWAS: genome-wide association studies; HDR: homology-directed repair; HMM: hidden Markov model; LTRs: long terminal repeats; MAS: marker-assisted selection; NHEJ: non-homologous end joining breaks; NGS: next-generation sequenceing; PTI: pathogen-associated molecular patterns (PAMP)-triggered immunity; QTL: quantitative trait loci; RE: restriction enzyme; RLK: receptor-like kinase; SNP: single nucleotide polymorphism; SSRs: simple sequence repeats; TEs: transposable elements; TPSs: terpene synthases; VOCs: volatile compounds; WGDs: whole genome duplications; YACs: yeast artificial chromosomes.

Authors' contributions

HHC, WCT, and YYH conceived the manuscript. WCT wrote the section of "Orchid genome evolution". CCH and SYC wrote the section of "Genome mapping". HHC, AD, and HH wrote the section of "Comparative genomics". YYH wrote the section of "Secondary metabolomics". HHC wrote the section of "Genome editing in orchids". All authors read and approved the final manuscript.

Author details

[1] Institute of Tropical Plant Sciences, National Cheng Kung University, Tainan 701, Taiwan. [2] Orchid Research and Development Center, National Cheng Kung University, Tainan 701, Taiwan. [3] Department of Life Sciences, National Cheng Kung University, Tainan 701, Taiwan. [4] CIRAD, UMR AGAP, TA A 108/03, Avenue Agropolis, 34398 Montpellier, France. [5] Present Address: School of Life Sciences and Biotechnology, Shanghai Jiao Tong University, 800 Dongchuan Road, Life Sciences Building, Room 3-117, Shanghai 200240, People's Republic of China.

Acknowledgements
We thank Ching-Jen Su's assistance in preparing the manuscript, Morris Ng provides the picture of *Anoectochilus formosanus*, and Dr. Yung-I Lee provides the picture of *Apostasia wallichii*.

Competing interests
The authors declare that they have no competing interests.

Funding
This work was supported by the Grants 105-2313-B-006-002-MY3, and 105-2321-B-006-026- to H.H.C. and W.C.T., respectively from Ministry of Science and Technology, Taiwan.

References

Aceto S, Sica M, De Paolo S, D'Argenio V, Cantiello P, Salvatore F, Gaudio L (2014) The analysis of the inflorescence miRNome of the orchid *Orchis italica* reveals a *DEF*-Like MADS-box gene as a new miRNA target. PLoS ONE 9:e97839

An FM, Chan MT (2012) Transcriptome-wide characterization of miRNA-directed and non-miRNA-directed endonucleolytic cleavage using degradome analysis under low ambient temperature in *Phalaenopsis aphrodite* subsp. *formosana*. Plant Cell Physiol 53:1737–1750

An FM, Hsiao SR, Chan MT (2011) Sequencing-based approaches reveal low ambient temperature-responsive and tissue-specific microRNAs in *Phalaenopsis* orchid. PLoS ONE 6:e18937

Antolín-Llovera M, Petutsching EK, Ried MK, Lipka V, Nürnberger T, Robatzek S, Parniske M (2014) Knowing your friends and foes–plant receptor-like kinases as initiators of symbiosis or defence. New Phytol 204:791–802

Bennetzen JL, Wang H (2014) The contributions of transposable elements to the structure, function, and evolution of plant genomes. Annu Rev Plant Biol 65:505–530

Bohlmann J, Meyer-Gauen G, Croteau R (1998) Plant terpenoid synthases: molecular biology and phylogenetic analysis. Proc Natl Acad Sci USA 95:4126–4133

Bolger ME, Weisshaar B, Scholz U, Stein N, Usadel B, Mayer KF (2014) Plant genome sequencing—applications for crop improvement. Curr Opin Biotechnol 26:31–37

Boller T, Felix G (2009) A renaissance of elicitors: perception of microbe-associated molecular patterns and danger signals by pattern-recognition receptors. Annu Rev Plant Biol 60:379–406

Boutanaev AM, Moses T, Zi J, Nelson DR, Mugford ST, Peters RJ, Osbourn A (2015) Investigation of terpene diversification across multiple sequenced plant genomes. Proc Natl Acad Sci USA 112:E81–E88

Bus A, Hecht J, Huettel B, Reinhardt R, Stich B (2012) High-throughput polymorphism detection and genotyping in *Brassica napus* using next-generation RAD sequencing. BMC Genomics 13:281

Cai J, Liu X, Vanneste K, Proost S, Tsai WC, Liu KW, Chen LJ, He Y, Xu Q, Bian C, Zheng Z, Sun F, Liu W, Hsiao YY, Pan ZJ, Hsu CC, Yang YP, Hsu YC, Chuang YC, Dievart A, Dufayard JF, Xu X, Wang JY, Wang J, Xiao XJ, Zhao XM, Du R, Zhang GQ, Wang M, Su YY, Xie GC, Liu GH, Li LQ, Huang LQ, Luo YB, Chen HH, Van de Peer Y, Liu ZJ (2015) The genome sequence of the orchid *Phalaenopsis equestris*. Nat Genet 47:65–72

Cao R, Zhang Y, Mann FM, Huang C, Mukkamala D, Hudock MP, Mead ME, Prisic S, Wang K, Lin F-Y, Chang T-K, Peters RJ, Oldfield E (2010) Diterpene cyclases and the nature of the isoprene fold. Proteins 78:2417–2432

Carroll D (2014) Genome engineering with targetable nucleases. Annu Rev Biochem 83:409–439

Chaney L, Sharp AR, Evans CR, Udall JA (2016) Genome mapping in plant comparative genomics. Trends Plant Sci 21:770–780

Chang CC, Lin HC, Lin IP, Chow TY, Chen HH, Chen WH, Cheng CH, Lin CY, Liu SM, Chang CC, Chaw SM (2006) The chloroplast genome of *Phalaenopsis aphrodite* (Orchidaceae): comparative analysis of evolutionary rate with that of grasses and its phylogenetic implications. Mol Biol Evol 23:279–291

Chang YY, Chu YW, Chen CW, Leu WM, Hsu HF, Yang CH (2011) Characterization of *Oncidium* 'Gower Ramsey' transcriptomes using 454 GS-FLX pyrosequencing and their application to the identification of genes associated with flowering time. Plant Cell Physiol 52:1532–1545

Chao YT, Su CL, Jean WH, Chen WC, Chang YCA, Shih MC (2014) Identification and characterization of the microRNA transcriptome of a moth orchid *Phalaenopsis aphrodite*. Plant Mol Biol 84:529–548

Chappell J (2002) The genetics and molecular genetics of terpene and sterol origami. Curr Opin Plant Biol 5:151–157

Chen F, Tholl D, Bohlmann J, Pichersky E (2011) The family of terpene synthases in plants: a mid-size family of genes for specialized metabolism that is highly diversified throughout the kingdom. Plant J 66:212–229

Chen WH, Kao YK, Tang CY, Tsai CC, Lin TY (2013) Estimating nuclear DNA content within 50 species of the genus *Phalaenopsis* Blume (Orchidaceae). Sci Hortic (Amsterdam) 161:70–75

Chen WH, Kao YK, Tang CY (2014) Variation of the genome size among *Phalaenospsis* species using DAPI. J Taiwan Soc Hort Sci 60:115–123

Chou ML, Shih MC, Chan MT, Liao SY, Hsu CT, Haung YT, Chen JJ, Liao DC, Wu FH, Lin CS (2013) Global transcriptome analysis and identification of a *CONSTANS*-like gene family in the orchid *Erycina pusilla*. Planta 237:1425–1441

Cozzolino S, Widmer A (2005) Orchid diversity: an evolutionary consequence of deception? Trends Ecol Evol 20:487–494

Croteau R, Kutchan TM, Lewis NG (2000) Natural products (secondary metabolites). Biochem Mol Biol Plants 24:1250–1319

Cui L, Wall PK, Leebens-Mack JH, Lindsay BG, Soltis DE, Doyle JJ, Soltis PS, Carlson JE, Arumuganathan K, Barakat A (2006) Widespread genome duplications throughout the history of flowering plants. Genome Res 16:738–749

Dardick C, Chen J, Richter T, Ouyang S, Ronald P (2007) The rice kinase database. A phylogenomic database for the rice kinome. Plant Physiol 143:579–586

De Paolo S, Salvemini M, Gaudio L, Aceto S (2014) De novo transcriptome assembly from inflorescence of *Orchis italica*: analysis of coding and non-coding transcripts. PLoS ONE 9:e102155

Delseny M, Han B, Hsing YI (2010) High throughput DNA sequencing: the new sequencing revolution. Plant Sci 179:407–422

Delteil A, Gobbato E, Cayrol B, Estevan J, Michel-Romiti C, Dievart A, Kroj T, Morel JB (2016) Several wall-associated kinases participate positively and negatively in basal defense against rice blast fungus. BMC Plant Biol 16:17

Deng H, Zhang GQ, Lin M, Wang Y, Liu ZJ (2015) Mining from transcriptomes: 315 single-copy orthologous genes concatenated for the phylogenetic analyses of Orchidaceae. Ecol Evol 5:3800–3807

Deng H, Zhang LS, Zhang GQ, Zheng BQ, Liu ZJ, Wang Y (2016) Evolutionary history of PEPC genes in green plants: implications for the evolution of CAM in orchids. Mol Phylogenet Evol 94:559–564

Dudareva N, Martin D, Kish CM, Kolosova N, Gorenstein N, Faldt J, Miller B, Bohlmann J (2003) (E)-beta-ocimene and myrcene synthase genes of floral scent biosynthesis in snapdragon: function and expression of three terpene synthase genes of a new terpene synthase subfamily. Plant Cell 15:1227–1241

Eckardt NA (2000) Sequencing the rice genome. Plant Cell 12:2011–2017

Eddy SR (2009) A new generation of homology search tools based on probabilistic inference. Genome Inform 23:205–211

Edwards D, Batley J (2010) Plant genome sequencing: applications for crop improvement. Plant Biotechnol J 8:2–9

Elshire RJ, Glaubitz JC, Sun Q, Poland JA, Kawamoto K, Buckler ES, Mitchell SE (2011) A robust, simple genotyping-by-sequencing (GBS) approach for high diversity species. PLoS ONE 6:e19379

Felsenstein J (1985) Confidence limits on phylogenies: an approach using the bootstrap. Evolution 39:783–791

Feng S, Zhao H, Lu J, Liu J, Shen B, Wang H (2013) Preliminary genetic linkage maps of Chinese herb *Dendrobium nobile* and *D. moniliforme*. J Genet 92:205–212

Fochi V, Chitarra W, Kohler A, Voyron S, Singan VR, Lindquist EA, Barry KW, Girlanda M, Grigoriev IV, Martin F, Balestrini R, Perotto S (2017) Fungal and plant gene expression in the *Tulasnella calospora–Serapias vomeracea* symbiosis provides clues about nitrogen pathways in orchid mycorrhizas. New Phytol 213:365–379

Fu CH, Chen YW, Hsiao YY, Pan ZJ, Liu ZJ, Huang YM, Tsai WC, Chen HH (2011) OrchidBase: a collection of sequences of the transcriptome derived from orchids. Plant Cell Physiol 52:238–243

Gaj T, Gersbach CA, Barbas Iii CF (2013) ZFN, TALEN, and CRISPR/Cas-based methods for genome engineering. Trends Biotechnol 31:397–405

Gebauer G, Preiss K, Gebauer AC (2016) Partial mycoheterotrophy is more widespread among orchids than previously assumed. New Phytol 211:11–15

Gershenzon J, Dudareva N (2007) The function of terpene natural products in the natural world. Nat Chem Biol 3:408–414

Gill DE (1989) Fruiting failure, pollinator inefficiency, and speciation in orchids. Speciat Conseq 458:481

Gish LA, Clark SE (2011) The RLK/Pelle family of kinases. Plant J 66:117–127

Guindon S, Delsuc F, Dufayard JF, Gascuel O (2009) Estimating maximum likelihood phylogenies with PhyML. In: Posada D (ed) Bioinformatics for DNA sequence analysis. Humana Press, Totowa, pp 113–137

Guo X, Li Y, Li C, Luo H, Wang L, Qian J, Luo X, Xiang L, Song J, Sun C, Xu H, Yao H, Chen S (2013) Analysis of the *Dendrobium officinale* transcriptome reveals putative alkaloid biosynthetic genes and genetic markers. Gene 527:131–138

Gustafsson ALS, Verola CF, Antonelli A (2010) Reassessing the temporal evolution of orchids with new fossils and a Bayesian relaxed clock, with implications for the diversification of the rare South American genus *Hoffmannseggella* (Orchidaceae: Epidendroideae). BMC Evol Biol 10:177

Haas BJ, Kamoun S, Zody MC, Jiang RH, Handsaker RE, Cano LM, Grabherr M, Kodira CD, Raffaele S, Torto-Alalibo T, Bozkurt TO, Ah-Fong AM, Alvarado L, Anderson VL, Armstrong MR, Avrova A, Baxter L, Beynon J, Boevink PC, Bollmann SR, Bos JI, Bulone V, Cai G, Cakir C, Carrington JC, Chawner M, Conti L, Costanzo S, Ewan R, Fahlgren N, Fischbach MA, Fugelstad J, Gilroy EM, Gnerre S, Green PJ, Grenville-Briggs LJ, Griffith J, Grunwald NJ, Horn K, Horner NR, Hu CH, Huitema E, Jeong DH, Jones AM, Jones JD, Jones RW, Karlsson EK, Kunjeti SG, Lamour K, Liu Z, Ma L, Maclean D, Chibucos MC, McDonald H, McWalters J, Meijer HJ, Morgan W, Morris PF, Munro CA, O'Neill K, Ospina-Giraldo M, Pinzon A, Pritchard L, Ramsahoye B, Ren Q, Restrepo S, Roy S, Sadanandom A, Savidor A, Schornack S, Schwartz DC, Schumann UD, Schwessinger B, Seyer L, Sharpe T, Silvar C, Song J, Studholme DJ, Sykes S, Thines M, van de Vondervoort PJ, Phuntumart V, Wawra S, Weide R, Win J, Young C, Zhou S, Fry W, Meyers BC, van West P, Ristaino J, Govers F, Birch PR, Whisson SC, Judelson HS, Nusbaum C (2009) Genome sequence and analysis of the Irish potato famine pathogen *Phytophthora infestans*. Nature 461:393–398

Hanks SK, Hunter T (1995) Protein kinases 6. The eukaryotic protein kinase superfamily: kinase (catalytic) domain structure and classification. FASEB J 9:576–596

Haruta M, Sussman MR (2017) Chapter ten—ligand receptor-mediated regulation of growth in plants. In: Andreas J (ed) Current topics in developmental biology, vol 123. Academic Press, Cambridge, pp 331–363

He G, Chen B, Wang X, Li X, Li J, He H, Yang M, Lu L, Qi Y, Wang X, Wang Deng X (2013) Conservation and divergence of transcriptomic and epigenomic variation in maize hybrids. Genome Biol 14:R57

Hearnden PR, Eckermann PJ, McMichael GL, Hayden MJ, Eglinton JK, Chalmers KJ (2007) A genetic map of 1000 SSR and DArT markers in a wide barley cross. Theor Appl Genet 115:383

Hruz T, Laule O, Szabo G, Wessendorp F, Bleuler S, Oertle L, Widmayer P, Gruissem W, Zimmermann P (2008) Genevestigator V3: a reference expression database for the meta-analysis of transcriptomes. Adv Bioinform 2008:5

Hsiao YY, Tsai WC, Kuoh CS, Huang TH, Wang HC, Wu TS, Leu YL, Chen WH, Chen HH (2006) Comparison of transcripts in *Phalaenopsis bellina* and *Phalaenopsis equestris* (Orchidaceae) flowers to deduce monoterpene biosynthesis pathway. BMC Plant Biol 6:14

Hsiao YY, Chen YW, Huang SC, Pan ZJ, Fu CH, Chen WH, Tsai WC, Chen HH (2011a) Gene discovery using next-generation pyrosequencing to develop ESTs for *Phalaenopsis* orchids. BMC Genomics 12:360

Hsiao YY, Pan ZJ, Hsu CC, Yang YP, Hsu YC, Chuang YC, Shih HH, Chen WH, Tsai WC, Chen HH (2011b) Research on orchid biology and biotechnology. Plant Cell Physiol 52:1467–1486

Hsiao YY, Huang TH, Fu CH, Huang SC, Chen YJ, Huang YM, Chen WH, Tsai WC, Chen HH (2013) Transcriptomic analysis of floral organs from *Phalaenopsis* orchid by using oligonucleotide microarray. Gene 518:91–100

Hsieh MH, Lu HC, Pan ZJ, Yeh HH, Wang SS, Chen WH, Chen HH (2013a) Optimizing virus-induced gene silencing efficiency with *Cymbidium* mosaic virus in *Phalaenopsis* flower. Plant Sci 201–202:25–41

Hsieh MH, Pan ZJ, Lai PH, Lu HC, Yeh HH, Hsu CC, Wu WL, Chung MC, Wang SS, Chen WH, Chen HH (2013b) Virus-induced gene silencing unravels multiple transcription factors involved in floral growth and development in *Phalaenopsis* orchids. J Exp Bot 64:3869–3884

Hsu CC, Chung YL, Chen TC, Lee YL, Kuo YT, Tsai WC, Hsiao YY, Chen YW, Wu WL, Chen HH (2011) An overview of the *Phalaenopsis* orchid genome through BAC end sequence analysis. BMC Plant Biol 11:3

Hsu CC, Chen YY, Tsai WC, Chen WH, Chen HH (2015) Three R2R3-MYB transcription factors regulate distinct floral pigmentation patterning in *Phalaenopsis* spp. Plant Physiol 168:175–191

Huang JZ, Lin CP, Cheng TC, Chang BC, Cheng SY, Chen YW, Lee CY, Chin SW, Chen FC (2015) A *de novo* floral transcriptome reveals clues into *Phalaenopsis* orchid flower development. PLoS ONE 10:e0123474

ICGMC (2015) High-resolution linkage map and chromosome-scale genome assembly for cassava (Manihot esculenta Crantz) from 10 populations. G3 (Bethesda) 5(1):133–144

Ishino Y, Shinagawa H, Makino K, Amemura M, Nakata A (1987) Nucleotide sequence of the iap gene, responsible for alkaline phosphatase isozyme conversion in *Escherichia coli*, and identification of the gene product. J Bacteriol 169:5429–5433

Jersáková J, Trávníček P, Kubátová B, Krejčíková J, Urfus T, Liu ZJ, Lamb A, Ponert J, Schulte K, Čurn V, Vrána J, Leitch IJ, Suda J (2013) Genome size variation in Orchidaceae subfamily Apostasioideae: filling the phylogenetic gap. Bot J Linn Soc 172:95–105

Jiao Y, Wickett NJ, Ayyampalayam S, Chanderbali AS, Landherr L, Ralph PE, Tomsho LP, Hu Y, Liang H, Soltis PS, Soltis DE, Clifton SW, Schlarbaum SE, Schuster SC, Ma H, Leebens-Mack J, de Pamphilis CW (2011) Ancestral polyploidy in seed plants and angiosperms. Nature 473:97–100

Jiao Y, Leebens-Mack J, Ayyampalayam S, Bowers JE, McKain MR, McNeal J, Rolf M, Ruzicka DR, Wafula E, Wickett NJ, Wu X, Zhang Y, Wang J, Zhang Y, Carpenter EJ, Deyholos MK, Kutchan TM, Chanderbali AS, Soltis PS, Stevenson DW, McCombie R, Pires JC, Wong GKS, Soltis DE, dePamphilis CW (2012) A genome triplication associated with early diversification of the core eudicots. Genome Biol 13:R3

Jin M, Liu H, He C, Fu J, Xiao Y, Wang Y, Xie W, Wang G, Yan J (2016) Maize pan-transcriptome provides novel insights into genome complexity and quantitative trait variation. Sci Rep 6:18936

Jinek M, Chylinski K, Fonfara I, Hauer M, Doudna JA, Charpentier E (2012) A programmable dual-RNA-guided DNA endonuclease in adaptive bacterial immunity. Science 337:816–821

Jo K, Schramm TM, Schwartz DC (2009) A single-molecule barcoding system using nanoslits for DNA analysis: nanocoding. Methods Mol Biol 544:29–42

Jones DT, Taylor WR, Thornton JM (1992) The rapid generation of mutation data matrices from protein sequences. Bioinformatics 8:275–282

Kao YY, Chang SB, Lin TY, Hsieh CH, Chen YH, Chen WH, Chen CC (2001) Differential accumulation of heterochromatin as a cause for karyotype variation in *Phalaenopsis* orchids. Ann Bot 87:387–395

Katoh K, Standley DM (2013) MAFFT multiple sequence alignment software version 7: improvements in performance and ssability. Mol Biol Evol 30:772–780

Katoh K, Standley DM (2014) MAFFT: iterative refinement and additional methods. In: Russell DJ (ed) Multiple sequence alignment methods. Humana Press, Totowa, pp 131–146

Kawakatsu T, Huang SC, Jupe F, Sasaki E, Schmitz RJ, Urich MA, Castanon R, Nery JR, Barragan C, He Y, Chen H, Dubin M, Lee C-R, Wang C, Bemm F, Becker C, O'Neil R, O'Malley RC, Quarless DX, Alonso-Blanco C, Andrade J, Becker C, Bemm F, Bergelson J, Borgwardt K, Chae E, Dezwaan T, Ding W, Ecker JR, Expósito-Alonso M, Farlow A, Fitz J, Gan X, Grimm DG, Hancock A, Henz SR, Holm S, Horton M, Jarsulic M, Kerstetter RA, Korte A, Korte P, Lanz C, Lee C-R, Meng D, Michael TP, Mott R, Muliyati NW, Nägele T, Nagler M, Nizhynska V, Nordborg M, Novikova P, Picó FX,

Platzer A, Rabanal FA, Rodriguez A, Rowan BA, Salomé PA, Schmid K, Schmitz RJ, Seren Ü, Sperone FG, Sudkamp M, Svardal H, Tanzer MM, Todd D, Volchenboum SL, Wang C, Wang G, Wang X, Weckwerth W, Weigel D, Zhou X, Schork NJ, Weigel D, Nordborg M, Ecker JR (2016) Epigenomic diversity in a global collection of *Arabidopsis thaliana* accessions. Cell 166:492–505

Keeling CI, Weisshaar S, Lin RPC, Bohlmann J (2008) Functional plasticity of paralogous diterpene synthases involved in conifer defense. Proc Natl Acad Sci USA 105:1085–1090

Kui L, Chen H, Zhang W, He S, Xiong Z, Zhang Y, Yan L, Zhong C, He F, Chen J, Zeng P, Zhang G, Yang S, Dong Y, Wang W, Cai J (2016) Building a genetic manipulation tool box for orchid biology: identification of constitutive promoters and application of CRISPR/Cas9 in the orchid, *Dendrobium officinale*. Front Plant Sci 7:2036

Külheim C, Padovan A, Hefer C, Krause ST, Köllner TG, Myburg AA, Degenhardt J, Foley WJ (2015) The *Eucalyptus* terpene synthase gene family. BMC Genomics 16:450

Kumar S, Stecher G, Tamura K (2016) MEGA7: molecular evolutionary genetics analysis version 7.0 for bigger datasets. Mol Biol Evol 33:1870–1874

Kurata N, Nagamura Y, Yamamoto K, Harushima Y, Sue N, Wu J, Antonio BA, Shomura A, Shimizu T, Lin SY, Inoue T, Fukuda A, Shimano T, Kuboki Y, Toyama T, Miyamoto Y, Kirihara T, Hayasaka K, Miyao A, Monna L, Zhong HS, Tamura Y, Wang ZX, Momma T, Umehara Y, Yano M, Sasaki T, Minobe Y (1994) A 300 kilobase interval genetic map of rice including 883 expressed sequences. Nat Genet 8:365–372

Lee YI, Yang CK, Gebauer G (2015) The importance of associations with saprotrophic non-Rhizoctonia fungi among fully mycoheterotrophic orchids is currently under-estimated: novel evidence from sub-tropical Asia. Ann Bot 116:423–435

Lehti-Shiu MD, Zou C, Hanada K, Shiu SH (2009) Evolutionary history and stress regulation of plant receptor-like kinase/pelle genes. Plant Physiol 150:12–26

Leitch IJ, Kahandawala I, Suda J, Hanson L, Ingrouille MJ, Chase MW, Fay MF (2009) Genome size diversity in orchids: consequences and evolution. Ann Bot 104:469–481

Li X, Luo J, Yan T, Xiang L, Jin F, Qin D, Sun C, Xie M (2014) Deep sequencing-based analysis of the *Cymbidium ensifolium* floral transcriptome. PLoS ONE 8:e85480

Li C, Bai G, Chao S, Wang Z (2015a) A high-density SNP and SSR consensus map reveals segregation distortion regions in wheat. Biomed Res Int 2015:10

Li D, Zhao C, Liu X, Liu X, Lin Y, Liu J, Chen H, Lǚ F (2015b) *De novo* assembly and characterization of the root transcriptome and development of simple sequence repeat markers in *Paphiopedilum concolor*. Genet Mol Res 14:6189–6201

Li X, Jin F, Jin L, Jackson A, Ma X, Shu X, Wu D, Jin G (2015c) Characterization and comparative profiling of the small RNA transcriptomes in two phases of flowering in *Cymbidium ensifolium*. BMC Genomics 16:622

Lin S, Lee HC, Chen WH, Chen CC, Kao YY, Fu YM, Chen YH, Lin TY (2001) Nuclear DNA contents of *Phalaenopsis* sp. and *Doritis pulcherrima*. J Am Soc Hort Sci 126:195–199

Lin CS, Chen JJW, Huang YT, Hsu CT, Lu HC, Chou ML, Chen LC, Ou CI, Liao DC, Yeh YY, Chang SB, Shen SC, Wu FH, Shih MC, Chan MT (2013a) Catalog of *Erycina pusilla* miRNA and categorization of reproductive phase-related miRNAs and their target gene families. Plant Mol Biol 82:193–204

Lin W, Ma X, Shan L, He P (2013b) Big roles of small kinases: the complex functions of receptor-like cytoplasmic kinases in plant immunity and development. J Integr Plant Biol 55:1188–1197

Lin YF, Chen YY, Hsiao YY, Shen CY, Hsu JL, Yeh CM, Mitsuda N, Ohme-Takagi M, Liu ZJ, Tsai WC (2016) Genome-wide identification and characterization of TCP genes involved in ovule development of *Phalaenopsis equestris*. J Exp Bot 67:5051–5066

Liu P, Wei W, Ouyang S, Zhang JS, Chen SY, Zhang WK (2009) Analysis of expressed receptor-like kinases (RLKs) in soybean. J Genet Genomics 36:611–619

Liu SS, Chen J, Li SC, Zeng X, Meng ZX, Guo SX (2015) Comparative transcriptome analysis of genes involved in GA-GID1-DELLA regulatory module in symbiotic and asymbiotic seed germination of *Anoectochilus roxburghii* (Wall.) Lindl. (Orchidaceae). Int J Mol Sci 16:26224

Lu HC, Chen HH, Tsai WC, Chen WH, Su HJ, Chang DC, Yeh HH (2007) Strategies for functional validation of genes involved in reproductive stages of orchids. Plant Physiol 143:558–569

Lu JJ, Wang S, Zhao HY, Liu JJ, Wang HZ (2012a) Genetic linkage map of EST-SSR and SRAP markers in the endangered Chinese endemic herb *Dendrobium* (Orchidaceae). Genet Mol Res 11:4654–4667

Lu JJ, Zhao HY, Suo NN, Wang S, Shen B, Wang HZ, Liu JJ (2012b) Genetic linkage maps of *Dendrobium moniliforme* and *D. officinale* based on EST-SSR, SRAP, ISSR and RAPD markers. Sci Hortic (Amsterdam) 137:1–10

Lu F, Lipka AE, Glaubitz J, Elshire R, Cherney JH, Casler MD, Buckler ES, Costich DE (2013) Switchgrass genomic diversity, ploidy, and evolution: novel insights from a network-based SNP discovery protocol. PLoS Genet 9:e1003215

Mali P, Yang L, Esvelt KM, Aach J, Guell M, DiCarlo JE, Norville JE, Church GM (2013) RNA-guided human genome engineering via Cas9. Science 339:823–826

Manning G, Plowman GD, Hunter T, Sudarsanam S (2002) Evolution of protein kinase signaling from yeast to man. Trends Biochem Sci 27:514–520

Martin DM, Faldt J, Bohlmann J (2004) Functional characterization of nine Norway spruce TPS genes and evolution of gymnosperm terpene synthases of the TPS-d subfamily. Plant Physiol 135:1908–1927

Martin DMA, Miranda-Saavedra D, Barton GJ (2009) Kinomer v. 1.0: a database of systematically classified eukaryotic protein kinases. Nucleic Acids Res 37:D244–D250

Mascher M, Gundlach H, Himmelbach A, Beier S, Twardziok SO, Wicker T, Radchuk V, Dockter C, Hedley PE, Russell J, Bayer M, Ramsay L, Liu H, Haberer G, Zhang XQ, Zhang Q, Barrero RA, Li L, Taudien S, Groth M, Felder M, Hastie A, Simkova H, Stankova H, Vrana J, Chan S, Munoz-Amatriain M, Ounit R, Wanamaker S, Bolser D, Colmsee C, Schmutzer T, Aliyeva-Schnorr L, Grasso S, Tanskanen J, Chailyan A, Sampath D, Heavens D, Clissold L, Cao S, Chapman B, Dai F, Han Y, Li H, Li X, Lin C, McCooke JK, Tan C, Wang P, Wang S, Yin S, Zhou G, Poland JA, Bellgard MI, Borisjuk L, Houben A, Dolezel J, Ayling S, Lonardi S, Kersey P, Langridge P, Muehlbauer GJ, Clark MD, Caccamo M, Schulman AH, Mayer KFX, Platzer M, Close TJ, Scholz U, Hansson M, Zhang G, Braumann I, Spannagl M, Li C, Waugh R, Stein N (2017) A chromosome conformation capture ordered sequence of the barley genome. Nature 544:427–433

Matsuba Y, Nguyen TT, Wiegert K, Falara V, Gonzales-Vigil E, Leong B, Schafer P, Kudrna D, Wing RA, Bolger AM, Usadel B, Tissier A, Fernie AR, Barry CS, Pichersky E (2013) Evolution of a complex locus for terpene biosynthesis in solanum. Plant Cell 25:2022–2036

Meng Y, Yu D, Xue J, Lu J, Feng S, Shen C, Wang H (2016) A transcriptome-wide, organ-specific regulatory map of *Dendrobium officinale*, an important traditional Chinese orchid herb. Sci Rep 6:18864

Niu SC, Xu Q, Zhang GQ, Zhang YQ, Tsai WC, Hsu JL, Liang CK, Luo YB, Liu ZJ (2016) De novo transcriptome assembly databases for the butterfly orchid *Phalaenopsis equestris*. Sci Data 3:160083

Ogura T, Busch W (2015) From phenotypes to causal sequences: using genome wide association studies to dissect the sequence basis for variation of plant development. Curr Opin Plant Biol 23:98–108

Oliver KR, McComb JA, Greene WK (2013) Transposable elements: powerful contributors to angiosperm evolution and diversity. Genome Biol Evol 5:1886–1901

Otero JT, Flanagan NS (2006) Orchid diversity—beyond deception. Trends Ecol Evol 21:64–65 **(author reply 65–66)**

Pan IC, Liao DC, Wu FH, Daniell H, Singh ND, Chang C, Shih MC, Chan MT, Lin CS (2012) Complete chloroplast genome sequence of an orchid model plant candidate: *Erycina pusilla* apply in tropical Oncidium breeding. PLoS ONE 7:e34738

Paterson AH, Bowers JE, Chapman BA (2004) Ancient polyploidization predating divergence of the cereals, and its consequences for comparative genomics. Proc Natl Acad Sci USA 101:9903–9908

Pawełkowicz M, Zieliński K, Zielińska D, Pląder W, Yagi K, Wojcieszek M, Siedlecka E, Bartoszewski G, Skarzyńska A, Przybecki Z (2016) Next generation sequencing and omics in cucumber (*Cucumis sativus* L.) breeding directed research. Plant Sci 242:77–88

Perotto S, Rodda M, Benetti A, Sillo F, Ercole E, Rodda M, Girlanda M, Murat C, Balestrini R (2014) Gene expression in mycorrhizal orchid protocorms suggests a friendly plant–fungus relationship. Planta 239:1337–1349

Piegu B, Guyot R, Picault N, Roulin A, Saniyal A, Kim H, Collura K, Brar DS, Jackson S, Wing RA (2006) Doubling genome size without polyploidization: dynamics of retrotransposition-driven genomic expansions in *Oryza australiensis*, a wild relative of rice. Genome Res 16:1262–1269

Poland JA, Rife TW (2012) Genotyping-by-sequencing for plant breeding and genetics. Plant Genome 5:92–102

Poland J, Endelman J, Dawson J, Rutkoski J, Wu S, Manes Y, Dreisigacker S, Crossa J, Sánchez-Villeda H, Sorrells M, Jannink JL (2012a) Genomic selection in wheat breeding using genotyping-by-sequencing. Plant Genome 5:103–113

Poland JA, Brown PJ, Sorrells ME, Jannink JL (2012b) Development of high-density genetic maps for barley and wheat using a novel two-enzyme genotyping-by-sequencing approach. PLoS ONE 7:e32253

Price MN, Dehal PS, Arkin AP (2009) FastTree: computing large minimum evolution trees with profiles instead of a distance matrix. Mol Biol Evol 26:1641–1650

Ramirez SR, Gravendeel B, Singer RB, Marshall CR, Pierce NE (2007) Dating the origin of the Orchidaceae from a fossil orchid with its pollinator. Nature 448:1042–1045

Ramsay L, Macaulay M, Ivanissevich Sd, MacLean K, Cardle L, Fuller J, Edwards KJ, Tuvesson S, Morgante M, Massari A, Maestri E, Marmiroli N, Sjakste T, Ganal M, Powell W, Waugh R (2000) A simple sequence repeat-based linkage map of barley. Genetics 156:1997–2005

Rao X, Krom N, Tang Y, Widiez T, Havkin-Frenkel D, Belanger FC, Dixon RA, Chen F (2014) A deep transcriptomic analysis of pod development in the vanilla orchid (*Vanilla planifolia*). BMC Genomics 15:964

Rasmussen HN, Rasmussen FN (2009) Orchid mycorrhiza: implications of a mycophagous life style. Oikos 118:334–345

Romay MC, Millard MJ, Glaubitz JC, Peiffer JA, Swarts KL, Casstevens TM, Elshire RJ, Acharya CB, Mitchell SE, Flint-Garcia SA, McMullen MD, Holland JB, Buckler ES, Gardner CA (2013) Comprehensive genotyping of the USA national maize inbred seed bank. Genome Biol 14:R55

Saitou N, Nei M (1987) The neighbor-joining method: a new method for reconstructing phylogenetic trees. Mol Biol Evol 4:406–425

Sakamoto T, Deguchi M, Brustolini OJ, Santos AA, Silva FF, Fontes EP (2012) The tomato RLK superfamily: phylogeny and functional predictions about the role of the LRRII-RLK subfamily in antiviral defense. BMC Plant Biol 12:229

Sattler MC, Carvalho CR, Clarindo WR (2016) The polyploidy and its key role in plant breeding. Planta 243:281–296

Sedeek KEM, Qi W, Schauer MA, Gupta AK, Poveda L, Xu S, Liu ZJ, Grossniklaus U, Schiestl FP, Schlüter PM (2013) Transcriptome and proteome data reveal candidate genes for pollinator attraction in sexually deceptive orchids. PLoS ONE 8:e64621

Sedeek KEM, Scopece G, Staedler YM, Schönenberger J, Cozzolino S, Schiestl FP, Schlüter PM (2014) Genic rather than genome-wide differences between sexually deceptive *Ophrys* orchids with different pollinators. Mol Ecol 23:6192–6205

Shimura K, Okada A, Okada K, Jikumaru Y, Ko KW, Toyomasu T, Sassa T, Hasegawa M, Kodama O, Shibuya N, Koga J, Nojiri H, Yamane H (2007) Identification of a biosynthetic gene cluster in rice for momilactones. J Biol Chem 282:34013–34018

Shiu SH, Bleecker AB (2001) Receptor-like kinases from *Arabidopsis* form a monophyletic gene family related to animal receptor kinases. Proc Natl Acad Sci USA 98:10763–10768

Shiu SH, Bleecker AB (2003) Expansion of the receptor-like kinase/Pelle gene family and receptor-like proteins in *Arabidopsis*. Plant Physiol 132:530–543

Shiu SH, Karlowski WM, Pan R, Tzeng YH, Mayer KF, Li WH (2004) Comparative analysis of the receptor-like kinase family in *Arabidopsis* and rice. Plant Cell 16:1220–1234

Silvera K, Santiago LS, Cushman JC, Winter K (2009) Crassulacean acid metabolism and epiphytism linked to adaptive radiations in the Orchidaceae. Plant Physiol 149:1838–1847

Slate J, Gratten J, Beraldi D, Stapley J, Hale M, Pemberton JM (2009) Gene mapping in the wild with SNPs: guidelines and future directions. Genetica 136:97–107

Sonah H, O'Donoughue L, Cober E, Rajcan I, Belzile F (2015) Identification of loci governing eight agronomic traits using a GBS-GWAS approach and validation by QTL mapping in soya bean. Plant Biotechnol J 13:211–221

Sonnhammer EL, Eddy SR, Birney E, Bateman A, Durbin R (1998) Pfam: multiple sequence alignments and HMM-profiles of protein domains. Nucleic Acids Res 26:320–322

Stankova H, Hastie AR, Chan S, Vrana J, Tulpova Z, Kubalakova M, Visendi P, Hayashi S, Luo M, Batley J, Edwards D, Dolezel J, Simkova H (2016) BioNano genome mapping of individual chromosomes supports physical mapping and sequence assembly in complex plant genomes. Plant Biotechnol J 14:1523–1531

Su CL, Chao YT, Alex Chang YC, Chen WC, Chen CY, Lee AY, Hwa KT, Shih MC (2011) *De novo* assembly of expressed transcripts and global analysis of the *Phalaenopsis aphrodite* transcriptome. Plant Cell Physiol 52:1501–1514

Su CL, Chao YT, Yen SH, Chen CY, Chen WC, Chang YC, Shih MC (2013) Orchidstra: an integrated orchid functional genomics database. Plant Cell Physiol 54:e11

Suetsugu K, Yamato M, Miura C, Yamaguchi K, Takahashi K, Ida Y, Shigenobu S, Kaminaka H (2017) Comparison of green and albino individuals of the partially mycoheterotrophic orchid *Epipactis helleborine* on molecular identities of mycorrhizal fungi, nutritional modes and gene expression in mycorrhizal roots. Mol Ecol 26:1652–1669

Tan J, Wang HL, Yeh KW (2005) Analysis of organ-specific, expressed genes in *Oncidium* orchid by subtractive expressed sequence tags library. Biotechnol Lett 27:1517–1528

Teague B, Waterman MS, Goldstein S, Potamousis K, Zhou S, Reslewic S, Sarkar D, Valouev A, Churas C, Kidd JM, Kohn S, Runnheim R, Lamers C, Forrest D, Newton MA, Eichler EE, Kent-First M, Surti U, Livny M, Schwartz DC (2010) High-resolution human genome structure by single-molecule analysis. Proc Natl Acad Sci USA 107:10848–10853

Teh SL, Chan WS, Abdullah JO, Namasivayam P (2011) Development of expressed sequence tag resources for *Vanda* Mimi Palmer and data mining for EST-SSR. Mol Biol Rep 38:3903–3909

Truong HT, Ramos AM, Yalcin F, de Ruiter M, van der Poel HJA, Huvenaars KHJ, Hogers RCJ, van Enckevort LJG, Janssen A, van Orsouw NJ, van Eijk MJT (2012) Sequence-based genotyping for marker discovery and co-dominant ccoring in germplasm and populations. PLoS ONE 7:e37565

Tsai WC, Hsiao YY, Lee SH, Tung CW, Wang DP, Wang HC, Chen WH, Chen HH (2006) Expression analysis of the ESTs derived from the flower buds of *Phalaenopsis equestris*. Plant Sci 170:426–432

Tsai WC, Fu CH, Hsiao YY, Huang YM, Chen LJ, Wang M, Liu ZJ, Chen HH (2013) OrchidBase 2.0: comprehensive collection of Orchidaceae floral transcriptomes. Plant Cell Physiol 54:e7

Tsai CC, Wu KM, Chiang TY, Huang CY, Chou CH, Li SJ, Chiang YC (2016) Comparative transcriptome analysis of *Gastrodia elata* (Orchidaceae) in response to fungus symbiosis to identify gastrodin biosynthesis-related genes. BMC Genomics 17:212

Uitdewilligen JGAML, Wolters A-MA, D'hoop BB, Borm TJA, Visser RGF, van Eck HJ (2013) A next-generation sequencing method for genotyping-by-sequencing of highly heterozygous autotetraploid potato. PLoS ONE 8:e62355

Vaid N, Macovei A, Tuteja N (2013) Knights in action: lectin receptor-like kinases in plant development and stressr esponses. Mol Plant 6:1405–1418

Valadares RBS, Perotto S, Santos EC, Lambais MR (2014) Proteome changes in *Oncidium sphacelatum* (Orchidaceae) at different trophic stages of symbiotic germination. Mycorrhiza 24:349–360

Van de Peer Y, Maere S, Meyer A (2009) The evolutionary significance of ancient genome duplications. Nat Rev Genet 10:725–732

van Poecke RMP, Maccaferri M, Tang J, Truong HT, Janssen A, van Orsouw NJ, Salvi S, Sanguineti MC, Tuberosa R, van der Vossen EAG (2013) Sequence-based SNP genotyping in durum wheat. Plant Biotechnol J 11:809–817

Varshney RK, Graner A, Sorrells ME (2005) Genic microsatellite markers in plants: features and applications. Trends Biotechnol 23:48–55

Walsh RM, Hochedlinger K (2013) A variant CRISPR–Cas9 system adds versatility to genome engineering. Proc Natl Acad Sci USA 110:15514–15515

Wen W, Liu H, Zhou Y, Jin M, Yang N, Li D, Luo J, Xiao Y, Pan Q, Tohge T, Fernie AR, Yan J (2016) Combining quantitative genetics approaches with regulatory network analysis to dissect the complex metabolism of the maize kernel. Plant Physiol 170:136–146

Wendel JF, Jackson SA, Meyers BC, Wing RA (2016) Evolution of plant genome architecture. Genome Biol 17:37

Wiedenheft B, Sternberg SH, Doudna JA (2012) RNA-guided genetic silencing systems in bacteria and archaea. Nature 482:331–338
Wisecaver JH, Borowsky AT, Tzin V, Jander G, Kliebenstein DJ, Rokas A (2017) A global co-expression network approach for connecting genes to specialized metabolic pathways in plants. Plant Cell 29:944–959
Woo JW, Kim J, Kwon SI, Corvalan C, Cho SW, Kim H, Kim SG, Kim ST, Choe S, Kim JS (2015) DNA-free genome editing in plants with preassembled CRISPR–Cas9 ribonucleoproteins. Nat Biotechnol 33:1162–1164
Wu Y, Zhou JM (2013) Receptor-like kinases in plant innate immunity. J Integr Plant Biol 55:1271–1286
Wyman C, Kanaar R (2006) DNA double-strand break repair: all's well that ends well. Annu Rev Genet 40:363–383
Xiao S, Li J, Ma F, Fang L, Xu S, Chen W, Wang ZY (2015) Rapid construction of genome map for large yellow croaker (*Larimichthys crocea*) by the whole-genome mapping in BioNano Genomics Irys system. BMC Genomics 16:670
Xu C, Zeng B, Huang J, Huang W, Liu Y (2015) Genome-wide transcriptome and expression profile analysis of *Phalaenopsis* during explant browning. PLoS ONE 10:e0123356
Xue D, Feng S, Zhao H, Jiang H, Shen B, Shi N, Lu J, Liu J, Wang H (2010) The linkage maps of *Dendrobium* species based on RAPD and SRAP markers. J Genet Genomics 37:197–204
Yan L, Wang X, Liu H, Tian Y, Lian J, Yang R, Hao S, Wang X, Yang S, Li Q, Qi S, Kui L, Okpekum M, Ma X, Zhang J, Ding Z, Zhang G, Wang W, Dong Y, Sheng J (2015) The genome of *Dendrobium officinale* illuminates the biology of the important traditional Chinese orchid herb. Mol Plant 8:922–934
Yang F, Zhu G (2015) Digital gene expression analysis based on *de novo* transcriptome assembly reveals new genes associated with floral organ differentiation of the orchid plant *Cymbidium ensifolium*. PLoS ONE 10:e0142434
Yang H, Tao Y, Zheng Z, Li C, Sweetingham MW, Howieson JG (2012) Application of next-generation sequencing for rapid marker development in molecular plant breeding: a case study on anthracnose disease resistance in *Lupinus angustifolius* L. BMC Genomics 13:318
Ye W, Shen CH, Lin Y, Chen PJ, Xu X, Oelmüller R, Yeh KW, Lai Z (2014) Growth promotion-related miRNAs in *Oncidium* orchid roots colonized by the endophytic fungus *Piriformospora indica*. PLoS ONE 9:e84920
Ye Y, Ding Y, Jiang Q, Wang F, Sun J, Zhu C (2017) The role of receptor-like protein kinases (RLKs) in abiotic stress response in plants. Plant Cell Rep 36:235–242
Yu H, Goh CJ (2000) Identification and characterization of three orchid *MADS*-box genes of the *AP1/AGL9* subfamily during floral transition. Plant Physiol 123:1325–1336
Zan Y, Ji Y, Zhang Y, Yang S, Song Y, Wang J (2013) Genome-wide identification, characterization and expression analysis of populusleucine-rich repeat receptor-like protein kinase genes. BMC Genomics 14:318
Zhang J, Wu K, Zeng S, Teixeira da Silva JA, Zhao X, Tian C-E, Xia H, Duan J (2013) Transcriptome analysis of *Cymbidium sinense* and its application to the identification of genes associated with floral development. BMC Genomics 14:279
Zhang GQ, Xu Q, Bian C, Tsai WC, Yeh CM, Liu KW, Yoshida K, Zhang LS, Chang SB, Chen F, Shi Y, Su YY, Zhang YQ, Chen LJ, Yin Y, Lin M, Huang H, Deng H, Wang ZW, Zhu S, Zhao X, Deng C, Niu SC, Huang J, Wang M, Liu GH, Yang HJ, Xiao XJ, Hsiao YY, Wu WL, Chen YY, Mitsuda N, Ohme-Takagi M, Luo YB, Van de Peer Y, Liu ZJ (2016a) The *Dendrobium catenatum* Lindl. genome sequence provides insights into polysaccharide synthase, floral development and adaptive evolution. Sci Rep 6:19029
Zhang J, He C, Wu K, Teixeira da Silva JA, Zeng S, Zhang X, Yu Z, Xia H, Duan J (2016b) Transcriptome analysis of *Dendrobium officinale* and its application to the identification of genes associated with polysaccharide synthesis. Front Plant Sci 7:5
Zhang L, Chen F, Zhang GQ, Zhang YQ, Niu SC, Xiong JS, Lin Z, Cheng ZM, Liu ZJ (2016c) Origin and mechanism of crassulacean acid metabolism in orchids as implied by comparative transcriptomics and genomics of the carbon fixation pathway. Plant J 86:175–185
Zhang GQ, Liu KW, Li Z, Lohaus R, Hsiao YY, Niu SC, Wang JY, Lin YC, Xu Q, Chen LJ, Yoshida K, Fujiwara S, Wang ZW, Zhang YQ, Mitsuda N, Wang M, Liu GH, Pecoraro L, Huang HX, Xiao XJ, Lin M, Wu XY, Wu WL, Chen YY, Chang SB, Sakamoto S, Ohme-Takagi M, Yagi M, Zeng SJ, Shen CY, Yeh CM, Luo YB, Tsai WC, Van de Peer Y, Liu ZJ (2017) The apostasia genome and the evolution of orchids. Nature 549:379–383
Zhao X, Zhang J, Chen C, Yang J, Zhu H, Liu M, Lv F (2014) Deep sequencing-based comparative transcriptional profiles of *Cymbidium hybridum* roots in response to mycorrhizal and non-mycorrhizal beneficial fungi. BMC Genomics 15:747
Zhou S, Kile A, Bechner M, Place M, Kvikstad E, Deng W, Wei J, Severin J, Runnheim R, Churas C, Forrest D, Dimalanta ET, Lamers C, Burland V, Blattner FR, Schwartz DC (2004) Single-molecule approach to bacterial genomic comparisons via optical mapping. J Bacteriol 186:7773–7782
Zhu G, Yang F, Shi S, Li D, Wang Z, Liu H, Huang D, Wang C (2015) Transcriptome characterization of *Cymbidium sinense* 'Dharma' using 454 pyrosequencing and its application in the identification of genes associated with leaf color variation. PLoS ONE 10:e0128592

13

Plant regeneration via direct somatic embryogenesis from leaf explants of *Tolumnia* Louise Elmore 'Elsa'

Hui-Ju Shen[1], Jen-Tsung Chen[2], Hsiao-Hang Chung[3*] and Wei-Chin Chang[1]

Abstract

Background: *Tolumnia* genus (equitant *Oncidium*) is a group of small orchids with vivid flower color. Thousands of hybrids have been registered on Royal Horticulture Society and showed great potential for ornamental plant market. The aim of this study is to establish an efficient method for in vitro propagation.

Results: Leaf explants taken from in vitro-grown plants were used to induce direct somatic embryogenesis on a modified 1/2 MS medium supplemented with five kinds of cytokinins, 2iP, BA, kinetin, TDZ and zeatin at 0.3, 1 and 3 mg l^{-1} in darkness. TDZ at 3 mg l^{-1} gave the highest percentage of explants with somatic globular embryos after 90 days of culture. It was found that 2,4-D and light regime highly retarded direct somatic embryogenesis and showed 95–100% of explant browning. Histological observations revealed that the leaf cells divided into meristematic cells firstly, followed by somatic proembryos, and then somatic globular embryos. Eventually, somatic embryos developed a bipolar structure with the shoot apical meristem and the root meristem. Scanning electron microscopy observations showed that the direct somatic embryogenesis from leaf explants was asynchronously. The somatic embryos were found on the leaf tip, the adaxial surface and also the mesophyll through a cleft, and it reflected the heterogeneity of the explant. The 90-day-old globular embryos were detached from the parent explants and transferred onto a hormone-free 1/2 MS medium in light condition for about 1 month to obtain 1-cm-height plantlets. After another 3 months for growth, the plantlets were potted with Sphagnum moss and were acclimatized in a shaded greenhouse. After 1 month of culture, the survival rate was 100%.

Conclusions: In this report, a protocol for efficient regenerating a *Tolumnia* orchid, Louise Elmore 'Elsa', was established via direct somatic embryogenesis and might reveal an alternative approach for mass propagation of *Tolumnia* genus in orchid industry.

Keywords: Cytokinin, Direct somatic embryogenesis, Leaf explant, Light regime

Background

Tolumnia, previously referred to as "the equitant oncidiums", is small and epiphytic with unapparent pseudobulbs that usually covered by leaves. The leaves arranged in pairs overlapping or straddling one another at the base. In cross section, the leaves showed circular or triangular shape. The inflorescence of *Tolumnia* orchids bears attractive and colorful flowers with large labella, and therefore became a popular pot plants in flower market worldwide. Based on our knowledge, there was no reliable protocol had been published for propagating *Tolumnia* in vitro.

In recent years, plant regeneration via direct somatic embryogenesis had been successfully established in various orchids including *Dendrobium* (Chung et al. 2005), *Epipactis* (Moradi et al. 2017), *Malaxis* (Mahendran and Bai 2016), *Phalaenopsis* (Chen and Chang 2006; Gow et al. 2010; Feng and Chen 2014), *Rhynchostylis* (Islam and Bhattacharjee 2015) and also a *Tolumnia* related genus, *Oncidium* (Chen et al. 1999; Chen and Chang

*Correspondence: hhchung@niu.edu.tw
[3] Department of Horticulture, National Ilan University, Yilan 260, Taiwan, ROC
Full list of author information is available at the end of the article

2002; Su et al. 2006; Hong et al. 2008a, b). In these previous studies, cytokinins, especially thidiazuron, were found to be effective to induce direct somatic embryogenesis from the leaf explants (Chen et al. 1999; Chen and Chang 2002; Su et al. 2006; Hong et al. 2008a, b). Therefore, to establish an efficient method for regenerating *Tolumnia* orchids, leaf explants taken from a popular cultivar, Louise Elmore 'Elsa', was used to test the effects of cytokinins on direct somatic embryo formation.

Methods

Plant materials

In vitro grown plants of *Tolumnia* Louise Elmore 'Elsa' were purchased from Taida Horticultural Co. Ltd., Chunghua County, Taiwan. These plants were maintained on a hormone-free 1/2 MS (Murashige and Skoog 1962) medium with sucrose (20 g l^{-1}) and Gelrite (2.4 g l^{-1}) at pH 5.2 in 250 ml flasks and subcultured every 2 months period. All of the cultures were incubated under a 16/8-h (light/dark) photoperiod at irradiance of 28–36 μmol $m^{-2}s^{-1}$ (daylight fluorescent tubes FL-30D/29, 40 W, China Electric Co., Taipei, Taiwan) and temperature of 26 ± 1 °C. Plants with several roots and leaves (about 2–4 cm in height) were used as donor plants.

Embryo induction

The basal medium for embryo induction is a modified MS medium containing 1/2-strength macro- and full-strength micro-elements of MS salts supplemented with [mg l^{-1}]: myo-inositol (100), niacin (0.5), pyridoxine HCl (0.5), thiamine HCl (0.1), glycine (2.0), peptone (1000), NaH_2PO_4 (170), sucrose (20,000) and Gelrite (2400). The pH of the media was adjusted to 5.2 with 1 M KOH or HCl prior to autoclaving at 121 °C for 15 min. One-year-old in vitro grown plants of *Tolumnia* Louise Elmore 'Elsa' were utilized as donor plantlets. Leaf tip segments (about 10 mm in length) excised from the top first leaves of donor plantlets were used to induce direct somatic embryogenesis on different media supplemented with five cytokinins, 2ip, BA, kinetin, zeatin and thidiazuron (TDZ) at 0, 0.3, 1 or 3 mg l^{-1}, 2,4-D (0, 1, 3 mg l^{-1}) combined with TDZ (0, 0.3, 1, 3 mg l^{-1}). The leaf explants were placed on the culture and incubated in 90 × 15 mm^2 Petri dishes under darkness or a light condition at irradiance of 28–36 μmol $m^{-2}s^{-1}$ as mention above in an incubator at temperature of 26 ± 1 °C.

Plantlet development

Plantlets derived from somatic embryos with 3–4 leaves were transferred onto a hormone-free 1/2 MS medium with sucrose (20 g l^{-1}) and Gelrite (2.4 g l^{-1}) in 250 ml flasks with a 1-months-interval subculture period. All of the cultures were incubated under a light condition with 16/8-h (light/dark) photoperiod at irradiance of 28–36 μmol $m^{-2}s^{-1}$ and temperature of 26 ± 1 °C for further development. After 2 months of culture, the plantlets were transferred into plastic pots with Sphagnum moss in a shaded greenhouse for the further growth.

Histological analysis

The tissue samples for histological observations were fixed in FAA (95% ethyl alcohol:glacial acetic acid:formaldehyde:water, 10:1:2:7) for 48 h and dehydrated in a tertiary-butyl-alcohol series (Liao and Wu 2011). Tissue samples were then infiltrated by liquid paraffin in the oven at 65 °C and embedded in paraffin wax. The wax blocks containing tissue samples were sectioned to a thickness of 10 μm using a Leica RM2125 RTS rotary microtome (Leica Microsystems) and stained with 0.5% safranin-O (Sigma-Aldrich Inc., USA) followed by 0.1% fast green (Sigma-Aldrich Inc., USA) (Jensen 1962). After staining, sections were permanently mounted on slides for observation.

Scanning electron microscopy (SEM) observations

Samples for SEM were fixed in 2.5% glutaraldehyde in 0.1 M phosphate buffer (pH 7.0) for 4 h at 4 °C, and then dehydrated in ethanol (Dawes 1971), dried using a critical point dryer (HCP-2, Hitachi), and coated with gold in an ion coater (IB-2, Giko Engineering Co.). A SEM (DSM-950, Carl Zeiss) was used for examination and photography of the specimens.

Data analysis

All the experiments were designed with a randomized complete block design. The percentages of explants forming somatic proembryos (smaller than 0.5 mm in diameter) and somatic globular embryos (larger than 0.5 mm in diameter) were counted under a stereomicroscope (SZH, Olympus, Tokyo, Japan). Data were scored after 75 and 90 days of culture. Five replicates (dishes) and each with four leaf explants were provided for each treatment. The data expressed as percentages were transformed using arc sine prior to ANOVA and then converted back to the original scale (Compton 1994). All treatment means were compared by following Duncan's Multiple Range Test (Duncan 1955). Significant differences between means were presented at the level of $p \leq 0.05$.

Results and discussion

The developmental pathway of direct somatic embryogenesis

When the leaf explants were cultured on 1/2 MS medium supplemented with suitable concentrations of cytokinins (the details will be discussed in the following paragraph) for 60 days in darkness, transparent somatic proembryos directly formed from the epidermal cells (Fig. 1a) or from the mesophyll cells through a cleft (Fig. 1b). The SEM observations revealed that the size of these somatic proembryos was less than 0.5 mm in diameter (Fig. 1a, b). After another 15–30 days of culture on the same medium in darkness, some of the somatic proembryos developed into somatic globular embryos that showed whitish to light yellowish in color (Fig. 2a, b). Most of the somatic embryos were found on the leaf tips and the cut ends (Fig. 2a), but less than 10% of the explants (data not shown) formed somatic embryos from the adaxial sides (Fig. 2b). In general, the size of somatic globular embryos was about 0.5–2 mm in diameter (Fig. 1a).

Somatic embryo development and plantlet establishment

The 90-day-old somatic globular embryos could be easily detached from the donor leaf explants and transferred onto hormone-free 1/2 MS medium in light condition. After 1–2 weeks of culture, the somatic embryos developed shoots and subsequent turned into 1-cm-height plantlets for another 2 weeks (Fig. 2c). After another 3 months of culture in light condition, morphological normal plantlets were ready to be transplanted into pots with Sphagnum moss (Fig. 2d). Dozens of plantlets were acclimatization in a shaded greenhouse for 1 month, and the survival percentage was 100% (data not shown).

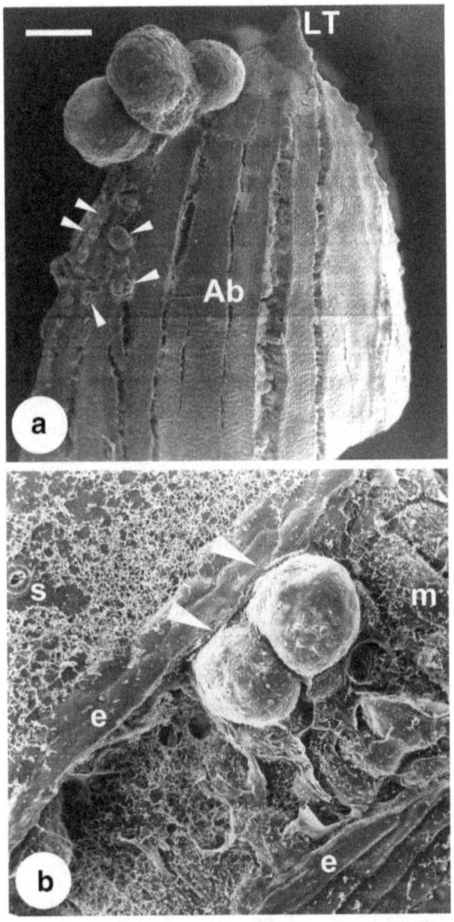

Fig. 1 Scanning electron microscopic observations on direct somatic embryogenesis from leaf explants of *Tolumnia* Louise Elmore 'Elsa'. **a** somatic proembryos (SPE, < 0.5 mm in diameter) and globular embryos (SGE, > 0.5 mm in diameter) formed from the leaf cells (*Scale bar* 1.5 mm); **b** a SPEs formed from the mesophyll cells (*Scale bar* 100 μm)

Histology of somatic embryo development

The somatic embryos were originated from the meristematic cells of the leaf explants (Fig. 3a). When compared to the original leaf cells, the meristematic cells showed smaller size and obviously densely stained (Fig. 3a). These meristematic cells were divided from the epidermal layers or together with the subepidermal layers of the explants (Fig. 3a). The somatic proembryo developed the typical bipolar structure with the putative shoot apical meristem and the putative root meristem (Fig. 3b). The somatic proembryo could be found on the same explant with the mature embryo, and thus the direct somatic embryogenesis showed an asynchronous behavior (Fig. 3b). On the posterior side of the mature embryos, the cells developed into meristematic cells, and thus suggested that secondary somatic embryogenesis will be in progress (Fig. 3b).

Effects of cytokinins on direct somatic embryogenesis

Different concentrations of five kinds of cytokinins, 2iP, BA, kinetin, TDZ and zeatin, were tested for their effects on direct somatic embryogenesis (Table 1). After 75 days of culture in darkness, 2iP at 1 and 3 mg l^{-1}, BA at 1 and 3 mg l^{-1}, TDZ at 0.3, 1 and 3 mg l^{-1}, zeatin at 0.3, 1 and 3 mg l^{-1} induced the leaf explants to form somatic proembryos (Table 1). The best result was found at 1 mg l^{-1} zeatin and 65% of the leaf explants were induced to form somatic proembryos in this treatment (Table 1). In the presence of TDZ at 0.3, 1 and 3 mg l^{-1}, the somatic globular embryos were obtained from the leaf explants (Table 1). Other cytokinins were not able to induce the somatic globular embryos after 75 days of culture in darkness (Table 1). It may partially due to the explant browning that delayed or inhibited the transition from somatic proembryos to globular embryos (Table 1). High browning percentage (75–100%) of explants were obtained at the hormone-free control treatment, 0.3, 1 and 3 mg l^{-1} 2iP and 0.3, 1 and 3 mg l^{-1} kinetin, and it reflected the relatively lower percentages of somatic embryogenesis from the explants (Table 1). Therefore, these treatments were not suitable for inducing direct somatic embryogenesis (Table 1).

Fig. 2 Somatic embryo development and plant regeneration from leaf explants of *Tolumnia* Louise Elmore 'Elsa'. **a** somatic embryos formed on the leaf tip and the cut end of a explant (*Scale bar* 3.5 mm); **b** somatic embryos formed on the edge of the adaxial surface of the leaf explant (*Scale bar* 2 mm); **c** multiple shoots developed from leaf-derived somatic embryos (*Scale bar* 5 mm); **d** plantlets converted from leaf-derived somatic embryos (*Scale bar* 2.5 cm)

After 90 days of culture in darkness, TDZ at 1 and 3 mg l^{-1}, zeatin at 0.3, 1 and 3 mg l^{-1} induced significantly higher percentages of the leaf explants with somatic proembryos, and TDZ at 3 mg l^{-1} gave the significantly highest percentages of the leaf explants with somatic globular embryos (Table 2). In the presence of TDZ and zeatin, the browning percentages of explants were lower than 20%, and it reflected the relatively higher percentages of somatic embryogenesis from the explants (Table 2). In a previous report on *Oncidium* 'Gower Ramsey', a close relative species to *Tolumnia*, 0.3, 1 and 3 mg l^{-1} kinetin induced 40% of the leaf explants to form somatic embryos (Chen and Chang 2001). However, in this present study on the *Tolumnia* leaf culture, the result showed that kinetin could not induce the direct embryogenesis (Table 2). In the previous report, 1 and 3 mg l^{-1} TDZ induced 75% of the leaf explants to form somatic embryos after 40 days of culture in darkness, and it seems these embryos had no boundary to get maturity (Chen and Chang 2001). In this study, although higher percentages of explants with somatic proembryos (75 and 90%) were induced by 1 and 3 mg l^{-1} TDZ after 90 days of culture in darkness (Table 2). However, only lower percentages (20 and 35%, respectively) of explants with somatic proembryos got through into somatic globular embryos (Table 2). It was suggested that the timing and requirements of growth regulators on embryo development was vary between the orchid species.

Effects of combinations of 2,4-D and TDZ on direct somatic embryogenesis

In previous reports, when 2,4-D combined with TDZ at appropriate concentrations, it was effective on embryogenic callus induction in *Oncidium* (Chen et al. 1999;

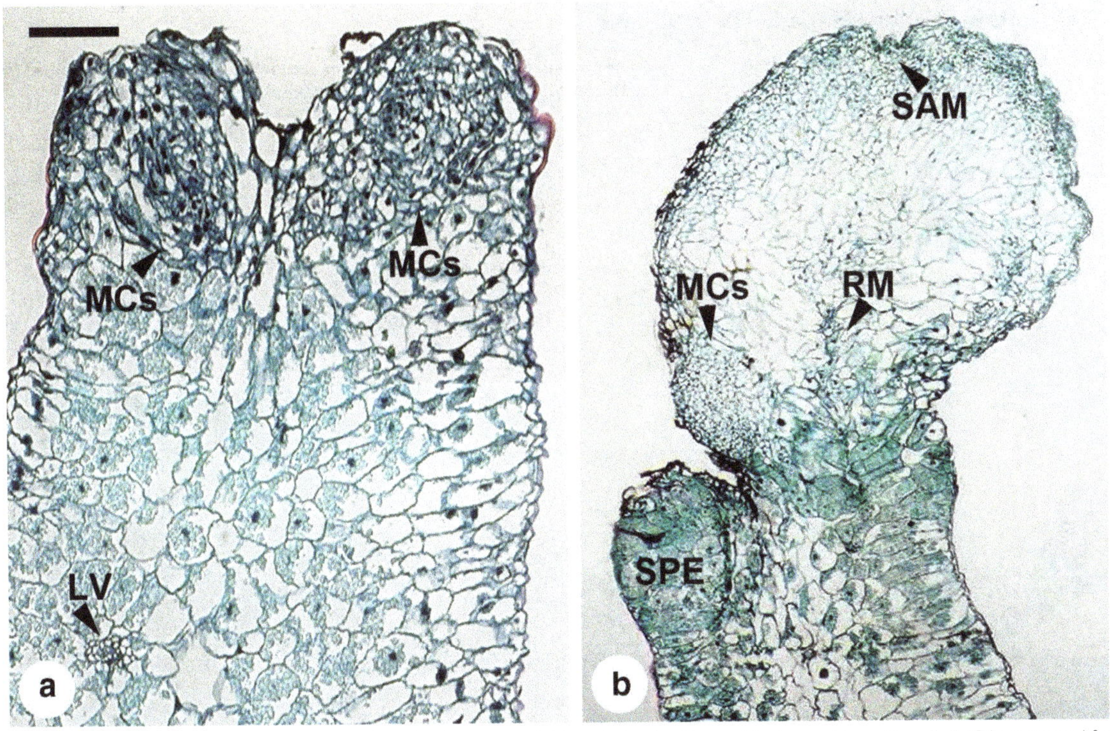

Fig. 3 Histology of direct somatic embryogenesis from leaf explants of *Tolumnia* Louise Elmore 'Elsa'. **a** meristematic cells (MCs) originated from the mesophyll cells of a leaf explant (*Scale bar* 0.1 mm); leaf vein (LV); **b** a somatic proembryo (SPE) and a somatic embryo developed the bipolar structure with putative shoot apical meristem (SAM) and putative root meristem (RM) (*Scale bar* 0.5 mm)

Table 1 Effects of cytokinins on direct somatic embryogenesis from leaf explants of *Tolumnia* Louise Elmore 'Elsa', when the explants were cultured in darkness for 75 days

Cytokinins	Concentration (mg l^{-1})	% of explants with somatic proembryos	% of explants with somatic globular embryos	% of browning
Hormone-free control	0	0 f	0 a	85.0 abc
2iP	0.3	0 f	0 a	90.0 abc
	1	20.0 def	0 a	75.0 c
	3	5.0 ef	0 a	80.0 bc
BA	0.3	0 f	0 a	30.0 d
	1	35.0 abcd	0 a	20.0 de
	3	15.0 def	0 a	10.0 de
Kinetin	0.3	0 f	0 a	100.0 a
	1	0 f	0 a	80.0 bc
	3	0 f	0 a	95.0 ab
TDZ	0.3	50.0 abc	5.0 a	5.0 de
	1	45.0 abc	10.0 a	0 e
	3	30.0 bcde	5.0 a	10.0 de
Zeatin	0.3	55.0 ab	0 a	10.0 de
	1	65.0 a	0 a	10.0 de
	3	30.0 cde	0 a	5.0 de

Means within a column followed by the same letter are not significantly different according Duncan's multiple range test at $p \leq 0.05$ (Duncan 1955)

Table 2 Effects of cytokinins on direct somatic embryogenesis from leaf explants of *Tolumnia* Louise Elmore 'Elsa', when the explants were cultured in darkness for 90 days

Cytokinins	Concentration (mg l^{-1})	% of explants with somatic proembryos	% of explants with somatic globular embryos	% of browning
Hormone-free control	0	0 d	0 d	100.0 a
2iP	0.3	0 d	0 d	95.0 ab
	1.0	20.0 cd	0 d	85.0 ab
	3.0	5.0 d	0 d	85.0 ab
BA	0.3	5.0 d	0 d	70.0 b
	1.0	60.0 ab	10.0 cd	35.0 c
	3.0	30.0 bc	5.0 cd	10.0 cd
Kinetin	0.3	0 d	0 d	100.0 a
	1.0	0 d	0 d	90.0 ab
	3.0	0 d	0 d	95.0 ab
TDZ	0.3	42.5 ab	15.0 bcd	5.0 d
	1.0	75.0 a	20.0 abc	20.0 cd
	3.0	90.0 a	35.0 a	15.0 cd
Zeatin	0.3	70.0 a	5.0 cd	10.0 cd
	1.0	90.0 a	20.0 abc	10.0 cd
	3.0	70.0 a	30.0 ab	10.0 cd

Means within a column followed by the same letter are not significantly different according Duncan's multiple range test at $p \leq 0.05$ (Duncan 1955)

Chen and Chang 2000). However, 2,4-D showed highly inhibitory on direct somatic embryogenesis in *Oncidium* (Chen et al. 1999; Chen and Chang 2001), *Dendrobium* (Chung et al. 2005, 2007) and *Phalaenopsis* (Kuo et al. 2005; Gow et al. 2008). In this present study, we combined 1 mg l^{-1} 2,4-D with 0, 0.3 and 3 mg l^{-1} TDZ to test the effect on direct somatic embryogenesis. Table 3 showed that at 1 mg l^{-1} 2,4-D plus 0.3 mg l^{-1} TDZ and 1 mg l^{-1} 2,4-D plus 3 mg l^{-1} TDZ, 5 and 10% of explants formed somatic proembryos, respectively. However, no globular embryos were found at these two treatments and it may due to the 100% of browning percentage (Table 3). In addition, the leaf explants did not form callus at all of the treatments.

Table 3 Effects of 2,4-D plus TDZ on direct somatic embryogenesis from leaf explants of *Tolumnia* Louise Elmore 'Elsa', when the explants were cultured in darkness for 90 days

2,4-D (mg l^{-1})	TDZ (mg l^{-1})	% of explants with somatic proembryos	% of explants with somatic globular embryos	% of browning
0	0	0 a	0	100.0 a
1	0	0 a	0	100.0 a
1	0.3	5.0 a	0	100.0 a
1	3	10.0 a	0	100.0 a

Means within a column followed by the same letter are not significantly different according Duncan's multiple range test at $p \leq 0.05$ (Duncan 1955)

Effect of light regime on direct somatic embryogenesis

In previous reports, light regime was a negative factor that highly retarded the embryogenic response of leaf explants in *Oncidium, Dendrobium* and *Phalaenopsis* (Chen et al. 1999; Chung et al. 2005, 2007; Gow et al. 2009). However, in the root-derived callus culture of *Oncidium* 'Gower Ramsey', light regime was a crucial factor for inducing of indirect embryogenesis (Chen and Chang 2000; Wu et al. 2004). In this study, Table 4 showed that light regime highly retarded direct somatic embryogenesis and resulted in 95–100% of explants browning.

Conclusion

This present study is the first to establish an efficient protocol for regenerating a *Tolumnia* orchid via direct somatic embryogenesis. It took about 4 months to obtain the regenerated plantlets with normal morphogenesis. Plant regeneration was derived from leaf cultures via an in vitro morphogenetic pathway that meristematic cells developed initially, and followed by somatic proembryos, somatic globular embryos, somatic embryos with the bipolar structure, and eventually plantlets. This protocol provided the basis for further investigation on micropropagation, in vitro

Table 4 Effects of TDZ and zeatin on direct somatic embryogenesis from leaf explants of *Tolumnia* Louise Elmore 'Elsa', when the explants were cultured in light condition for 90 days

Cytokinins	Concentration (mg l^{-1})	% of explants with somatic pro-embryos	% of explants with somatic globular embryos	% of browning
Hormone-free control	0	0 c	0	100.0 a
TDZ	1	10.0 abc	0	100.0 a
	3	5.0 bc	0	100.0 a
Zeatin	1	25.0 ab	0	100.0 a
	3	40.0 a	0	95.0 a

Means within a column followed by the same letter are not significantly different according Duncan's multiple range test at $p \leq 0.05$ (Duncan 1955)

morphogenesis, gene transfer or breeding programs in *Tolumnia* orchids.

Abbreviations

2,4-D: 2,4-dichlorophenoxyacetic acid; 2iP: N^6-2-isopentenyl adenine; BA: N^6-benzyladenine; Kinetin: N^6-furfuryladenine; TDZ: 1-phenyl-3-(1,2,3-thiadiazol-5-yl)-urea; Zeatin: 6-[4-hydroxy-3-methylbut-2-enylamino]purine.

Authors' contributions

JTC and WCC conceived and designed the experiments; HJS carried out the experimental work and data collection; JTC helped in the data analysis; JTC and HHC wrote the manuscript; HHC reviewed and finalized the manuscript; WCC provided laboratory facilities and important comments. All authors read and approved the final manuscript.

Author details

[1] Institute of Plant and Microbial Biology, Academia Sinica, Taipei 115, Taiwan, ROC. [2] Department of Life Sciences, National University of Kaohsiung, Kaohsiung 811, Taiwan, ROC. [3] Department of Horticulture, National Ilan University, Yilan 260, Taiwan, ROC.

Competing interests

The authors declare that they have no competing interests.

Funding

No funding was received.

References

Chen JT, Chang WC (2000) Efficient plant regeneration through somatic embryogenesis from callus cultures of *Oncidium* (Orchidaceae). Plant Sci 160:87–93

Chen JT, Chang WC (2001) Effects of auxins and cytokinins on direct somatic embryogenesis in *Oncidium* 'Gower Ramsey'. Plant Growth Regul 34:229–232

Chen JT, Chang WC (2002) Effects of tissue culture conditions and explant characteristics on direct somatic embryogenesis in *Oncidium* 'Gower Ramsey'. Plant Cell Tiss Org 69:41–44

Chen JT, Chang WC (2006) Direct somatic embryogenesis and plant regeneration from leaf explants of *Phalaenopsis amabilis*. Biol Plant 50:169–173

Chen JT, Chang C, Chang WC (1999) Direct somatic embryogenesis on leaf explants of *Oncidium* 'Gower Ramsey' and subsequent plant regeneration. Plant Cell Rep 19:143–149

Chung HH, Chen JT, Chang WC (2005) Cytokinins induce direct somatic embryogenesis of *Dendrobium* Chiengmai Pink and subsequent plant regeneration. Vitro Cell Dev Biol Plant 2005(41):765–769

Chung HH, Chen JT, Chang WC (2007) Plant regeneration through direct somatic embryogenesis from leaf explants of *Dendrobium*. Biol Plant 51:346–350

Compton ME (1994) Statistical methods suitable for the analysis of plant tissue culture data. Plant Cell Tissue Org 37:217–242

Dawes CJ (1971) Biological techniques in electron microscopy. Barnes & Noble, New York

Duncan DB (1955) Multiple range and multiple F test. Biometrics 11:1–42

Feng JH, Chen JT (2014) A novel in vitro protocol for inducing direct somatic embryogenesis in *Phalaenopsis aphrodite* without taking explants. Sci World J. Article ID 263642: 7. https://doi.org/10.1155/2014/263642

Gow WP, Chen JT, Chang WC (2008) Influence of growth regulators on direct embryo formation from leaf explants of *Phalaenopsis* orchids. Acta Physiol Plant 30:507–512

Gow WP, Chen JT, Chang WC (2009) Effects of genotype, light regime, explant position and orientation on direct embryo formation from leaf explants of *Phalaenopsis* orchids. Acta Physiol Plant 31:363–369

Gow WP, Chen JT, Chang WC (2010) Enhancement of direct somatic embryogenesis and plantlet growth from leaf explants of *Phalaenopsis* by adjusting culture period and explant length. Acta Physiol Plant 32:621–627

Hong PI, Chen JT, Chang WC (2008a) Effects of salicylic and acetylsalicylic acid on direct somatic embryogenesis in *Oncidium*. J Plant Biochem Biotechnol 17:149–153

Hong PI, Chen JT, Chang WC (2008b) Promotion of direct somatic embryogenesis of *Oncidium* by adjusting carbon sources. Biol Plant 52:597–600

Islam SS, Bhattacharjee B (2015) Plant regeneration through somatic embryogenesis from leaf and root explants of *Rhynchostylis retusa* (L.) Blume. Appl Biol Res 17:158–165

Jensen WA (1962) Botanical histochemistry: principles and practice. Freeman, San Francisco

Kuo HL, Chen JT, Chang WC (2005) Efficient plant regeneration through direct somatic embryogenesis from leaf explants of *Phalaenopsis* 'Little Steve'. Vitro Cell Dev Biol Plant 41:453–456

Liao YK, Wu YH (2011) *In vitro* propagation of *Platycerium bifurcatum* (Cav.) C. Chr. green globular body initiation. Bot Stud 52:455–463

Mahendran G, Bai VN (2016) Direct somatic embryogenesis of *Malaxis densiflora* (A. Rich.) Kuntze. J Genet Eng Biotechnol 14:77–81

Moradi S, Dianati Daylami S, Arab M, Vahdati K (2017) Direct somatic embryogenesis in *Epipactis veratrifolia*, a temperate terrestrial orchid. J Hortic Sci Biotechnol 92:88–97

Murashige T, Skoog F (1962) A revised medium for rapid growth and bioassays with tobacco tissue cultures. Physiol Plant 15:495–497

Su YJ, Chen JT, Chang WC (2006) Efficient and repetitive production of leaf-derived somatic embryos of *Oncidium*. Biol Plant 50:107–110

Wu IF, Chen JT, Chang WC (2004) Effects of auxins and cytokinins on embryo formation from root-derived callus of *Oncidium* 'Gower Ramsey'. Plant Cell Tissue Org 77:107–109

14

Morphology, taxonomy and mating-type loci in natural populations of *Volvox carteri* in Taiwan

Hisayoshi Nozaki[1*], Noriko Ueki[2], Mari Takusagawa[3], Shota Yamashita[1], Osami Misumi[4], Ryo Matsuzaki[5], Masanobu Kawachi[5], Yin-Ru Chiang[6] and Jiunn-Tzong Wu[6]

Abstract

Background: *Volvox carteri* f. *nagariensis* is a model taxon that has been studied extensively at the cellular and molecular level. The most distinctive morphological attribute of *V. carteri* f. *nagariensis* within *V. carteri* is the production of sexual male spheroids with only a 1:1 ratio of somatic cells to sperm packets or androgonidia (sperm packet initials). However, the morphology of male spheroids of *V. carteri* f. *nagariensis* has been examined only in Japanese strains. In addition, *V. carteri* f. *nagariensis* has heterothallic sexuality; male and female sexes are determined by the sex-determining chromosomal region or mating-type locus composed of a > 1 Mbp linear chromosome. Fifteen sex-specific genes and many sex-based divergent shared genes (gametologs) are present within this region. Thus far, such genes have not been identified in natural populations of this species.

Results: During a recent fieldwork in Taiwan, we encountered natural populations of *V. carteri* that had not previously been recorded from Taiwan. In total, 33 strains of this species were established from water samples collected in Northern Taiwan. Based on sequences of the internal transcribed spacer 2 region of nuclear ribosomal DNA and the presence of asexual spheroids with up to 16 gonidia, the species was clearly identified as *V. carteri* f. *nagariensis*. However, the sexual male spheroids of the Taiwanese strains generally exhibited a 1:1 to > 50:1 ratio of somatic cells to androgonidia. We also investigated the presence or absence of several sex-specific genes and the sex-based divergent genes *MAT3m*, *MAT3f* and *LEU1Sm*. We did not identify recombination or deletion of such genes between the male and female mating-type locus haplotypes in 32 of the 33 strains. In one putative female strain, the female-specific gene *HMG1f* was not amplified by genomic polymerase chain reaction. When sexually induced, apparently normal female sexual spheroids developed in this strain.

Conclusions: Male spheroids are actually variable within *V. carteri* f. *nagariensis*. Therefore, the minimum ratio of somatic cells to androgonidia in male spheroids and the maximum number of gonidia in asexual spheroids may be diagnostic for *V. carteri* f. *nagariensis*. *HMG1f* may not be directly related to the formation of female spheroids in this taxon.

Keywords: Mating-type locus, Morphology, Sexual reproduction, Taxonomy, *Volvox*, *Volvox carteri*, *Volvox carteri* f. *nagariensis*

Background

The genus *Volvox* represents the most complex member of volvocine green algae, especially in terms of multicellularity and sex (Kirk 1998; Hiraide et al. 2013). In recent years, *Volvox carteri* f. *nagariensis* has been studied extensively at the cellular and molecular level (e.g., Kirk et al. 1999; Ferris et al. 2010). Although this taxon or form was originally described based on a natural sample collected in Nagari, India (Iyengar 1933), most of the studied strains are Japanese strains of *V. carteri* f.

*Correspondence: nozaki@bs.s.u-tokyo.ac.jp
[1] Department of Biological Sciences, Graduate School of Science, University of Tokyo, Hongo, Bunkyo-ku, Tokyo 113-0033, Japan
Full list of author information is available at the end of the article

nagariensis such as Eve (Starr 1969; Kirk et al. 1999; Ferris et al. 2010; Geng et al. 2014). Six forms are recognized within *V. carteri*, and the most distinctive morphological attribute for distinguishing *V. carteri* f. *nagariensis* from other forms within this species is the production of sexual male spheroids with only a 1:1 ratio of somatic cells to sperm packets or androgonidia (sperm packet initials) (Nozaki 1988). However, this ratio has been examined only in Japanese strains (Starr 1969; Nozaki 1988; Geng et al. 2014) irrespective of the presence of Indian strains from Poona (Adams et al. 1990).

Volvox carteri f. *nagariensis* is an oogamous species that has heterothallic sexuality determined by the sex-determining chromosomal region or mating-type locus where the presence of sex-specific genes (genes harboring in only one of the two sexes) and sex-based divergent shared genes (gametologs; two homologous genes harboring in this regions of both of the two sexes) was recently demonstrated (Ferris et al. 2010). Sex in *V. carteri* f. *nagariensis* is determined based on the presence or absence of such genes. However, sex identification in natural populations of this species has not previously been conducted by using molecular markers. Furthermore, the mating-type locus of *V. carteri* f. *nagariensis* is composed of a > 1 Mbp linear chromosome where recombination is suggested to be repressed (Ferris et al. 2010). Although the mating-type locus or sex chromosomal region does not exhibit recombination in *Chlamydomonas reinhardtii* under the laboratory conditions, recombination and gene conversion of *C. reinhardtii* mating-type locus genes were demonstrated in natural populations of this species (De Hoff et al. 2013). However, recombination of the mating-type locus genes has not been examined in natural populations of *V. carteri*.

During a recent field survey of the freshwater green algae in Taiwan, we encountered natural populations of *V. carteri*, which have not previously been recorded in Taiwan (Yamagishi 1992). Based on the internal transcribed spacer 2 region of nuclear ribosomal DNA (nuclear rDNA ITS-2), the populations were clearly identified as *V. carteri* f. *nagariensis*. This study was undertaken to reveal morphological details of *V. carteri* f. *nagariensis* originating from Taiwan and to examine presence or absence of several sex-specific and sex-based divergent genes in these natural populations of *V. carteri* f. *nagariensis*.

Methods

Table 1 shows the origins of the 33 Taiwanese strains used in the present study. Clones were established by the pipette-washing method (Pringsheim 1946). The cultures were maintained in screw-cap tubes (18 × 150 mm) containing 11 mL AF-6 or VTAC medium (Kawachi et al. 2013) at 25 °C with a 14:10 h light:dark (LD) schedule under cool-white fluorescent lamps at an intensity of 80–130 μmol m^{-2} s^{-1}. Several of these new wild strains are available from the Microbial Culture Collection at the National Institute for Environmental Studies (Kawachi et al. 2013; http://mcc.nies.go.jp/index_en.html) as NIES-4205–4210 (Table 1). To observe the morphology of asexual spheroids, the cultures were grown in 11 mL VTAC medium in Petri dishes (10 × 60 mm) at 25 °C with a 14:10 h LD cycle. For comparison, we also used Eve [i.e., UTEX 1885, obtained from the Culture Collection of Algae at the University of Texas at Austin (UTEX,

Table 1 List of Taiwanese strains of *Volvox carteri* f. *nagariensis*

Habitat/date [locality designation]	Longitude, latitude [temperature/pH]	Number of strains	Sex based on genomic PCR	Strains deposited in NIES-collection and their DDBJ/EMBL/GenBank accession numbers of nuclear rDNA ITS-2
Rice paddy/25 May 2016 [2]	N 25°13′02.76″, E 121°36′22.44″ [26.1 °C, pH 6.7]	0	Male	
		1	Female	2016-tw-nuk-Chg (=NIES-4205), LC376031
Fallow rice paddy/9 Jun 2016 [4]	N 25°13′51.27″, E 121°37′53.78″ [36.2 °C, pH 6.8]	4	Male	
		12	Female	2016-0609-v-1 (=NIES-4206), LC376032
Fallow rice paddy/10 Jun 2016 [6]	N 25°00′52.51″, E 121°55′33.38″ [29.7 °C, pH 6.7]	14	Male	2016-0610-v-11 (=NIES-4207), LC376033; 2016-tw-nuk-6-1 (=NIES-4208), LC376034
		0	Female	
Rice paddy/10 Jun 2016 [8]	N 25°01′09.82″, E 121°55′21.76″ [29.0 °C, pH 5.7]	1	Male	2016-tw-nuk-8-1 (=NIES-4209), LC376035
		1	Female	2016-tw-nuk-8-2 (=NIES-4210), LC376036

https://utex.org)], a typical female strain of *V. carteri* f. *nagariensis* (Ferris et al. 2010; Geng et al. 2014).

Morphological details of asexual and sexual spheroids were observed mainly in three strains, 2016-0609-v-1 (female), 2016-tw-nuk-6-1 (male) and 2016-tw-nuk-8-1 (male). To observe asexual spheroids and reproduction, we examined small aliquots of asexual spheroids, in actively grown 2- to 5-day-old cultures in the Petri dishes (10 × 60 mm). Sexual male spheroids developed spontaneously in a culture of male strains grown in VTAC medium at 20 °C. To induce sexual spheroids, 0.1–0.2 mL of the sexual inducer (supernatant of the male culture after the formation of sperm packets) and ca. 0.5 mL of an actively growing female or male strain were added to 11 mL of fresh VTAC medium in a tube and placed at 25 °C with a 14:10 h LD cycle. After 2–4 days, sexual spheroids developed. To observe zygotes, sexual male and female spheroids induced in tubes (each 11 mL) were mixed with fresh VTAC medium (11 mL) in Petri dishes (20 × 150 mm) that were placed at 25 °C with a 14:10 h LD cycle.

Light microscopy was carried out using a BX60 microscope (Olympus, Tokyo, Japan) equipped with Nomarski optics. The cells in spheroids were counted as previously described (Nozaki 1988). Individual cellular sheaths of the gelatinous matrix of the spheroids were examined after mixing approximately 10 μL cultured material with 2–5 μL of 0.002% (w/v in distilled water) methylene blue (1B-429 WALDECK GmbH & Co Division Chroma, Münster, Germany).

Extraction of total DNA and sequencing of nuclear rDNA ITS-2 were performed as previously described (Nozaki et al. 2000, 2006; Hayama et al. 2010). Nuclear rDNA ITS-2 sequences of the Taiwanese strains (Table 1) and a Chinese strain "Wang (Shenzhen)" (DDBJ/EMBL/GenBank Accession Number HG422808) and the sequences aligned with the data set in Kawafune et al. (2015) were used for phylogenetic analyses. The alignment of the whole data matrix was refined as described by Kawafune et al. (2015). Since the ITS-2 sequences from all of the 33 Taiwanese strains were identical to those of strains Eve and NIES-397 of *V. carteri* f. *nagariensis*, these 35 strains were treated as a single operational taxonomic unit (OTU). The alignment of 10 OTUs of *V. carteri* and *V. obversus* (Additional file 1: Figure S1) was subjected to the maximum-likelihood (ML) analysis, based on the selected model (T92 + G) with 1000 bootstrap replicates (Felsenstein 1985) using MEGA7 (Kumar et al. 2016). In addition, 1000 bootstrap replicates were performed using the maximum-parsimony (MP) method, based on a branch-and-bound search by PAUP 4.0b10 (Swofford 2002). *V. obversus* was treated as the outgroup because it is sister to *V. carteri* (Kawafune et al. 2015; Nozaki et al. 2015).

We performed genomic polymerase chain reaction (PCR) using the primers listed in Table 2 to check the presence or absence of several sex-specific genes and the sex-based divergent genes *MAT3m*, *MAT3f* and *LEU1Sm* (Ferris et al. 2010) in the 33 Taiwanese strains and Eve. PCR was carried out using KOD FX Neo DNA polymerase (TOYOBO, Osaka, Japan) according to the manufacturer's instructions.

Results

Morphology

Asexual spheroids of Taiwanese strains were ovoid or spherical in shape and measured up to 920 μm long; each spheroid contained 1400–3000 somatic cells embedded within the periphery of the gelatinous matrix and large reproductive cells or gonidia in the posterior two-thirds (Fig. 1a, b). Each somatic cell was nearly spherical or ovoid in shape, up to 10 μm in diameter, and enclosed by a rectangular to hexagonal space formed by individual sheaths of gelatinous matrix (Fig. 1c, d). The cells lacked adjoining cytoplasmic bridges (Fig. 1c), and exhibited cup-shaped chloroplasts with a single stigma and a basal pyrenoid. The spheroid exhibited a gradual decrease in stigma size from the anterior to posterior pole. Gonidia were spherical in shape and had large vacuoles, measuring up to 92 μm in diameter (Fig. 1e). There were up to 16 gonidia per spheroid (Fig. 1a, b).

Asexual reproduction occurred as previously described (Starr 1969; Nozaki 1988). Successive divisions of each gonidium resulted in formation of a plakea. Gonidia of the next generation were evident outside the plakea because of the unequal cytokinesis (Fig. 1f, g). The plakea then inverted to form a compact, spheroidal daughter spheroid.

Sexual male spheroids were ellipsoidal or ovoid and 128- or 256-celled, containing biflagellate somatic cells and androgonidia that developed into sperm packets (Fig. 2a–d). The number of androgonidia or sperm packets in male spheroids varied even within the same culture. The ratio of somatic cells to androgonidia (sperm packets) in male spheroids was generally 4–12:1 (Fig. 2a, b). However, male spheroids sometimes exhibited a > 50:1 ratio (Fig. 2c) or approximately 1:1 ratio (Fig. 2d) of somatic cells to androgonidia. Female spheroids were ellipsoidal or ovoid in shape, and contained 1000–2000 biflagellate somatic cells and 22–28 eggs (Fig. 2e). The zygotes developed within the female spheroid after possible fertilization (Fig. 2f). The mature zygotes had a reticulate cell wall and were reddish brown in color, measuring 34–40 μm in diameter (Fig. 2g).

Table 2 Primer sequences and product sizes of sex-specific genes, sex-based divergent genes *MAT3* and *LEU1S* (Fig. 4b), and *ACTIN* gene

Gene name [Accession number]	Primer name	Primer sequence (5′ to 3′)	Product size (bp)
MID1m [GU784916]	MID1-F1	GCTGAAGGAGTGTATCGACGCATT	287
	MID1-R1[a]	GCTGCCTGCAAATTCGCTTAAGGT	
HMG1f [GU784915]	HMG1-1F	GACTGCTCACCTATCAGTTGGTCA	326
	HMG1-1R[a]	TTGCCACCTCGTATGCATCGAAGT	
MAT3f [GU784915]	MAT3f-F	AATCGCATGCAGCGCGCACTTTCT	404
	MAT3f-R[a]	TCAGCTGCCAGGCTGGCAGTTATG	
MAT3m [GU784916]	MAT3 m-F	GTGCGCATGCAACGTGCGCTTGAC	401
	MAT3 m-R[a]	TCAGCTGCAAGGTTAGCCGTCCTC	
FSI1f [GU784915]	FSI-1-F	ATGCAGCGGATTGAGATGGCCTTC	428
	FSI-1-R[a]	ATCACCTCTGTACAGGTCGCCTCC	
MTM0946 [GU784916]	MTM-946-F	GCCAACCAACCCGTTCCCAGCATA	420
	MTM-946-R[a]	GCCCTGGACGACTCTAGGGGTCTT	
MTF0821 [GU784915]	MTF-0821-F	GTCGGGCTCATCAGTAAAAGCTGG	358
	MTF-0821-R[a]	GTTCGCCAAGGAACCGAGCTTAAC	
LEU1Sm [GU784916]	LEU1Sm-F	ATGCAGTTTCAGGTCCTAAGGGCT	451
	LEU1Sm-R[a]	CACGTACAGAGTTTCCAAGCCGCT	
MTF0991 [GU784915]	MTF0991-1F	ATGAACACATGCCCATCCTGGTGC	396
	MTF0991-1R[a]	TCAAGGCGCTGCATCGCTGCTACT	
MTF2030 [GU784915]	MTF2030-1F	ATGGCAAACGCAGACCCTGGTACA	464
	MTF2030-2R[a]	CCCTTGCGCCTGCCAGCAGACCCA	
ACTIN [XM_002955490]	VxCnAct-F2	GAAGCTCTGCTATGTGGCGCTGGA	410[b]
	VxCnAct-R3[a]	ATGGTCGTTCCACCAGAGAGCACG	

[a] Reverse primer

[b] Determined in the present study

Morphology of asexual and sexual spheroids in the present Taiwanese strains was essentially consistent with that of the Japanese strains of *V. carteri* f. *nagariensis* except for male spheroids (Starr 1969; Nozaki 1988). The Japanese strains produce male spheroids with only a 1:1 ratio of somatic cells to androgonidia (Starr 1969; Nozaki 1988). In contrast, male spheroids of the present Taiwanese strains exhibited variability in the ratios (Fig. 2).

Molecular phylogeny of nuclear rDNA ITS-2

All 33 strains from Taiwan had the same nuclear rDNA ITS-2 sequences, which were also identical to the *V. carteri* f. *nagariensis* strain Eve (UTEX 1885). The phylogenetic analysis clearly demonstrated that the Taiwanese strains belonged to the *V. carteri* f. *nagariensis* clade (Fig. 3).

Genomic PCR of 33 strains from Taiwan

We screened the 33 Taiwanese strains for the presence or absence of five female-specific genes, two male-specific genes, the female and male types of the sex-based divergent gene *MAT3*, and the male type of *LEU1S* (Ferris et al. 2010) (Tables 1, 2).

Thirty-two of the Taiwanese strains and Eve could be clearly subdivided into female and male types (Table 1). The female type was composed of 13 Taiwanese strains and Eve, which shared the presence of five female-specific genes and *MAT3f* and the absence of two male-specific genes and male types of two sex-based divergent genes (*MAT3m* and *LEU1Sm*) (Table 1, Fig. 4; Ferris et al. 2010). Nineteen other strains belonged to the male type, based on the presence of two male-specific genes, *MAT3m* and *LEU1Sm* and the absence of five female-specific genes and *MAT3f* (Table 1, Fig. 4; Ferris et al. 2010). Thus, there seemed to be no recombination or deletion of such genes between male and female mating-type locus haplotypes in 32 of the 33 Taiwanese strains. However, in the remaining strain (2016-tw-nuk-8-2), one female-specific gene (*HMG1f*) was not amplified whereas this strain contained four other female-specific genes and *MAT3f* and showed no amplifications of all four of the male genes examined (Fig. 4). Thus, this strain seemed to be a female strain without amplification of *HMG1f* in genomic PCR.

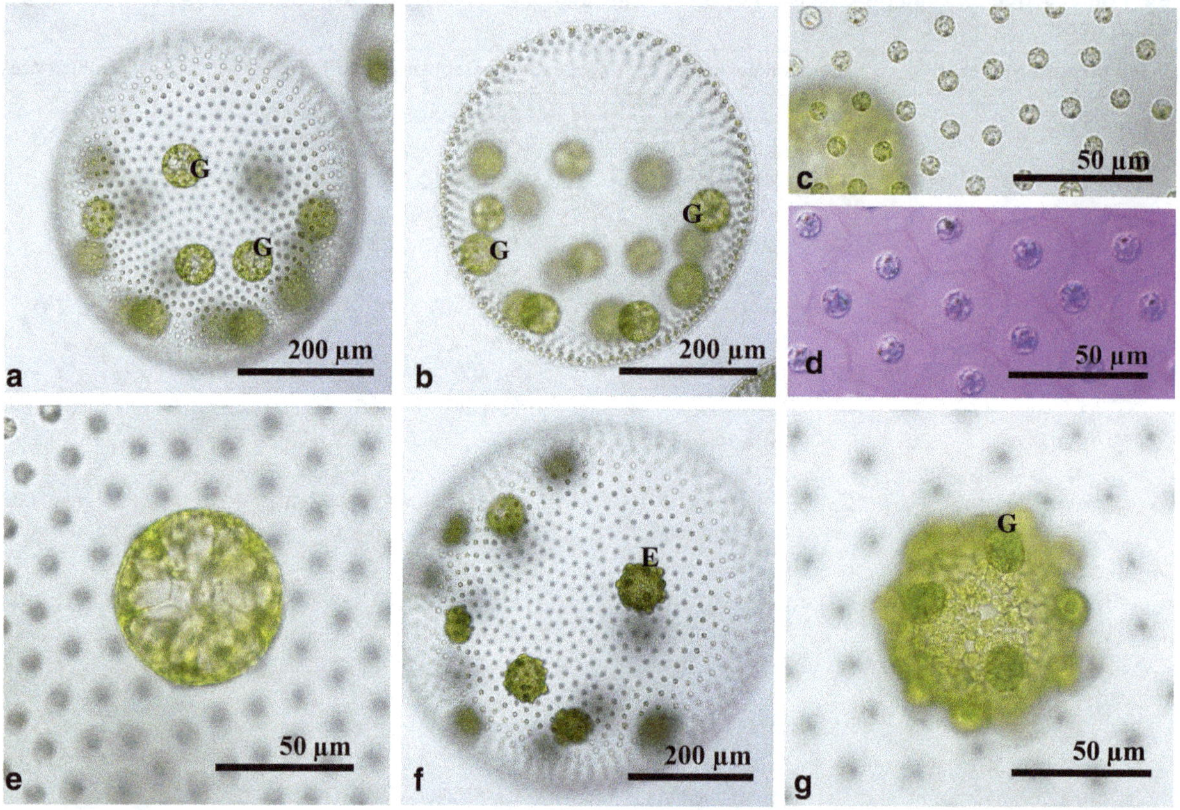

Fig. 1 Light microscopy of asexual spheroids in Taiwanese strains of *Volvox carteri* f. *nagariensis*. **a** Surface view of a spheroid showing undivided gonidia (G). 2016-tw-nuk-6-1. **b** Optical section of a spheroid in (**a**) with gonidia (G). **c** Surface view of spheroid. Note no cytoplasmic bridges between somatic cells. 2016-tw-nuk-6-1. **d** Surface view of spheroid showing individual sheaths of the gelatinous matrix. Stained with methylene blue. 2016-tw-nuk-8-1. **e** Optical section of gonidium. 2016-tw-nuk-6-1. **f, g** Pre-inversion plakea or embryo (E) showing gonidia (G) of the next generation outside. 2016-tw-nuk-8-1

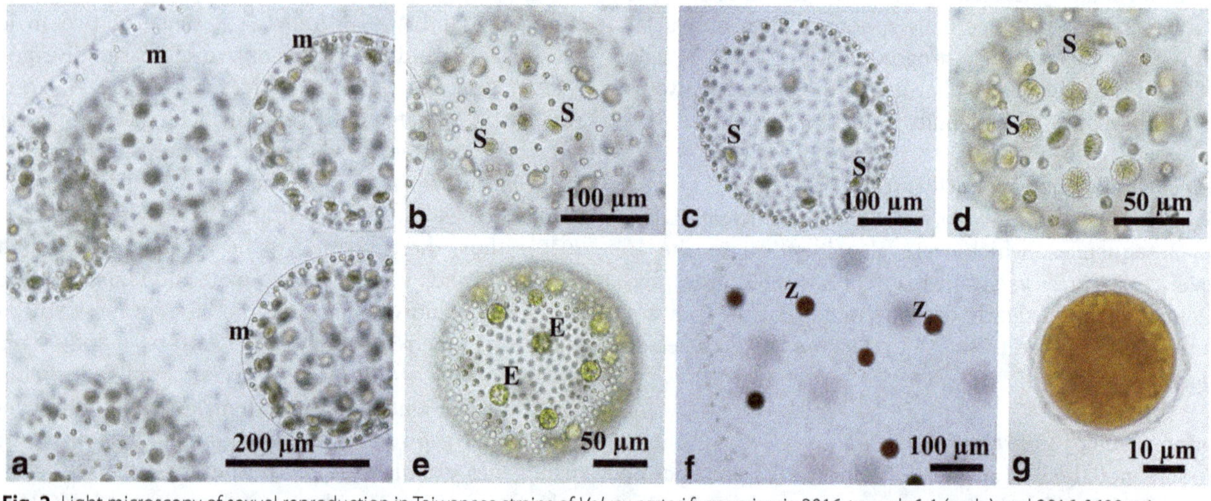

Fig. 2 Light microscopy of sexual reproduction in Taiwanese strains of *Volvox carteri* f. *nagariensis*. 2016-tw-nuk-6-1 (male) and 2016-0609-v-1 (female). **a** Male spheroids (m) developing within parental spheroid. **b–d** Various ratios of somatic cells to sperm packets (S) in male spheroids. Note 4–12:1 ratio in (**a** and **b**), > 50:1 ratio in (**c**), and 1:1 ratio in (**d**). **e** Female spheroids with eggs (E). **f** Mature zygotes (Z) in female spheroid. **g** Optical section of mature zygote with reticulate wall

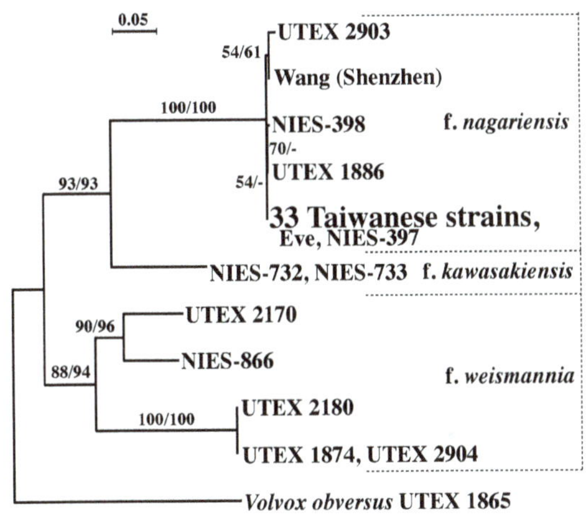

Fig. 3 Phylogenetic positions of 33 Taiwanese strains within three forms of *Volvox carteri*. The tree was based on the maximum likelihood (ML) analysis of the data matrix of Kawafune et al. (2015) plus a Chinese strain "Wang (Shenzhen)" (Additional file 1: Figure S1). Numbers at left and right above the branches represent bootstrap values (50% or more) of ML and maximum parsimony analyses, respectively, based on 1000 replicates. The branch lengths are proportional to the evolutionary distances used in the ML analysis and indicated by bar above the tree. For details, see "Methods"

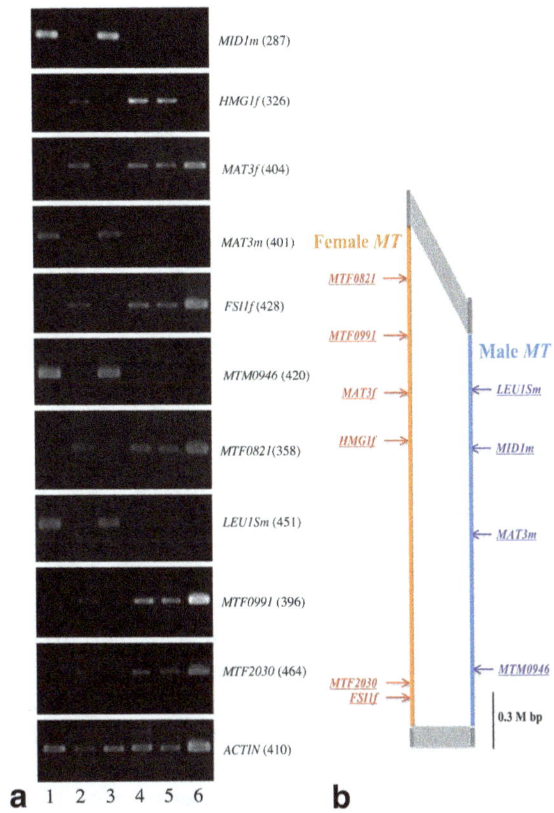

Fig. 4 Mating-type locus (*MT*) genes of *Volvox carteri* f. *nagariensis*. **a** Results of genomic PCR of five Taiwanese strains (lanes 1–4 and 6) and Eve (UTEX 1885) (lane 5) using primers listed in Table 2. Numbers within the parentheses just right to the gene designations represent expected lengths of the PCR products. Lane 1: 2016-tw-nuk-6-1. Lane 2: 2016-0609-v-1. Lane 3: 2016-0610-v-11. Lane 4: 2016-tw-nuk-Chg. Lane 6: 2016-tw-nuk-8-2. Note that both male genes and *HMG1f* were not amplified in 2016-tw-nuk-8-2. **b** Diagram showing positions of sex-specific and sex-based divergent genes in the mating-type locus analyzed by genomic PCR in (**a**). Based on Ferris et al. (2010)

Asexual spheroids and sexual induction of the HMG1f-lacking putative female strain

Asexual spheroids of the *HMG1*-lacking strain (2016-tw-nuk-8-2) were essentially the same as those of other Taiwanese strains, with up to 16 gonidia that developed into daughter spheroids via successive divisions and inversion (Fig. 5a–c). When sexually induced, apparently normal female sexual spheroids developed in this strain. The number of eggs produced was generally 20 or more (Fig. 5d). No morphological differences were recognized between these female spheroids and the female spheroids of other female stains. Attempts to cross 2016-tw-nuk-8-2 with male Taiwanese strains were unsuccessful, because the male Taiwanese strains suddenly lost their ability to form male spheroids during the study (from April to June 2017).

Discussion

Nozaki (1988) recognized six forms within *V. carteri*, and characterized *V. carteri* f. *nagariensis* as having male spheroids with only a 1:1 ratio of somatic cells to androgonidia (sperm packets). However, this ratio was based on observations of only Japanese strains studied by Starr (1969), and subsequent studies have also used only these strains and/or their progenies (e.g., Kirk et al. 1999; Ferris et al. 2010). Based on the ITS-2 phylogeny, the present Taiwanese strains are clearly identified as *V. carteri* f. *nagariensis* because these have the exactly the same ITS-2 sequence as that of the Japanese strains (Fig. 3). However, the male spheroids of the Taiwanese strains exhibited a 1:1 to >50:1 ratio of somatic cells to androgonidia (Fig. 2). Thus, the morphology of male spheroids is actually variable within *V. carteri* f. *nagariensis*. On the other hand, *V. carteri* f. *nagariensis* is similar to *V. carteri* f. *kawasakiensis* with respect to the production of 16 or more gonidia in asexual spheroids. However, the minimum ratio of somatic cells to androgonidia differs between these two forms. The minimum ratio in *V. carteri* f. *nagariensis* is 1:1, whereas that of *V. carteri* f. *kawasakiensis* is 0:1. Thus, *V. carteri* f. *nagariensis* can be distinguished from other forms within *V. carteri* in producing asexual spheroids with 16 or more gonidia and lacking male spheroids without somatic cells.

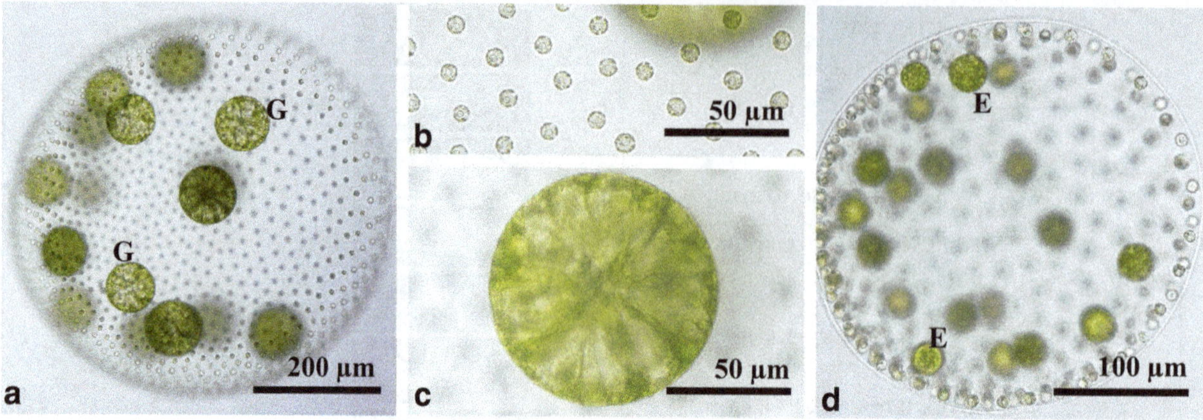

Fig. 5 Light microscopy of a putative female strain (2016-tw-nuk-8-2) with no amplification of *HMG1f* in genomic PCR (see Fig. 3 and Additional file 1: Figure S2). **a** Asexual spheroid with 16 gonidia (G). **b** Surface view of asexual spheroid showing no cytoplasmic bridges between cells. **c** Optical section of gonidium in asexual spheroid. **d** Female spheroid with eggs (E)

HMG1f is one of five female-specific genes of *V. carteri* f. *nagariensis* (Ferris et al. 2010), and encodes a putative HMG-box DNA binding protein (Umen 2011). In Opisthokonta (Metazoa and Fungi), sex determination involves HMG-box domain proteins (Waters et al. 2007; Idnurm et al. 2008). However, it has been suggested that the female-specific gene *HMG1f* is a possible negative regulator of sexual function in females because its transcription is downregulated in the sexual phase (Umen 2011). The present genomic PCR analyses resolved a possible female strain of *V. carteri* f. *nagariensis* (2016-tw-nuk-8-2) that did not exhibit amplification of *HMG1f* (Fig. 4; Additional file 1: Figures S2, S3). This strain exhibited normal asexual and female spheroids that could not be distinguished from those of other female strains (Fig. 5). Thus, *HMG1f* may not contribute to the formation of sexual female spheroids. Further analyses of this strain with sexually active males and other females of *V. carteri* f. *nagariensis* are needed to determine the function of *HMG1f*. However, we cannot completely rule out the alternative explanation of polymorphisms in *HMG1f* that precluded amplification. Thus, further analyses are needed for evidencing the complete absence of *HMG1f* in this interesting strain.

Conclusions

Given that the Japanese and Taiwanese populations of *V. carteri* f. *nagariensis* grow in two geographically separated regions (temperate and subtropical, respectively), they can be presumed to be independent lineages derived from a common ancestor. However, the present genomic analyses of 33 Taiwanese strains could not demonstrate the existence of recombination within the >1 Mbp mating-type locus. This is consistent with previous suggestions that recombination is suppressed within the mating-type locus (Ferris et al. 2010). This may indicate that the >1 Mbp mating-type locus in *V. carteri* f. *nagariensis* is biologically important. Alternatively, such an expanded mating-type locus may result from only passive accumulation of mutations in the duplicated (male and female) genome regions that are divergent enough to inhibit gene conversion (Teshima and Innan 2004). Further studies are needed to resolve this problem, especially in other forms of *V. carteri* (Fig. 3).

Abbreviations

LD: light:dark; nuclear rDNA ITS2: internal transcribed spacer 2 region of nuclear ribosomal DNA; OTU: operational taxonomic unit; PCR: polymerase chain reaction.

Authors' contributions

HN, OM and JTW conceived the study, participated in its design and coordination. NU, MT, SY, RM, MK and YRC collected, analyzed, and interpreted the data. HN wrote the manuscript. All authors read and approved the final manuscript.

Author details

[1] Department of Biological Sciences, Graduate School of Science, University of Tokyo, Hongo, Bunkyo-ku, Tokyo 113-0033, Japan. [2] Department of Biology, Brooklyn College, City University of New York, Brooklyn, NY 11210, USA. [3] Department of Botany, Graduate School of Science, Kyoto University, Sakyo-ku, Kyoto 606-8502, Japan. [4] Department of Biological Science and Chemistry, Faculty of Science, Graduate School of Medicine, Yamaguchi University, Yoshida, Yamaguchi 753-8512, Japan. [5] Center for Environmental Biology and Ecosystem Studies, National Institute for Environmental Studies, Onogawa, Tsukuba, Ibaraki 305-8506, Japan. [6] Biodiversity Research Center, Academia Sinica, Nankang, Taipei 11529, Taiwan.

Acknowledgements

We thank Dr. Chao-Jen Shih and Lai-Chu Kow at Biodiversity Research Center, Academia Sinica, Taiwan, for their help with field collection.

Competing interests

The authors declare that they have no competing interests.

Funding

This work was supported by Grants-in-Aid for Scientific Research (Nos. 25304012 and 16H02518 to HN) from MEXT/JSPS KAKENHI.

References

Adams CR, Stamer KA, Miller JK, McNally JG, Kirk MM, Kirk DL (1990) Patterns of organellar and nuclear inheritance among progeny of two geographically isolated strains of *Volvox carteri*. Curr Genet 18:141–153

De Hoff PL, Ferris P, Olson BJSC, Miyagi A, Geng S, Umen JG (2013) Species and population level molecular profiling reveals cryptic recombination and emergent asymmetry in the dimorphic mating locus of *C. reinhardtii*. PLoS Genet 9:e1003724. https://doi.org/10.1371/journal.pgen.100372

Felsenstein J (1985) Confidence limits on phylogenies: an approach using the bootstrap. Evolution 39:783–791. https://doi.org/10.1111/j.1558-5646.1985.tb00420.x

Ferris P, Olson BJ, De Hoff PL, Douglass S, Casero D, Prochnik S, Geng S, Rai R, Grimwood J, Schmutz J, Nishii I, Hamaji T, Nozaki H, Pellegrini M, Umen JG (2010) Evolution of an expanded sex determining locus in *Volvox*. Science 328:351–354. https://doi.org/10.1126/science.1186222

Geng S, De Hoff P, Umen JG (2014) Evolution of sexes from an ancestral mating-type specification pathway. PLoS Biol 12:e1001904. https://doi.org/10.1371/journal.pbio.1001904

Hayama M, Nakada T, Hamaji T, Nozaki H (2010) Morphology, molecular phylogeny and taxonomy of *Gonium maiaprilis* sp. nov. (Goniaceae, Chlorophyta) from Japan. Phycologia 49:221–234. https://doi.org/10.2216/PH09-56.1

Hiraide R, Kawai-Toyooka H, Hamaji T, Matsuzaki R, Kawafune K, Abe J, Sekimoto H, Umen J, Nozaki H (2013) The evolution of male–female sexual dimorphism predates the sex-based divergence of the mating locus gene *MAT3/RB*. Mol Biol Evol 30:1038–1040. https://doi.org/10.1093/molbev/mst018

Idnurm A, Walton FJ, Floyd A, Heitman J (2008) Identification of the sex genes in an early diverged fungus. Nature 451:193–196. https://doi.org/10.1038/nature06453

Iyengar MOP (1933) Contribution to our knowledge of the colonial Volvocales of South India. J Linn Soc Bot 49:323–373. https://doi.org/10.1111/j.1095-8339.1933.tb00391.x

Kawachi M, Ishimoto M, Mori F, Yumoto K, Sato M, Noël M-H (2013) MCC-NIES list of strains, 9th Edition, microbial culture collection at National Institute for Environmental Studies, Tsukuba, Japan. http://mcc.nies.go.jp/download/list9th_e.pdf. Accessed 9 Jan 2018

Kawafune K, Hongoh Y, Hamaji T, Sakamoto T, Kurata T, Hirooka S, Miyagishima S, Nozaki H (2015) Two different rickettsial bacteria invading *Volvox carteri*. PLoS ONE 10:e0116192. https://doi.org/10.1371/journal.pone.0116192

Kirk DL (1998) *Volvox*: molecular genetic origins of multicellularity and cellular differentiation. Cambridge University Press, Cambridge

Kirk MM, Stark K, Miller SM, Müller W, Taillon BE, Gruber H, Schmitt R, Kirk DL (1999) *regA*, a *Volvox* gene that plays a central role in germ-soma differentiation, encodes a novel regulatory protein. Development 126:639–647

Kumar S, Stecher G, Tamura K (2016) MEGA7: molecular evolutionary genetics analysis version 7.0 for bigger datasets. Mol Biol Evol 30:2725–2729. https://doi.org/10.1093/molbev/msw054

Nozaki H (1988) Morphology, sexual reproduction and taxonomy of *Volvox carteri* f. *kawasakiensis* f. nov. (Chlorophyta) from Japan. Phycologia 27:209–220. https://doi.org/10.2216/i0031-8884-27-2-209.1

Nozaki H, Misawa K, Kajita T, Kato M, Nohara S, Watanabe MM (2000) Origin and evolution of the colonial Volvocales (Chlorophyceae) as inferred from multiple, chloroplast gene sequences. Mol Phylogenet Evol 17:256–268. https://doi.org/10.1006/mpev.2000.0831

Nozaki H, Ott FD, Coleman AW (2006) Morphology, molecular phylogeny and taxonomy of two new species of *Pleodorina* (Volvoceae, Chlorophyceae). J Phycol 42:1072–1080. https://doi.org/10.1111/j.1529-8817.2006.00255.x

Nozaki H, Matsuzaki R, Yamamoto K, Kawachi M, Takahashi F (2015) Delineating a new heterothallic species of *Volvox* (Volvocaceae, Chlorophyceae) using new strains of "*Volvox africanus*". PLoS ONE 10:e0142632. https://doi.org/10.1371/journal.pone.0142632

Pringsheim EG (1946) Pure cultures of algae. Cambridge University Press, London

Starr RC (1969) Structure, reproduction and differentiation in *Volvox carteri* f. *nagariensis* Iyengar, strains HK9 & 10. Arch Protistenkd 111:204–222

Swofford DL (2002) PAUP*: phylogenetic analysis using parsimony (* and other methods), version 40b10. Sinauer, Sunderland

Teshima KM, Innan H (2004) The effect of gene conversion on the divergence between duplicated genes. Genetics 166:1553–1560

Umen JG (2011) Evolution of sex and mating loci: an expanded view from Volvocine algae. Curr Opin Microbiol 14:634–641. https://doi.org/10.1016/j.mib.2011.10.005

Waters PD, Wallis MC, Marshall Graves JA (2007) Mammalian sex—origin and evolution of the Y chromosome and SRY. Semin Cell Dev Biol 18:389–400. https://doi.org/10.1016/j.semcdb.2007.02.007

Yamagishi T (1992) Plankton algae in Taiwan (Formosa). Uchida Rokakuho, Tokyo

15

Expression shifts of floral symmetry genes correlate to flower actinomorphy in East Asia endemic *Conandron ramondioides* (Gesneriaceae)

Kuan-Ting Hsin[1] and Chun-Neng Wang[1,2*]

Abstract

Background: Bilateral symmetry flower (zygomorphy) is the ancestral state for Gesneriaceae species. Yet independent reversions to actinomorphy have been parallelly evolved in several lineages. *Conandron ramondioides* is a natural radially symmetrical species survived in dense shade mountainous habitats where specialist pollinators are scarce. Whether the mutations in floral symmetry genes such as *CYC*, *RAD* and *DIV* genes, or their expression pattern shifts contribute to the reversion to actinomorphy in *C. ramondioides* thus facilitating shifts to generalist pollinators remain to be investigated. To address this, we isolated putative orthologues of these genes and relate their expressions to developmental stages of flower actinomorphy.

Results: Tissue specific RT-PCR found no dorsal identity genes *CrCYCs* and *CrRADs* expression in petal and stamen whorls, while the ventral identity gene *CrDIV* was expressed in all petals. Thus, ventralized actinomorphy is evolved in *C. ramondioides*. However, *CrCYCs* still persists their expression in sepal whorl. This is congruent with previous findings that *CYC* expression in sepals is an ancestral state common to both actinomorphic and zygomorphic core Eudicot species.

Conclusions: The loss of dorsal identity genes *CrCYCs* and *CrRADs* expression in petal and stamen whorl without mutating these genes specifies that a novel regulation change, possibly on cis-elements of these genes, has evolved to switch zygomorphy to actinomorphy.

Keywords: Peloria, Reversal, Zygomorphy, *CYCLOIDEA*, *DIVARICATA*, *RADIALIS*, Bilateral symmetry

Background

Evolutionary reversal to actinomorphy (flower radial symmetry) from zygomorphy (flower bilateral symmetry) have occurred multiple times independently across flowering plant diversification (Hileman 2014). Although zygomorphy enhances pollination specialization, these reversals may have evolved in a benefit of increased pollinator generalization when pollinators are scarce or in harsh conditions (Cronk and Möller 1997; Donoghue et al. 1998). The frequent transitions of floral symmetry reversals along many flowering plant lineages imply that a similar or modified developmental program has been independently recruited for floral symmetry transition (Zhang et al. 2013; Hileman 2014).

Gesneriaceae species are predominantly bilateral flower symmetry (zygomorphy) and exhibit a great diversity of floral forms (Endress 1999, 2001; Weber 2004). Their flowers have been adaptively evolved with bee, fly, moth, birds and even bat into a variety of pollination syndromes (Harrison et al. 1999; Perret et al. 2007). However, there are also certain lineages evolved with flower reversion to actinomorphy. There are at least five and four independent reversals to actinomorphy events

*Correspondence: leafy@ntu.edu.tw
[1] Institute of Ecology and Evolutionary Biology, National Taiwan University, Taipei, Taiwan
Full list of author information is available at the end of the article

occurred in old world and new world Gesneriaceae lineages, respectively (Wang et al. 2010; Smith et al. 2004; Clark et al. 2011). This frequency of reversals to actinomorphy in Gesneriaceae is the highest among Lamiales, probably because it is the basal most lineage of Lamiales which is just derived from the ourgroup actinomorphic Oleaceae species (Endress 1999). The repeated reversals to actinomorphy in Gesneriaceae species also implied there are perhaps similar yet modified developmental programs repeatedly recruited among these reversals.

It has been argued that the reversals to actinomorphy may contain selection disadvantage by losing its specific pollinators (Cronk and Möller 1997). However, in the aforementioned examples, the actinomorphic floral forms caused by reversals in Gesneriaceae species are usually compromised by pollinator shifts and pollination strategies switched from nectar to pollen rewards (Weber 2004). The reversal to actinomorphy with corolla tube fully opened to attract every kinds of general pollinators may be selected for when only few pollinators are available (Cronk and Möller 1997). For example, the European relict actinomorphic species in Pyrenees, *Ramonda myconi*, has been inferred that reversion to actinomorphy by opening of the corolla tube provides adaptive advantage to attract more available general pollinators in harsh alpine habitat (Cronk and Möller 1997; Picó et al. 2002). Additionally, its pollination syndrome has shifted from nectar reward which is usually favored by specific pollinators to pollen reward allowing general pollinators to visit (Weber 2004; Wang et al. 2010).

Conandron ramondioides is another relic and paleoendemic genus in East Asia which is also apparently evolved as an actinomorphic reversal from zygomorphic ancestor (Kokubugata and Peng 2004; Wang et al. 2010; Xiao et al. 2012). Like *R. myconi*, its corolla tube is lost and five petals are equally large with five stamens fully developed. These indicate they are natural peloria with complete actinomorphy. Nectary glands of *C. ramondioides* are lost. Also, its stamens are dehisced by apical pores and pollens are powdery suggest they are pollinated by pollen-collecting bees (Wang, pers. obs.). *Conandron ramondioides* can only survive in limestone cliff, often in deep shade forest, where pollinators are scarce. Taken together, the reversal to actinomorphy in *C. ramondioides* could prevent it from relying only on specific pollinators because insect activity is relative low under this habitat.

The establishment of flower zygomorphy requires *CYCLOIDEA* (*CYC*) gene specifically expressed in dorsal side of the flower to promote growth difference between dorsal and ventral petals and retarding dorsal stamen (Luo et al. 1996, 1999; Corley et al. 2005). Thus the reversal to actinomorphy in *C. ramondioides* could probably results from loss of *CYC* function or a *CYC* expression shifts in flower bud. In *Antirrhinum majus*, mutation of *CYC* and its paralog *DICH* can result in complete actinomorphy with all petals equal in size resembling the ventral one and no retardation on stamens. *CYC* in *C. ramondioides* may also be mutated thus becoming actinomorphy. However, *CYC* and *DICH*'s effect is through activating a downstream MYB family gene *RADIALIS* (*RAD*) at dorsal side, whose encoded protein restricts another MYB-like protein *DIVARICATA* (*DIV*) to the ventral region (Corley et al. 2005; Costa et al. 2005; Raimundo et al. 2013). Thus in addition to *CYC*, fully functional *RAD* is also needed to develop flower zygomorphy. In *cyc dich* double mutant, *RAD* could not be activated to restrict ventral identity gene *DIV* to ventral region resulting in all petals resembling to ventral petal of wild type. This type of actinomorphic reversal due to mutation of *CYC* and/or its downstream *RAD* therefore is often called abaxialized (ventralized) effect (Cronk 2006; Zhang et al. 2013; Hileman 2014; Spencer and Kim 2017).

In contrast to ventralization, actinomorphy can be established through expanded expression of *CYC* and its homologues in petal whorl, an adaxialized (dorsalized) effect. In actinomorphic species such as *Cadia* of legumes and certain Malpighiaceae species, the expression of *CYC* extended from dorsal regions to lateral and ventral regions of the corolla (Citerne et al. 2006; Zhang et al. 2013). It thus appears that both the loss of *CYC*-like gene expression (ventralization) and the expansion (dorsalization) of *CYC*-like gene expression are two major mechanisms in creating flower actinomorphic reversal in angiosperm.

In Gesneriaceae, reversal to actinomorphy through both ventralization or dorsalization were reported (e.g. *Bournea leiophylla, Tengia scopulorum and Saintpaulia ionantha*) (Zhou et al. 2008; Pang et al. 2010; Hsu et al. 2018). In *B. leiophylla*, the *BlCYC1* and *BlRAD* genes were transiently expressed in floral meristem initiation stage and then quickly vanished at latter developmental stages. The loss of *CYC* expression at later stages correlates with the fact that all petals resembling ventral ones, demonstrating a ventralization form of reversal (Zhou et al. 2008). Unlike *B. leiophylla*, *CYC*-like expression in *T. scopulorum* has ubiquitously expressed in all petals across dorsoventral axis, a dorsalized form of actinomorphic reversal (Pang et al. 2010). Partial *CYC* sequences have been isolated from both *C. ramondioides* and *R. myconi* but no apparent SNP mutations result in premature stop codon in their coding regions (Xiao and Wang 2007; Picó et al. 2002). These imply *CYC* in *C. ramondioides* may still function. It would therefore essential to investigate whether expression patterns of *CYC* together with downstream *RAD* and *DIV* shifts, which may correlate and explain the developmental switch of actinomorphic reversal in *C. ramondioides.*

To examine possible roles of floral symmetry genes (homologues of *CYC, RAD* and *DIV*) involving in establishment of actinomorphic in *C. ramondioides,* we examined their expression patterns along floral developmental stages and separated floral organs. In order to ascertain the developmental process of actinomorphy in *C. ramondioides,* we also observed the bud development using scanning electron microscope. From these results, we hope to find whether there is a correlation between shifts of expression patterns among these floral symmetry genes and corresponded floral symmetry transition.

Materials and methods

Floral development in *C. ramondioides*

Floral buds of *C. ramondioides* were collected from top of inflorescence containing developing floral organ primordia for SEM examination. The materials were fixed in FAA overnight then transfer to 70% EtOH for preservation. Fixed materials were pre-dissected under stereo microscope S8APO (Leica) then dehydrated through an ethanol series (85%, 95%, and 100% twice) with each step for 20 min. After dehydration in 100% ethanol, materials were dehydrated through ascending gradients of acetone, dried with molecular sieve, and finally dried in a critical point dryer (Hitachi E101). Dried samples were mounted on aluminum stubs and then coated with gold–palladium using Hitachi E1011I sputter. Specimens were viewed using FEI SEM at working distance at 10 mm, and operating at 15 kV. Stages of flower development were summarized in Additional file 1: Table S1.

Isolation and characterization of *CYC, RAD* and *DIV* homologues

CYC, RAD and *DIV* homologues were isolated from *C. ramondioides* total genomic DNA by using degenerate primer pairs. We used either previously published primers which have been claimed able to amplify all possible gene copies, or newly designed primers located at conserve domain of each gene to isolate these floral symmetry gene copies. For example, for isolating *CYC* homologues from *C. ramondioides,* a pair of Geseneriaceae specific primers, GcycFS and GcycR, were used for amplifying *CYC* homologues (Möller et al. 1999). For *RAD,* primers were design from conserved MYB domain with all available GenBank sequences from Gesneriaceae species and *Antirrhinum.* Similarly, *DIV* primers were design from conserved R2 and R3 domain. To isolate *CYC* homologues from *C. ramondioides,* The PCR products were then cloned into pGEM-T easy vector system (Promega, USA) and 8 clones were sequenced for checking numbers of *CYC* homologues of *C. ramondioides.* Then, to isolate *RAD* homologues of *C. ramondioides,* degenerate primer pair QAL-F (5′-RTTRGC RGTKTAYGACA-3′) and FPN-R (5′-TTYCCYAAYTAC WGGACCA-3′) locating at conserve MYB domain and conserve 3′ end were designed according to available *RAD* homologues of other Gesneriaceae species. Last, to isolate *DIV* homologues of *C. ramondioides,* degenerate primer pair DIV-F MEI (5′-ATGGAGATTTTRDCM-CCAAGTT-3′) and DIV-YGK-R1 (5′-CTCCARTCYCCY TTYCCATA-3′) locating at R2 end and R3 domain were designed based on available Gesneriaceae sequences and *DIV* form *A. majus* respectively. Both *RAD* and *DIV* homologue PCR products were cloned and sequenced (8 clones for examining *RAD* homologue; 7 clones for examining *DIV* homologue) following the same process as previously mentioned above. To extend into the 5′ and 3′ unknown sequence region of amplified *RAD* and *DIV* partial sequence above, rapid amplification of cDNA ends (5′- and 3′-RACE, SMART RACE cDNA amplification kit, Clontech) technique is applied for obtaining full length cDNA according to manufacturer suggestion. To investigate the homology of isolated *CYC, RAD* and *DIV* of *C. ramondioides,* we aligned full length sequences of them with their genbank available homologs from subfamily Didymocarpoideae where it belongs to and those from closely related model species *Antirrhinum majus* (Scrophulariaceae) (Luo et al. 1996; Almeida et al. 1997; Zhou et al. 2008; Yang et al. 2012).

Phylogenetic analysis of isolated *CYC, RAD* and *DIV* homologues in *C. ramondioides*

To check the homology of these isolated *CYC, RAD* and *DIV* genes from *C. ramondioides,* available NCBI homologues of Gesneriaceae species, *Antirrhinum majus* and *Arabidopsis thaliana* were downloaded and used for reconstructing their phylogenies respectively. Sequences used to reconstruct phylogeny were listed in Additional file 2: Table S2. Nucleotide sequences were first translated into amino acid and aligned using default settings in CLUSTALX (Thompson et al. 1997) with major domains specified, then manually aligned afterward. We apply both neighbor joining (NJ) and maximum likelihood (ML) algorithm for testing the robustness of reconstructed *CYC, RAD* and *DIV* phylogeny. The NJ tree of each gene dataset was reconstructed using MEGA 6 (Tamura et al. 2013). For ML tree, the web interface PhyML 3.0 was applied (Guindon et al. 2010). Best-fit nucleotide substitution model of each dataset was evaluated by smart model selection (SMS) which is implementing in PhyML 3.0 (Lefort et al. 2017). For *CYC*-like gene dataset, the best-fit model is HKY+G model. For

RAD, TN93+G model is suggested and GTR+I+G model for *DIV*.

Locus-specific RT-PCR

Flower buds were categorized into three stages (Fig. 1): early stage (Stage 10, 2–4 mm in diameter, sepal longer than petal), middle stage (Stage 13, 5–7 mm in diameter, petal longer than sepal), and late stage (Stage 15, anthesis) (see Additional file 1: Table S1). They were collected, in the field through fixing in RNAlater (Ambion, Life technologies, USA), or freshly collected from individuals grown in the greenhouse. Next, to detect expression locations of *CYC*, *RAD* and *DIV* homologues on petals, a single petal was dissected from flower buds at early developmental stage. Because RNA yield may be low in single petal, pooled sepals, petals, stamen and gynoecium were dissected from flower buds at early stage to confirm the expression locations of *CYC*, *RAD* and *DIV* homologues in *C. ramondioides*. Total RNA of floral buds and dissected floral organs were extracted following TRIzol® Reagent (Invitrogen, USA) protocol. Single-strand cDNA (20 ng/μl) were reverse transcribed from total RNA of these samples by SuperscriptIV Reverse Transcriptase (Invitrogen). Gene specific primer pairs were used to examine each candidate gene's expression level: *CrCYC1C* (forward: 5′-AGACATGCTTTCTGG CCACT-3′, Reverse: 5′-CTTCTTCGCCTTCTGAAT GC-3′), *CrCYC1D* (Forward: 5′-CAGGTGCAGATT CGATGAGA-3′, Reverse: 5′-GTTCCATTGCAGTCT CCCAT-3′), *CrCYC2* (Forward: 5′-TCTTGCTTCATC AGCACCAC-3′, Reverse: 5′-GTGATGCCCCTACTT GCACT-3′), *CrRAD1* (Forward: 5′-ATGGCCTCAAGT TCCTTGACCGCT, Reverse: 5′-GTGGTCCAGTAG TTGGGAAATGGT), *CrRAD2* (Forward: 5′-GCAATG AGCTCCATGTCTAGTGGT, Reverse: 5′-CCCACT TCGATATTCTTCACATCCTCC-3′) and *CrDIV* (Forward: 5′-CGATCGATGGCAAAGAGTGG-3′, Reverse: 5′-TACTCGAACACCAACCCAGG-3′). The RT-PCR mixture contains 12 μl of Ampliqon master mix Red III (Denmark), 6.25 mM of each primer, 9.5 μl ddH_2O and 1 μl first-strand cDNA. The PCR condition used for amplifying *CrCYC1C*, *CrCYC1D* and *CrCYC2* genes were 94 °C for 3 min followed by 35 cycles of 94 °C for 30 s, 53 °C for 40 s, 72 °C for 20 s and a 2 min final extension at 72 °C. To amplify higher Tm *CrRAD1* and *CrRAD2* genes, the program is 94 °C for 3 min followed by 35 cycles of 94 °C for 30 s, 57 °C for 40 s, 72 °C for 20 s and a 2 min final extension at 72 °C. For *CrDIV* gene, the program is 94 °C for 3 min followed by 35 cycles of 94 °C for 30 s, 53 °C for 40 s, 72 °C for 20 s and a 2 min final extension at 72 °C. *Cr18S* was used as positive control. Two biological replicates were carried out to validate the reproducibility of the results (see Additional file 3: Figure S1).

Results

Floral development

Development of flower in *C. ramondioides* can be divided into 15 milestone stages (Additional file 1: Table S1). From Stage 3 SEM picture (Fig. 2a), all five sepals were already initiated but dorsal and lateral sepals were slightly smaller than ventral ones (a residual zygomorphy). During Stage 7 (sepal removed), petals and stamens appeared to be equaled in size when initiated (Fig. 2c). This is more evident during Stage 7A when gynoecium started to emerged, in which all petals and stamens were grown in equal size (Fig. 2d). Petals and stamens continued to grow in equal rate thus all 5 petals and stamens are the same size toward anthesis (Fig. 2e). The floral diagram of fully developed *C. ramondioides* thus can be drawn as Fig. 2f, showing complete actinomorphy of *C. ramondioides* flower at anthesis.

Characterization and phylogeny of *CrCYC*, *CrRAD* and *CrDIV* genes

Among *CYC* clones, three out of 8 belonged to *CrCYC1C*, two belongs to *CrCYC1D* and three belongs to *CrCYC2*. For *CrRADs*, four clones belonged to *CrRAD1*, while the other four were *CrRAD2*. For *CrDIV*, all seven clones belonged to *CrDIV*. We believed our approach can effectively isolate all possible copies of each gene because we designed the primers in most conserved domain of each gene (see "Material and methods" section). Full-length cDNA of *CrCYC* genes, *CrRAD* genes and *CrDIV* were isolated from developing floral tissues and dissected tissues. The *CrCYC*, *CrRAD* and *CrDIV* sequences we isolated from *C. ramondioides* have been deposited in NCBI database (Accession number MH366524 to MH366529, detailed information see Additional file 2: Table S2). There are three *CYC* homologs, *CrCYC1C*, *CrCYC1D*, *CrCYC2* identified in *C. ramondioides* (Fig. 3a). Their full length amino acid sequences are 339, 338, and 335, respectively. According to phylogeny (Fig. 4a), we designated them as *CrCYC1C*, *CrCYC1D* and *CrCYC2*. Sequence analysis shows that *CrCYC1C*, *CrCYC1D* and *CrCYC2* are 26.1%, 17% and 23% identical to *Antirrhinum CYC*, respectively. When comparing the TCP domain, R domain, and ECE domains, *CrCYC1C*, *CrCYC1D*, and *CrCYC2* shared 92.6%, 90.2% and 92.6% amino acid sequence identity with *Antirrhinum CYC* , suggesting these genes are functionally related. When compared with *CYC*-like genes from available closely related Didymocarpoid Gesneriaceae species, *CrCYC1C*

Fig. 1 The flower of *C. ramondioides* from early to late stage. Early, middle and late stage corresponding to Stage 10, 13, 15 in Additional file 1: Table S1. Actinomorphy was observed in late stage *C. ramondioides* flower. Scale bar represented 1 cm

and *CrCYC1D* are 86% and 78.4% identical to *BlCYC1* of *Bournea leiophylla*, respectively, and *CrCYC2* is 88.1% identical to *BlCYC2*.

The sequence difference between these *CYC*-like genes are mainly located in the intervening regions of *CYC* domains mentioned above. Phylogenetic analysis shows that *CrCYC1C*, *CrCYC1D* (first isolated in this study) and *CrCYC2* formed three monophyletic clades (*GCYC1C*, *GCYC1D* and *GCYC2*) with other Gesneriaceae *CYC* homologs at amino acid level, confirming to previous phylogenetic trees (Fig. 4a).

Next, two *RAD* homologues isolated from *C. ramondioides* were *CrRAD1* and *CrDAD2*. They shared 76% and 68% amino acid identity with *RAD* from *A. majus*, respectively. Both *CrRAD1* and *CrDAD2* have one conserved 55aa-MYB-domain as *RAD* does (Fig. 3b). Phylogenetic analyses based on neighbor-joining method showed that *CrRAD1* and *CrRAD2* formed two distinct clades with high support (bootstrap/ML: 96/99 for *RAD1* clade; 100/100 for *RAD2* clade) (Fig. 4b). With *A. majus RAD* as outgroup, *CrRAD1* formed one monophyletic clade with Didymocarpoid Generiaceae species such as *RAD1* of *Saintpaulia ionantha* with high support, while *CrRAD2* formed another monophyletic clade with *Saintpaulia RAD2* and *RAD*-like gene of *B. leiophylla* according to nucleotide and amino acid NJ tree (Fig. 4b).

We only isolated one *DIV* homolog (*CrDIV*) from *C. ramondioides* which encode a protein of 296 amino acids. *CrDIV* is 54% identical to *DIV* from *A. majus* at amino acid level, and 90% and 89% identical to *BlDIV1* and *BlDIV2* from *B. leiophylla*. Conserved R2 and R3 domain and MYB-specific motif "SHAQKY" were found in *CrDIV* (Fig. 3c). Phylogenetic trees reconstructed from nucleotide and amino acid showed that *CrDIV* forms a monophyletic clade with *BlDIV1* (Fig. 4c).

Tissue-specific expressions of *CrCYC, CrRAD and CrDIV*

Transcripts of *CrCYC1C* and *CrCYC1D* were first detected in the early developmental stage (E) of the floral bud and then the expression of *CrCYC1C* and *CrCYC1D* were gradually reduced to almost invisible from middle (M) to late stage (L) of entire floral buds (Fig. 5). To further confirm organ expression pattern of *CrCYC1C* and *CrCYC1D*, pooled tissues including petals, gynoecium, stamens and sepals dissected from early development stage were used. Surprisingly, transcripts of *CrCYC1C* and *CrCYC1D* were both only detected in sepals (Fig. 5). This unique expression pattern of *CrCYC1C* and *CrCYC1D* in *C. ramondioides* seems to correlate with the residual zygomorphy found in sepal whorl (see Fig. 2a). As to *CrCYC2*, its expression was undetectable throughout all flower developmental stages and dissected organs. In summary, our RT-PCR results showed that no *CrCYCs* transcripts were detectable at petal and stamen whorl.

As to *CrRADs*, *CrRAD1* expression can be detected throughout all flower developmental stages but *CrRAD2* could only be detected in the early floral development stage (Fig. 5). Their organ expression pattern revealed that both *CrRAD1* and *CrRAD2* expressed in stamens, while *CrRAD2* also expressed in gynoecium (Fig. 5). The *CrDIV* was continuously expressed throughout all flower developmental stages. Specifically, *CrDIV* expressed in petals, stamens and gynoecium but not in sepals (Fig. 5).

Discussion

Reversal to actinomorphy correlate to diverse *CYC* expression shifts in Gesneriaceae and other eudicot lineages

Our SEM pictures showed that all five petals of *C. ramondioides* seems to grow in consistent rate of enlargement ever since petal primordia initiation stages (Fig. 2). Similarly, this was also observed in the stamen whorl. Development of actinomorphy in *C. ramondioides* therefore resembles to that in *cyc/dich* mutant of *A. majus* in which all petal primordia maintain equal growth rate along flower developmental stages and all stamens are fully developed (Luo et al. 1996).

From RT-PCR results, *CrCYCs* and *CrRADs* have no sign of expression in petal whorl (Fig. 5). This suggests *C. ramondioides* is a ventralized actinomorphy. In Gesneriaceae, ventralized actinomorphy has been reported in the peloric cultivar of *Sinningia speciosa* where *CYC* is not expressed in petal whorl because a deletion in *CYC* coding region (Hsu et al. 2015, 2017; Wang et al. 2015). Loss of *CYC* expression in late petal developmental stage but not in early stage can also

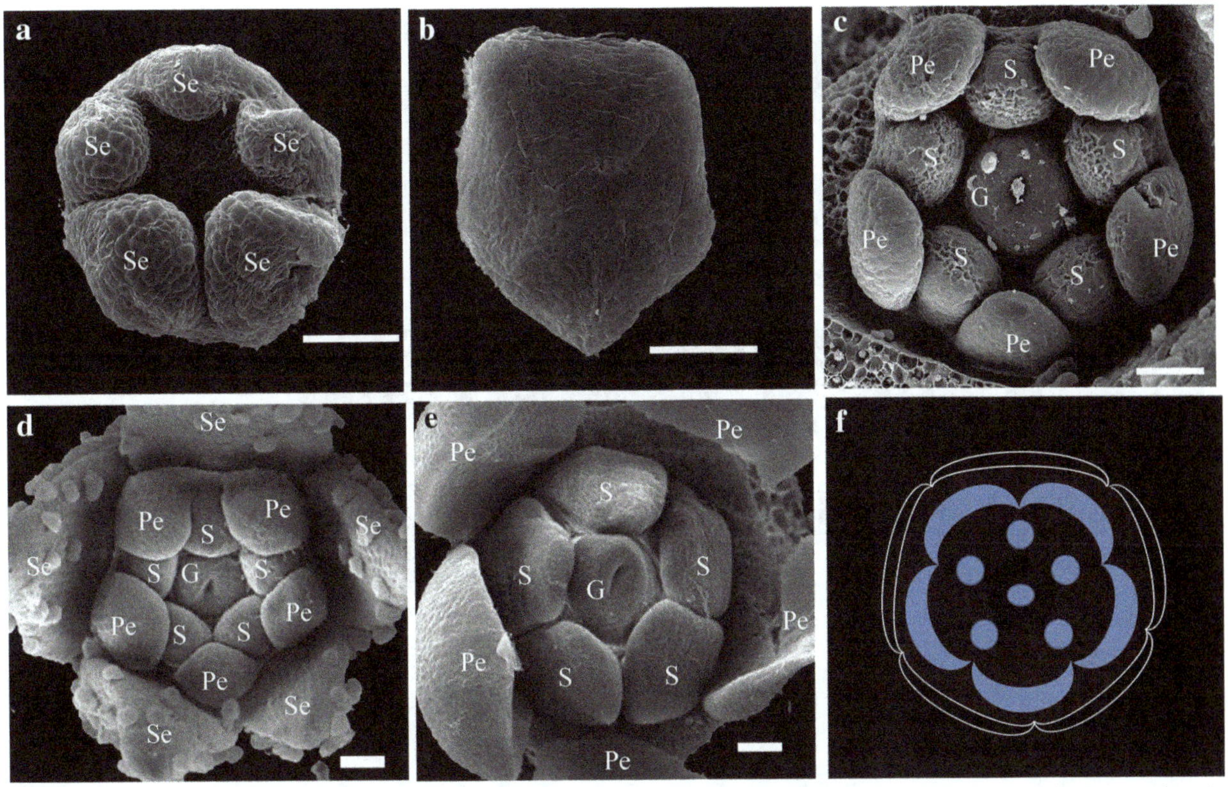

Fig. 2 The SEM photos of morphological development process of *Conandron ramondioides* flowers. Zygomorphy was observed at Stage 3 of sepal whorl. Definition of developmental stages of *C. ramondioides* based on (Harrison et al. 1999). **a** Stage 3, the sepals form as bulges at the points of the pentagon. **b** Stage 4, the sepals grow, while the floral meristem remains undifferentiated. **c** Stage 6, the corolla and androecium grow, and the gynoecium initiates. **d** Stage 7A, petal growth. **e** Stage 7B, androecium and gynoecium development. **f** Floral diagram of *C. ramondioides*. *Se* Sepals, *Pe* petals, *S* stamen, *G* gynoecium. Scale bar represents 50 μm

contribute to actinomorphic reversal in *B. leiophylla* (Zhou et al. 2008). On the other hand, dorsalized actinomorphy due to ubiquitous *CYC* expression in entire petal whorl has been reported in *T. scopulorum* and occasionally in *Petrocosmea* hybrids (Pang et al. 2010; Yang et al. 2015). Other than Gesneriaceae, examples of dorsalized actinomorphy have been reported in *Cardia purpurea* (Leguminosae), in *Aquilegia alpina* (Ranunculaceae), and in certain Malpighiaceae lineages (Citerne et al. 2006; Zhang et al. 2010, 2012, 2013; Jabbour et al. 2014). While ventralized actinomorphy where the loss of *CYC* expression has been documented in *Tradescantia* (Commelinaceae), in *Nigella damascene* (Ranunculaceae), in *Plantago lanceolata* (Plantaginaceae), in Malpighiaceae lineages and in *Arabidopsis* (Cubas et al. 2001; Reardon et al. 2009; Preston et al. 2011; Preston and Hileman 2012; Zhang et al. 2013; Jabbour et al. 2014).

Although we did not detect any *CrCYC1C* and *CrCYC1D* expression in petal and stamen whorls, we did find them distinctively expressed in sepals. Sepal specific expression of *CYC* has been inferred the ancestral state among Ranunculales including both actinomorphic Papaveraceae species and zygomorphic Fumariaceae species, the basal most core Eudicot lineage (Damerval et al. 2007). Thus, the shifts of *CrCYC1C/1D* back to ancestral expression in sepals and cease of expression in petals/stamens seem to associate with the actinomorphic reversal in *C. ramondioides*. Indeed, the recruitment of *CYC* expression (ECE clade) in sepal was evolved earlier than expression in dorsal-specific manner (Preston and Hileman 2009). It would be interesting to examine whether all those reversals to actinomorphy species (Wang et al. 2010; Clark et al. 2011) in Geseneriaceae also have their *CYC* expression return to ancestral state of sepal expression.

Duplications of *CYC* may associate to expression shift and flower shape variation

Our phylogenetic analysis revealed that the three *CrCYCs* we isolated (*CrCYC1C, CrCYC1D* and *CrCYC2*) were resulted from at least two duplication events among

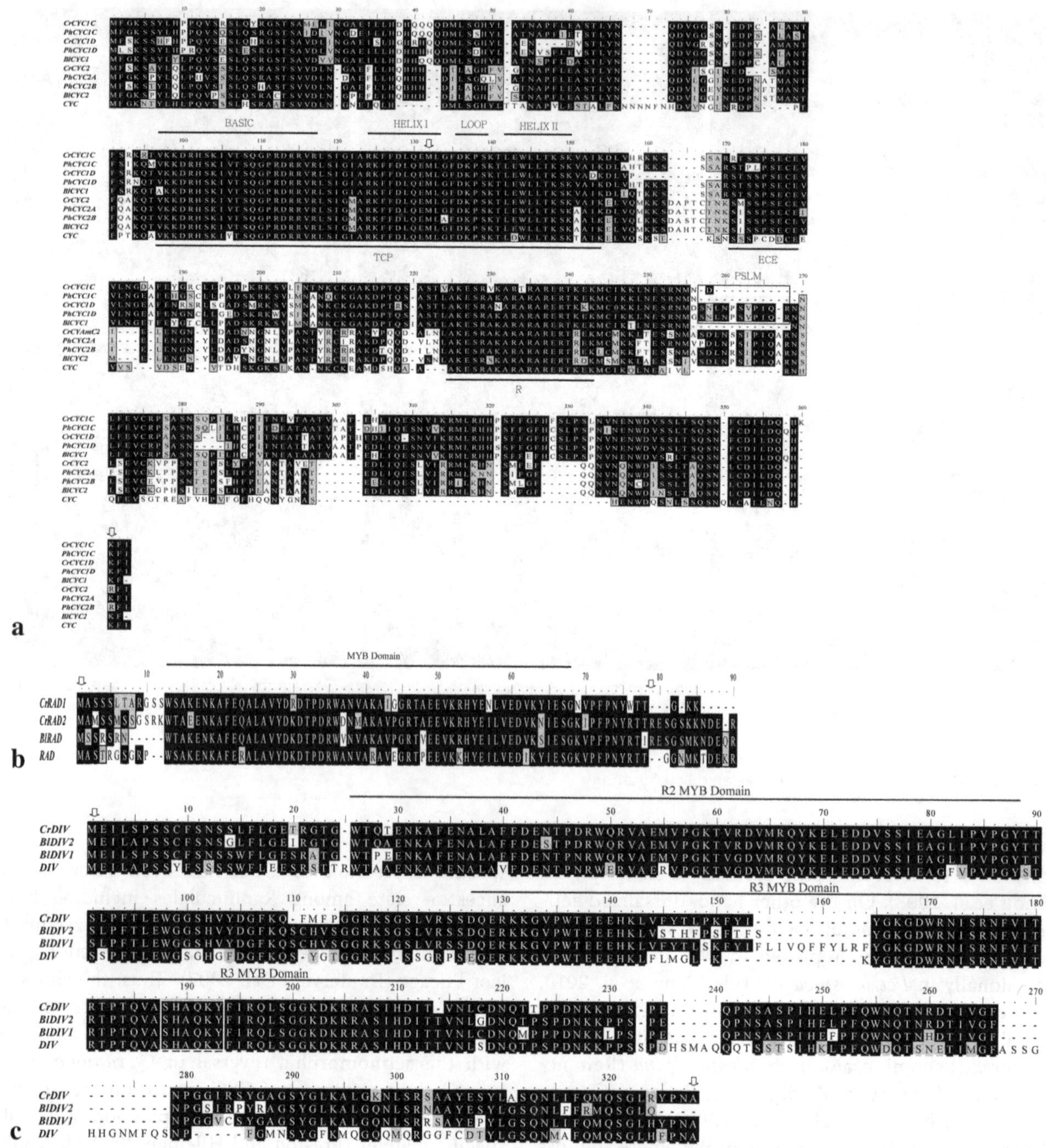

Fig. 3 Alignments of protein sequences of *CrCYC*, *CrRAD* and *CrDIV* genes with homologs from *Antirrhinum majus* (*CYC*, *RAD* and *DIV*) and *Bournea leiophylla* (*BlCYC1*, *BlCYC2*, *BlRAD*, *BlDIV1* and *BlDIV2*). **a** Alignmentes of *CYC* homologs. TCP, ECE and R domain are outlined. Identical amino acids are in black and similar amino acids are in gray. **b** Alignments of *RAD* homologs. MYB domain are outlined. **c** Alignments of *DIV* homologs. Two MYB domains (R2 and R3) are outlined. A highly conserved SHAQKY motif in R3 MYB domain is identified and labelled in white box. Arrows indicate sequence region used in phylogeny analysis. Sequence region used in phylogenetic analysis coving almost all important domains in all three gene dataset except for *PlDIV* in *DIV* dataset

Gesneriaceae species, congruent to previous findings (Möller et al. 1999; Citerne et al. 2000; Wang et al. 2004; Du and Wang 2008; Song et al. 2009; Pang et al. 2010; Yang et al. 2015). Although these *CrCYCs* have lost their expression in petal and stamen whorls, which is correlate to the reversal to actinomorphy in *C. ramondioides*, their

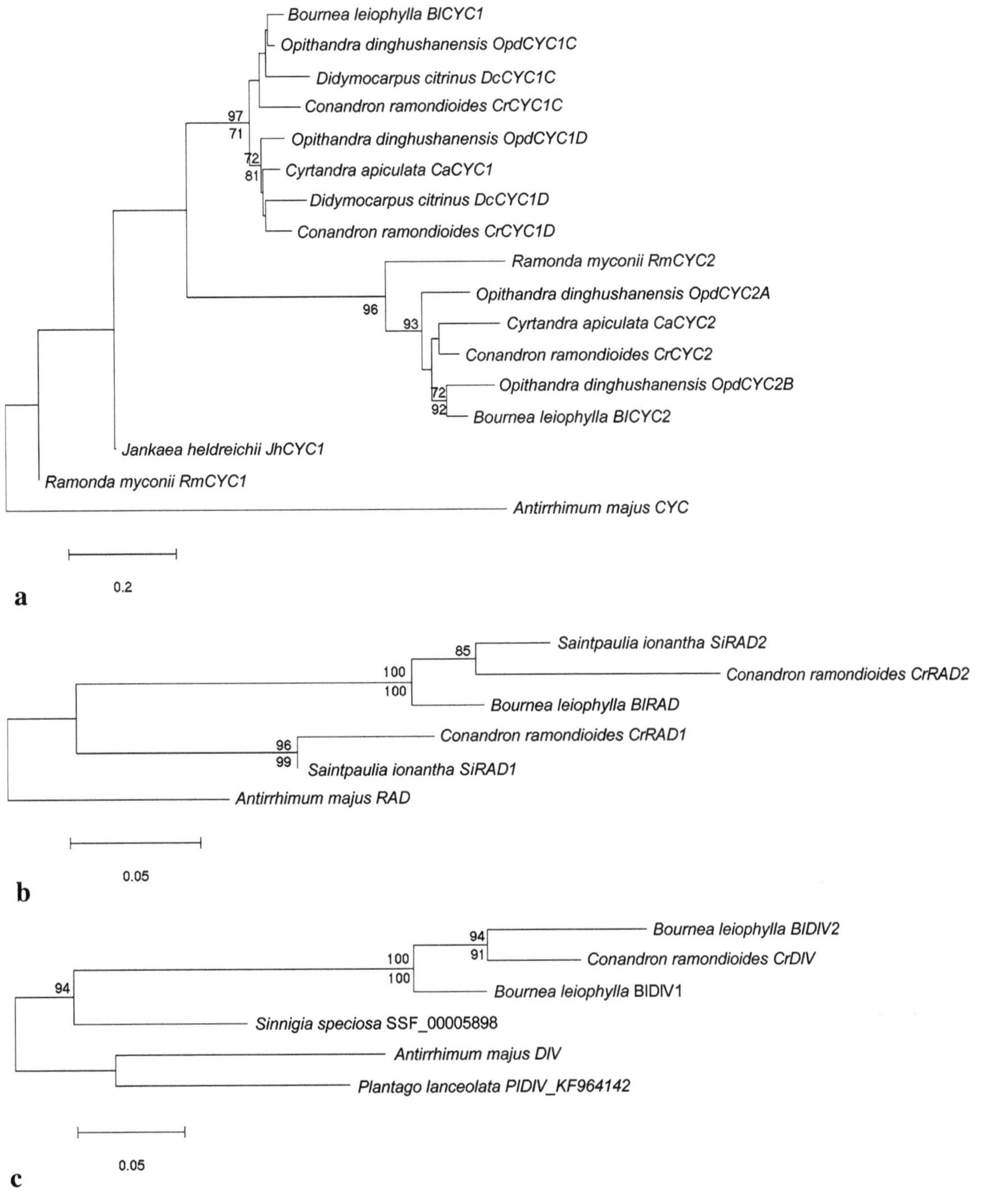

Fig. 4 Neighbour-joining trees of *CYC*-like, *RAD*-like and *DIV*-like genes. Trees from **a** to **c** are reconstructed based on amino acid sequences. **a** *CYC* is from *A. majus*, others are *CYC*-like genes from Gesneriaceae species (see Additional file 2: Table S2). Trees show *CrCYCs* cluster into three groups with high support. **b** The tree shows *CrRADs* cluster into two distinct clade with high support. **c** Bootstrap values from NJ and ML are listed above and below branches respectively. Bootstrap values > 70 are shown. Detailed sequences information is listed in Additional file 2: Table S2

coding sequences contain no frame shift or nonsense mutations. This implies these *CrCYCs* may still function yet the regulation controls on cis-elements in their promoter regions could have been mutated to become mis-expression (i.e. sepal only expression). It would be interesting to test this hypothesis by ectopically express these *CrCYCs* separately in *Arabidopsis* and compare their phenotypic effects.

In snapdragon, *CYC* evolved as major effect copy (higher and broader dorsal-specific expression) and its duplicates *DICH* as helper function (Luo et al. 1996, 1999). Similarly, *CYC* tend to duplicate in most Angiosperm lineages with either both copies retain similar expression pattern (major/helper) or expressions become diversified thus under different selection pressures (Ree et al. 2004; Chapman et al. 2008; Bello et al. 2017). In *C. ramondioides*, *CrCYC1C* and *CrCYC1D* both have sepal only expression but expression level of *CrCYC1C* is higher than *CrCYC1D*. This is similar to the case of *CYC* and *DICH* in snapdragon. Duplication could allow one copy to maintain the essential function but the other to evolve into novel or modified function. *CYC* duplications may therefore link to the evolution of diverse floral shape in angiosperms, although yet to be determined. There are reports, however, indicating shifts of expression between *CYC* paralogs correlate to floral symmetry transitions and/or flower shape variations (Bartlett and Specht 2011; Zhang et al. 2012; Jabbour et al. 2014).

Specific expression patterns of *CrRADs* and *CrDIV* may suggest loss of antagonistic expression pattern between *CrRADs* and *CrDIV* following loss of *CYC* expression in *C. ramondioides*

In *cyc*/*dich* double mutant *A. majus*, *RAD* does not express in dorsal region, thus, allowing ventral region restricted *DIV* to spread to whole flower (Corley et al. 2005). In *B. leiophylla*, once dorsal region restricted *BlCYC1* and *BlRAD* are downregulated, *BlDIV* spread to corolla and stamen whorl (Zhou et al. 2008). To sum up, *RAD*-like genes and *DIV*-like genes seem have antagonistic function in *A. majus* and *B. leiophylla*. However, expression patterns of *CrRADs* and *CrDIV* are different from those gene expression pattern in *A. majus* and *B. leiophylla*. In *C. ramondioides*, both *CrRADs* and *CrDIV* expressed in stamens (including *CrRAD1*, *CrRAD2* and *CrDIV*) and gynoecium (including *CrRAD2* and *CrDIV*) at the same development stage (Fig. 5). Comparing expression patterns of *RAD*-like and *DIV*-like genes among these three species, antagonistic function seem have lost in *C. ramondioides*. Based on expression patterns from these three species, we postulate that loss of expression of upstream gene (e.g. *CYC* and its homologue) may provide opportunity for its downstream gene (e.g. *RAD* and its homologue, *DIV* and its homologue) releasing from genetic constraint. To conclude, antagonistic expression pattern between *RAD*-like and *DIV*-like genes was maintained in *B. leiophylla*, which resembling to *CYC*-mediated regulatory pathway in *A. majus*. However, this antagonistic expression pattern was lost in *C. ramondioides*. Since the *RAD*-like and *DIV*-like genes were rarely studied in Gesneriaceae, it would be interesting to examine whether the maintaining or loss of antagonistic function between *RAD*-like and *DIV*-like genes is a common pattern or not in Gesneriaceae.

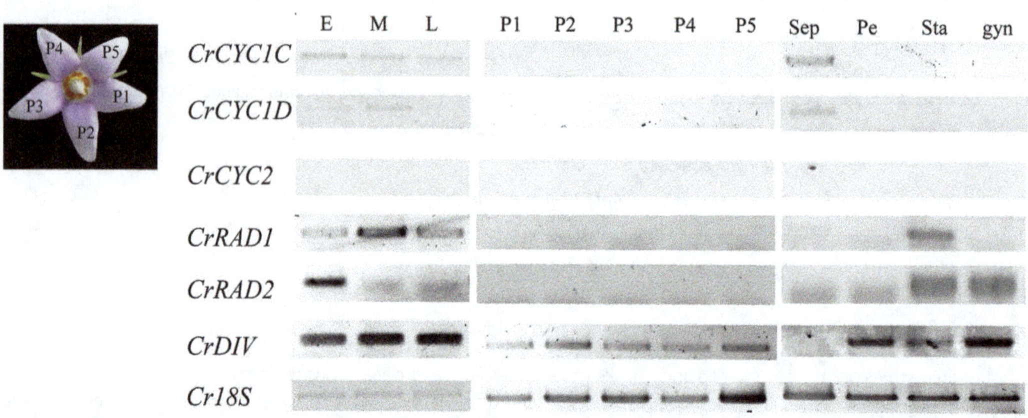

Fig. 5 Gene-specific reverse transcriptase polymerase chain reaction (RT-PCR) analysis of *CrCYC*, *CrRAD* and *CrDIV* genes from *C. ramondioides* buds and dissected flower tissues. E, M, L represent three flower development stage defined in this study. E stands for early flower development stage; M stands for middle flower development stage; L stands for anthesis stage. P1 to P5 represent dissected petal from flower bud at early (E) flower developmental stage. Sep, Pe, Sta and gyn denote pooled sepals, pooled petals, pooled stamens and gynoecium dissected from early flower developmental stage. *CrCYC1C*, *CrCYC1D* and *CrCYC2* indicate expression of *CrCYC1C*, *CrCYC1D* and *CrCYC2*. *CrRAD1* and *CrRAD2* indicate expression of *CrRAD1* and *CrRAD2*. *CrDIV* indicates expression of *CrDIV*. *18S* is included as a positive control. *CrCYC1C*, *CrCYC1D*, *CrRAD1*, *CrRAD2* and *CrDIV* are detected through flower development stages, only *CrCYC2* is restricted through flower development stages. *CrDIV* is detected in all petals, whereas *CrCYC1C*, *CrCYC1D*, *CrCYC2*, *CrRAD1* and *CrRAD2* are restricted in petals. *CrCYC1C* and *CrCYC2* were detected in pooled sepal tissue

Reversal to actinomorphy may help to attract general pollinator visiting *C. ramondioides*

In *C. ramondioides*, reversal to actinomorphy coupling with very short corolla tube at anthesis may facilitate generalist (e.g. bees, small beetles) visitation because they can obtain pollen from any direction (Fig. 1). Similarly, it has been postulated that reversal to actinomorphy in *R. myconii* could allow visits from a wider range of pollinators in alpine extreme habitats (Cronk and Möller 1997). In alpine or harsh conditions, plants may suffer low pollinator visiting. If certain zygomorphic species still rely on their specific pollinator visiting in harsh condition, their reproductive success may be low. But if species which can reverse to actinomorphy by opening the corolla, such as *C. ramondioides* and *R. myconii*, they could have pollinator shifts to a variety of pollinators to increase visiting rate. Although *C. ramondioides* is not distributed in alpine environment like *R. myconii*, its deep-forest dense shape habitat may discourage insect pollinator visiting (Peat and Goulson 2005). Reversal to actinomorphy to attract more general pollinators in pollinator scarce habitats may actually compensate for maintaining the reproductive success in *C. ramondioides*. To support this idea, detailed pollination experiment and pollinator observation in the field are necessary in the future.

Abbreviations
Sep: pooled sepals; Pe: pooled petals; Sta: pooled stamens; gyn: gynoecium.

Authors' contribution
CNW conceived the context of the manuscript. CNW and KTH designed the experiment. KTH performed the experiment. Both authors read and approved the final manuscript.

Author details
[1] Institute of Ecology and Evolutionary Biology, National Taiwan University, Taipei, Taiwan. [2] Department of Life Science, National Taiwan University, Taipei, Taiwan.

Acknowledgements
We are grateful to the staffs including Dr. Shiang-Jiuun Chen, Ms. Yi-Chun Chuang, Ms. Ya-Chan Yang and Ms. Pei-Yin Wu of TC5 Bio-image Tools, Technology Commons, College of Life science, National Taiwan University for help with SEM operation and observation. We also thank Mr. Chun-Hsien Wu for help with isolating target genes in this study and Mr. Hong-Wen Ma for help with collecting floral buds of *C. ramondioides* in the field.

Competing interests
The authors declare that they have no competing interests. And there have neither financial competing interests nor other competing interests.

Funding
We thank fundings from Ministry of Science and Technology of Taiwan MOST-106-2313-B-002-035-MY3 and National Science Council of Taiwan 95-2311-B-002-014-MY3 were granted to Chun-Neng Wang.

References

Almeida J, Rocheta M, Galego L (1997) Genetic control of flower shape in *Antirrhinum majus*. Development 124:1387

Bartlett ME, Specht CD (2011) Changes in expression pattern of the teosinte branched1-like genes in the Zingiberales provide a mechanism for evolutionary shifts in symmetry across the order. Am J Bot 98:227–243

Bello MA, Cubas P, Álvarez I, Sanjuanbenito G, Fuertes-Aguilar J (2017) Evolution and expression patterns of CYC/TB1 genes in anacyclus: phylogenetic insights for floral symmetry genes in Asteraceae. Front Plant Sci 8:589

Chapman MA, Leebens-Mack JH, Burke JM (2008) Positive selection and expression divergence following gene duplication in the sunflower cycloidea gene family. Mol Biol Evol 25:1260–1273

Citerne HL, Möller M, Cronk QCB (2000) Diversity of cycloidea-like genes in gesneriaceae in relation to floral symmetry. Ann Bot 86:167–176

Citerne HL, Pennington RT, Cronk QCB (2006) An apparent reversal in floral symmetry in the legume Cadia is a homeotic transformation. Proc Natl Acad Sci 103:12017–12020

Clark JL, Roalson EH, Pritchard RA, Coleman CL, Teoh V-H, Matos J (2011) independent origin of radial floral symmetry in the Gloxinieae (Gesnerioideae: gesneriaceae) is Supported by the rediscovery of *Phinaea pulchella* in Cuba. Syst Bot 36:757–767

Corley SB, Carpenter R, Copsey L, Coen E (2005) Floral asymmetry involves an interplay between TCP and MYB transcription factors in Antirrhinum. Proc Natl Acad Sci USA 102:5068–5073

Costa MMR, Fox S, Hanna AI, Baxter C, Coen E (2005) Evolution of regulatory interactions controlling floral asymmetry. Development 132:5093

Cronk QCB (2006) Legume flowers bear fruit. Proc Natl Acad Sci USA 103:4801–4802

Cronk Q, Möller M (1997) Genetics of floral symmetry revealed. Trends Ecol Evol 12:85–86

Cubas P, Coen E, Zapater JMMN (2001) Ancient asymmetries in the evolution of flowers. Curr Biol 11:1050–1052

Damerval C, Guilloux ML, Jager M, Charon C (2007) Diversity and evolution of CYCLOIDEA-Like TCP genes in relation to flower development in Papaveraceae. Plant Physiol 143:759

Donoghue MJ, Ree RH, Baum DA (1998) Phylogeny and the evolution of flower symmetry in the Asteridae. Trends Plant Sci 3:311–317

Du Z-Y, Wang Y-Z (2008) Significance of RT-PCR expression patterns of CYC-like genes in *Oreocharis benthamii* (Gesneriaceae). J Syst Evol. https://doi.org/10.1007/s00427-008-0227-y

Endress PK (1999) Symmetry in flowers: diversity and evolution. Int J Plant Sci 160:S3–S23

Endress PK (2001) Evolution of floral symmetry. Curr Opin Plant Biol 4:86–91

Guindon S, Dufayard J-F, Lefort V, Anisimova M, Hordijk W, Gascuel O (2010) New algorithms and methods to estimate maximum-likelihood phylogenies: assessing the performance of PhyML 3.0. Syst Biol 59:307–321

Harrison CJ, MÖller M, Cronk QCB (1999) Evolution and development of floral diversity in *Streptocarpus* and *Saintpaulia*. Ann Bot 84:49–60

Hileman LC (2014) Trends in flower symmetry evolution revealed through phylogenetic and developmental genetic advances. Philos Trans R Soc B Biol Sci. https://doi.org/10.1098/rstb.2013.0348

Hsu H-C, Chen C-Y, Lee T-K, Weng L-K, Yeh D-M, Lin T-T, Wang C-N, Kuo Y-F (2015) Quantitative analysis of floral symmetry and tube dilation in an F2 cross of *Sinningia speciosa*. Sci Hortic 188:71–77

Hsu H-C, Wang C-N, Liang C-H, Wang C-C, Kuo Y-F (2017) Association between petal form variation and CYC2-like genotype in a hybrid line of *Sinningia speciosa*. Front Plant Sci 8:558

Hsu HJ, He CW, Kuo WH, Hsin KT, Lu JY, Pan ZJ, Wang CN (2018) Genetic analysis of floral symmetry transition in African violet suggests the involvement of trans-acting factor for CYCLOIDEA expression shifts. Front Plant Sci. 9:1008. https://doi.org/10.3389/fpls.2018.01008

Jabbour F, Cossard G, Le Guilloux M, Sannier J, Nadot S, Damerval C (2014) Specific duplication and dorsoventrally asymmetric expression patterns of cycloidea-like genes in Zygomorphic species of Ranunculaceae. PLoS ONE 9:e95727

Kokubugata G, Peng CI (2004) Floral morphology and recognition of varieties in *Conandron ramondioides* (Gesneriaceae) from Japan and Taiwan. Edinburgh J Bot 61:21–30

Lefort V, Longueville J-E, Gascuel O (2017) SMS: smart model selection in PhyML. Mol Biol Evol 34:2422–2424
Luo D, Carpenter R, Vincent C, Copsey L, Coen E (1996) Origin of floral asymmetry in Antirrhinum. Nature 383:794
Luo D, Carpenter R, Copsey L, Vincent C, Clark J, Coen E (1999) Control of organ asymmetry in flowers of *Antirrhinum*. Cell 99(4):367–376
Möller M, Clokie M, Cubas P, Cronk QCB (1999) Integrating molecular and developmental genetics: a Gesneriaceae case study. In: Hollingsworth PM, Bateman RJ, Gornall RJ (eds) Molecular systematics and plant evolution. Taylor & Francis, London, pp 375–402
Pang H-B, Sun Q-W, He S-Z, Wang Y-Z (2010) Expression pattern of CYC-like genes relating to a dorsalized actinomorphic flower in *Tengia* (Gesneriaceae). J Syst Evol 48:309–317
Peat J, Goulson D (2005) Effects of experience and weather on foraging rate and pollen versus nectar collection in the bumblebee, *Bombus terrestris*. Behav Ecol Sociobiol 58:152–156
Perret M, Chautems A, Spichiger R, Barraclough TG, Savolainen V (2007) The geographical pattern of speciation and floral diversification in the neotropics: THE Tribe Sinningieae (Gesneriaceae) as a case study. Evolution 61:1641–1660
Picó FX, Möller M, Ouborg NJ, Cronk QCB (2002) Single nucleotide polymorphisms in the coding region of the developmental gene Gcyc in natural populations of the relict *Ramonda myconi* (Gesneriaceae). Plant Biol 4:625–629
Preston JC, Hileman LC (2009) Developmental genetics of floral symmetry evolution. Trends Plant Sci 14(3):147–154
Preston JC, Hileman LC (2012) Parallel evolution of TCP and B-class genes in Commelinaceae flower bilateral symmetry. EvoDevo 3:6
Preston JC, Martinez CC, Hileman LC (2011) Gradual disintegration of the floral symmetry gene network is implicated in the evolution of a wind-pollination syndrome. Proc Natl Acad Sci 108:2343–2348
Raimundo J, Sobral R, Bailey P, Azevedo H, Galego L, Almeida J, Coen E, Costa Maria Manuela R (2013) A subcellular tug of war involving three MYB-like proteins underlies a molecular antagonism in Antirrhinum flower asymmetry. Plant J 75:527–538
Reardon W, Fitzpatrick DA, Fares MA, Nugent JM (2009) Evolution of flower shape in *Plantago lanceolata*. Plant Mol Biol 71:241–250
Ree RH, Citerne HL, Lavin M, Cronk QCB (2004) Heterogeneous selection on LEGCYC paralogs in relation to flower morphology and the phylogeny of Lupinus (Leguminosae). Mol Biol Evol 21:321–331
Smith JF, Hileman LC, Powell MP, Baum DA (2004) Evolution of GCYC, a Gesneriaceae homolog of CYCLOIDEA, within Gesnerioideae (Gesneriaceae). Mol Phylogenet Evol 31:765–779
Song C-F, Lin Q-B, Liang R-H, Wang Y-Z (2009) Expressions of ECE-CYC2 clade genes relating to abortion of both dorsal and ventral stamens in Opithandra (Gesneriaceae). BMC Evol Biol 9:244
Spencer V, Kim M (2017) Re"CYC"ling molecular regulators in the evolution and development of flower symmetry. Semin Cell Dev Biol. https://doi.org/10.1016/j.semcdb.2017.08.052
Tamura K, Stecher G, Peterson D, Filipski A, Kumar S (2013) MEGA6: molecular evolutionary genetics analysis version 6.0. Mol Biol Evol 30:2725–2729
Thompson JD, Gibson TJ, Plewniak F, Jeanmougin F, Higgins DG (1997) The CLUSTAL_X windows interface: flexible strategies for multiple sequence alignment aided by quality analysis tools. Nucleic Acids Res 25:4876–4882
Wang C-N, Möller M, Cronk QCB (2004) Phylogenetic position of *Titanotrichum oldhamii* (Gesneriaceae) inferred from four different gene regions. Syst Bot 29:407–418
Wang Y-Z, Liang R-H, Wang B-H, Li J-M, Qiu Z-J, Li Z-Y, Weber A (2010) Origin and phylogenetic relationships of the Old World Gesneriaceae with actinomorphic flowers inferred from ITS and trnL-trnF sequences. Taxon 59:1044–1052
Wang C-N, Hsu H-C, Wang C-C, Lee T-K, Kuo Y-F (2015) Quantifying floral shape variation in 3D using microcomputed tomography: a case study of a hybrid line between actinomorphic and zygomorphic flowers. Front Plant Sci 6:724
Weber A (2004) Gesneriaceae. In: Kubitzki K, Kadereit JW (eds) The families and genera of vascular plants. vol. 7. Flowering plants, Dicotyledons: Lamiales (except Acanthaceae including Avicenniaceae). Springer, Berlin, pp 63–158
Xiao LH, Wang YZ (2007) Single nucleotide polymorphisms of Gcyc1 (Cycloidea) in *Conandron ramondioides* (Gesneriaceae) from Southeast China. Plant Syst Evol 269:145–157
Xiao L-H, Li Z, Wang R, Wang Y-Z (2012) Population differentiation and phylogeographic pattern of a relict species, *Conandron ramondioides* (Gesneriaceae), revealed from sequence polymorphism and haplotypes of the CYCLOIDEA gene. J Syst Evol 50:45–57
Yang X, Pang H-B, Liu B-L, Qiu Z-J, Gao Q, Wei L, Dong Y, Wang Y-Z (2012) Evolution of double positive autoregulatory feedback loops in CYCLOIDEA2 clade genes is associated with the origin of floral zygomorphy. Plant Cell 24:1834
Yang X, Zhao X-G, Li C-Q, Liu J, Qiu Z-J, Dong Y, Wang Y-Z (2015) Distinct regulatory changes underlying differential expression of TEOSINTE BRANCHED1-CYCLOIDEA-PROLIFERATING CELL FACTOR genes associated with petal variations in zygomorphic flowers of *Petrocosmea* spp. of the family Gesneriaceae. Plant Physiol 169:2138–2151
Zhang W, Kramer EM, Davis CC (2010) Floral symmetry genes and the origin and maintenance of zygomorphy in a plant-pollinator mutualism. Proc Natl Acad Sci USA 107:6388–6393
Zhang W, Kramer EM, Davis CC (2012) Similar genetic mechanisms underlie the parallel evolution of floral phenotypes. PLoS ONE 7:e36033
Zhang W, Steinmann VW, Nikolov L, Kramer EM, Davis CC (2013) Divergent genetic mechanisms underlie reversals to radial floral symmetry from diverse zygomorphic flowered ancestors. Front Plant Sci 4:302
Zhou X-R, Wang Y-Z, Smith JF, Chen R (2008) Altered expression patterns of TCP and MYB genes relating to the floral developmental transition from initial zygomorphy to actinomorphy in Bournea (Gesneriaceae). New Phytol 178:532–543

16

Genome-wide association study of rice genes and loci conferring resistance to *Magnaporthe oryzae* isolates from Taiwan

Heng-An Lin[1†], Szu-Yu Chen[1†], Fang-Yu Chang[2], Chih-Wei Tung[3], Yi-Chia Chen[1], Wei-Chiang Shen[1], Ruey-Shyang Chen[4], Chih-Wen Wu[2] and Chia-Lin Chung[1*]

Abstract

Background: Rice blast, caused by *Magnaporthe oryzae*, is an important rice disease occurring in all rice-growing areas. To manage blast disease effectively and in an environmentally friendly way, it is important to continually discover diverse resistant resources for breeding. In this study, genome-wide association study (GWAS) was used to map genes/loci resistant to rice blast in the open-access rice diversity panel 1 (RDP1), previously genotyped with a 44K single-nucleotide polymorphism array. Two geographically and genetically different *M. oryzae* isolates from Taiwan, D41-2 and 12YL-DL3-2, were used to challenge RDP1. Infected leaves were visually rated for lesion type (LT) and evaluated for proportion of diseased leaf area (%DLA) by image analysis software.

Results: A total of 32 quantitative trait loci (QTLs) were identified, including 6 from LT, 30 from DLA, and 4 from both LT and DLA. In all, 22 regions co-localized with previously reported resistance (*R*) genes and/or QTLs, including two cloned *R* genes, *Pita* and *Ptr*; 19 mapped *R* loci, and 20 QTLs. We identified 100 candidate genes encoding leucine-rich repeat-containing proteins, transcription factors, ubiquitination-related proteins, and peroxidases, among others, in the QTL intervals. Putative resistance and susceptibility haplotypes of the 32 QTL regions for each tested rice accessions were also determined.

Conclusions: By using Taiwanese *M. oryzae* isolates and image-based phenotyping for detailed GWAS, this study offers insights into the genetics underlying the natural variation of blast resistance in RDP1. The results can help facilitate the selection of desirable donors for gene/QTL validation and blast resistance breeding.

Keywords: Rice blast, Quantitative trait loci (QTLs), Genome-wide association study (GWAS), Haplotype analysis, Rice diversity panel 1 (RDP1)

Background

Rice is considered the major staple food of over half of the world's population. In 2014–2015, the global rice yield was 494.7 million tonnes (milled basis) and the cultivated area ~ 162 million hectares, mostly in Asia (Food and Agriculture Organization of the United Nations 2016). Rice blast is a devastating disease occurring in all rice cultivated areas. This polycyclic disease is caused by the filamentous ascomycete fungus *Magnaporthe oryzae* (anamorph *Pyricularia oryzae*), which can infect all parts of the rice plant at all growth stages (Wilson and Talbot 2009). Blast disease can be managed well with fungicides (e.g., isoprothiolane, thiophanatemethyl) and antibiotics (e.g., kasugamycin) applied at the proper time. However, abuse of agricultural chemicals can cause additional costs and undesirable environmental side effects. An economic and eco-friendly alternative is to use resistant rice varieties. However, owing to the rapid evolution of pathogen *avirulence* (*Avr*) genes in field populations of *M. oryzae*, resistant varieties can be overcome in a few years after

*Correspondence: clchung@ntu.edu.tw
†Heng-An Lin and Szu-Yu Chen contributed equally to this work
[1] Department of Plant Pathology and Microbiology, National Taiwan University, No. 1, Sec. 4, Roosevelt Rd., Taipei 10617, Taiwan
Full list of author information is available at the end of the article

their release and large-scale cultivation (Chen et al. 2009; Zeigler et al. 1994). To more effectively control rice blast, it is important to continually discover new resistant-related genes for breeding.

Both qualitative and quantitative types of resistance against rice blast have been reported. To date, more than 500 quantitative trait loci (QTLs) have been identified in ~ 30 different studies, mostly using populations derived from *indica/japonica* crosses (Ashkani et al. 2016; Miah et al. 2013; Sharma et al. 2012). At least 100 resistance (*R*) genes have been mapped in the rice genome. Most *R* genes are from *japonica* and *indica* lines and 4% are from wild species (Sharma et al. 2012). A total of 31 *R* genes (*Pi1, Pi2/Pi9/Pi50/Pigm/Piz-t, Pi5, Pi21, Pi25/Pid3, Pi35/Pish, Pi36, Pi37, Pi54rh, Pi56, Pi63, Pi64, Pia/PiCO39, Pib, Pid2, Pi-k/Pik-m/Pik-p, Pik-h/Pi54, Pit, Pita, NLS1* and *Pb1*) have been cloned and characterized (Ashkani et al. 2016; Das et al. 2012; Deng et al. 2009; Fukuoka et al. 2014; Ma et al. 2015; RiceData 2012; Tang et al. 2011). Most blast resistance genes encode nucleotide binding site–leucine-rich repeat (NBS–LRR) proteins (Koide et al. 2009; McDowell and Woffenden 2003). The NBS domain functions in signaling, and the LRR domain usually plays an important role in R-AVR protein–protein interaction (Marone et al. 2013). There are three non-NBS–LRR blast *R* genes: *Pid2*, encoding a B-lectin kinase protein (Chen et al. 2006); *pi21*, encoding a proline-rich protein (Fukuoka et al. 2009); and *Ptr*, encoding two isoforms each with four Armadillo repeats (Zhao et al. 2018).

The success of molecular breeding for developing new resistant varieties for blast management depends on fine-scale localization of *R* genes and QTLs. The conventional approach to mapping QTLs is linkage analysis, which involves an experimental population of F_2, back-crossed, double haploid, or recombinant inbred lines derived from two parental lines with contrasting phenotypes (Collard et al. 2005). Linkage analysis with a bi-parental population is ideal for low-resolution mapping and allows for testing a maximum of two alleles at a locus. Another mapping approach, genome-wide association study (GWAS), is used to correlate patterns of genomic variation with phenotype(s) in a collection of diverse genotypes (Buckler et al. 2009; Huang and Han 2014). Because of the numerous historic recombination events resulting from a long natural evolution (Soto-Cerda and Cloutier 2012), GWAS can identify QTLs at high resolution, possibly to the gene level, and does not require the time and labor needed for constructing a mapping population (Oraguzie et al. 2007; Pasam and Sharma 2014). With the development of high-throughput genotyping techniques, GWAS has also been used to dissect complex traits in crops such as rice, maize, and wheat (Buckler et al. 2009; Huang et al. 2010; Huang and Han 2014; Rasheed et al. 2014).

Several populations of rice have been established to investigate traits related to agronomic properties and stress tolerance by GWAS (Begum et al. 2015; Famoso et al. 2011; Huang et al. 2010, 2012; Norton et al. 2014; Wang et al. 2014; Zhao et al. 2011). Among those, Wang et al. (2014) inoculated a subset of 366 *indica* rice accessions from a population of 517 China landraces with 16 local *M. oryzae* strains and detected 30 blast-associated loci and several candidate genes (Wang et al. 2014). Another established population, rice diversity panel 1 (RDP1), is a collection of 421 rice accessions containing landraces and cultivars from 82 countries (Eizenga et al. 2014). The open-access RDP1 was genotyped with 44,100 single nucleotide polymorphisms (SNPs) and used to discover loci associated with agronomic traits (Zhao et al. 2011); aluminum tolerance (Famoso et al. 2011); grain concentrations of arsenic, copper, molybdenum, and zinc (Norton et al. 2014); and blast resistance (Kang et al. 2016; Mgonja et al. 2016; Zhao et al. 2011). Recently, a panel of 1568 diverse rice accessions (RDP1 included) and high-density 700,000 SNP genotyping data were released to the rice community (McCouch et al. 2016).

The current research aimed to use GWAS to search for blast resistance genes and QTLs in RDP1. RDP1 has been used to locate blast resistance loci in several previous studies. A major QTL was identified with 413 accessions by using a mixture of three *M. oryzae* isolates from the United States (Zhao et al. 2011); 97 QTLs were mapped with 390 accessions by using five *M. oryzae* isolates from South Korea, China, Columbia, India, and the Philippines (Kang et al. 2016); 31 QTLs were mapped with 162 accessions by using four *M. oryzae* isolates from Tanzania, Uganda, Kenya, and Burkina Faso (Mgonja et al. 2016); and 16 QTLs were mapped with 413 accessions by using natural infection in three blast nurseries in Shanghang, Wuchang, and Taojiang in China (Zhu et al. 2016). Here we selected two Taiwan isolates to challenge the RDP1 accessions. Image-based phenotyping and detailed GWAS offered more insights into the genetics underlying the natural variation of blast resistance in RDP1. Analysis of haplotypes helped identify rice accessions with favorable alleles at candidate QTL regions. The results can help facilitate marker development and the selection of desirable donors for blast resistance breeding.

Methods

Plant and fungal materials

Seeds of RDP1 were acquired from the Genetics Stocks *Oryza* (GSOR) germplasm collection (Agricultural Research Service, US Department of Agriculture). Sufficient seeds from 314 of 421 diverse accessions were successfully reproduced in a greenhouse at Kaohsiung District Agriculture Research and Extension Station and in the Phytotron at National Taiwan University during 2011–2013. The 314 accessions contained 55 *indica*, 51 *aus*, 83 *temperate japonica* (TEJ), 75 *tropical japonica* (TRJ), 5 *aromatic*, and 45 admixed varieties (Additional file 1: Table S1). Principle component analysis (PCA) defined the five subpopulations *indica*, *aus*, TEJ, TRJ, and *aromatic* according to clear clustering based on the top four principal components (PCs) explaining ~ 50% of the genetic variation in RDP1, and the admixed accession did not fit well into any clusters (Zhao et al. 2011). Because *aus* is closely related to *indica* and *aromatic* is related to *japonica* (Kovach et al. 2007), the full population could be further divided into *indica* (*indica* and *aus*) and *japonica* (TEJ, TRJ, and *aromatic*) varietal subgroups. Two susceptible varieties, LTH and Lomello, were included as positive controls.

From the collection of nationwide surveys of *M. oryzae* isolates in Taiwan during 2009–2013 (Shen, W.-C. and Chen, R.-S, unpublished), two geographically and genetically different isolates, D41-2 and 12YL-DL3-2, were chosen for blast inoculation. D41-2, isolated from Chiayi in 2009, by Chen, belongs to a dominant *Pot2* fingerprint lineage in Taiwan; 12YL-DL3-2 was isolated from Yilan in 2012, by Shen, belongs to a minor *Pot2* lineage. Both isolates grow and sporulate well on artificial media and showed high virulence in LTH and Lomello. D41-2 and 12YL-DL3-2 were inoculated in 8 international standard blast differential varieties and 16 Taiwan blast differential varieties. The reaction patterns indicate their difference in pathogenicity (Additional file 2: Table S2).

Evaluation of blast resistance

Resistance of rice accessions to D41-2 and 12YL-DL3-2 was evaluated in three and two independent inoculation trials, respectively, with two replications per trial and 6–7 seedlings per rice accession per replication. Inoculations with D41-2 and 12YL-DL3-2 were conducted from December 2013 to March 2014 and February to July 2015, respectively. Inoculation followed methods modified from Azizi et al. (2015) and Valent et al. (1991). *M. oryzae* was cultured on oat meal agar (OMA) for 2 weeks at 26 °C in a 12/12-h light/dark photoperiod in a growth chamber. Conidia were dislodged by using a glass rod and 0.05% Tween 20 (Sigma, USA), filtered through sterilized double-layered cheesecloth, and adjusted to 2×10^5 conidia/ml by using a haemocytometer. Rice seeds were sown in plug trays (5 seeds per plug). In each tray, 48 rice accessions (2 plugs each) and 2 susceptible control varieties (4 plugs each) were arranged in a randomized complete block design (RCBD). Rice seedlings were grown at 28/26 °C day/night temperature and 16/8-h light/dark photoperiod (luminous intensity 7000–8000 lx) in a growth chamber. Seedlings were fertilized with a 500X dilution of HYPONeX No.5 (N:P:K = 30:10:10) (HYPONEX Corp., USA) at 7 and 20 days after planting. Three- to four-leaf-stage seedlings were inoculated by spraying with 2×10^5 conidia/ml suspension (50 ml per tray) with use of an airbrush (Ming Yang, Taiwan) at 10–15 psi. Inoculated seedlings were maintained at 26 °C and 95–100% relative humidity in a sealed plastic storage box, with its interior covered with wet paper towels. After 36 h of incubation, the seedlings were grown at 28/26 °C day/night temperature and 16/8-h light/dark photoperiod (7000–8000 lx) for disease development. Seven days after inoculation, the second or third leaves were excised, placed on a light box (Chartmat, Taiwan), flattened with a transparent slide and photographed (Canon EOS 700D, Japan; ISO: 400, F8.0, shutter rate 1/100). Each digital image contained 3–7 diseased leaves from one rice accession in a replication. Diseased leaf area (DLA) was analyzed by using Assess 2.0 (Lamari 2008), with the color threshold manually adjusted to correctly differentiate lesions from healthy tissue. Predominant lesion type (LT) was visually rated according to the Standard Evaluation System for Rice (IRRI 2014), with scores 0, 1, and 3 considered resistant (R) and 5, 7, and 9 considered susceptible (S). Only rice accessions showing consistent results in all replications were included for subsequent analyses. Pearson correlation analysis was used to analyze the correlation between DLA and LT by using SAS 6.1 (SAS Inst. Inc., Cary, NC, USA).

GWAS

The GWAS involved use of the 44K SNP dataset (36,775 high-quality SNPs) (Zhao et al. 2011). The physical map positions of the SNPs were converted from the MSU v6.0 Nipponbare rice reference genome (MSU6) to the MSU v7.0 Nipponbare rice reference genome (MSU7) by using the assembly converter tool in Gramene (http://www.gramene.org/). The phenotypic datasets were LT and DLA. The LT scores were averaged over different replications and trials. Accessions showing resistant-type lesions (LT in each trial = 0, 1, 3) and susceptible-type lesions (LT in each trial = 5, 7, 9) were defined as R and S accessions, respectively. To control the variation among blocks in different inoculation experiments, best linear unbiased estimates (BLUEs) for DLA data were generated by using TASSEL 5.0.5 (Bradbury et al. 2007).

The generalized linear model (GLM) and mixed linear model (MLM) were used for analyzing different traits (LT, DLA) and populations (full population, *indica*, *japonica*) by using TASSEL 5.0.5. The formulas for the GLM and MLM were $y = X\beta + e$ and $y = X\beta + Zu + e$, respectively, where y is the vector of phenotypic data, β is the vector of genotypic data and population structure (Q), u is the vector estimated from kinship matrix (K), e is vector of residuals, X is the matrix of genotype and population structure, and Z is the known design matrices (Bradbury et al. 2007). The population structure was obtained by PCA with TASSEL 5.0.5. We tested GLM and MLM with/without the population structure as covariates (GLM, GLM-Q, MLM-K, MLM-K+Q). Quantile–quantile (Q–Q) plots were produced by using TASSEL 5.0.5 to assess the fitness of different models for each phenotype dataset. Q–Q plots showing less deviation from the y (observed test statistics) = x (expected test statistics) lines suggest less systemic bias (Reed et al. 2015). Manhattan plots were generated by using the qqman package in R (R Development Core Team 2011).

To understand the linkage disequilibrium (LD) structure, pairwise LD analysis of SNPs involved using TASSEL 5.0.5, and LD blocks were defined by using Haploview (Barrett et al. 2005) with default settings. The LD blocks containing more than three SNPs with $P < 3.1 \times 10^{-4}$ [$-\text{Log}_{10}(P) > 3.5$] were considered QTLs significantly associated with the traits. The QTL interval was the size of the significant LD block. Each candidate QTL (significant LD block) was checked for known *R* genes/loci in review papers (Ashkani et al. 2016; Sharma et al. 2012) and recent literature (Kang et al. 2016; Liu et al. 2015; Ma et al. 2014, 2015; Mgonja et al. 2016; Su et al. 2015; Xu et al. 2014; Zhao et al. 2011; Zhu et al. 2016). Previously identified blast QTLs and defense-related candidate genes were also checked according to the reference Nipponbare genomic sequences (MSU7) in the Gramene database (http://www.gramene.org/), Rice Genome Annotation Project website (http://rice.plantbiology.msu.edu/) (Kawahara et al. 2013), and The Rice Annotation Project website (http://rapdb.dna.affrc.go.jp/index.html) (Kawahara et al. 2013; Sakai et al. 2013). Candidate genes were determined by their gene ontology (GO) terms and gene descriptions.

Haplotype analysis

To identify possible resistance (R) and susceptible (S) haplotypes for the regions significantly associated with resistance, haplotype association analysis was performed with Haploview. Phenotype data were converted to case–control datasets. According to the frequency distributions of LT data (Fig. 1a, c), the accessions showing LT > 3 (S lesions) and ≤ 3 (R lesions) were assigned as "cases" and "controls", respectively. There is no standard way to determine a threshold for the quantitative DLA data, so DLA = 15% was arbitrarily set as the threshold for DLA data conversion. For rice accessions inoculated with D41-2, the ratios of "cases (S): controls (R)" for LT and DLA were 168:136 and 210:94, respectively, and with 12YL-DL-3-2, the ratios were 142:86 and 142:86, respectively. In each LD block, the resistance/susceptibility haplotype was determined by Chi-square analysis and *P* value for haplotype frequencies in cases versus controls. Putative R and S haplotypes in all candidate regions were identified by using the SNP genotypes of the 44K SNP data. Haplotype identity was assigned if the SNP genotypes were 100% identical in length and sequences. Pearson correlation analysis was used for analysis of correlation between the resistance performance (LT and DLA) and total number of non-overlapped R/S haplotypes (in each rice accession) by using SAS 6.1 (SAS Inst. Inc., Cary, NC, USA). The total number of non-overlapped R or S haplotypes was defined as the sum of the number of R or S haplotypes identified from different phenotype datasets. The R or S haplotype identified from more than one phenotype was counted only once.

Results

Phenotypic variation of blast resistance in RDP1

Phenotyping data consistent in all replications were obtained from 304 rice accessions [104 *indica* (54 *indica* and 50 *aus*), 155 *japonica* (78 TEJ, 72 TRJ, 5 *aromatic*), and 45 admixed] inoculated with *M. oryzae* isolate D41-2, and 228 rice accessions [80 *indica* (48 *indica* and 32 *aus*), 121 *japonica* (56 TEJ, 63 TRJ, 2 *aromatic*), and 27 admixed] inoculated with *M. oryzae* isolate 12YL-DL3-2 (Additional file 1: Table S1). Most of the eliminated accessions did not grow well in all the replications. The distributions of LT and DLA are shown for the full population (Fig. 1a–d) and for *indica* and *japonica* subgroups (Fig. 1e–h). The distributions of LT were bimodal with discrete resistance (LT ≤ 3) and susceptibility (LT > 3) categories (Fig. 1a, c). LT and DLA were positively correlated (full population with D41-2: $r = 0.58$, $P < 0.001$; full population with 12YL: $r = 0.66$, $P < 0.001$). Overall, 55 accessions were resistant (LT in each trial = 0, 1, 3) and 89 accessions were susceptible (LT in each trial = 5, 7, 9) to both isolates. The mean of LT and DLA were lower and the resistance was relatively greater for the *indica* than *japonica* subgroup (Fig. 1e–h).

Identification of loci associated with resistance to two *M. oryzae* isolates

Manhattan plots and Q–Q plots for LT and DLA in the full population and *indica* and *japonica* subgroups are in Figs. 2, 3. For each experimental dataset, the optimal model for GWAS was selected from GLM, GLM-Q, MLM-K, and MLM-K+Q. MLM-K+Q had the best explanatory power for global accessions, and GLM and MLM-K were optimal for GWAS at the subgroup level. The strategy of using different optimal models for different traits in the same populations was also adopted by

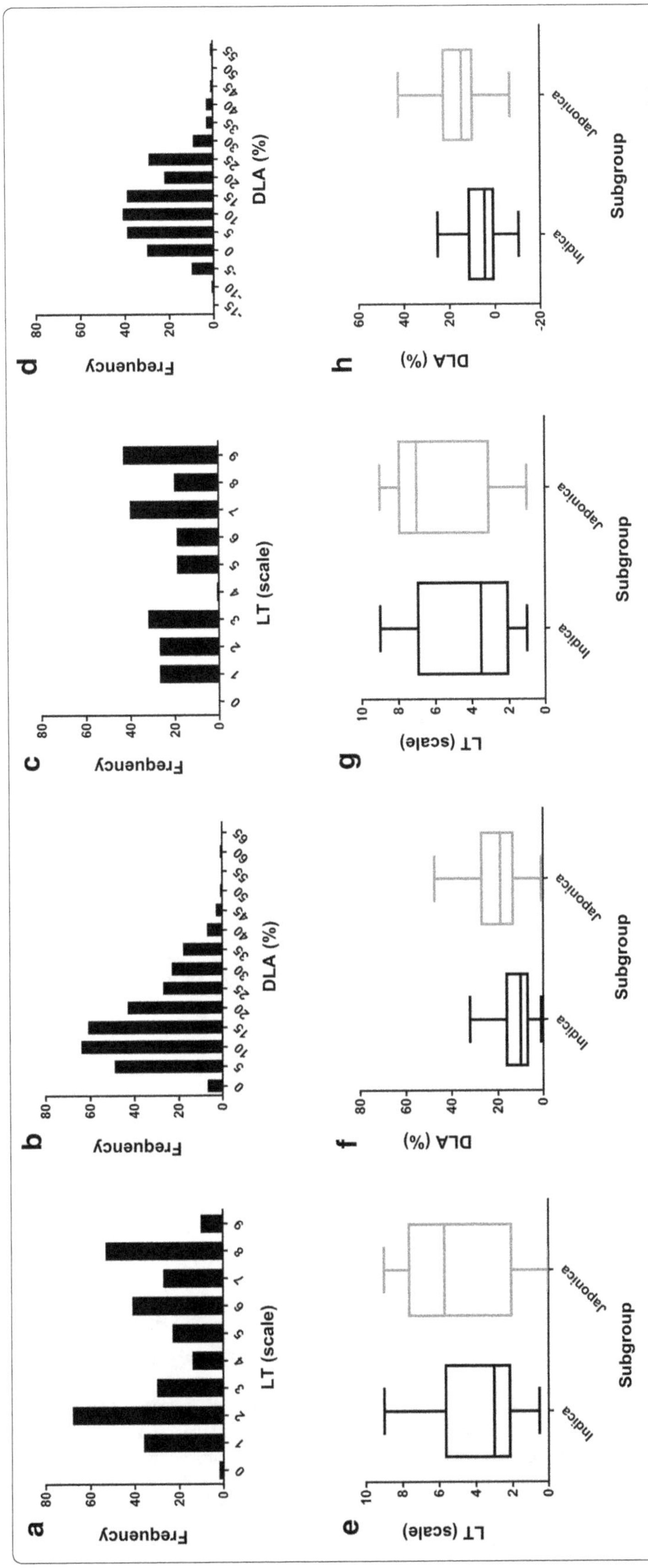

Fig. 1 Distribution of blast resistance with *Magnaporthe oryzae* isolates D41-2 and 12YL-DL-3-2. The histograms and boxplots show the phenotype distributions in the full population and the two subgroups, respectively. **a**, **e** Lesion type (LT) from inoculation with D41-2. **b**, **f** Diseased leaf area (DLA) from inoculation with D41-2. **c**, **g** Lesion type (LT) from inoculation with 12YL-DL-3-2. **d**, **h** Diseased leaf area (DLA) from inoculation with 12YL-DL-3-2

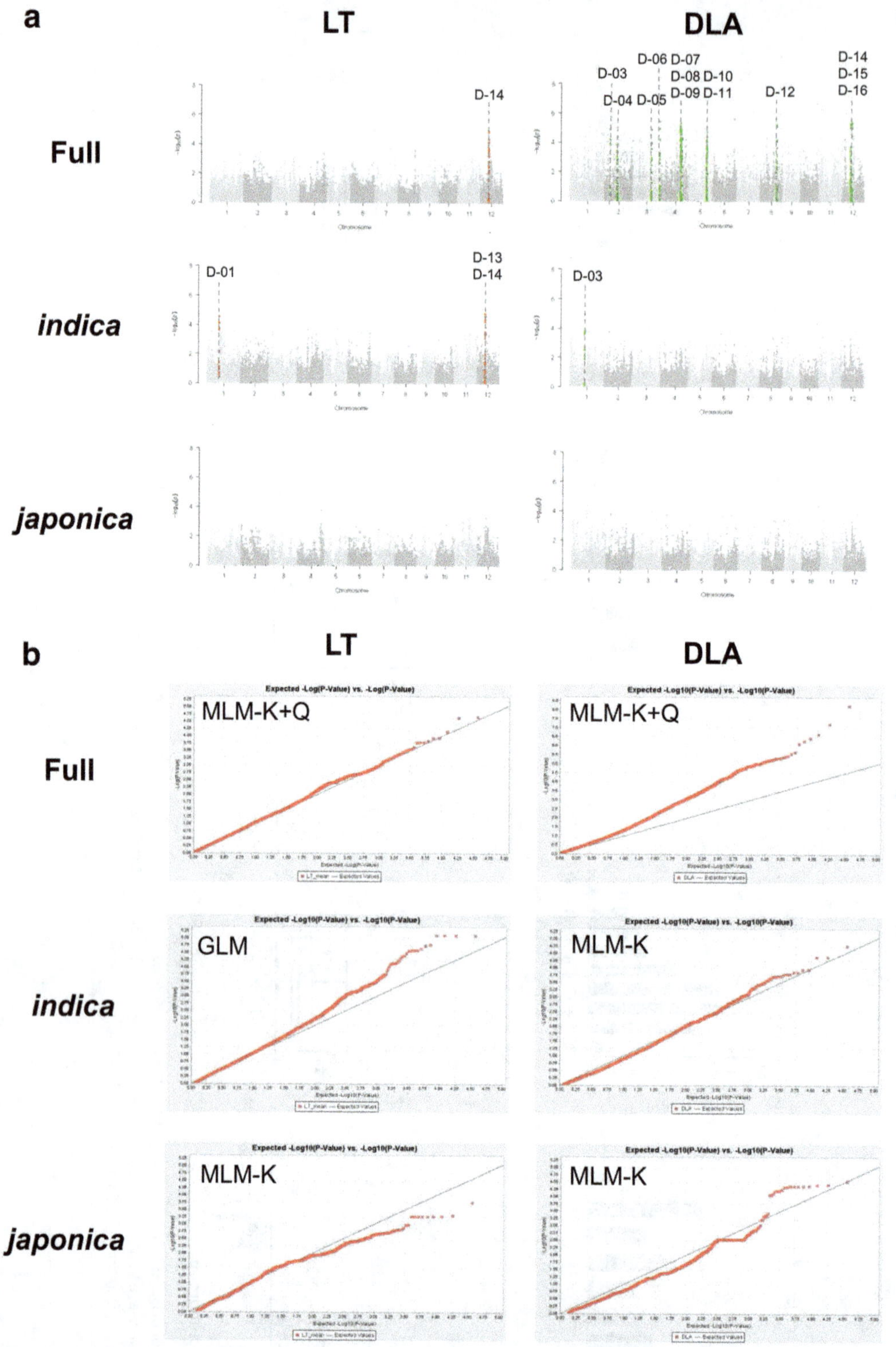

Fig. 2 Genome-wide association study (GWAS) of loci associated with resistance to *M. oryzae* isolate D41-2. **a** Manhattan plots show significant genomic regions (D-01 to D-16) identified by using lesion type (LT) and diseased leaf area (DLA) dataset. X axis: rice chromosomes; Y axis: $-\text{Log}_{10}(P)$. **b** Quantile–quantile (Q–Q) plots show the fitness of the selected models used for different traits in the full population or subgroups. X axis: expected $-\text{Log}_{10}(P)$; Y axis: observed $-\text{Log}_{10}(P)$

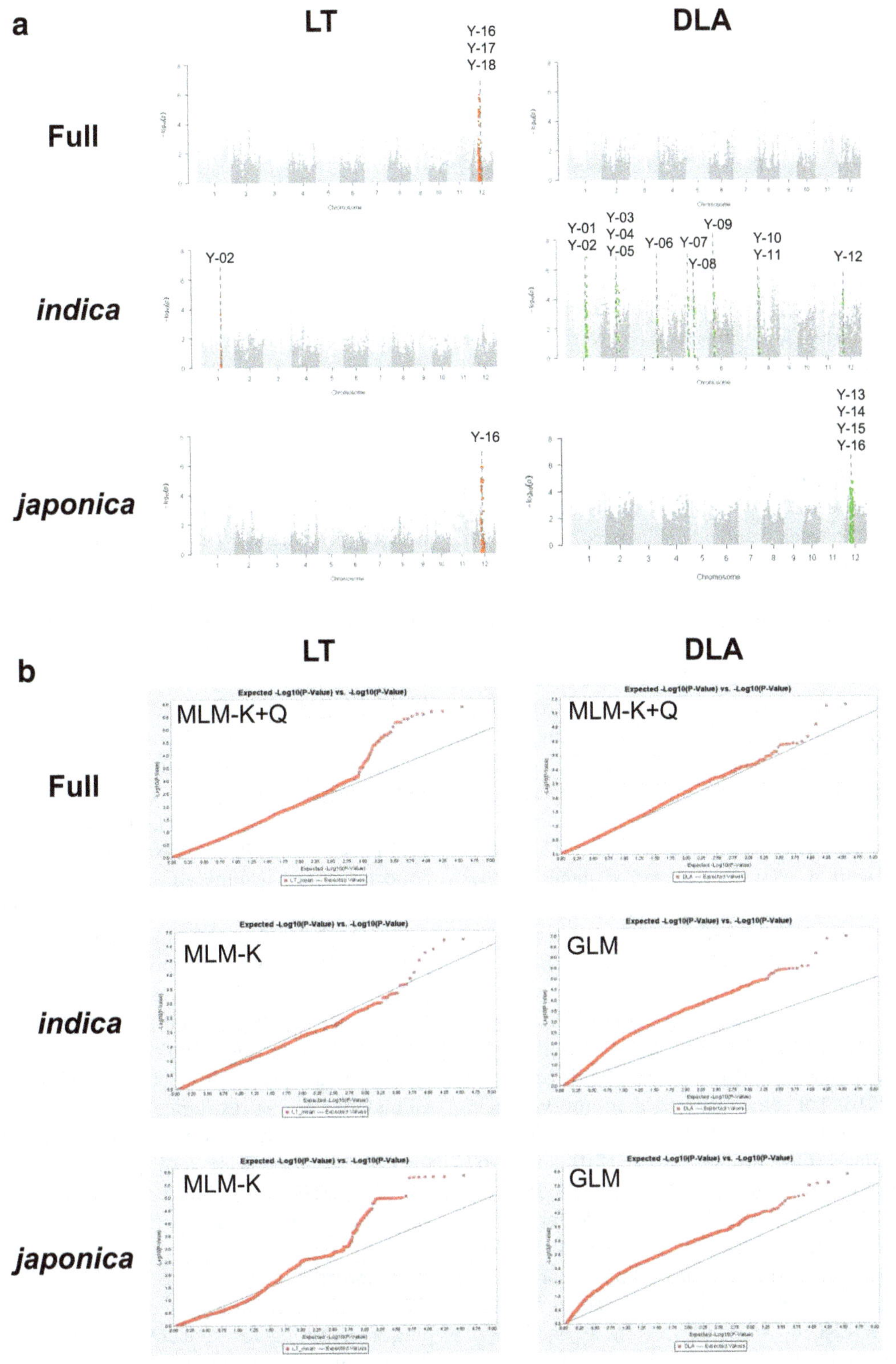

Fig. 3 GWAS of loci associated with resistance to *M. oryzae* isolate 12YL-DL-3-2. **a** Manhattan plots show significant genomic regions (Y-01 to Y-18) identified by using lesion type (LT) and diseased leaf area (DLA) datasets. X axis: rice chromosomes; Y axis: $-\text{Log}_{10}(P)$. **b** Quantile-quantile (Q-Q) plots show the fitness of the selected models used for different traits in the full population or subgroups. X axis: expected $-\text{Log}_{10}(P)$; Y axis: observed $-\text{Log}_{10}(P)$

Gajardo et al. (2015), Jaiswal et al. (2016), and Wan et al. (2017).

A total of 32 non-overlapped loci associated with blast resistance were identified across 8 of the 12 rice chromosomes (no QTL identified in chromosomes 7, 9, 10 and 11). The number of QTLs identified from different phenotypes, populations, and *M. oryzae* isolates are summarized in Additional file 3: Table S3. Overall, 16 and 18 QTLs were detected with D41-2 (D-01 to D-16; Table 1 and Fig. 2) and 12YL-DL3-2 (Y-01 to Y-18; Table 2 and Fig. 3), respectively. Four QTLs (D-13/Y-16, D-14, D-15/Y-17, and Y-02) were identified for both DLA and LT, and more QTLs were mapped from the *indica* than *japonica* subpopulation. Two QTLs were identified with both isolates: D-13 and Y-16 overlapped at 10.21–10.61 Mb, and D-15 and Y-17 were both located at 11.06–11.56 Mb in chromosome 12. Detailed information on the QTL intervals, the most significantly associated markers, and the co-localized *R* genes/QTLs are in Tables 1 and 2. In total, 6 and 30 QTLs were detected for LT and DLA, respectively, with $-\ Log_{10}(P)$ values ranging from 3.72 to 6.83 and the proportion of phenotypic variation explained (R^2) from 0.02 to 0.31. GWAS of global accessions and within varietal subgroups revealed different QTLs. In all, 13 of the 16 QTLs against D41-2 were detected in the full population, and 12 of the 18 QTLs against 12YL-DL3-2 were detected specifically in the *indica* subgroup. D-13/Y-16 and D-14 were the common QTLs identified at both the full-population and subgroup levels.

Among the 32 identified candidate QTLs, 22 regions co-localized with previously reported *R* genes and/or QTLs, including two cloned *R* genes, *Pita* (D-13/Y-16) and *Ptr* (D-14); 19 mapped *R* loci and 20 mapped QTLs. Only 3 regions co-localized with the QTLs identified in other GWAS of RDP1: D-13/Y-16 co-localized with *Pita* (Zhao et al. 2011), LABR_87, LABR_88, and LABR_89 (Kang et al. 2016); D-15/Y-17 co-localized with LABR_90 (Kang et al. 2016); and Y-01 co-localized with LABR_10 (Kang et al. 2016). For some reported large-interval *R* loci and QTLs, more than one candidate QTL was detected within the ranges. On chromosome 1, D-01 and D-02 were detected within AQCT001, AQEN001, AQEN011, and AQEN018, and Y-01 and Y-02 were detected within AQCT001 and AQEN001. On chromosome 4, D-07 to D-09 were detected within CQAC2; on chromosome 12, D-13 to D-15 and Y-14 to Y-18 were detected within AQCT008, AQEN010, AQEN017, *Pi6(t)*, *Pi12*, *Pi19(t)*, *Pi62(t)*, *Pi157*, and *Pita2*. In addition to known *R* genes or QTLs, we detected 10 new genomic regions associated with blast resistance on chromosome 2 (D-03, D-04, Y-03, Y-05), chromosome 3 (D-06, Y-06), chromosome 5 (D-11, Y-07), and chromosome 8 (Y-10, Y-11).

A total of 100 defense-related genes were identified within the intervals of the 32 QTLs (Additional file 4: Table S4). These included NBS–LRR genes, receptor-like protein kinase (RLK) genes, transcription factor genes, ubiquitination-related genes, and oxidase/oxidoreductase genes.

Resistance and susceptibility haplotypes in RDP1

LD analyses defined a total of 3261 (full population), 3226 (*indica* subgroup) and 1644 (*japonica* subgroup) LD blocks in the rice genome. Each block contained 3–13 haplotypes. Putative R haplotypes (frequency of control > frequency of case) and S haplotypes (frequency of case > frequency of control) of the 32 QTL regions in all the tested rice accessions are shown in Additional file 5: Table S5 and Additional file 6: Table S6. For each accession, R haplotypes were observed from 0 to 10 QTL regions and S haplotypes were observed from 0 to 11 QTL regions. For both isolates, the total number of non-overlapped R haplotypes was negatively correlated with disease severity ($r = -\ 0.29$ to $-\ 0.41$, $P < 0.001$), but the total number of non-overlapped S haplotypes was positively correlated with disease severity ($r = 0.28$ to 0.41, $P < 0.001$) (Additional file 7: Table S7).

To apply the haplotype information in Additional file 5: Table S5 and Additional file 6: Table S6 for resistance breeding, one can first choose resistance donors based on the resistance performance data in rows 3–6 (DLA and LT from D41-2 and 12YL-DL3-2 inoculations), then scroll down to find the QTLs likely contributing resistance in each accession. For example, NSFTV 17 exhibited good resistance to both *M. oryzae* isolates (DLA = 8.95 and 2.58; LT = 1.7 and 3), and it may contain resistant QTLs D-08, D-09, and D-14 against D41-2 (Additional file 5: Table S5) and resistant QTLs Y-02, Y-03, Y-07, Y-08, Y-09, and Y-11 against 12YL-DL3-2 (Additional file 6: Table S6). If a certain QTL is of interest, one can easily find accessions containing putative R or S haplotypes in this region. For instance, NSFTV 6 and NSFTV 10 may be used as resistance and susceptible donors for D-05, respectively (Additional file 5: Table S5). With the open access 44K or 700K genotypic data, molecular markers targeting candidate QTL can be designed and applied in marker-assisted selection programs.

Discussion

GWAS has been widely used for high-resolution mapping and mining for resistance resources in plants. In this research, we identified 32 non-overlapped regions associated with resistance to blast infection in rice by inoculating the open-access RDP1 with two representative *M. oryzae* isolates from Taiwan. Most of the QTLs were identified by inoculation of D41-2 or 12YL-DL3-2,

Table 1 Genetic regions significantly associated with resistance to the *Magnaporthe oryzae* isolate D41-2

No.	Population	Phenotype	Chr.	Range[a]	Physical position (bp)[b]	Highest marker information					Known *R* genes/QTLs
						Marker ID	Allele	MAF (%)[c]	$-\mathrm{Log}_{10}(P)$	R^2[d]	
D-01	*indica*	LT	1	id1010103–id1010205	15,438,604–15,649,252	id1010164	T/C	19.6	4.55	0.16	AQCT001, AQEN001, AQEN011, AQEN018
D-02	*indica*	DLA	1	id1011342–id1011401	18,821,356–18,925,916	id1011349	C/A	9.28	3.96	0.18	AQCT001, AQEN001, AQEN011, AQEN018
D-03	Full	DLA	2	id2003636–id2003653	7,104,193–7,129,000	id2003651	C/T	11.59	4.88	0.07	–
D-04	Full	DLA	2	id2006540-id2006687	16,209,277–16,648,368	id2006548	G/T	8.28	4.38	0.06	–
D-05	Full	DLA	3	id3010758–ud3001344	23,818,976–24,280,279	id3010813	C/T	32.01	4.86	0.06	AQAF029
						ud3001342	T/C	31.91		0.06	
D-06	Full	DLA	3	id3016979–id3017136	34,609,998–34,809,786	id3017114	G/A	12.37	5.41	0.08	–
D-07	Full	DLA	4	id4008106–id4008335	24,437,545–24,950,637	id4008148	T/A	46.8	5.30	0.07	CQAC2, *Pikur1*
D-08	Full	DLA	4	id4008360–id4008522	25,043,111–25,696,775	id4008464	G/A	7.01	4.87	0.07	CQAC2, *Pikur1*
D-09	Full	DLA	4	id4008601–id4008681	26,300,907–26,575,531	id4008636	G/A	5.92	4.29	0.06	CQAC2, *Pikur1*
D-10	Full	DLA	5	id5009568–id5009958	22,335,609–22,650,069	id5009818	G/T	5.96	4.13	0.06	AQAQ024
D-11	Full	DLA	5	id5009967–id5010294	22,654,409–23,102,774	id5010176	A/T	10.2	3.72	0.05	–
D-12	Full	DLA	8	id8005634–ud8001357	20,708,815–20,937,247	id8005717	A/G	34.65	5.21	0.07	*Pizh*
D-13[e]	*indica*	LT	12	ud12000652–id12004102	10,218,678–10,611,754	wd12001025	T/A	38.04	4.77	0.19	AQCT008, AQEN010, AQEN017, LABR_87, LABR_88, LABR_89, *Pi6(t)*, *Pi12*, *Pi19(t)*, *Pi20*, *Pi31(t)*, *Pi58(t)*, *Pi62(t)*, *Pi157*, *Pita2*, *Pita*
D-14	Full	LT	12	id12004113–id12004234	10,669,467–10,900,950	id12004183	A/G	35.64	4.64	0.07	AQCT008, AQEN010, AQEN017, *Pi6(t)*, *Pi12*, *Pi19(t)*, *Pi31(t)*, *Pi58(t)*, *Pi62(t)*, *Pi157*, *Pita2*, *Ptr*
	Full	DLA									
	indica	LT				id12004196	G/A	15.46	5.03	0.07	
						id12004196	G/A	37.5	5.09	0.12	
D-15[f]	Full	DLA	12	id12004251–wd12001183	11,060,008–11,555,556	id12004303	A/G	15.43	5.31	0.08	AQCT008, AQEN010, AQEN017, CQAC4, LABR_90, *Pi6(t)*, *Pi12*, *Pi19(t)*, *Pi31(t)*, *Pi58(t)*, *Pi62(t)*, *Pi157*, *Pita2*
D-16	Full	DLA	12	id12004432–id12004540	12,065,937–12,565,383	id12004538	G/A	16.33	5.37	0.08	AQCT008, AQCT009, AQEN010, AQEN017, CQAC4, *Pi6(t)*, *Pi12*, *Pi19(t)*, *Pi51(t)*, *Pi62(t)*, *Pi157*, *Pita2*

[a] The left-border marker and right-border marker of the QTL

[b] The physical map position is based on MSU version 7.0 Nipponbare rice reference genome

[c] MAF: minor allele frequency

[d] R^2 represents the proportion of the phenotypic variation explained by the single nucleotide polymorphism

[e,f] Same or largely overlapped genetic region identified by using D41-2 and 12YL-DL3-2

Table 2 Genetic regions significantly associated with resistance to the *Magnaporthe oryzae* isolate 12YL-DL3-2

No.	Population	Phenotype	Chr.	Range[a]	Physical position (bp)[b]	Highest marker information[b]					Known *R* genes/QTLs
						Marker ID	Allele	MAF (%)[c]	$-\log_{10}(P)$	R^{2d}	
Y-01	*indica*	DLA	1	id1013754–id1014034	23,732,729–23,934,283	id1013942	T/G	36	4.47	0.21	AQCT001, AQEN001, LABR_10
Y-02	*indica*	DLA	1	id1015101–id1015282	25,284,622–25,681,092	id1015101	A/G	35.06	6.83	0.31	AQAH002, AQCT001, AQEN001, *Pitp(t)*
		LT						35.06	4.85	0.28	
Y-03	*indica*	DLA	2	id2007485–id2007516	18,977,320–19,192,422	id2007494	G/T	41.09	5.37	0.26	–
Y-04	*indica*	DLA	2	id2008594–id2008644	21,493,161–21,607,823	id2008601	A/G	39.24	4.88	0.22	*Pid(t)1*
Y-05	*indica*	DLA	2	id2009379–id2009403	23,397,267–23,466,199	id2009379	T/C	34.21	4.42	0.21	–
Y-06	*indica*	DLA	3	id3017643–id3017725	35,480,048–35,558,427	id3017643	C/G	33.33	4.19	0.19	–
Y-07	*indica*	DLA	5	id5002016-id5002075	3,483,482-3,586,560	id5002075	C/T	30	4.06	0.18	–
Y-08	*indica*	DLA	5	wd5001198–wd5001344	10,351,569–10,848,969	id5004839, id5004859	C/T	22.79	4.15	0.19	AQEN004, *Pi23*
							A/G	23.75		0.18	
Y-09	*indica*	DLA	6	id6004112–id6004235	6,458,990–6,636,955	id6004212	C/T	46.15	4.37	0.20	AQCT004, AQEN005, *Pii1*, *Pi8*, *Pi59(t)*, *Pi27(t)*
Y-10	*indica*	DLA	8	id8000511–id8000544	1,744,391–1,839,119	id8000544	A/G	32.88	4.66	0.23	–
Y-11	*indica*	DLA	8	id8000658–id8000695	2,130,616–2,184,609	id8000663	C/T	26.67	4.47	0.21	–
Y-12	*indica*	DLA	12	id12002136–id12002219	4,731,647–4,872,049	id12002210	A/G	11.54	4.52	0.21	*Pi6(t)*, *Pi62(t)*
Y-13	*japonica*	DLA	12	id12003134–id12003432	7,827,050–8,296,350	id12003144	T/C	7.64	4.09	0.10	AQCT008, AQEN010, AQEN017, *Pi6(t)*, *Pi12*, *Pi20*, *Pi31(t)*, *Pi58(t)*, *Pi62(t)*
Y-14	*japonica*	DLA	12	ud12000485–id12003608	8,383,596–8,874,826	id12003547	C/T	7.9	4.04	0.10	AQCT008, AQEN010, AQEN017, *Pi6(t)*, *Pi12*, *Pi19(t)*, *Pi20*, *Pi31(t)*, *Pi58(t)*, *Pi62(t)*, *Pi157*
						id12003562	A/T	7.9	4.04	0.10	
Y-15	*japonica*	DLA	12	id12003642–ud12000566	8,914,661–9,231,272	id12003728	T/A	8.1	3.93	0.10	AQCT008, AQEN010, AQEN017, *Pi6(t)*, *Pi12*, *Pi19(t)*, *Pi20*, *Pi31(t)*, *Pi58(t)*,, *Pi62(t)*, *Pi157*
Y-16[e]	Full	LT	12	id12003919–id12004113	10,058,758–10,669,467	id12004093	G/A	10	5.67	0.11	AQCT008, AQEN010, AQEN017, LABR_87, LABR_88, LABR_89, *Pi6(t)*, *Pi12*, *Pi19(t)*, *Pi20*, *Pi31(t)*, *Pi58(t)*, *Pi62(t)*, *Pi157*, *Pita2*, *Pita*
	japonica	LT				id12004007	A/T	7	4.82	0.02	
	japonica	DLA				wd12001025	T/A	8.2	4.62	0.13	
Y-17[f]	Full	LT	12	id12004251–wd12001183	11,060,008–11,555,556	id12004284	C/G	8.52	5.87	0.11	AQCT008, AQEN010, AQEN017, CQAC4, LABR_90, *Pi6(t)*, *Pi12*, *Pi19(t)*, *Pi31(t)*, *Pi58(t)*, *Pi62(t)*, *Pi157*, *Pita2*
Y-18	Full	LT	12	id12004546–id12004685	12,567,686–13,060,942	id12004554	C/A	42.11	5.4	0.10	AQCT008, AQCT009, AQEN010, AQEN017, CQAC4, *Pi6(t)*, *Pi12*, *Pi19(t)*, *Pi51(t)*, *Pi62(t)*, *Pi157*, *Pita2*

[a] The left-border marker and right-border marker of the QTL

[b] The physical map position is based on MSU version 7.0 Nipponbare rice reference genome

[c] MAF: minor allele frequency

[d] R^2 represents the proportion of the phenotypic variation explained by the single nucleotide polymorphism

[e,f] Same or largely overlapped genetic region identified by using D41-2 and 12YL-DL3-2

but two QTLs, D-13/Y-16 and D-15/Y-17, on chromosome 12, were resistant to both isolates. The blast resistance QTLs mapped in our and other studies using RDP1 largely differ (Kang et al. 2016; Mgonja et al. 2016; Zhao et al. 2011; Zhu et al. 2016). In agreement with Kang et al. (2016), we identified some QTLs specifically in the *indica* or *japonica* subgroup, which may be associated with the frequencies of different resistant alleles in the two subgroups.

Among the QTLs detected in the present study, D-13/Y-16 on chromosome 12 showed the strongest resistance to both isolates. D-13/Y-16 had the greatest number of significant SNPs, high -$\log_{10}$ (*P*-value), and high R^2 value (Tables 1 and 2). The QTL is located at 10.05–10.67 Mb on chromosome 12, a region containing *Pita* (10.60–10.62 Mb), and co-localized with numerous *R* loci [*Pi6(t), Pi12, Pi19(t), Pi20, Pi31(t), Pi58(t), Pi62(t), Pi157, Pita2*] and reported QTLs. The neighboring D-14 (10.67–10.90 Mb) contains *Ptr* (LOC_Os12g18729; 10.82–10.84 Mb), an atypical *R* gene required for *Pita*- and *Pita-2*-mediated resistance (Zhao et al. 2018). The *Pita* locus was also identified as a major QTL in RDP1 by Kang et al. (2016) and Zhao et al. (2011). D41-2 and 12YL-DL3-2 are compatible with IRBLTA-K1, IRBLTA-CT2, and IRBLTA-CP1, the monogenic lines carrying *Pita* in the LTH background (unpublished data), so the major QTL we detected is likely allelic or linked to *Pita*. In addition to *Pita* (LOC_Os12g18360), LOC_Os12g17550 (10.05–10.07 Mb), encoding an LRR-containing protein, and LOC_Os12g18374 (10.62–10.64 Mb), encoding an NB-ARC-containing protein, are potential causal genes. As well, D-13 was identified from the *indica* subgroup, and Y-16 was identified from the full population and the *japonica* subgroup, which suggests the presence of diverse resistant variants of *Pita* and/or other linked *R* genes in *indica* and *japonica* rice accessions.

Previous GWAS of blast resistance relied on the 0-9 scoring systems that combine the evaluation of lesion size, LT and DLA by the naked eye (IRRI 2014; Kang et al. 2016; Mgonja et al. 2016; Zhao et al. 2011). In this study, we photographed the diseased samples, and LT and DLA were independently assessed by using image analysis. The use of LT for GWAS revealed only 6 QTLs on chromosome 1 (D-01: 15.43–15.65 Mb; Y-02: 25.28–25.68 Mb) and chromosome 12 (D-13/Y-16: 10.05–10.67 Mb; D-14: 10.66–10.90 Mb; Y-17: 11.06–11.56 Mb; Y-18: 12.56–13.06 Mb). An additional 26 QTLs were identified when the DLA trait was analyzed. Hence, the detailed phenotype data and accurate quantification of DLA can improve the power of QTL detection.

Our association mapping results can serve as a good reference for finer delimitation of previously reported large-interval *R* loci and QTLs. Considering that significant SNP signals may not locate nearby or within the coding regions of causative genes (Kang et al. 2016), we present the candidate QTLs from this study based on LD blocks. The mean interval of our candidate QTLs was ~ 309 kb (range 25 to 654 kb), which is higher than the resolution from conventional linkage mapping studies. In fact, many previously identified blast *R* loci were located at > 1000 kb genetic regions (Sharma et al. 2012).

A variety of candidate genes were identified within the QTL regions. According to the annotation, these genes may be involved in recognition, signaling, and/or antimicrobial activities in effector-triggered immunity (ETI) or pathogen-associated molecular pattern (PAMP)-triggered immunity (PTI) in plants (Jones and Dangl 2006). For instance, in D-05, the − $\text{Log}_{10}(P)$ peak and four significant SNPs were detected within a gene encoding an LRR protein (LOC_Os03g43390; 24.19–24.20 Mb on chromosome 3) (Additional file 8: Fig. S1A). This *R*-like gene is worthy of further investigation. A cluster of 21 significant SNPs with strong LD (LD parameter $r^2 > 0.8$) was found across D-08 (Additional file 8: Fig. S1B), which co-localizes with a large-interval *R* locus *Pikur1* on chromosome 4 (Goto 1970). Genes encoding LRR proteins (LOC_Os04g42470, LOC_Os04g42670, LOC_Os04g43340) and ubiquitin-related proteins were found in the D-08 interval. Ubiquitination is known to be involved in the modulation of (compatible and incompatible) plant–pathogen interactions via posttranslational modifications of the protein components of PTI and ETI (Park et al. 2012; Shirsekar et al. 2010). Fine-mapping and functional assays will be needed to validate the causal genes underlying the identified QTLs.

A number of rice accessions in RDP1 have great potential for resistance breeding and further exploitation. LAC 23 (NSFTV 99) and OS 6 (WC 10296, NSFTV 395) showed resistance to both Taiwanese isolates (LT < 3 and DLA < 10%); they were also reported to be resistant against isolates from China, South Korea, Columbia, Philippines, India, and the United States (Kang et al. 2016; Zhao et al. 2011; Zhu et al. 2016). The mechanisms of broad-spectrum resistance are worthy of further exploration. Our haplotype analysis revealed that in general, the more R haplotypes detected in an accession, the greater the degree of resistance. Eight accessions (NSFTV 17, NSFTV 135, NSFTV 161, NSFTV 183, NSFTV 202, NSFTV 235, NSFTV 642) carried ≥ 7 R haplotypes and were resistant to both isolates. These accessions are particularly useful for cultivar improvement in Taiwan. Moreover, *Pi27(t)* was originally identified from the resistance donor IR64 (NSFTV 644). This accession was included in RDP1 and the resistance haplotype was accurately assigned to the QTL region (Y-09) corresponding to *Pi27(t)*, which suggests the validity of the haplotype identification analysis in this study.

Conclusions

This study used the GWAS approach to explore blast resistance in the open-access RDP1. Although RDP1 had been used to locate blast resistance QTLs in a few recent studies (Zhao et al. 2011; Kang et al. 2016; Mgonja et al. 2016; Zhu et al. 2016), the QTLs mapped with geographically distinct *M. oryzae* isolates were largely different from those in our study. This observation indicates the diverse composition of *Avr* genes in geographically distinct *M. oryzae* isolates and the richness of qualitative and quantitative resistance genes in RDP1. By conventional visual rating and the use of image analysis for accurate quantitative assessment of disease severity, we delineated 32 known and new genomic regions controlling blast resistance and identified 100 candidate genes encoding leucine-rich repeat-containing proteins, transcription factors, ubiquitination-related proteins, and peroxidases. We also determined putative resistance and susceptibility haplotypes of the 32 QTLs for each tested rice accession. The information provided in Additional file 5: Table S5 and Additional file 6: Table S6 will aid in selecting suitable resistance donor lines for gene/QTL validation and further application.

Abbreviations

%DLA: proportion of diseased leaf area; BLUEs: best linear unbiased estimates; ETI: effector-triggered immunity; GLM: generalized linear model; GO: gene ontology; GSOR: Genetics Stocks *Oryza*; GWAS: genome-wide association study; LD: linkage disequilibrium; LT: lesion type; MLM: mixed linear model; NBS–LRR: nucleotide binding site–leucine-rich repeat; OMA: oat meal agar; PAMP: pathogen-associated molecular pattern; PTI: pathogen-associated molecular pattern-triggered immunity; Q–Q: quantile–quantile; QTL: quantitative trait loci; RCBD: randomized complete block design; RDP1: rice diversity panel 1; *R* genes: resistance genes; RLK: receptor-like protein kinase; SNP: single-nucleotide polymorphism; TEJ: *temperate japonica*; TRJ: *tropical japonica*.

Authors' contributions

HAL, SYC, FYC, and CLC conceived and designed the experiments. HAL performed the experiments. CWT helped in GWAS. YCC performed the haplotype analysis. WCS and RSC provided *M. oryzae* isolates and helped in fungal culture. FYC and CWW amplified the seeds of RDP1. HAL, SYC, and CLC analyzed the data and wrote the paper. All authors read and approved the final manuscript.

Author details

[1] Department of Plant Pathology and Microbiology, National Taiwan University, No. 1, Sec. 4, Roosevelt Rd., Taipei 10617, Taiwan. [2] Kaohsiung District Agricultural Research and Extension Station, No. 2-6, Dehe Rd., Pingtung County 90846, Taiwan. [3] Department of Agronomy, National Taiwan University, No. 1, Sec. 4, Roosevelt Rd., Taipei 10617, Taiwan. [4] Department of Biochemical Science and Technology, National Chiayi University, No. 300, Syuefu Rd., Chiayi City 60004, Taiwan.

Acknowledgements

We thank Khong-Sam Chia, Yi-Jui Lo, and Hao Li for help with seed amplification; Jui-Yu Liao, Yu-Chia Chen, Wei-Lun Chen, An-Po Cheng, Yu-Cyuan Shih, Qiao-Juan Lai, and Shun-Yuan Huang for participating in phenotyping; and Wei-Bin Chang and Hsin-Han Lee for assisting data analysis.

Competing interests

The authors declare that they have no competing interests.

Funding

The work was supported by the Ministry of Science and Technology of Taiwan (102-2313-B-002-065-MY3; 105-2313-B-002-014-; 106-2313-B-002-021-MY3) and Bureau of Animal and Plant Inspection and Quarantine (BAPHIQ), Council of Agriculture, Taiwan [104AS-10.6.1-BQ-B1(1); 105AS-10.5.4-BQ-B1(1); 106AS-9.5.4-BQ-B1(1); 107AS-8.4.4-BQ-B1(1)].

References

Ashkani S, Rafii MY, Shabanimofrad M, Ghasemzadeh A, Ravanfar SA, Latif MA (2016) Molecular progress on the mapping and cloning of functional genes for blast disease in rice (*Oryza sativa* L.): current status and future considerations. Crit Rev Biotechnol 36:353–367

Azizi P, Rafii MY, Mahmood M, Abdullah SN, Hanafi MM, Nejat N, Latif MA, Sahebi M (2015) Differential gene expression reflects morphological characteristics and physiological processes in rice immunity against blast pathogen *Magnaporthe oryzae*. PLoS ONE 10(5):e0126188

Barrett JC, Fry B, Maller J, Daly MJ (2005) Haploview: analysis and visualization of LD and haplotype maps. Bioinformatics 21:263–265

Begum H, Spindel JE, Lalusin A, Borromeo T, Gregorio G, Hernandez J, Virk P, Collard B, McCouch SR (2015) Genome-wide association mapping for yield and other agronomic traits in an elite breeding population of tropical rice (*Oryza sativa*). PLoS ONE 10:e0119873

Bradbury PJ, Zhang Z, Kroon DE, Casstevens TM, Ramdoss Y, Buckler ES (2007) TASSEL: software for association mapping of complex traits in diverse samples. Bioinformatics 23:2633–2635

Buckler ES, Holland JB, Bradbury PJ, Acharya CB, Brown PJ, Browne C, Ersoz E, Flint-Garcia S, Garcia A, Glaubitz JC, Goodman MM, Harjes C, Guill K, Kroon DE, Larsson S, Lepak NK, Li H, Mitchell SE, Pressoir G, Peiffer JA, Rosas MO, Rocheford TR, Romay MC, Romero S, Salvo S, Sanchez Villeda H, da Silva HS, Sun Q, Tian F, Upadyayula N, Ware D, Yates H, Yu J, Zhang Z, Kresovich S, McMullen MD (2009) The genetic architecture of maize flowering time. Science 325:714–718

Chen X, Shang J, Chen D, Lei C, Zou Y, Zhai W, Liu G, Xu J, Ling Z, Cao G, Ma B, Wang Y, Zhao X, Li S, Zhu L (2006) A B-lectin receptor kinase gene conferring rice blast resistance. Plant J 46:794–804

Chen LC, Huang SH, Cheng CH (2009) Review of the screening tests for rice varietal resistance to major diseases and insect pests in Taiwan. In: Ni HF, Yang HR (eds) Proceedings of symposium on achievements and perspectives of rice protection in Taiwan, 161p, 1st ed. Chiayi Agricultural Experiment Branch, Taiwan Agricultural Research Institute, Chiayi, Taiwan. pp 83–103

Collard BCY, Jahufer MZZ, Brouwer JB, Pang ECK (2005) An introduction to markers, quantitative trait loci (QTL) mapping and marker-assisted selection for crop improvement: the basic concepts. Euphytica 142:169–196

Das A, Soubam D, Singh PK, Thakur S, Singh NK, Sharma TR (2012) A novel blast resistance gene, *Pi54rh* cloned from wild species of rice, *Oryza rhizomatis* confers broad spectrum resistance to *Magnaporthe oryzae*. Funct Integr Genomics 12:215–228

Deng Y, Zhu X, Xu J, Chen H, He Z (2009) Map-based cloning and breeding application of a broad-spectrum resistance gene *Pigm* to rice blast. Advances in genetics, genomics and control of rice blast disease. Springer, Dordrecht, pp 161–171

Eizenga GC, Ali M, Bryant RJ, Yeater KM, McClung AM, McCouch SR (2014) Registration of the rice diversity panel 1 for genomewide association studies. J Plant Regist 8:109–116

Famoso AN, Zhao K, Clark RT, Tung C-W, Wright MH, Bustamante C, Kochian LV, McCouch SR (2011) Genetic architecture of aluminum tolerance in rice

(*Oryza sativa*) determined through genome-wide association analysis and QTL mapping. PLoS Genet 7:e1002221

Food and Agriculture Organization of the United Nations (2016) FAO Rice Market Monitor 19:1–31. http://www.fao.org/economic/RMM/en/. Accessed 5 Feb 2017

Fukuoka S, Saka N, Koga H, Ono K, Shimizu T, Ebana K, Hayashi N, Takahashi A, Hirochika H, Okuno K, Yano M (2009) Loss of function of a proline-containing protein confers durable disease resistance in rice. Science 325:998–1001

Fukuoka S, Yamamoto SI, Mizobuchi R, Yamanouchi U, Ono K, Kitazawa N, Yasuda N, Fujita Y, Thanh Nguyen TT, Koizumi S, Sugimoto K, Matsumoto T, Yano M (2014) Multiple functional polymorphisms in a single disease resistance gene in rice enhance durable resistance to blast. Sci Rep 4:4550

Gajardo HA, Wittkop B, Soto-Cerda B, Higgins EE, Parkin IA, Snowdon RJ, Federico ML, Iniguez-Luy FL (2015) Association mapping of seed quality traits in *Brassica napus* L. using GWAS and candidate QTL approaches. Mol Breed 35:143

Goto I (1970) Genetic studies on the resistance of rice plant to the blast fungus. I. Inheritance of resistance in crosses Sensho x H-79 and Imochi-shirazu x H-79. Ann Phytopathol Soc Jpn 36:304–312

Huang X, Han B (2014) Natural variations and genome-wide association studies in crop plants. Annu Rev Plant Biol 65:531–551

Huang X, Wei X, Sang T, Zhao Q, Feng Q, Zhao Y, Li C, Zhu C, Lu T, Zhang Z, Li M, Fan D, Guo Y, Wang A, Wang L, Deng L, Li W, Lu Y, Weng Q, Liu K, Huang T, Zhou T, Jing Y, Li W, Lin Z, Buckler ES, Qian Q, Zhang Q-F, Li J, Han B (2010) Genome-wide association studies of 14 agronomic traits in rice landraces. Nat Genet 42:961–967

Huang X, Zhao Y, Wei X, Li C, Wang A, Zhao Q, Li W, Guo Y, Deng L, Zhu C, Fan D, Lu Y, Weng Q, Liu K, Zhou T, Jing Y, Si L, Dong G, Huang T, Lu T, Feng Q, Qian Q, Li J, Han B (2012) Genome-wide association study of flowering time and grain yield traits in a worldwide collection of rice germplasm. Nat Genet 44:32–39

IRRI (2014) Standard evaluation system for rice, 5th edn. International Rice Research Institute, Manila, p 57

Jaiswal V, Gahlaut V, Meher PK, Mir RR, Jaiswal JP, Rao AR, Balyan HS, Gupta PK (2016) Genome wide single locus single trait, multi-locus and multi-trait association mapping for some important agronomic traits in common wheat (*T. aestivum* L.). PLoS ONE 11:e0159343

Jones JDG, Dangl JL (2006) The plant immune system. Nature 444:323–329

Kang H, Wang Y, Peng S, Zhang Y, Xiao Y, Wang D, Qu S, Li Z, Yan S, Wang Z, Liu W, Ning Y, Korniliev P, Leung H, Mezey J, McCouch SR, Wang GL (2016) Dissection of the genetic architecture of rice resistance to the blast fungus *Magnaporthe oryzae*. Mol Plant Pathol 17:959–972

Kawahara Y, de la Bastide M, Hamilton JP, Kanamori H, McCombie WR, Ouyang S, Schwartz DC, Tanaka T, Wu J, Zhou S, Childs KL, Davidson RM, Lin H, Quesada-Ocampo L, Vaillancourt B, Sakai H, Lee SS, Kim J, Numa H, Itoh T, Buell CR, Matsumoto T (2013) Improvement of the *Oryza sativa* Nipponbare reference genome using next generation sequence and optical map data. Rice 6:4

Koide Y, Kobayashi N, Xu D, Fukuta Y (2009) Resistance genes and selection DNA markers for blast disease in rice (*Oryza sativa* L.). Jpn Agric Res Q 43:255–280

Kovach MJ, Sweeney MT, McCouch SR (2007) New insights into the history of rice domestication. Trends Genet 23:578–587

Lamari L (2008) Assess 2.0: image analysis software for plant disease quantification. American Phytopathological Society, St. Paul, p 125

Liu Y, Qi X, Young ND, Olsen KM, Caicedo AL, Jia Y (2015) Characterization of resistance genes to rice blast fungus *Magnaporthe oryzae* in a "Green Revolution" rice variety. Mol Breed 35:52

Ma J, Jia MH, Jia Y (2014) Characterization of rice blast resistance gene *Pi61(t)* in rice germplasm. Plant Dis 98:1200–1204

Ma J, Lei C, Xu X, Hao K, Wang J, Cheng Z, Ma X, Ma J, Zhou K, Zhang X, Guo X, Wu F, Lin Q, Wang C, Zhai H, Wang H, Wan J (2015) *Pi64*, encoding a novel CC–NBS–LRR protein, confers resistance to leaf and neck blast in rice. Mol Plant Microbe Interact 28:558–568

Marone D, Russo MA, Laidò G, De Leonardis AM, Mastrangelo AM (2013) Plant nucleotide binding site–leucine-rich repeat (NBS–LRR) genes: active guardians in host defense responses. Int J Mol Sci 14:7302–7326

McCouch SR, Wright MH, Tung C-W, Maron LG, McNally KL, Fitzgerald M, Singh N, DeClerck G, Agosto-Perez F, Korniliev P, Greenberg AJ, Naredo ME, Mercado SM, Harrington SE, Shi Y, Branchini DA, Kuser-Falcão PR, Leung H, Ebana K, Yano M, Eizenga G, McClung A, Mezey J (2016) Open access resources for genome-wide association mapping in rice. Nat Commun 7:10532

McDowell JM, Woffenden BJ (2003) Plant disease resistance genes: recent insights and potential applications. Trends Biotechnol 21:178–183

Mgonja EM, Balimponya EG, Kang H, Bellizzi M, Park CH, Li Y, Mabagala R, Sneller C, Correll J, Opiyo S, Talbot NJ, Mitchell T, Wang GL (2016) Genome-wide association mapping of rice resistance genes against *Magnaporthe oryzae* isolates from four African countries. Phytopathology 106:1359–1365

Miah G, Rafii MY, Ismail MR, Puteh AB, Rahim HA, Asfaliza R, Latif MA (2013) Blast resistance in rice: a review of conventional breeding to molecular approaches. Mol Biol Rep 40:2369–2388

Norton GJ, Douglas A, Lahner B, Yakubova E, Guerinot ML, Pinson SR, Tarpley L, Eizenga GC, McGrath SP, Zhao F-J, Islam MR, Islam S, Duan G, Zhu Y, Salt DE, Meharg AA, Price AH (2014) Genome wide association mapping of grain arsenic, copper, molybdenum and zinc in rice (*Oryza sativa* L.) grown at four international field sites. PLoS ONE 9:e89685

Oraguzie NC, Gardiner SE, Rikkerink EH, Silva HN (2007) Association mapping in plants. Springer, New York, p 277

Park CH, Chen S, Shirsekar G, Zhou B, Khang CH, Songkumarn P, Afzal AJ, Ning Y, Wang R, Bellizzi M, Valent B, Wang GL (2012) The *Magnaporthe oryzae* effector AvrPiz-t targets the RING E3 ubiquitin ligase APIP6 to suppress pathogen-associated molecular pattern–triggered immunity in rice. Plant Cell 24:4748–4762

Pasam RK, Sharma R (2014) Association mapping: a new paradigm for dissection of complex traits in crops. In: Kishor PBK, Bandopadhyay R, Suravajhala P (eds) Agricultural bioinformatics. Springer India, New Delhi, pp 1–20

R Development Core Team (2011) R: a language and environment for statistical computing. The R Foundation for Statistical Computing, Vienna. ISBN: 3-900051-07-0. http://www.R-project.org/. Accessed 7 Feb 2017

Rasheed A, Xia X, Ogbonnaya F, Mahmood T, Zhang Z, Mujeeb-Kazi A, He Z (2014) Genome-wide association for grain morphology in synthetic hexaploid wheats using digital imaging analysis. BMC Plant Biol 14:128

Reed E, Nunez S, Kulp D, Qian J, Reilly MP, Foulkes AS (2015) A guide to genome-wide association analysis and post-analytic interrogation. Stat Med 34:3769–3792

RiceData (2012) http://www.ricedata.cn/gene/gene_pi.htm. Accessed 22 Oct 2018

Sakai H, Lee SS, Tanaka T, Numa H, Kim J, Kawahara Y, Wakimoto H, Yang C, Iwamoto M, Abe T, Yamada Y, Muto A, Inokuchi H, Ikemura T, Matsumoto T, Sasaki T, Itoh T (2013) Rice Annotation Project Database (RAP-DB): an integrative and interactive database for rice genomics. Plant Cell Physiol 54:e6

Sharma TR, Rai AK, Gupta SK, Vijayan J, Devanna BN, Ray S (2012) Rice blast management through host-plant resistance: retrospect and prospects. Agric Res 1:37–52

Shirsekar G, Dai L, Hu Y, Wang X, Zeng L, Wang GL (2010) Role of ubiquitination in plant innate immunity and pathogen virulence. J Plant Biol 53:10–18

Soto-Cerda BJ, Cloutier S (2012) Association mapping in plant genomes. In: Caliskan M (ed) Genetic diversity in plants, 498p, 1st edn. InTech, Rijek, pp 29–54

Su J, Wang W, Han J, Chen S, Wang C, Zeng L, Feng A, Yang J, Zhou B, Zhu X (2015) Functional divergence of duplicated genes results in a novel blast resistance gene *Pi50* at the *Pi2/9* locus. Theor Appl Genet 128:2213–2225

Tang J, Zhu X, Wang Y, Liu L, Xu B, Li F, Fang J, Chu C (2011) Semi-dominant mutations in the CC-NB-LRR-type *R* gene, *NLS1*, lead to constitutive activation of defense responses in rice. Plant J 66:996–1007

Valent B, Farrall L, Chumley FG (1991) *Magnaporthe grisea* genes for pathogenicity and virulence identified through a series of backcrosses. Genetics 127:87–101

Wan H, Chen L, Guo J, Li Q, Wen J, Yi B, Ma C, Tu J, Fu T, Shen J (2017) Genome-wide association study reveals the genetic architecture underlying salt tolerance-related traits in rapeseed (*Brassica napus* L.). Front Plant Sci 8:593

Wang C, Yang Y, Yuan X, Xu Q, Feng Y, Yu H, Wang Y, Wei X (2014) Genome-wide association study of blast resistance in indica rice. BMC Plant Biol 14:311

Wilson RA, Talbot NJ (2009) Under pressure: investigating the biology of plant infection by *Magnaporthe oryzae*. Nat Rev Microbiol 7:185–195

Xu X, Hayashi N, Wang CT, Fukuoka S, Kawasaki S, Takatsuji H, Jiang CJ (2014) Rice blast resistance gene *Pikahei-1(t)*, a member of a resistance gene cluster on chromosome 4, encodes a nucleotide-binding site and leucine-rich repeat protein. Mol Breed 34:691–700

Zeigler RS, Leong SA, Teng PS (1994) Rice blast disease, 1st edn. CAB International, Wallingford, p 626

Zhao K, Tung C-W, Eizenga GC, Wright MH, Ali ML, Price AH, Norton GJ, Islam MR, Reynolds A, Mezey J, McClung AM, Bustamante CD, McCouch SR (2011) Genome-wide association mapping reveals a rich genetic architecture of complex traits in *Oryza sativa*. Nat Commun 2:467

Zhao H, Wang X, Jia Y, Minkenberg B, Wheatley M, Fan J, Jia MH, Famoso A, Edwards JD, Wamishe Y, Valent B, Wang GL, Yang Y (2018) The rice blast resistance gene *Ptr* encodes an atypical protein required for broad-spectrum disease resistance. Nat Commun 9:2039

Zhu D, Kang H, Li Z, Liu M, Zhu X, Wang Y, Wang D, Wang Z, Liu W, Wang GL (2016) A genome-wide association study of field resistance to *Magnaporthe oryzae* in rice. Rice 9:44

17

Begonia myanmarica (Begoniaceae), a new species from Myanmar and molecular phylogenetics of *Begonia* sect. *Monopteron*

Yu-Hsin Tseng[1], Young-Dong Kim[2*], Ching-I Peng[1], Khin Myo Htwe[3], Seong-Hyun Cho[4], Yoshiko Kono[1] and Kuo-Fang Chung[1*]

Abstract

Background: A new species, *Begonia myanmarica*, was discovered from Myanmar and herein documented. Characterized by a single developed wing in the ovary/fruit, this species would be assigned to sect. *Monopteron* (sensu Doorenbos et al. in The sections of *Begonia* including descriptions, keys and species lists: studies in Begoniaceae VI. Wageningen Agricultural University, Wageningen, 1998) that is known by *B. griffithiana* and *B. nepalensis* from the Himalaya. To confirm its sectional assignment, we conducted morphological, phylogenetic and cytological studies.

Results: Morphological observations indicated that *B. myanmarica* was distinguishable from the two known species of sect. *Monopteron* by the leaf shape and size, 1-locular ovary, parietal placentation and chromosome number. Molecular phylogenetic analysis using nrITS sequences showed that *B. myanmarica* was not allied with the clade of sect. *Monopteron*, though both were nested within sect. *Platycentrum*-sect. *Sphenanthera* clade.

Conclusions: Studies of morphology, molecular phylogenetics and cytology support the recognition of the new species, *Begonia myanmarica*, which is fully described and illustrated. Our results also indicate that *B. myanmarica* is not closely related to species previously assigned to sect. *Monopteron*, suggesting that the fruit morphology of a single developed wing in the ovary/fruit characterizing sect. *Monopteron* is homoplasious.

Keywords: *Begonia griffithiana*, *Begonia nepalensis*, Chromosome, Morphology

Background

Begonia L. (Begoniaceae), comprising more than 1800 species classified into 68 sections (Doorenbos et al. 1998; Hughes et al. 2015; Christenhusz and Byng 2016), is one of the largest genera of vascular plants. With more than 760 *Begonia* species in Asia, Doorenbos et al. (1998) recognized 18 sections [*Alicida* C.B. Clarke, *Apterobegonia* Warb., *Baryandra* A. DC., *Bracteibegonia* A. DC., *Coleocentrum* Irmscher, *Diploclinium* (Lindl.) A. DC., *Haagea* (Klotzsch) A. DC., *Heeringia* Irmscher, *Lauchea* (Klotzsch) A. DC., *Monophyllon* A. DC., *Monopteron* (A. DC.) Warb., *Parvibegonia* A. DC., *Petermannia* (Klotzsch) A. DC., *Platycentrum* (Klotzsch) A. DC., *Putzeysia* (Klotzsch) A. DC., *Reichenheimia* (Klotzsch) A. DC., *Ridleyella* Irmscher, and *Sphenanthera* (Hassk.) Warb.]. Thereafter, four additional Asian sections were proposed [*Leprosae* (T.C. Ku) Y.M. Shui, *Monolobium* T.C. Ku, *Pleiothece* T.C. Ku, and *Symbegonia* (Warb.) G. Forrest & Hollingsw.] (Ku 1999; Shui et al. 2002; Forrest and Hollingsworth 2003; Ku et al. 2007). These 22 Asian sections are highly unequal in species numbers: eight of the large sections (*Petermannia*, *Platycentrum*, *Diploclinium*, *Reichenheimia*, *Coleocentrum*, *Parvibegonia*, *Sphenanthera*, and *Symbegonia*) comprise 95% of Asian *Begonia* species and the rest 14 sections each with less than five species (Thomas 2010). Several molecular phylogenetic studies have demonstrated the paraphyly or polyphyly of these large sections, suggesting homoplasy

*Correspondence: ydkim@hallym.ac.kr; bochung@gate.sinica.edu.tw
[1] Research Museum and Herbarium (HAST), Biodiversity Research Center, Academia Sinica, Taipei 11542, Taiwan
[2] Department of Life Science, Hallym University, Chuncheon, Gangwon 24252, Republic of Korea
Full list of author information is available at the end of the article

of morphological characters used for current sectional delimitations (Tebbitt et al. 2006; Thomas et al. 2011; Chung et al. 2014). However, few studies have tested the monophyly of small Asian section of *Begonia* thus far [but see Rajbhandary (2010); Rubite (2010); Thomas (2010)].

Myanmar is botanically a most interesting country, but there have been no critical floristic surveys for nearly half a century. Thus far about 60 species of *Begonia* have been recorded from Myanmar (Hughes 2008; Tanaka and Hughes 2007; Tanaka and Hayami 2011; Peng et al. 2014b; Tanaka and Peng 2016). During the fieldwork in western Myanmar on 2 February 2012, the second author (YDK) collected an unknown *Begonia* with only one developed wing in ovary/fruit, which is the key character of *Begonia* sect. *Monopteron* sensu Doorenbos et al. (1998) first delimited by de Candolle (1864) as *Mezierea* sect. *Monopteron*. Presently, only two species, *B. griffithiana* Warb. and *B. nepalensis* Warb., are recognized in sect. *Monopteron* (de Candolle 1864; Doorenbos et al. 1998). *Begonia nepalensis*, the type species of sect. *Monopteron*, is native to Bhutan, Nepal and India (Fig. 5; Doorenbos et al. 1998; Hughes et al. 2015). Its chromosome number was reported to be $2n = 16$ (Legro and Doorenbos 1971), with an uncertain chromosome count $2n = 28–42$ by Sharma and Bhattacharyya (1961). *Begonia griffithiana*, occurring in Bhutan and India (Fig. 5), is characterized by lanceolate to oblong leaves with subcordate base. Chromosome number of *B. griffithiana* was documented as $2n = 22$ (Doorenbos et al. 1998). Based on recent systematics and phylogenetics of *Begonia*, sect. *Monopteron* is nested within the *Platycentrum-Sphenanthera* clade (Rubite 2010; Thomas 2010; Rajbhandary et al. 2011; Leong 2017).

Although morphology of the 1-winged ovary/capsule of the undescribed *Begonia* should be assigned to sect. *Monopteron*, it differs from *B. griffithiana* and *B. nepalensis* significantly the leaf shape, leaf size and distribution. In this study, we described it as a new species. We also provide detailed morphological data and molecular phylogenetic analysis to elucidate the sectional assignment for this species.

Methods

Morphological observations

Rhizomes of *Begonia myanmarica* collected by YDK from Myanmar were cultivated in the experimental greenhouse of the Biodiversity Research Center, Academia Sinica, Taipei, Taiwan. Fully grown plants with flowers and fruits (*Peng 23565, 23566*) were used for morphological observation. The two species of sect. *Monopteron*, *B. griffithiana* (*Peng 20851*) and *B. nepalensis* (*Peng 20854*), cultivated in the greenhouse were also studied as a comparison.

Chromosome preparations

Root tips were obtained from cultivated materials from greenhouse of Academic Sinica. Somatic chromosome of the new species, *B. myanmarica* (*Peng 23566*), and two species of sect. *Monopteron*: *B. griffithiana* (*Peng 20851*) and *B. nepalensis* (*Peng 20854*), were examined using root tips following the methods by Peng et al. (2014a).

Phylogenetic analyses

DNA sequences of the nuclear ribosomal internal transcribed spacer (nrITS) were used to evaluate the phylogenetic relationship among new species and the two species of sect. *Monopteron*. DNA extraction, PCR amplification and DNA sequencing followed Chung et al. (2014). To test the monophyly of sect. *Monopteron* and sectional assignment of new species, nrITS of 96 species used in Chung et al. (2014) were adopted for phylogenetic analysis (see Appendix for details). Alignment was conducted using MUSCLE implemented in MEGA5.2 (Tamura et al. 2011) and verified in Mesquite v3.03 (Maddison and Maddison 2015). Phylogenetic relationships were constructed by Bayesian Inference (BI) method. The best nucleotide substitution models were determined by Modeltest v2.7 (Posada and Crandall 1998). For BI analysis, the consensus topology was based on Markov chains algorithm implemented in MRBAYES 3.0b4 (Huelsenbeck and Ronquist 2001). Four chains of Markov chain Monte Carlo (MCMC) simulation were carried out for 1,500,000 generations each with trees sampled per 500 generations. The first 500 trees of sampled trees were discarded before the node probability was calculated (posterior probability: PP).

Taxonomic treatment

Begonia myanmarica C.-I Peng & Y. D. Kim, sp. *nov.* (Figs. 1, 2)

Type

MYANMAR. Sagaing Region, Alangdaw Kathapa National Park, 22°18′ 47.7″ N, 94°28′32.7″ E, alt. 438 m, mixed deciduous forest, along the stream. Living collection made by *Seong-hyun Cho, Young-dong Kim, Yong-in Kim & Jeong-hun Lee MM-0611*, 2 Feb 2012; type specimens (with flowers and fruits) pressed from plants cultivated in the experimental greenhouse, Academia Sinica, Taipei, Taiwan, 20 Mar 2016, *Ching-I Peng* 23566 (holotype: RAF; isotypes: HAST, KB).

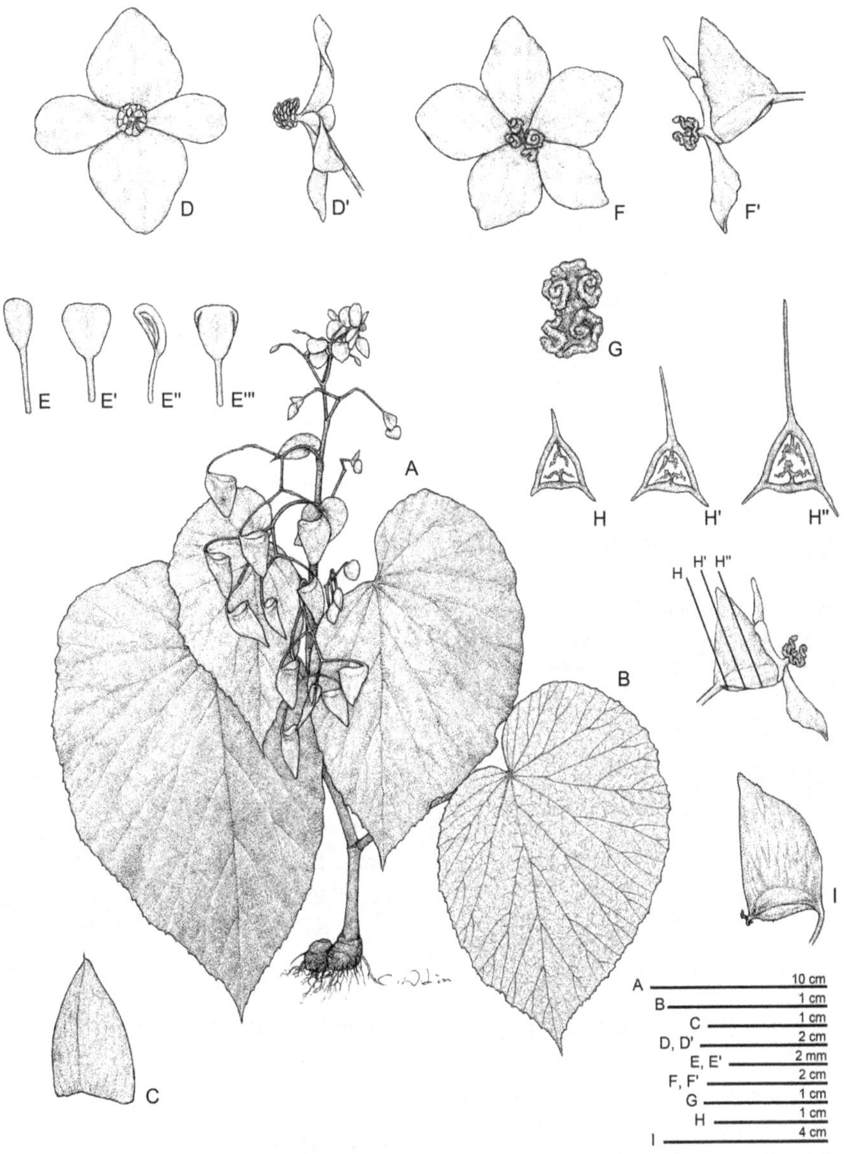

Fig. 1 *Begonia myanmarica* C.-I Peng & Y. D. Kim. **A** Habit. **B** Leaf adaxial surface. **C** Stipule. **D** Male flower, face view, **D′** Male flower, side view. **E, E′, E″, E‴** Stamen. **F** Female flower, face view, **F′** Female flower, side view. **G** Style and stigma. **H, H′, H″, H‴** Cross section of ovary. **I** Capsule

Diagnosis

Begonia myanmarica is a unique species with an erect habit; large, ovate to broadly ovate leaves (ca. 20–40 cm long, 22–30 mm across); sole, much protruded wing in ovary/fruit; 1-locular ovary with parietal placentation and 2 placentae; and the somatic chromosomes are determined as $2n = 38$.

Herbs monoecious, perennial. Rhizomes stout, 2–4 cm across, to 9 cm long; erect stem 50–90 cm tall, 1–2 cm thick, internodes 5–15 cm, glabrous. Stipules caducous, glabrous, triangular, apex aristate or apiculate, margin entire, 7–15 mm long, 6–15 mm wide. Leaves alternate, green with slightly paler veins; petiole 20–40 cm long, 2.2–3 cm across, glabrous; leaf blade fleshy, asymmetric,

Fig. 2 *Begonia myanmarica* C.-I Peng & Y. D. Kim. **a** Habit and habitat. **b** Cultivated plant at anthesis. **c** Leaf abaxial view. **d** Stipule. **e** Bract. **f** Male flower, face view. **g** Male flower, side view. **h** Female flower, face view. **i** Female flower, side view. **j** Cross section of ovaries. **k** Capsule

ovate to broadly ovate, 21–35 cm long, 15–28 cm wide, upper surface glabrous, underside slightly hairy on veins, base obliquely cordate, margin irregularly loosely serrulate or denticulate, apex acute or short acuminate; venation 7-or 8-palmate. Inflorescence mainly terminal but also axillary racemes of dichasial cymes, bisexual, protandrous; cyme 10–15 cm long, with 2 female flowers at apex and 7–9 male flowers at base, peduncle 2–5 cm long, glabrous; bracts deciduous, ovate to triangular, apex acuminate, margin entire, 0.7–1.5 cm long, 0.4–0.6 cm wide. Male flower: pedicel 2–2.8 cm long, glabrous; tepals 4, white to pinkish, outer 2 ovate or orbicular, 1.9–2.3 cm long, 1.5–2 cm wide, inner 2, broadly oblong, 1.8–2.3 cm long, 0.5–1.5 cm wide, glabrous; androecium actinomorphic, subglobose, ca. 0.7 cm across, stamens 60–80, yellow, clavate; filaments ca. 2 mm long, fused to a short central column; anthers 1–1.2 mm long, apex truncate. Female flower: pedicel 1.7–3.5 cm long, glabrous; ovary white, wings 3, manifestly unequal, 2 side wings almost undeveloped, abaxial wing much protruded, white to pale greenish or pinkish, 1-locular; placentation parietal, placentae 2, each bilamellate; tepals 5, white in the greenhouse (pinkish in the wild), unequal, elliptic to obovate, 1.5–2 cm long, 0.5–1.1 cm wide, apex obtuse; styles 2, ca. 5 mm long, 2- or 3-cleft, fused at base, stigmatic band wavy-twisted and spiralled. Capsule nodding, stalk 3.5–5.5 cm long, abaxial wing triangular to rectangular, 2.4–3.8 cm tall, 2.0–2.4 cm wide, lateral two wings barely developed, rounded, 0.2–0.4 cm tall, 1.8–2.2 cm wide. Seeds barrel-shaped, 0.25–0.3 mm long.

Distribution and habit

Known only from the type locality.

Etymology

The epithet refers to Myanmar (formerly Burma) where it was discovered.

Additional specimens examined

MYANMAR. Sagaing Region: Alang Daw Kathapa National Park, 22°18′47.7″ N, 94°28′32.7″ E, 438 m, 2 Feb 2012, *Peng 23565* (HAST); 22°18′49.5″ N, 94°28′30.2″ E, 434 m, 2 Feb 2012, *MM-0556* (KB); 22°18′47.7″ N, 94°28′32.7″ E, 438 m, 2 Feb 2012, *MM-0611* (KB, HHU); 22°18′44.2″ N, 94°28′28.7″ E, 380 m, 2 Feb 2012, *MM-0616* (KB); 22°19′25″ N, 94°29′37.7″ E, 380 m, 5 Feb 2012, *MM-0848* (KB, RAF).

Chromosome cytology

Somatic chromosome at metaphase of *B. myanmarica* were shown to be $2n = 38$ in this study (Fig. 4c).

Discussion

Begonia myanmarica has only one developed wing in ovary/fruit (Figs. 2k, 3c, d), the key character of sect. *Monopteron* in *Begonia* (Doorenbos et al. 1998). The new species, however, deviates from sect. *Monopteron* with axillary placentation and two locules in ovary (Fig. 3e, f) in having 1-locular ovary and parietal placentation (Fig. 2j). Additionally, *B. myanmarica* has ovate to broadly ovate leaves and large leaves (ca. 20–40 cm long, 22–30 mm across) (Fig. 2b, c), whereas leaves of *B. griffithiana* and *B. nepalensis* are lanceolate to oblong and no longer than 20 × 10 cm (Fig. 3a, b). Cytologically,

Fig. 3 Overview of leaves, fruit and cross section of ovary in *Begonia griffithiana* and *B. nepalensis*. **a**, **b** Leaves. **c**, **d** Fruit. **e**, **f** Cross section of ovary. *Begonia griffithiana* (**a**, **c**, **e**). *Begonia nepalensis* (**b**, **d**, **f**)

somatic chromosome of *B. myanmarica* is determined to be $2n = 38$ (Fig. 4c), while chromosomes of *B. griffithiana* and *B. nepalensis* are both $2n = 16$ (Fig. 4a, b) in our study. Geographically, *B. myanmarica* is endemic to Myanmar while *B. griffithiana* and *B. nepalensis* are distributed in India, Nepal and Bhutan (Fig. 5). We concluded that *B. myanmarica* is sharply distinct from *B. griffithiana* and *B. nepalensis*.

In our molecular phylogenetic study, *B. griffithiana* and *B. nepalensis* form a strongly supported clade (posterior probability, PP = 1) nested within the clade dominated by sect. *Platycentrum*-sect. *Sphenanthera* clade [Fig. 6; Clade *PLA-SPH* in Chung et al. (2014)], congruent with the studies of Rubite (2010), Thomas (2010), Rajbhandary et al. (2011), and Leong (2017). Two sampled individuals of *B. myanmarica* also fall within the Clade *PLA-SPH*

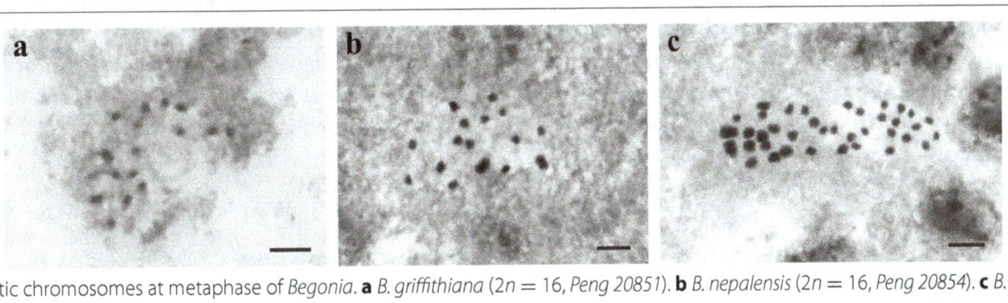

Fig. 4 Somatic chromosomes at metaphase of *Begonia*. **a** *B. griffithiana* ($2n = 16$, *Peng 20851*). **b** *B. nepalensis* ($2n = 16$, *Peng 20854*). **c** *B. myanmarica* ($2n = 38$, *Peng 23566*). *Scale bar* 5 μm

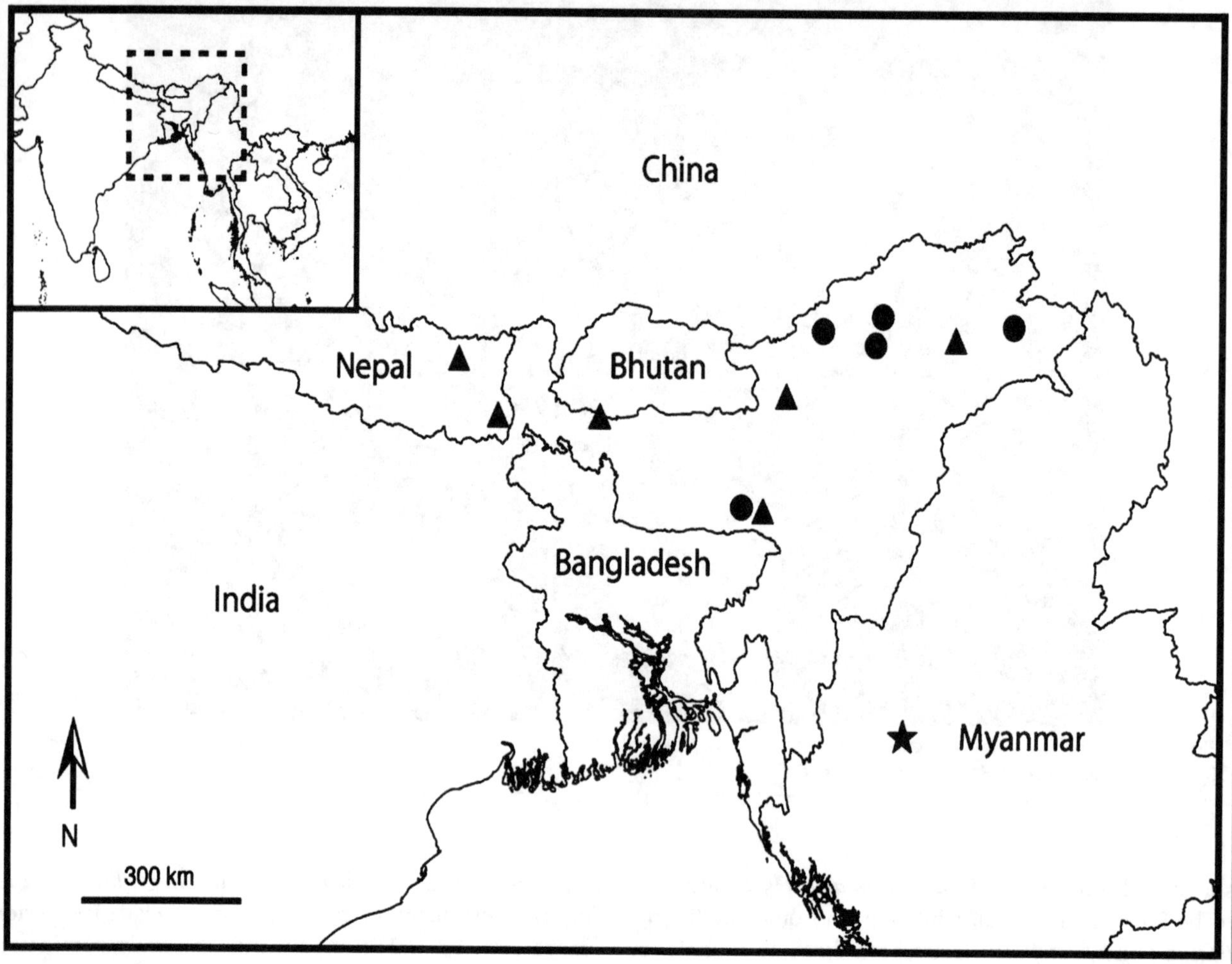

Fig. 5 Distribution map of *Begonia griffithiana* (*circle*), *B. myanmarica* (*star*), *B. nepalensis* (*trangle*). Distribution data of *B. griffithiana* and *B. nepalensis* is based on GPS data in the Begonia Resource Center (Hughes et al. 2015)

but not clustered within sect. *Monopteron* clade (Fig. 6), suggesting that the fruit morphology of a single developed wing in the ovary/fruit is homoplasious. Morphologically, the key characters for sect. *Platycentrum*-sect. *Sphenanthera* clade are evergreen, rhizomatous and two locules in ovary (Leong 2017). *Begonia myanmarica* is evergreen with stout rhizome, but having 1-locular ovary. Compared with other species in sect. *Platycentrum*-sect. *Sphenanthera, B. myanmarica* is unique with a single developed wing and having 1-locular ovary not known in any other taxa in this clade. Further studies with increasing sampling of Myanmar *Begonia* are needed to place *B. myanmarica* in its proper infrageneric position.

Conclusion

Studies of morphology, molecular phylogenetics and cytology support the recognition of the new species, *Begonia myanmarica*, which is fully described and illustrated. Our results also indicate that *B. myanmarica* is not closely related to species previously assigned to sect. *Monopteron*, suggesting that the fruit morphology of a single developed wing in the ovary/fruit characterizing sect. *Monopteron* is homoplasious.

Authors' contributions

YDK, KMH, and SHC conducted the fieldwork and collected the new species from Myanmar; YDK and CIP took color photographs of *B. myanmarica* from the wild and the experimental greenhouse respectively; YHT carried out the morphological observation and undertook laboratory analyses; YK carried out the cytological study; YHT, CIP and KFC prepared the manuscript. All authors read and approved the final manuscript.

Author details

[1] Research Museum and Herbarium (HAST), Biodiversity Research Center,

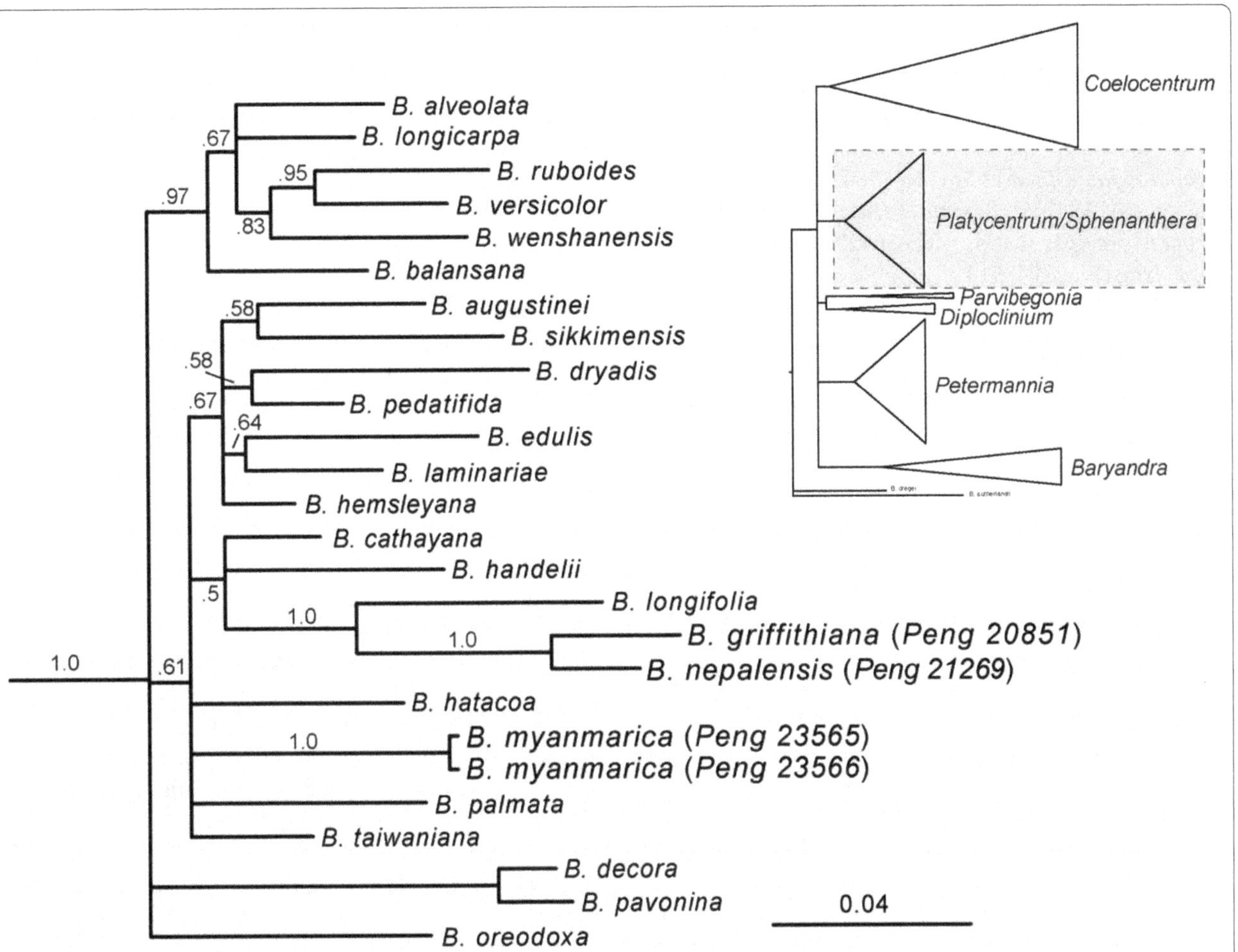

Fig. 6 Phylogenetic tree of section *Platycentrum*-section *Sphenanthera* in *Begonia* generated from Bayesian analysis of nrITS sequence data. *Numbers on the branches* indicate posterior probability of Bayesian inference analysis. The *inset* is simplified phylogenetic tree based on the nrITS dataset and sectional classification of Chung et al. (2014)

Academia Sinica, Taipei 11542, Taiwan. [2] Department of Life Science, Hallym University, Chuncheon, Gangwon 24252, Republic of Korea. [3] Popa Mountain Park, Nature and Wildlife Conservation Division, Ministry of Natural Resources and Environmental Conservation, Kyaukpadaung Township, Mandalay Region, Myanmar. [4] International Biological Material Research Center, Korea Research Institute of Bioscience and Biotechnology, Daejeon 34141, Republic of Korea.

Acknowledgements

This work was supported by grants from the National Institute of Biological Resources (NIBR), the Ministry of Environment (MOE) of the Republic of Korea (NIBR2016 04201) and Biodiversity Research Center, Academia Sinica, Taiwan. Our gratitude extends to the Ministry of Natural Resources and Environmental Conservation of Myanmar for the continuous guidance and support for this project. We are indebted to Dr. Rekha Morris who provided research materials of *B. griffithiana* and *B. nepalensis* for cytological study; Mr. Ku-Feng Lin, manager of Academia Sinica greenhouse, for providing facilities and assistance to maintain the *Begonia* living collections.

Competing interests

The authors declare that they have no competing interests.

Appendix

Voucher information and GenBank accession numbers in newly generated sequences and morphological observation are listed here. Voucher data is given using following format: Taxon name, Collection locality, collector(s) and collector number (herbarium for voucher specimen), GenBank accession numbers for nrITS. Other nrITS sequences refer to Chung et al. (2014).

Newly generated sequences

Begonia myanmarica C.I Peng & Y.D. Kim, Myanmar, *Peng 23565* (HAST), KY088184; *Peng 23566* (HAST), KY088185. *Begonia griffithiana* Warb., India, *Peng 20851* (HAST), KY088186. *Begonia nepalensis* Warb., India, *Peng 20854* (HAST), KY088187.

Other sequences from Chung et al. (2014)

Begonia aequata A. Gray, AF485147; *Begonia alicida* C.B.Clarke ex Hook. f., KF636419; *Begonia alveolata* Yu, AY048977, *Begonia amphioxus* Sands, AF485150; *Begonia arachnoidea* C.I Peng, Yan Liu & S.M. Ku, KF636420; *Begonia augustinei* Hemsl., KF636421; *Begonia aurit-*

istipula Y.M. Shui & W.H. Chen, KF636422; *Begonia austroguangxiensis* Y.M. Shui & W.H. Chen, KF636423; *Begonia balansana* Gagnep., AF485091; *Begonia bamaensis* Yan Liu & C.I Peng, KF636424; *Begonia bataiensis* Kiew, KF636425; *Begonia berhamanii* Kiew, KF636426; *Begonia bipinnatifida* J.J.Sm., KF636427; *Begonia boisiana* Gagnep., AF534719; *Begonia bolsteri* Merr., KF636428; *Begonia brevipes* Merr., KF636429; *Begonia brevirimosa* Irmsch., AF485145; *Begonia cathayana* Hemsl., AF280106; *Begonia cavaleriei* H.Lév., KF636430; *Begonia chingii* Irmsch., KF636432; *Begonia cirrosa* L.B.Sm. & Wassh., AY048979; *Begonia contracta* Warb., KF636433; *Begonia cylindrica* D.R.Liang & X.X.Chen, KF636434; *Begonia decora* Stapf, KF636435; *Begonia dipetala* Graham, AF469124; *Begonia dregei* Otto & A.Dietr., AY429336; *Begonia dryadis* Irmsch., KF636436; *Begonia edulis* H.Lév., KF636437; *Begonia erythrogyna* Sands, KF636438; *Begonia fimbristipula* Hance, KF636439; *Begonia fuscisetosa* Sands, KF636440; *Begonia goegoensis* N.E.Br., AF485138; *Begonia grandis* subsp. *holostyla* Irmsch., AF485088; *Begonia griffithiana* Warb., KY088186; *Begonia gueritziana* Gibbs, KF636441; *Begonia guixiensis* Yan Liu, S.M. Ku, C.I Peng, KF636442; *Begonia hainanensis* Chun & F.Chun, KF636443; *Begonia handelii* Irmsch., AY048982; *Begonia hatacoa* Buch.-Ham. ex D. Don, KF636444; *Begonia hemsleyana* Hook. f., AF485099; *Begonia hernandioides* Merr., KF636445; *Begonia inostegia* Stapf, KF636446; *Begonia isoptera* Dryand. ex Sm., KF636447; *Begonia jingxiensis* D. Fang & Y.G. Wei, KF636448; *Begonia kinabaluensis* Sands, KF636450; *Begonia kingiana* Irmsch., KF636451; *Begonia labordei* H.Lév., KF636452; *Begonia lagunensis* Elmer, KF636453; *Begonia lambii* Kiew, KF636454; *Begonia laminariae* Irmsch., KF636455; *Begonia lanternaria* Irmsch., KF636456; *Begonia leprosa* Hance, KF636457; *Begonia liuyanii* C.I Peng, S.M. Ku & W.C. Leong, KF636458; *Begonia longa* C.I Peng & W.C. Leong, KF636459; *Begonia longicarpa* K.Y. Guan & D.K. Tian, AY048985; *Begonia longifolia* Blume, AF485105; *Begonia longistyla* Y.M.Shui & W.H.Chen, KF636460; *Begonia luzhaiensis* T.C. Ku, KF636461; *Begonia madaiensis* Kiew, KF636462; *Begonia masoniana* Irmsch. ex Ziesenh., KF636463; *Begonia merrittii* Merr., KF636464; *Begonia nigritarum* Steud., KF636465; *Begonia ningmingensis* D. Fang, Y.G. Wei & C.I Peng, KF636466; *Begonia oreodoxa* Chun & F.Chun ex C.Y.Wu & T.C.Ku, KF636467; *Begonia oxysperma* A.DC., AF485131; *Begonia palmata* D.Don, KF636468; *Begonia panayensis* Merr., KF636469; *Begonia paracauliflora* sp. ined., KF636470; *Begonia parvula* H. Lév. & Vaniot, KF636471; *Begonia pavonina* Ridl., KF636472; *Begonia pedatifida* H.Lév., KF636473; *Begonia peltatifolia* H.L. Li, KF636474; *Begonia pengii* S.M. Ku & Yan Liu, KF636475; *Begonia pseudolateralis* Warb., KF636476; *Begonia pulvinifera* C.I Peng & Yan Liu, KF636477; *Begonia ramosii* Merr., KF636478; *Begonia ravenii* C.I Peng & Y.K.Chen, KF636479; *Begonia retinervia* D.Fang, D.H.Qin & C.I Peng, KF636480; *Begonia ruboides* C.M. Hu ex C.Y.Wu & T.C.Ku, KF636481; *Begonia rufipila* Merr., KF636482; *Begonia semiparietalis* Yan Liu, S.M.Ku & C.I Peng, KF636483; *Begonia serratipetala* Irmsch., KF636484; *Begonia sikkimensis* A.DC., KF636485; *Begonia sinofloribunda* Dorr, KF636486; *Begonia subnummularifolia* Merr., KF636487; *Begonia sutherlandii* Hook. f., AF485215; *Begonia symsanguinea* L.L. Forrest & Hollingsw., AF485151; *Begonia taiwaniana* Hayata, KF636488; *Begonia variabilis* Ridl., AY753732; *Begonia variegata* Y.M. Shui & W.H.Chen, KF636489; *Begonia versicolor* Irmsch., AF485090; *Begonia wadei* Merr. & Quisumb., KF636490; *Begonia wenshanensis* C.M. Hu ex C.Y. Wu & T.C. Ku, AY048974; *Begonia yappii* Ridl., KF636491.

Morphological observation

Begonia myanmarica C.I Peng & Y.D. Kim, Myanmar, *Peng 23565* (HAST); *Peng 23566* (HAST). *Begonia griffithiana* Warb., India, *Peng 20851* (HAST). *Begonia nepalensis* Warb., India, *Peng 20854* (HAST).

Begonia cathayana Hemsl., China, *Peng 20288* (HAST). *Begonia decora* Stapf, Malaysia, *Peng 20261* (HAST). *Begonia dryadis* Irmsch., China, *Peng 18016* (HAST). *Begonia edulis* H.Lév., China, *Peng 18747* (HAST). *Begonia hatacoa* Buch.-Ham. ex D. Don, *Peng 20861* (HAST). *Begonia hemsleyana* Hook. f., China, *Peng 17590* (HAST). *Begonia laminariae* Irmsch., China, *Peng 17447* (HAST). *Begonia oreodoxa* Chun & F. Chun ex C.Y. Wu & T.C. Ku, China, *Peng 20454* (HAST). *Begonia palmata* D. Don, Taiwan, *Peng 20993* (HAST). *Begonia pavonina* Ridl., Malaysia, *Peng 20239* (HAST). *Begonia pedatifida* H. Lév., China, *Peng 18779* (HAST). *Begonia sikkimensis* A. DC., China, *Peng 20848* (HAST). *Begonia augustinei* Hemsl., China, *Peng 20759* (HAST). *Begonia versicolor* Irmsch., China, *Peng 18688* (HAST). *Begonia balansana* Gagnep., Vietnam, *Peng 21928* (HAST). *Begonia handelii* Irmsch., China, *Peng 17513* (HAST). *Begonia longifolia* Blume, Taiwan, *Peng 16795* (HAST).

References

Christenhusz MJM, Byng JW (2016) The number of known plants species in the world and its annual increase. Phytotaxa 261(3):201–217. doi:10.11646/phytotaxa.261.3.1

Chung K-F, Leong W-C, Rubite RR, Repin R, Kiew R, Liu Y, Peng C-I (2014) Phylogenetic analyses of *Begonia* sect. *Coelocentrum* and allied limestone species of China shed light on the evolution of Sino-Vietnamese karst flora. Bot Stud 55:1. doi:10.1186/1999-3110-55-1

de Candolle A (1864) Begoniaceae. In: de Candolle A (ed) Prodromus systematis naturalis regni vegetabilis. Treuffel et Würtz, Masson, pp 266–408

Doorenbos JM, Sosef MS, de Wilde JJFE (1998) The sections of *Begonia* including descriptions, keys and species lists: studies in Begoniaceae VI.

Wageningen Agricultural University, Wageningen
Forrest LL, Hollingsworth PM (2003) A recircumscription of *Begonia* based on nuclear ribosomal sequences. Plant Syst Evol 241(3):193–211. doi:10.1007/s00606-002-0033-y
Huelsenbeck JP, Ronquist F (2001) MRBAYES: Bayesian inference of phylogenetic trees. Bioinformatics 17(8):754–755. doi:10.1093/bioinformatics/17.8.754
Hughes M (2008) An annotated checklist of Southeast Asian Begonia. Royal Botanic Garden Edinburgh, Edinburgh
Hughes M, Moonlight PW, Jara A, Pullan M (2015) Begonia Resource Centre. http://padme.rbge.org.uk/begonia. Accessed 15 July 2016
Ku T-C (1999) Begoniaceae. In: Ku T-C (ed) Flora Reipublicae Popularis Sinicae, vol 52, no 1. Science Press, Beijing, pp 126–269
Ku C-Z, Peng C-I, Turland NJ (2007) Begoniaceae. In: Raven PH, Hong D-Y (eds) Flora of China, vol 13., Science Press, Beijing & Missouri Botanical Garden Press, St. Louis, pp 153–207
Legro RAH, Doorenbos J (1971) Chromosome numbers in *Begonia* II. Neth J Agric Sci 19:176–183
Leong, W-C (2017) Molecular phylogenetics of Asian *Begonia* (Begoniaceae): systematics and biogeographic implications. PhD Dissertation, National Taiwan University
Maddison WP, Maddison DR (2015) Mesquite: a modular system for evolutionary analysis, version 3.03 ed. http://mesquiteproject.org. Accessed 7 Mar 2015
Peng C-I, Jin X-H, Ku S-M, Kono Y, Huang H-Y, Yang H-A (2014a) *Begonia wuzhishanensis* (sect. *Diploclinium*, Begoniaceae), a new species from Hainan Island, China. Bot Stud 55:24. doi:10.1186/1999-3110-55-24
Peng C-I, Wang H, Kono Y, Yang H-A (2014b) *Begonia wui-senioris* (sect. *Platycentrum*, Begoniaceae), a new species from Myanmar. Bot Stud 55:13. doi:10.1186/1999-3110-55-13
Posada D, Crandall KA (1998) MODELTEST: testing the model of DNA substitution. Bioinformatics 14(9):817–818. doi:10.1093/Bioinformatics/14.9.817
Rajbhandary S (2010) Systematic revision of the genus *Begonia* L. (Begoniaceae) in the Himalayas. PhD dissertation, Tribhuvan University
Rajbhandary S, Hughes M, Phutthai T, Thomas DC, Krishna KS (2011) Asian *Begonia*: out of Africa via the Himalayas? Gard Bull (Singapore) 63(1 & 2):277–286
Rubite RR (2010) Systematic studies on Philippine *Begonia* L. secion *Diploclinium* (Lindl.) A.D.C. (Begoniaceae). PhD dissertation, De La Salle University
Sharma AK, Bhattacharyya UC (1961) Cytological studies in *Begonia* II. Caryologia 14:279–301
Shui Y-M, Peng C-I, Wu C-Y (2002) Synopsis of the Chinese species of *Begonia* (Begoniaceae), with a reappraisal of sectional delimitation. Bot Bull Acad Sin 43(4):313–327
Tamura K, Peterson D, Peterson N, Stecher G, Nei M, Kumar S (2011) MEGA5: molecular evolutionary genetics analysis using maximum likelihood, evolutionary distance, and maximum parsimony methods. Mol Biol Evol 28(10):2731–2739. doi:10.1093/molbev/msr121
Tanaka N, Hayami Y (2011) *Begonia kachinensis* (sect. *Sphenanthera*, Begoniaceae), a new species from Myanmar. Acta Phytotax Geobot 61(3):151–154
Tanaka N, Hughes M (2007) *Begonia* (Sect *Sphenanthera*) *hayamiana* (Begoniaceae), a new species of *Begonia* (Begoniaceae) from Northern Myanmar. Acta Phytotax Geobot 58(2–3):83–86
Tanaka N, Peng C-I (2016) *Begonia togashii* (Begoniaceae: sect. *Platycentrum*), a new species from central Myanmar. Acta Phytotax Geobot 67(3):191–197. doi:10.18942/apg.201608
Tebbitt MC, Lowe-Forrest L, Santoriello A, Clement WL, Swensen SM (2006) Phylogenetic relationships of Asian *Begonia*, with an emphasis on the evolution of rain-ballist and animal dispersal mechanisms in sections *Platycentrum*, *Sphenanthera* and *Leprosae*. Syst Bot 31(2):327–336. doi:10.1600/036364406777585784
Thomas DC (2010) Phylogenetics and historical biogeography of Southeast Asian *Begonia* L. (Begoniaceae). PhD dissertation, University of Glasgow
Thomas DC, Hughes M, Phutthai T, Rajbhandary S, Rubite R, Ardi WH, Richardson JE (2011) A non-coding plastid DNA phylogeny of Asian *Begonia* (Begoniaceae): evidence for morphological homoplasy and sectional polyphyly. Mol Phylogenet Evol 60(3):428–444. doi:10.1016/j.ympev.2011.05.006

18

Mitigation of salt stress in white clover (*Trifolium repens*) by *Azospirillum brasilense* and its inoculation effect

Muhammad Khalid[1], Muhammad Bilal[2], Danial Hassani[1], Hafiz M. N. Iqbal[3], Hang Wang[4] and Danfeng Huang[1*]

Abstract

Background: Salinity is one of the increasingly serious environmental problems worldwide for cultivating agricultural crops. The present study was aimed to ascertain the potential of beneficial soil bacterium *Azospirillum brasilense* to alleviate saline stress in *Trifolium repens*. Experimental plants (white clover) were grown from seeds and inoculated with or without *A. brasilense* bacterial strain supplemented with 0, 40, 80, or 120 mM NaCl into soil.

Results: The growth attributes including, shoot heights, root lengths, fresh and dry weights, leaf area and chlorophyll content were significantly enhanced in *T. repens* plants grown in *A. brasilense* inoculated soil than un-inoculated controls, particularly under elevated salinity conditions (40, 80 and 120 mM NaCl). Malondialdehyde content of leaf was recorded to be declined under saline conditions. Moreover, the K^+/Na^+ ratio was also improved in bacterium-inoculated plants, since *A. brasilense* significantly reduced the root and shoot Na^+ level under high salty environment.

Conclusions: Results revealed that soil inoculation with *A. brasilense* could significantly promote *T. repens* growth under both non-saline and saline environments, and this study might be extended to other vegetables and crops for the germination and growth enhancement.

Keywords: Salinity, Leaf physiology, Stem physiology, Salinity stress, K^+/Na^+ ratio

Background

Soil salinity is considered to be the most brutal environmental factor which caused a reduction in plant growth and crops productivity (Allakhverdiev et al. 2000). Cultivated soils all over the world are becoming more saline due to continuing use of poor quality water for irrigation, excessive fertilization, and desertification processes (Ramadoss et al. 2013). Currently, more than 20% of the worlds (26% of Pakistan) agricultural irrigated land is affected by levels of salt that could markedly downscale the net agriculture yield (Munns and Tester 2008). Excessive salt concentration augments the Na^+ and Cl^- ions level in different plants that negatively affect the plant survival by disrupting different plant metabolisms, cellular homeostasis, and uncoupling major biochemical as well as physiological processes (Mahajan and Tuteja 2005). Although both Na^+ and Cl^- ions have the capability to induce many physiological and/biochemical disorders in plants, nevertheless, Cl^- is among the most hazardous (Tavakkoli et al. 2010).

Effective strategies for alleviating salinity stress involve developing salt-tolerant cultivars, in situ soil flushing, leaching excess soluble salts from upper to lower soil depths, reducing salt by collecting salt-accumulating aerial plant parts, and improvement of alkaline soils under cropping and leaching (Bacilio et al. 2004). Inoculation of crop with plant growth promoting bacteria (PGPB) is becoming more imperative as an alternative to mitigate abiotic stresses (i.e., salinity) since PGPB have been

*Correspondence: hdf@sjtu.edu.cn
[1] School of Agriculture and Biology, Shanghai Jiao Tong University, Shanghai 200240, China[3] School of Engineering and Science, Tecnologico de Monterrey, Campus Monterrey, Ave. Eugenio Garza Sada 2501, Monterrey, N.L. 64849, Mexico
Full list of author information is available at the end of the article

reported to overcome stress effects. Further, the beneficial properties of microbes under salinity has been attributed to the hydraulic conductivity, sequestering toxic Na^+ ions, osmolyte accumulation, retaining higher stomatal conductance and photosynthetic activities (Dodd and Pérez-Alfocea 2012).

PGPB are a diverse group of bacteria that possess the remarkable capability to promote growth and yield of many crops and wild plants (de-Bashan et al. 2012). Several species of PGPB, particularly species of the genus *Azospirillum*, *Pseudomonas*, and *Arbuscular mycorrhizal* fungi can effectively alleviate salt stress in plants (Fasciglione et al. 2015). Among PGPB, *Azospirillum* spp. is the most investigated bacterium because of its potential to colonize a wide-range vegetable crops and to enhance the general plant performance under normal or stressed environment. Indeed, it has been reported that inoculation mitigated the undesirable effects of NaCl on wheat seedlings with *Azospirillum brasilense*. The benefits of *Azospirillum* inoculation have also been observed in chickpea irrigated with saline water (Hamaoui et al. 2001). These favorable effects of the *Azospirillum* strains might be attributed to the secretion of several types of phytohormones in the surroundings of the roots. These phytohormones are physiologically active and hence, potentially assist the plants to propagate in stress conditions (Fasciglione et al. 2015).

Trifolium repens, the white clover, has been considered an important forage crop throughout the world. *T. repens* is also deliberated as a folk medicine in India against intestinal helminthic worms, and an in vivo experimental study corroborated that its aerial shoots possess unique anti-custodial properties (Tangpu et al. 2005). Like several other forage crops, *T. repens* is susceptible to salinity stress; reducing its growth and number of root nodules, and as a consequence, nitrogen fixation, as well as soil fertility level, is compromised (Acharya et al. 2011). The present work was carried out to evaluate *A. brassilence* potential for growth promotion and salinity resistance in white clover.

Methods

Bacterial strain and growth conditions

Bacterial strain, *A. brasilense* was provided by "*The Leibniz Institute DSMZ-German Collection of Microorganisms and Cell Cultures*". The strain was streaked in Lauria-Bertani (LB) agar plates and incubated at 30 ± 0.2 °C for 24 h. From LB agar plates, the bacterial cells were then transferred into LB-broth and cultured at 30 ± 0.2 °C under an agitation speed of 250 rpm to obtain 1.0×10^9 colony-forming units (CFU)/mL.

Seed treatment and germination experiment

The healthy and mature seeds of *T. repens* were cleaned and disinfected with 3.5% sodium hypochlorite (NaClO) solution for 1 min followed by soaking in 70% ethanol for 10 min, and finally washed with distilled water for further use. Pre-sterilized plastic pots (20 cm diameter) comprising 800 g of soil were used to germinate the seeds by watering with Hoagland's solution (400 mL), which contained $NH_4H_2PO_4$, 1 mM; KNO_3, 5 mM; $MgSO_4$, 0.5 mM; Fe-citrate, 60 μM; $Ca(NO_3)_2$, 0.5 mM; $MnCl_2{\cdot}4H_2O$, 18 μM; H_3BO_3, 92 μM; $CuSO_4{\cdot}5H_2O$, 0.6 μM; $ZnSO_4{\cdot}7H_2O$, 1.6 μM and $(NH_4)_6\,Mo_7O_{24}{\cdot}4H_2O$, 0.7 μM. After germination, Three uniform seedlings per pot were selected for continued growth and were treated with 1 mL bacterial suspension (1.0×10^9 CFU/mL), per plant as inoculation treatment. The seedlings treated with 1.0 mL liquid LB medium were considered as control. For salt stress treatments, the seedlings were watered with Hoagland's nutrient solution supplemented with 0, 40, 80, 120 mM NaCl. Three plastic pots (Three plants/pot) were used as replications for each treatment. Plants were grown under controlled conditions at 28 °C during the day and 16 °C at night, irrigated daily through dropper with deionized water to maintain the moisture approximately at 60% water holding capacity of the soil, efforts were made to protect the leaching of salt and all pots were randomly placed in a greenhouse.

Physiological measurements

The plants were harvested after 60 days to determine physiological index and growth measurements including, leaf area and a whole number of leaves per plant. Measurement of leaf area was made through portable leaf area meter (Model YMJ-A, China). Leaf area/plant was divided by some leaves/plant to calculate the average leaf area. Once the plants were separated from pots, the growth parameters such as root and shoot lengths (cm) were measured by a ruler. Shoot fresh weight (g) was determined immediately, whereas the plant parts were oven-dried at 75 °C for 2-days after rinsing with deionized water for quantitative analysis of dry weights (g). Leaf chlorophyll content was analyzed following the reported method (Porra et al. 1989). Briefly, 80% acetone was used to pulverize fresh leaves sample and centrifuged at 4 °C for 10 min at 10,000 rpm. The clear supernatants, thus obtained, were collected and recording OD measured chlorophyll content at 645 and 663 nm. Thiobarbituric acid (TBA) protocol was adopted to probe the oxidative stress using the biomarker malondialdehyde (MDA) (Bao et al. 2009).

Ionic analysis

Ionic analysis (Na^+ and K^+) was made using an atomic absorption spectrophotometer available in Instrumental analysis center, SJTU following the method described previously (Wang et al. 2007). Ice-chilled $CaCl_2$ (20 mM) was used to wash the plant roots twice for 5 min to exchange cell wall bound Na^+. Surface salts from the shoots were removed by rinsing it in deionized water and the sample parts (root, shoot) were oven dried at 70 °C for three days to calculate dry weights. Dried plant tissues were used for Na^+ and K^+ analysis. These ions were extracted from dried plant samples for 2 h at 90 °C in 100 mM acetic acid.

Statistical analysis

All the data such as growth measurements, physiological index, ion contents, and ratio were analyzed and reported with standard deviations and means of three replicates. The means and standard errors were computed for each treatment, and the SE values were displayed as Y-error bars in the figures. The collected data was subjected to analysis of variance (ANOVA) and treatment means were compared by Duncan's New Multiple Range Test (DMRT) at $P < 0.05$ using a computer-based statistical software package (IBM SPSS Statistics 21).

Results

White clover growth (shoot height and root length) improvement by A. brasilense

Azospirillum brasilense possessed beneficial effects on white clover cultivation by increasing its growth and decreasing the susceptibilities to different salt concentration. Under the non-saline condition, the shoot height was improved to 60.86% in comparison to media control, whereas the shoot height was significantly increased in *A. brasilense* inoculated plants than that of un-inoculated plants. The enhancement in shoot height was recorded to be 58.82, 57.8 and 70% at 40, 80, and 120 mM NaCl treatments, respectively (Fig. 1A). Compared with media control, the root length of inoculated plants was improved to 62.06% under non-saline condition. *A. brasilense* significantly increased the root length by 70.83, 58.82, and 63.63% under at experimental concentration of 40, 80, and 120 mM NaCl, respectively (Fig. 1B).

Biomass measurement

Azospirillum brasilense inoculation led to increased biomass of the white clover in non-saline conditions and also alleviated the detrimental effect of different concentration of salts (NaCl). The shoot fresh weight of inoculated white clover was found to be 2.1-folds higher in non-saline conditions as compared to similar medium control. In the same way, shoot fresh weight of experimental plants treated with 40, 80, and 120 mM NaCl were increased to 2.66, 1.16-folds and 2.33-folds (Fig. 2A). In non-saline condition, shoot dry weight was recorded to be 4.2-folds higher with respect to the control medium. While in inoculated plants, the salt stress was significantly alleviated by *A. brasilense*, and therefore, 3.0-, 5.3- and 1.6-folds improvement in shoot dry weight was observed as compared to un-inoculated ones (Fig. 2B). Inoculated plants root dry weight was assessed to be 0.57% higher in non-saline condition. Increases in dry weight of other samples were also significant and recorded to be 0.55%, 0.56 and 0.6-folds greater than un-inoculated plants at 40, 80, 120 mM NaCl concentration (Fig. 2C). In conclusion, *A. brasilense* had a significant salinity stress relieving effect on the growth of plants under all the tested concentrations of NaCl.

Improvement in leaf number and area

Azospirillum brasilense supported the growth of plants under abiotic (salt etc.) stress and plays a significant

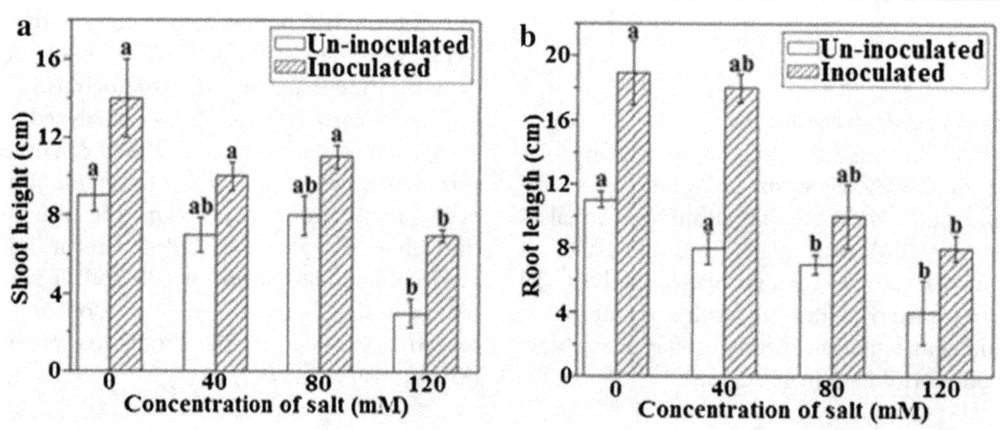

Fig. 1 Effects of *A. brasilense* inoculation on shoot height (**A**) and root length (**B**) of *T. repens* under varying concentrations of NaCl

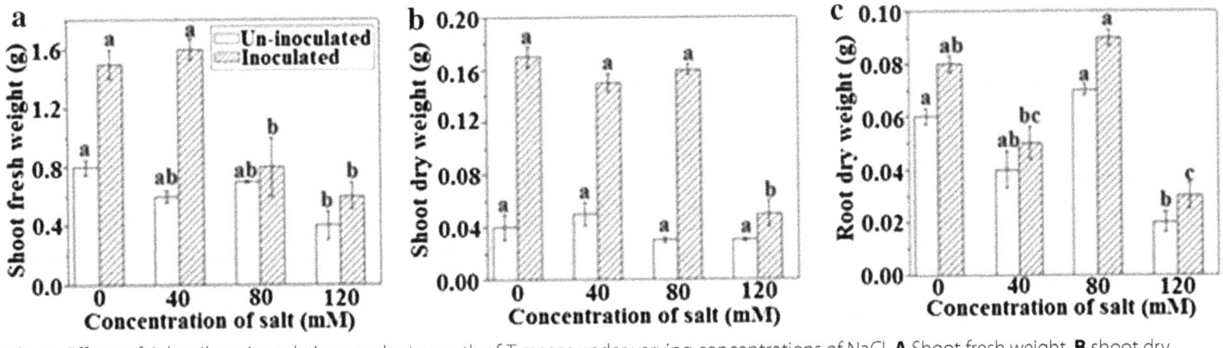

Fig. 2 Effects of *A. brasilense* inoculation on plant growth of *T. repens* under varying concentrations of NaCl. **A** Shoot fresh weight, **B** shoot dry weight, **C** root dry weight

role in the development of leaves and chlorophyll in the absence as well as the presence of salinity stress. Leaf area of bacterial inoculated plants with no salt treatment was increased to 1.4-folds in contrast to equivalent un-inoculated samples. Stress was alleviated in the rest of treatments by increasing the number of leaves to 2.27-, 1.25- and 1.83-folds relatively at 40, 80, 120 mM NaCl concentrations, correspondingly (Fig. 3A). Enhancement in some leaves per inoculated plant under non-saline condition was 54.5%. While under NaCl treatment of 40 and 120 mM, leaf numbers were increased to 68.1 and 62.5, respectively compared with un-inoculated plants samples. Nevertheless, this trend was observed to be contradicted for the plants treated with 80 mM NaCl concentration (Fig. 3B).

Determination of chlorophyll content

Besides increasing the number and area of leaves, *A. brasilense* also stimulated the amount of chlorophyll formation and supported its specific growth in bacterial inoculated plants. Leaf chlorophyll content in inoculated plants was increased to 61.5% in non-saline condition. Under appropriated saline condition, this ratio was 63.6, 75 and 0.62% grew at NaCl concentration of 40, 80, and 120 mM, accordingly (Fig. 4).

Measurement of leaf MDA, sodium and potassium content

Lipid peroxidation is an index of membrane damage and/or permeability, and ions leakage occurred in salinity stress environments. The results showed that MDA content in plants inoculated by *A. brasilense* decreased significantly to 64.8% in non-salinity conditions and 24.1, 20.90 and 27.5% in 40, 80, and 120 mM NaCl treatments, respectively. MDA biomarker assessment also demonstrated that *A. brasilense* could favorably regulate the oxidative stress and cell membrane integrity. The results from ion content analysis revealed that sodium content was remarkably reduced to 40% for non-saline conditions in plants treated with *A. brasilense*, whereas this ratio was 41.6, 46.66 and 43.33% increased for inoculated plants under 40, 80 and 120 mM NaCl level (Fig. 5A). Dry roots evaluation also showed that in all treatments

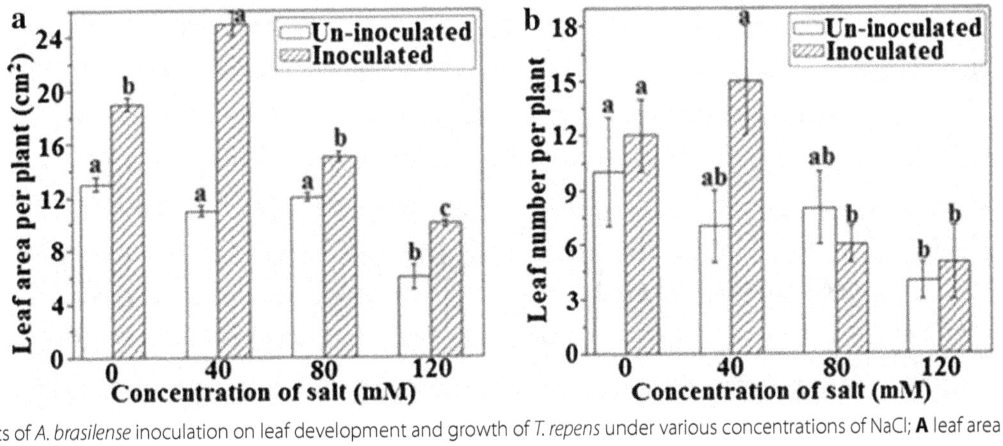

Fig. 3 Effects of *A. brasilense* inoculation on leaf development and growth of *T. repens* under various concentrations of NaCl; **A** leaf area per plant, and **B** leaf number per plant

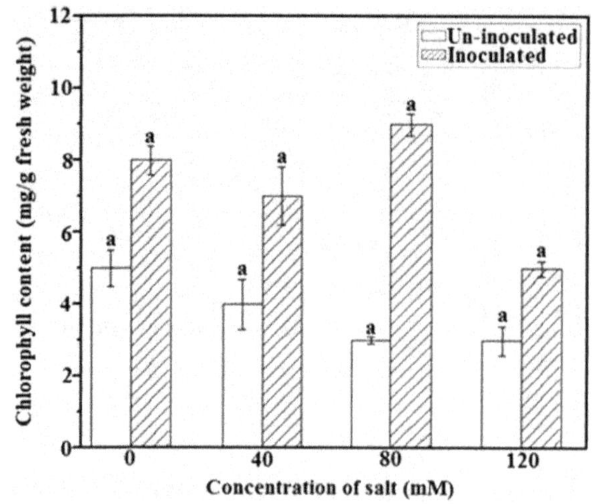

Fig. 4 Effects of *A. brasilence* inoculation on chlorophyll content of *T. repens* under different NaCl concentrations

including 0, 40, 80, 120 mM mM NaCl, the sodium content of inoculated plants decreased to 40, 55.5, 42.8 and 41.1% respectively (Fig. 5C). However, no significant reduction was recorded in potassium accumulation for all treatments (Fig. 5B, D).

Discussion

Salt stress negatively influences the morphological, ionic, and physiological as well as biochemical characteristics in plants (Chen et al. 2013; Abbasi et al. 2015). It has been reported that increasing the salt level, reduces the osmotic potential, which results in cell dehydration, due to increased water efflux from cell (Amjad et al. 2014). Beneficial soil bacterial mediated growth enhancing attributes have been documented in enormous plant species (Gray and Smith 2005; Xie et al. 2009; Paré et al. 2011). The growth amplification of several plants, such as wheat (*Triticum aestivum*), maize (*Zea mays*), tomato

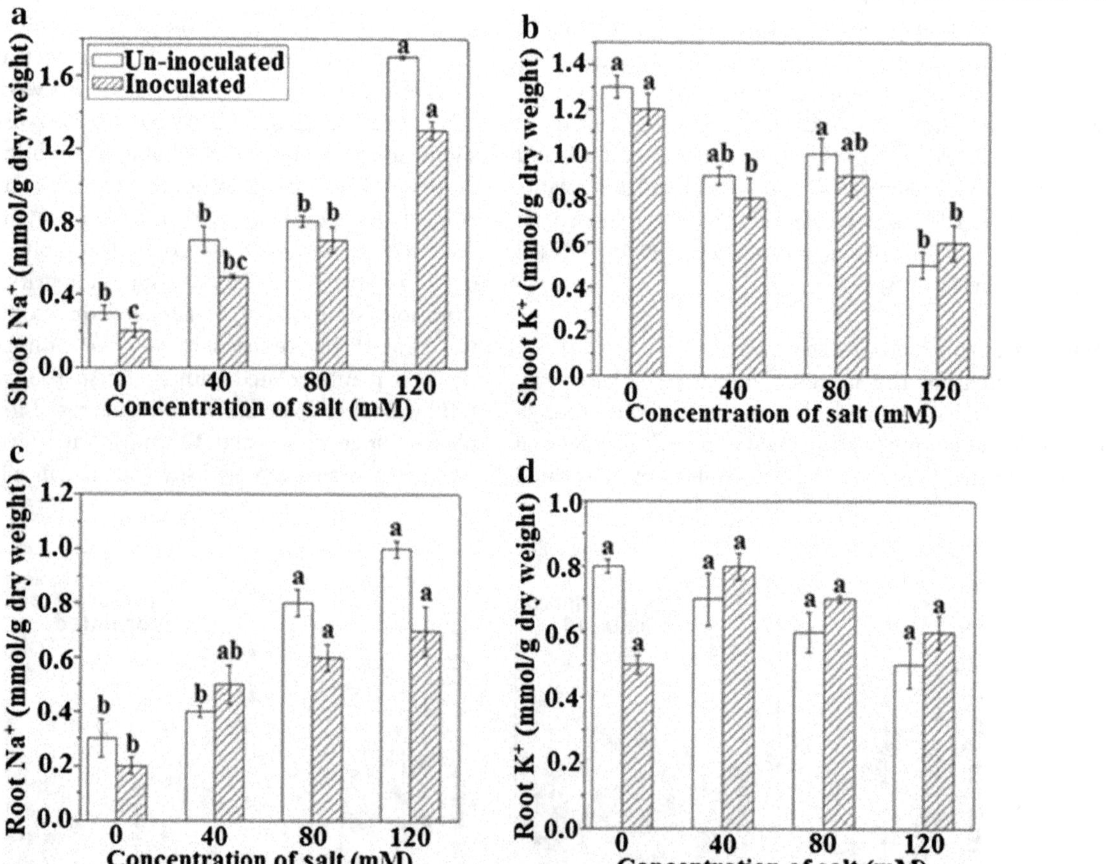

Fig. 5 Effects of *A. brasilence* inoculation on ion-contents of *T. repens* at varying concentrations of NaCl **A** shoot Na^+ content, **B** shoot K^+ content, **C** root Na^+ content, **D** root K^+ content

(*Lycopersicon esculentum*), dwarf salt wort (*Salicornia bigelovii*), chickpea (*Cicer arietinum*), and alfalfa (*Medicago sativa*) by beneficial soil bacteria have also been observed at varying salt exposures (Bashan et al. 2000; Mayak et al. 2004; Ibragimova et al. 2006; Bano and Fatima 2009; Tiwari et al. 2011). A significant growth improvement of white clover by *A. brasilense* inoculated soil recorded in the present study was comparable with previous reports in *Arabidopsis* (Zhang et al. 2007, 2008, Zhang et al. 2010a, b; Xie et al. 2009; Paré et al. 2011). Noticeably, our findings revealed that *A. brasilense* in white clover has more active shoot growth promoting role compared to its root growth, particularly under salt stress conditions (40, 80, and 120 mM NaCl concentration). Similar results for salt preventing potentiality of *Azospirillum* strains have been documented previously by several researchers evaluated the bio-control and salt stress alleviating perspective of *A. lipoferum* in wheat crop and found that *A. lipoferum* displayed the good potential to promote the growth of wheat under saline conditions (up to 150 mM of NaCl) (Bashan and Levanony 1990). Synthesis of growth promoting substances (GPS) and adaptation to current environmental conditions may confer increased germination percentage and alleviation under prevailing salt conditions (Rueda-Puente et al. 2007; Nadeem et al. 2013). Leaf development and/or enlargement play a significant role in plant production since it is strongly associated with plant growth and biomass accumulation (Gutiérrez-Boem and Thomas 1998; Bacilio et al. 2004; Battie-Laclau et al. 2013).

A significant augmentation in leaf area and leaf number per plant was also noted in *A. brasilence* inoculated white clover plants. In another study, Han et al. (2014) reported that GB03 inoculation significantly enlarged leaf area and number per plant, and thus average leaf area in white clover. Likewise, photosynthetic pigment, leaf chlorophyll content is also an important physiological/biological attribute directly related to photosynthesis rate in plants (Ma et al. 2012). Inoculation of *A. brasilence* considerably improved the leaf chlorophyll content in white clover plant, cultivated under both non-saline as well as variable salt concentrations of 40, 80, and 120 mM NaCl. Similar to our study, Zhang et al. (2008) described that *B. subtilis* GB03 enhances photosynthetic efficiency by increasing chlorophyll content in *Arabidopsis*. In contrast, some reports highlighted that plants grown under salinity environments produced a smaller amount of chlorophyll and dry matter than those without salt exposure presumably due to chlorophyll per-oxidation (Lunde et al. 2007; Tuna et al. 2008; Barry 2009).

Soils with salinity contain an array of several cation–anion pairs (i.e., $CaSO_4$, $MgSO_4$, $MgCl_2$, Na_2CO_3, and Na_2SO_4), with Na^+ ions being the predominant species (Zhang et al. 2010a). Growth is impeding, a common plant response to salt stress is often related to elevated intracellular Na^+ concentration and low K^+/Na^+ ratio in the plant (Zhang et al. 2010a). Researchers have reported that plants grown under saline conditions can curtail sodium toxicity by limiting Na^+ uptake, re-directing Na^+ from shoots to roots, and also extruding Na^+ loadings from root cells (Tester and Davenport 2003; Munns and Tester 2008; Zhang et al. 2010a, 2011; Kronzucker and Britto 2011). In a previous study, (Zhang et al. 2010a, b) reported that *B. subtilis* significantly reduced (54%) the plant Na^+ content in *Arabidopsis* compared with control plants by down-regulating and up-regulating *HKT1* expression in roots and shoots, respectively. In this study, soil inoculated with *A. brassilence* considerably reduced the Na^+ level and increased K^+ to Na^+ ratio in both roots and shoots of the experimental plant at all levels of tested salt stresses.

Conclusions

In conclusion, the results revealed that soil inoculation with *A. brasilense* bacterium remarkably augments plant development and biomass of the tested plant (white clover) under both salted as well as non-salted settings. The present findings offer an opportunity for the application of this beneficial soil bacterium to cultivate different plants to combat saline toxicity. Further, comprehensive investigations elucidating the mechanisms by which bacteria elicit salt tolerance would be the focus of future studies.

Authors' contributions

MK, and DH, performed the experiments, analyzed the data and wrote the paper; HW contributed in analytical analyses; MB, and HMNI contributed by analyzing the data, writing and reviewing the paper; DH provided financial support and supervised the project. All authors read and approved the final manuscript.

Author details

[1] School of Agriculture and Biology, Shanghai Jiao Tong University, Shanghai 200240, China. [2] State Key Laboratory of Microbial Metabolism, and School of Life Sciences and Biotechnology, Shanghai Jiao Tong University, Shanghai 200240, China. [3] School of Engineering and Science, Tecnologico de Monterrey, Campus Monterrey, Ave. Eugenio Garza Sada 2501, Monterrey, N.L. 64849, Mexico. [4] Instrumental Analysis Center, Shanghai Jiao Tong University, Shanghai 200240, China.

Acknowledgements

The authors are grateful to School of Agriculture and Biology, Shanghai Jiao Tong University, Shanghai 200240, China for providing experimental facilities. The technical and analytical help provided by Instrumental Analysis Center of SJTU is also thankfully acknowledged.

Competing interests

The authors declare that they have no competing interests.

References

Abbasi GH, Akhtar J, Ahmad R, Jamil M, Anwar-ul-Haq M, Ali S, Ijaz M (2015) Potassium application mitigates salt stress differentially at different growth stages in tolerant and sensitive maize hybrids. Plant Growth Regul 76:111–125

Acharya AR, Wofford DS, Kenworthy K, Quesenberry KH (2011) Combining ability analysis of resistance in white clover to southern root-knot nematode. Crop Sci 51:1928–1934

Allakhverdiev SI, Sakamoto A, Nishiyama Y, Inaba M, Murata N (2000) Ionic and osmotic effects of NaCl-induced inactivation of photosystems I and II in *Synechococcus* sp. Plant Physiol 123:1047–1056

Amjad M, Akhtar J, Anwar-ul-Haq M, Yang A, Akhtar SS, Jacobsen SE (2014) Integrating role of ethylene and ABA in tomato plants adaptation to salt stress. Sci Hortic 172:109–116

Bacilio M, Rodriguez H, Moreno M, Hernandez JP, Bashan Y (2004) Mitigation of salt stress in wheat seedlings by a gfp-tagged *Azospirillum lipoferum*. Biol Fert Soil 40:188–193

Bano A, Fatima M (2009) Salt tolerance in *Zea mays* (L). following inoculation with *Rhizobium* and *Pseudomonas*. Biol Fertil Soil 45:405–413

Bao AK, Wang SM, Wu GQ, Xi JJ, Zhang JL, Wang CM (2009) Overexpression of the *Arabidopsis* H + -PPase enhanced resistance to salt and drought stress in transgenic alfalfa (*Medicago sativa L.*). Plant Sci 176:232–240

Barry CS (2009) The stay-green revolution: recent progress in deciphering the mechanisms of chlorophyll degradation in higher plants. Plant Sci 176:325–333

Bashan Y, Levanony H (1990) Current status of *Azospirillum* inoculation technology: *Azospirillum* as a challenge for agriculture. Can J Microbiol 36:591–608

Bashan Y, Moreno M, Troyo E (2000) Growth promotion of the seawater-irrigated oilseed halophyte *Salicornia bigelovii* inoculated with mangrove rhizosphere bacteria and halotolerant *Azospirillum* spp. Biol Fertil Soil 32:265–272

Battie-Laclau P, Laclau JP, de Cassia Piccolo M, Arenque BC, Beri C, Mietton L, Ranger J (2013) Influence of potassium and sodium nutrition on leaf area components in *Eucalyptus grandis* trees. Plant Soil 371:19–35

Chen P, Wang H, Tang X, Brestic M, Shao H (2013) Comparative physiological study between cultivated and wild soybean species under salt stress. Jokull J 63:1–11

de-Bashan LE, Hernandez JP, Bashan Y (2012) The potential contribution of plant growth-promoting bacteria to reduce environmental degradation? A comprehensive evaluation. Appl Soil Ecol 61:171–189

Dodd IC, Pérez-Alfocea F (2012) Microbial amelioration of crop salinity stress. J Exp Bot 2012:Ers033

Fasciglione G, Casanovas EM, Quillehauquy V, Yommi AK, Goni MG, Roura SI, Barassi CA (2015) *Azospirillum* inoculation effects on growth, product quality and storage life of lettuce plants grown under salt stress. Sci Hortic 195:154–162

Gray E, Smith D (2005) Intracellular and extracellular PGPR: commonalities and distinctions in the plant–bacterium signaling processes. Soil Biol Biochem 37:395–412

Gutiérrez-Boem FH, Thomas GW (1998) Phosphorus nutrition affects wheat response to water deficit. Agron J 90:166–171

Hamaoui B, Abbadi JM, Burdman S, Rashid A, Sarig S, Okon Y (2001) Effects of inoculation with *Azospirillum brasilense* on chickpeas (*Cicer arietinum*) and faba beans (*Vicia faba*) under different growth conditions. Agronomie 21:553–560

Han QQ, Lü XP, Bai JP, Qiao Y, Paré PW, Wang SM, Wang ZL (2014) Beneficial soil bacterium *Bacillus subtilis* (GB03) augments salt tolerance of white clover. Front Plant Sci 5:525

Ibragimova MV, Rumyantseva ML, Onishchuk OP, Belova VS, Kurchak ON, Andronov EE, Simarov BV (2006) Symbiosis between the root-nodule bacterium *Sinorhizobium meliloti* and alfalfa (*Medicago sativa*) under salinization conditions. Microbiology 75:77–81

Kronzucker HJ, Britto DT (2011) Sodium transport in plants: a critical review. New Phytol 189:54–81

Lunde C, Drew DP, Jacobs AK, Tester M (2007) Exclusion of Na^+ via sodium ATPase (PpENA1) ensures normal growth of *Physcomitrella patens* under moderate salt stress. Plant Physiol 144:1786–1796

Ma Q, Yue LJ, Zhang JL, Wu GQ, Bao AK, Wang SM (2012) Sodium chloride improves photosynthesis and water status in the succulent xerophyte *Zygophyllum xanthoxylum*. Tree Physiol 32:4–13

Mahajan S, Tuteja N (2005) Cold, salinity and drought stresses: an overview. Arch Biochem Biophys 444:139–158

Mayak S, Tirosh T, Glick BR (2004) Plant growth-promoting bacteria confer resistance in tomato plants to salt stress. Plant Physiol Biochem 42:565–572

Munns R, Tester M (2008) Mechanisms of salinity tolerance. Annu Rev Plant Biol 59:651–681

Nadeem SM, Zahir ZA, Naveed M, Nawaz S (2013) Mitigation of salinity-induced negative impact on the growth and yield of wheat by plant growth-promoting rhizobacteria in naturally saline conditions. Ann Microbiol 63:225–232

Paré PW, Zhang H, Aziz M, Xie X, Kim MS, Shen X, Zhang J (2011) Beneficial rhizobacteria induce plant growth: mapping signaling networks in *Arabidopsis*. In: Biocommunication in soil microorganisms. Springer, Berlin, pp 403–412

Porra R, Thompson W, Kriedemann P (1989) Determination of accurate extinction coefficients and simultaneous equations for assaying chlorophylls a and b extracted with four different solvents: verification of the concentration of chlorophyll standards by atomic absorption spectroscopy. Biochim Biophys Acta Bioenerg 975:384–394

Ramadoss D, Lakkineni VK, Bose P, Ali S, Annapurna K (2013) Mitigation of salt stress in wheat seedlings by halotolerant bacteria isolated from saline habitats. Springer Plus 2:1–7

Rueda-Puente EO, García-Hernández JL, Preciado-Rangel P, Murillo-Amador B, Tarazón-Herrera MA, Flores-Hernández A, Mwandemele O (2007) Germination of *Salicornia bigelovii* ecotypes under stressing conditions of temperature and salinity and ameliorative effects of plant growth-promoting bacteria. J Agron Crop Sci 193:167–176

Tangpu V, Temjenmongla K, Yadav AK (2005) Anticestodal activity of *Trifolium repens*. extract. Pharm Biol 42:656–658

Tavakkoli E, Rengasamy P, McDonald GK (2010) High concentrations of Na^+ and Cl^- ions in soil solution have simultaneous detrimental effects on growth of faba bean under salinity stress. J Exp Bot 2010:erq25

Tester M, Davenport R (2003) Na^+ tolerance and Na^+ transport in higher plants. Ann Bot 91:503–527

Tiwari S, Singh P, Tiwari R, Meena KK, Yandigeri M, Singh DP, Arora DK (2011) Salt-tolerant rhizobacteria-mediated induced tolerance in wheat (*Triticum aestivum*) and chemical diversity in rhizosphere enhance plant growth. Biol Fertil Soil 47:907–916

Tuna AL, Kaya C, Higgs D, Murillo-Amador B, Aydemir S, Girgin AR (2008) Silicon improves salinity tolerance in wheat plants. Environ Exp Bot 62:10–16

Wang S-M, Zhang J-L, Flowers TJ (2007) Low-affinity Na^+ uptake in the halophyte *Suaeda maritima*. Plant Physiol 145:559–571

Xie X, Zhang H, Pare P (2009) Sustained growth promotion in *Arabidopsis* with long-term exposure to the beneficial soil bacterium *Bacillus subtilis* (GB03). Plant Sig Behav 4:948–953

Zhang H, Kim MS, Krishnamachari V, Payton P, Sun Y, Grimson M, Paré PW (2007) Rhizobacterial volatile emissions regulate auxin homeostasis and cell expansion in *Arabidopsis*. Planta 226:839–851

Zhang H, Kim MS, Sun Y, Dowd SE, Shi H, Paré PW (2008) Soil bacteria confer plant salt tolerance by tissue-specific regulation of the sodium transporter HKT1. Mol Plant Microb Interact 21:737–744

Zhang H, Murzello C, Sun Y, Kim MS, Xie X, Jeter RM, Paré PW (2010a) Choline and osmotic-stress tolerance induced in Arabidopsis by the soil microbe *Bacillus subtilis* (GB03). Mol Plant Microb Interact 23:1097–1104

Zhang J-L, Flowers TJ, Wang S-M (2010b) Mechanisms of sodium uptake by roots of higher plants. Plant Soil 326:45–60

Zhang N, Wu K, He X, Li SQ, Zhang ZH, Shen B, Shen QR (2011) A new bioorganic fertilizer can effectively control banana wilt by strong colonization with *Bacillus subtilis* N11. Plant Soil 344:87–97

19

Chloroplast proteome analysis of *Nicotiana tabacum* overexpressing TERF1 under drought stress condition

Wei Wu and Yanchun Yan*

Abstract

Background: Chloroplast is indispensable for plant response to environmental stresses, growth and development, whose function is regulated by different plant hormones. The chloroplast proteome is encoded by chloroplast genome and nuclear genome, which play essential roles in plant photosynthesis, metabolism and other biological processes. Ethylene response factors (ERFs) are key transcription factors in activating the ethylene signaling pathway and plant response to abiotic stress. But we know little about how ethylene regulates plastid function under drought stress condition. In this study we utilized tobacco overexpressing tomato ethylene responsive factor 1 (TERF1), an ERF transcription factor isolated from tomato, to investigate its effects on the plastid proteome under drought stress condition by method of iTRAQ technology.

Results: Results show that TERF1 represses the genes encoding the photosynthetic apparatus at both transcriptional and translational level, but the genes involved in carbon fixation are significantly induced by TERF1. TERF1 regulates multiple retrograde signaling pathways, providing a new mechanism for regulating nuclear gene expression. TERF1 also regulates plant utilization of phosphorus (Pi) and nitrogen (N). We find that several metabolic and signaling pathways related with Pi are significantly repressed and gene expression analysis shows that TERF1 significantly represses the Pi transport from root to shoot. However, the N metabolism is upregulated by TERF1 as shown by the activation of different amino acids biosynthesis pathways due to the induction of glutamine synthetase and stabilization of nitrate reductase although the root-to-shoot N transport is also reduced. TERF1 also regulates other core metabolic pathways and secondary metabolic pathways that are important for plant growth, development and response to environmental stresses. Gene set linkage analysis was applied for the upregulated proteins by TERF1, showing some new potential for regulating plant response to drought stress by TERF1.

Conclusions: Our research reveals effects of ethylene signaling on plastid proteome related with two key biological processes, including photosynthesis and nutrition utilization. We also provide a new mechanism to regulate nuclear gene expression by ERF1 transcription factor through retrograde signals in chloroplast. These results can enrich our knowledge about ERF1 transcription factor and function of ethylene signaling pathway.

Keywords: Chloroplast proteome, Drought stress, TERF1, Ethylene signaling pathway

*Correspondence: yanyanchun@caas.cn
Graduate School of Chinese Academy of Agricultural Sciences, No. 12 Zhongguancun South St., Haidian District, Beijing 100081, People's Republic of China

Background

Ethylene is a gaseous phytohormone that plays vital roles in plant growth, development as well as response to abiotic and biotic stresses. ERF1 protein is a key transcription factor for activating the ethylene signaling pathway by binding the GCC box and other *cis*-elements in the promoters of ethylene-responsive genes, which is an ideal candidate gene for improving plant tolerance of different abiotic stresses (Kavas et al. 2015).

Chloroplast is an organelle derived from the cyanobacteria and its proteome is encoded by nuclear genome and its own genome. Apart from the key function of photosynthesis, chloroplast is also responsible for the biosynthesis of different metabolites, including amino acids, vitamins, fatty acids, nucleotides, phytohormones as well as assimilation of sulfur and nitrogen (Daniell et al. 2016). Chloroplast has been found to regulate plants' response to different environmental stresses through organellar signaling pathways because it can act as environment sensor (Kmiecik et al. 2016).

At transcriptional level different phytohormones have been found to regulate chloroplast transcription. Abscisic acid (ABA) represses the plastid encoded RNA polymerase (PEP) and nuclear encoded RNA polymerase (NEP), leading to the reduced transcript level of most chloroplast genes, which can be minimized by cytokinin (Yamburenko et al. 2015). Methyl jasmonate (MeJA), gibberellic acid [GA(3)], an auxin (indole-3-acetic acid, IAA) repress both transcription and transcript accumulation of chloroplast genes, while a brassinosteroid (24-epibrassinolide, BR) can counteract these effects (Zubo et al. 2011).

Additionally, characterization of chloroplast proteome changes under environmental stress condition attracted the attention of the researchers. 31 chloroplast proteins were found to be differentially regulated by drought stress, which are mainly enriched in the Gene Ontology (GO) category of "energy" (Zubo et al. 2011). Kosmala et al. identified 10 differentially accumulated chloroplast proteins in high-drought-tolerant and low-drought-tolerant of *Festuca arundinacea* (Kosmala et al. 2012). Uberegui et al. revealed the function of Executer proteins in plant chloroplast response to increased light condition (Uberegui et al. 2015). Wang et al. studied the chloroplast proteome changes in mangrove under different salt stress conditions (Wang et al. 2013). Although these studies have revealed some mechanisms in chloroplast response to environmental stresses, most of these proteomics studies utilized 2D-Gel method, which restricted coverage of the chloroplast proteome and the accuracy of protein quantification. Additionally, there is little report on how ethylene regulate chloroplast proteome under environmental stress conditions.

TERF1, an ERF1 protein isolated from tomato, can activate ethylene signaling pathway through binding the GCC box and dehydration responsive element (DRE) in the promoters of the target genes (Huang et al. 2004). Overexpression of TERF1 can significantly improve tobacco tolerance of abiotic stress (Huang et al. 2004), however the detailed mechanism of TERF1 in improving the plant tolerance of drought stress is not clear. In this study we compared the chloroplast proteome between WT tobacco and tobacco overexpressing TERF1 under natural dehydration condition by the isobaric tags for relative and absolute quantification (iTRAQ) method, aiming to further reveal the function of TERF1 and ethylene in plant response to drought stress.

Results

TERF1 significantly changes the chloroplast proteome under drought stress condition

After drought stress was exerted we measured the relative water content (RWC) of the WT and transgenic tobacco and results showed that the RWC of WT tobacco decreased faster than the transgenic ones after 10 days of dehydration and at 20 days of dehydration the RWC of WT tobacco decreased by more than 35% and the TERF1 tobacco decreased only about 18% (Additional file 1: Figure S1), indicating that TERF1 significantly reduced water loss under drought stress condition. We compared the chloroplast proteome between TERF1 and WT tobacco after 20 days of dehydration when the RWC of WT tobacco was significantly lower than TERF1 tobacco.

We identified 13,049 unique peptides (Additional file 2: Table S9), which were searched against the UniProt database of *Nicotiana tabacum* and 4732 proteins were identified and quantified (at least two unique peptide with high confidence), among which 4694 proteins are encoded by nuclear genome (Additional file 3: Table S2) and 38 proteins are encoded by chloroplast genome (Additional file 9: Table S7). A significant difference is observed in chloroplast proteome between WT and TERF1 tobacco supported by principal component analysis (PCA) (Additional file 4: Figure S2).

We use more than 1.20-fold ($P < 0.05$) or less than 0.83-fold ($P < 0.05$) cutoff to identify the differentially expressed proteins (DEPs), of which 189 proteins are significantly induced (Additional file 5: Table S3) and 273 proteins are significantly repressed (Additional file 6: Table S4). In order to dissect the function of the DEPs the KEGG pathway enrichment analysis was applied and the significantly enriched pathways are shown in Tables 1 and 2. The results of gene ontology (GO) enrichment analysis for the DEPs are listed in Additional file 7: Table S5 and Additional file 8: Table S6, respectively.

Table 1 KEGG pathway enrichment for the upregulated DCPs

Term	ID	Number of the enriched proteins (%)	P-value
Metabolic pathways	ko01100	61 (32%)	0.009696835
Biosynthesis of secondary metabolites	ko01110	41 (21%)	0.005100797
Biosynthesis of amino acids	ko01230	19 (10%)	0.002111965
Terpenoid backbone biosynthesis	ko00900	5 (3%)	0.012298651
Fatty acid biosynthesis	ko00061	5 (3%)	0.013993754
mRNA surveillance pathway	ko03015	5 (3%)	0.05547601
Pantothenate and CoA biosynthesis	ko00770	4 (2%)	0.007031075
Valine, leucine and isoleucine biosynthesis	ko00290	4 (2%)	0.010395488
One carbon pool by folate	ko00670	4 (2%)	0.014658739
Cyanoamino acid metabolism	ko00460	4 (2%)	0.029672314
Isoquinoline alkaloid biosynthesis	ko00950	3 (2%)	0.021900026
Biotin metabolism	ko00780	3 (2%)	0.026505785

Table 2 KEGG pathway enrichment for the downregulated DCPs

TERM	ID	Number of the enriched proteins (%)	P-value
Metabolic pathways	ko01100	62 (23%)	0.007587671
Photosynthesis	ko00195	15 (5%)	1.66E−07
Photosynthesis-antenna proteins	ko00196	8 (3%)	1.22E−07
Glycerophospholipid metabolism	ko00564	6 (2%)	0.000985438
Phenylpropanoid biosynthesis	ko00940	6 (2%)	0.007947381
Phosphatidylinositol signaling system	ko04070	4 (1%)	0.004652457
Ether lipid metabolism	ko00565	3 (1%)	0.002725805
ABC transporters	ko02010	2 (1%)	0.005861127
Diterpenoid biosynthesis	ko00904	2 (1%)	0.002013018

TERF1 exerts different effects on the expression of the genes in photosynthetic light and dark reaction

Chloroplast proteome are encoded by nuclear genome and chloroplast genome together, among which 98 proteins are encoded by tobacco chloroplast genome. The genes encoded by chloroplast genome are related with photosynthetic apparatus, the chloroplast genetic system, and other functions (Shimada and Sugiura 1991). 38 chloroplast genome encoded proteins were identified and quantified in our proteomics analysis, among which 27 proteins were repressed by TERF1 and most of these proteins are components of photosynthesis apparatus (Additional file 9: Table S7).

We also assessed the transcript level of the 79 genes encoded by chloroplast genome with confirmed function, showing that 61 genes were repressed and only 18 were induced by TERF1 (Additional file 10: Table S8). The significantly repressed genes encode subunits of PSI, PSII, NADH dehydrogenase, ATP synthase, cytochrome b6/f and ribosome proteins (Additional file 10: Table S8). As the bacteria-type 70S ribosome is responsible for protein biosynthesis in chloroplast (Tiller and Bock 2014), the significant repression of the *rps* and *rpl* genes by TERF1 contribute to the repression of protein biosynthesis in chloroplast.

Additionally TERF1 negatively regulates RNA editing in chloroplast as shown by the significant repression of Multiple Organellar RNA Editing Factor 8 (MORF8) (No. 122, Additional file 6: Table S4). MORF8 participates in chloroplast RNA editing as well as 8 mitochondrial genes editing (Glass et al. 2015; Huang et al. 2017). Repression of MORF8 will negatively affect chloroplast and plant growth and development (Glass et al. 2015).

TERF1 also represses many genes encoded by nuclear genome participating in the process of light harvesting, light reaction and photosynthetic electron transport (Additional file 8: Table S6). Additionally, we also found FtsH5, a thylakoid membrane-bound metalloprotease, was significantly repressed (No. 32, Additional file 6: Table S4), which will impair the D1 protein degradation and finally leads to photoinhibition (Kato et al. 2009). Finally, two genes involved in the chlorophyll biosynthesis, Glutamate–tRNA ligase (GluRSAt) and Protoporphyrinogen oxidase (PPO2) (No. 91 and No. 200, Additional file 6: Table S4), were also significantly repressed by TERF1. Repression of GluRSAt and PPO2 will reduce the accumulation of Glutamyl-tRNA^{Glu}, starting substrate of tetrapyrroles biosynthesis, and Protoporphyrin IX, the last common precursor for chlorophyll and heme biosynthesis, respectively.

Although many genes involved in photosynthetic light reactions were repressed by TERF1, seven GO terms related with carbon fixation are significantly enriched for the upregulated DEPs, including "GO 0034637: cellular carbohydrate biosynthetic process", "GO 0033692: cellular polysaccharide biosynthetic process", "GO 0042732: D-xylose metabolic process", "GO 0009250: glucan biosynthetic process", "GO 0000271: polysaccharide biosynthetic process", "GO 0019252: starch biosynthetic process", GO 0005982: starch metabolic process" (Additional file 7: Table S5), indicating that carbon fixation in TERF1 tobacco is significantly upregulated.

According to the protein expression profile, several key genes involved in starch biosynthesis are significantly induced by TERF1. Protein Targeting To Starch (PTTS) is upregulated by more than 32% by TERF1, which localise granule-bound starch synthase to starch granules and is responsible for normal amylose synthesis (No. 125, Additional file 5: Table S3) (Seung et al. 2015). ADP-Glucose Pyrophosphorylase (ApL3), catalyzing the first and limiting step in starch biosynthesis (Sulmon et al. 2011), is significantly induced by 26% in TERF1 tobacco (No. 66, Additional file 5: Table S3). Starch Synthase 2 (SS2), involved in amylopectin biosynthesis (Zhang et al. 2008), is induced by more than 25% in TERF1 tobacco. The upregulation of the carbon fixation in TERF1 tobacco will also contribute to the repression of the genes involved in photosynthetic light reaction (Hausler et al. 2014).

TERF1 regulates four key chloroplast metabolic pathways

According to the KEGG pathway enrichment analysis, the pathway of "valine, leucine and isoleucine biosynthesis", "biosynthesis of amino acids" and "cyanoamino acid metabolism" are significantly enriched for the upregulated DEPs in TERF1 tobacco (Table 1). Another five GO terms related with amino acids metabolism (GO 1901607: alpha-amino acid biosynthetic process, GO 0006541: glutamine metabolic process, GO 0006545: glycine biosynthetic process, GO 0009070: serine family amino acid biosynthetic process, GO 0006556: S-adenosylmethionine biosynthetic process) are also significantly enriched for the upregulated DEPs (Additional file 7: Table S5), indicating that N assimilation is improved by TERF1 (Li et al. 2017).

Gene expression results show that TERF1 significantly represses *NRT2.4* and *NRT1.2* (Fig. 1a). NRT2.4 and NRT1.2 encode high-affinity and low-affinity nitrate transporter, respectively (Filleur et al. 2001), indicating that TERF1 restricts the nitrate uptake in the root under drought stress condition. Moreover, the significant induction of NLA in leaf will degrade NRT1.7 although *NRT1.7* transcript level is slightly induced (Fig. 1b). NRT1.7 is responsible for source-to-sink nitrate remobilization and negatively regulates plant adaptation to limited nitrogen (Liu et al. 2017a, b), indicating that TERF1 also limits the N remobilization under drought stress condition.

However, at protein level we find that TERF1 significantly upregulates the glutamine synthetase (GS) (No. 67, Additional file 5: Table S3), a key enzyme for improving N use efficiency (NUE) through converting ammonium into glutamine (Wang et al. 2018). And two 14-3-3 proteins are significantly repressed by TERF1 (No. 2 and No. 3, Additional file 6: Table S4), which can stabilize nitrate reductase (NR), a key enzyme for reducing nitrate in chloroplast (Miller et al. 2008). N assimilation in chloroplast also correlates with the TCA cycle in the mitochondria because it can provide C skeleton for de novo ammonium assimilation. We find that the dicarboxylate transporter (No. 54, Additional file 5: Table S3) is significantly induced by TERF1, which can promote the transport of 2-oxoglutarate into the chloroplast, an important C skeleton for ammonium assimilation by GS (Taniguchi and Taniguchi et al. 2002). 2-oxoglutarate also regulates GS activity at transcriptional or post-translational level (Galvez et al. 1999). Upregulation of GS and stabilization of NR provide a possible mechanism for the enhanced N

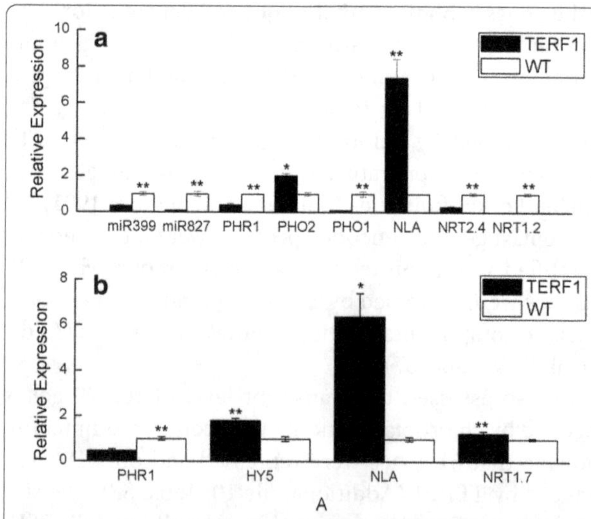

Fig. 1 Quantitative real-time polymerase chain reaction (qRT-PCR) for expression analysis of the genes related N and Pi metabolism in root (**a**) and leaf (**b**). Means ± SDs, n = 3; * and **significantly different at 5% and 1% level of probability, respectively

assimilation efficiency by TERF1. As NUE is composed of N uptake efficiecy and N utilization efficiency (Wang et al. 2018), we propose that TERF1 significantly improve N utilization efficiency.

As described above many genes related with photosynthesis are significantly repressed we speculate that TERF1 may repress Phosphorus (Pi) utilization because Pi is an essential nutrient for plant photosynthesis (Zhang et al. 2014). Additionally we also find several pathways related with Pi metabolism or signaling pathway are significantly enriched for the downregulated proteins, including phosphatidylinositol signaling system, glycerophospholipid metabolism, inositol phosphate metabolism (Table 2) and phosphatidylcholine metabolic process (GO: 0046470) (Additional file 8: Table S6). Additionally, repression of the genes encoding subunits of ATP synthase described above may also result from the repression of Pi deficiency (Zhang et al. 2014).

Based on the above results we compared the expression of genes responsible for root-to-shoot Pi transport between WT and TERF1 tobacco. Phosphate Starvation Response 1 (PHR1), a conserved MYB transcription factor for regulating plant Pi starvation response, is significantly repressed by TERF1, which can cause the repression of miR399 and miR827 (Fig. 1a) (Chien et al. 2017). Repression of miR399 can upregulate the expression of *Phosphate 2* (*PHO2*) (Fig. 1a), a ubiquitin-conjugating E2 enzyme for degrading Pi transporters phosphate 1 (PHO1) and phosphate transporter 1 (PHT1) (Chien et al. 2017). Moreover, the transcript level of *PHO1* is also significantly reduced by TERF1 (Fig. 1a). Repression of miR827 can lead to the upregulation of *NLA* (Fig. 1a), another ubiquitin E3 ligase for degrading the PHT1 (Chien et al. 2017). So TERF1 represses the expression of the Pi transporters at both transcriptional and posttranslational level, indicating that it can significantly reduce the root-to-shoot Pi uptake. In leaf *PHR1* is also significantly repressed by TERF1, which may result from the upregulation of HY5 as it can repress the expression of PHR1 activated by light and ethylene signaling pathway (Fig. 1b) (Liu et al. 2017a, b).

Sulfur assimilation is upregulated by TERF1 as shown by the two GO term, "sulfur compound biosynthetic process" and "sulfur compound metabolic process", are significantly enriched for the upregulated proteins in TERF1 tobacco. The enriched proteins for the sulfur metabolism correlates with the biosynthesis of S-adenosylmethionine (SAM) (No. 2 and No. 143, Additional file 5: Table S3), thiamine (vitamin B1) (No. 158, Additional file 5: Table S3), biotin (vitamin H) (No. 35, Additional file 5: Table S3). Accumulation of SAM provides substrates for the biosynthesis of plant hormones, polyamines, and defense metabolites (Kusano et al. 2010). Acetyl CoA carboxylase is significantly upregulated by TERF1 (No. 22 and No. 23, Additional file 5: Table S3) and biotin is an important co-factor for it (Nikolau et al. 2003), which contribute to the significant upregulation of fatty acid biosynthesis as shown in Table 1. Biosynthesis of thiamine contributes to the carbon fixation shown in Additional file 7: Table S5, glycolysis and citric acid cycle (Martinis et al. 2016).

Finally we find that TERF1 can increase the supply of one-carbon. Serine accumulation in TERF1 tobacco provides the substrate for one-carbon metabolism. And three serine hydroxymethyltransferase are significantly upregulated by TERF1 (No. 147–149, Additional file 5: Table S3), which convert serine into glycine and 5,10-methylenetetrahydrofolate that are important for plant growth and development (Zhang et al. 2010). And the upregulation of SAM also provide an important source of one-carbon.

TERF1 regulates the retrograde signals in chloroplast

Retrograde signaling is defined by chan et al. as a process in which a stimulus perturbs chloroplast homeostasis and gives rise to one or more retrograde signals that alter transcriptional through to posttranslational processes and ultimately regulate chloroplast function (Chan et al. 2016).

Transcription as well as protein biosynthesis in chloroplast can act as retrograde signals to regulate the nuclear genes expression, especially the photosynthesis-associated nuclear genes (PhANGs) (Chan et al. 2016). The repression of transcription and protein synthesis in chloroplast by TERF1 described above are important signals for the downregulation of proteins enriched with pathway of "photosynthesis" and "photosynthesis-antenna proteins" described in Table 2 and Additional file 6: Table S4 because chloroplast gene expression need to coordinate with their counterparts encoded by nuclear genome performing the same function (Nott et al. 2006).

Methylerythritol cyclodiphosphate (MEcPP), an isoprenoid precursor in chloroplast, is the second retrograde signal regulated by TERF1. TERF1 significantly induces 1-hydroxy-2-methyl-2-(E)-butenyl-4-diphosphate synthase (HDS) (No. 16, Additional file 5: Table S3), which can convert MEcPP into hydroxymethylbutenyl diphosphate (HMBPP) (Chan et al. 2016). The decrease of MEcPP can repress the expression of hydroxyperoxide hyase (HPL) protein (No. 109, Additional file 6: Table S4) and salicylic acid (SA) accumulation (Xiao et al. 2012). HPL protein also correlates with the chloroplast retrograde signal of oxylipins, derived from the oxidized linoleic acid and linolenic acid in chloroplast. The repression of HPL protein will cause the accumulation of 12-oxophytodienoic acid (12-OPDA) as well as the biosynthetic precursor for jasmonates (de Souza et al. 2017). Additionally 12-OPDA itself can also act as signal to regulate nuclear gene expression through interaction with the transcription factors binding the TGACG motif, which regulates the stomatal closure in response to drought (de Souza et al. 2017).

Thioredoxins (Trxs) are a large protein family acting as redox regulators to regulate the structure and function of plant proteins through catalyzing the reduction of disulfide bonds in them. (Holmgren 1985). As Trxs can lead to changes in sugars, intermediates of chlorophyll synthesis, ROS, chloroplast protein synthesis, they have been proposed to be involved in the retrograde signaling to nucleus (Brunkard et al. 2015). Two chloroplast-localized Trxs (Trx f1 and HCF164) are significantly repressed by TERF1 (No. 230 and No. 233, Additional file 6: Table S4), while CDSP32 is significantly induced (No. 159, Additional file 5: Table S3). Trx f1 regulates multiple enzymes involved in Calvin–Benson cycle and starch synthesis (Thormahlen et al. 2013). Repression of Trx f1 in TERF1 will attenuate the starch biosynthesis (Thormahlen et al. 2013), indicating that sucrose signaling may be activated by TERF1. HCF164 is involved in the biosynthesis and assembly of cyt b6/f complex (Gabilly et al. 2010), providing an important mechanism for the downregulation of the genes encoding cyt b6/f complex by TERF1 (Additional file 6: Table S4 and Additional file 10: Table S8), indicating that it is involved in the plastid gene expression retrograde signaling. CDSP32 protects photosynthetic apparatus against oxidative damage, showing that it may participate in the retrograde signaling of oxidative stress (Broin et al. 2002).

Tetrapyrroles are the fourth kind of retrograde signals regulated by TERF1. As described above the repression of GluRSAt and PPO2 can inhibit the biosynthesis of tetrapyrroles, which can also lead to the repression of PhANGs (Nott et al. 2006).

The last retrograde signal regulated by TERF1 results from the repression of STN7, which phosphorylates LHCII to mediate state transitions and acts as sensor of the plastoquinol (PQH2)/plastoquinone (PQ) redox state (Chan et al. 2016). PQ/PQH2 redox state can regulate at least 750 nuclear genes expression through STN7 (Chan et al. 2016), so the repression of STN7 partially blocks the redox cues from chloroplast.

Gene set linkage analysis of the DEPs

Gene set linkage analysis (GSLA) interpretates the differentially expressed genes functional impact on biological processes in Arabidopsis with the aim to solve the problems of "no annotation" or "only conceptually general terms (such as "GO 0007165: signal transduction")" in GO enrichment analysis (Yao et al. 2018). We first searched the corresponding orthologous genes of the DEPs in Arabidopsis and conducted GSLA analysis. As no significant GO term is enriched for the downregulated proteins, we only analyzed the PPI network of upregulated proteins based on GO enrichment and topology characteristics of the network (Fig. 2).

According to PPI network described in Fig. 2, most of the upregulated DEPs interact with more than 10 proteins, whose functions are summarized in Table 3. These proteins participate in many important biological processes, including salicylic acid signaling pathway, autophagy, chloroplast development, circadian rhythm as well as many important metabolic pathways, including biosynthesis of fatty acid, ethylene, alkaloid, lignin, demonstrating the important roles played by these proteins.

Discussion

As an important component of ethylene signaling pathway transcription factor of ERF1 plays important roles in activating the ethylene signaling pathway through regulating ethylene responsive genes expression in response to environmental stresses. Our research reveals several new mechanisms for regulating plant response to drought stress by ERF1 protein through regulating chloroplast function.

TERF1 negatively regulates the expression of genes related with photosynthesis

Our results show that overexpression of TERF1 negatively regulates the expression of genes related with photosynthetic light reaction at both transcriptional and translational level, which may result from several important mechanisms. Firstly, our previous research showed that TERF1 significantly repressed the genes encoding the chloroplast transcriptional apparatuses under drought stress condition (Wu et al. 2018). Secondly, a recent study proves that plant response to light needs the degradation of Ethylene-Insensitive3 (EIN3), an upstream activator for ERF1, by phyB (Liu et al. 2017a, b). So overexpression of TERF1 may constitutively repress plant response to light, leading to the repression of genes encoding PhANGs, as it acts downstream of EIN3. Thirdly, TERF1 has also been found to activate the ABA signaling pathway (Zhang et al. 2005) and induce MPK6 (Additional file 11: Figure S3), which also contribute to the repression of the genes related with photosynthetic light reaction (Yamburenko et al. 2015). Finally, the upregulation of carbon fixation and repression of Pi uptake can also lead to the repression of PhANGs (Hausler et al. 2014; Zhang et al. 2014).

Under stress condition the repression of PhANGs in light will lead to the accumulation of reactive oxygen species (ROS), which can induce the expression of stress-related genes (Su et al. 2018). The interaction between

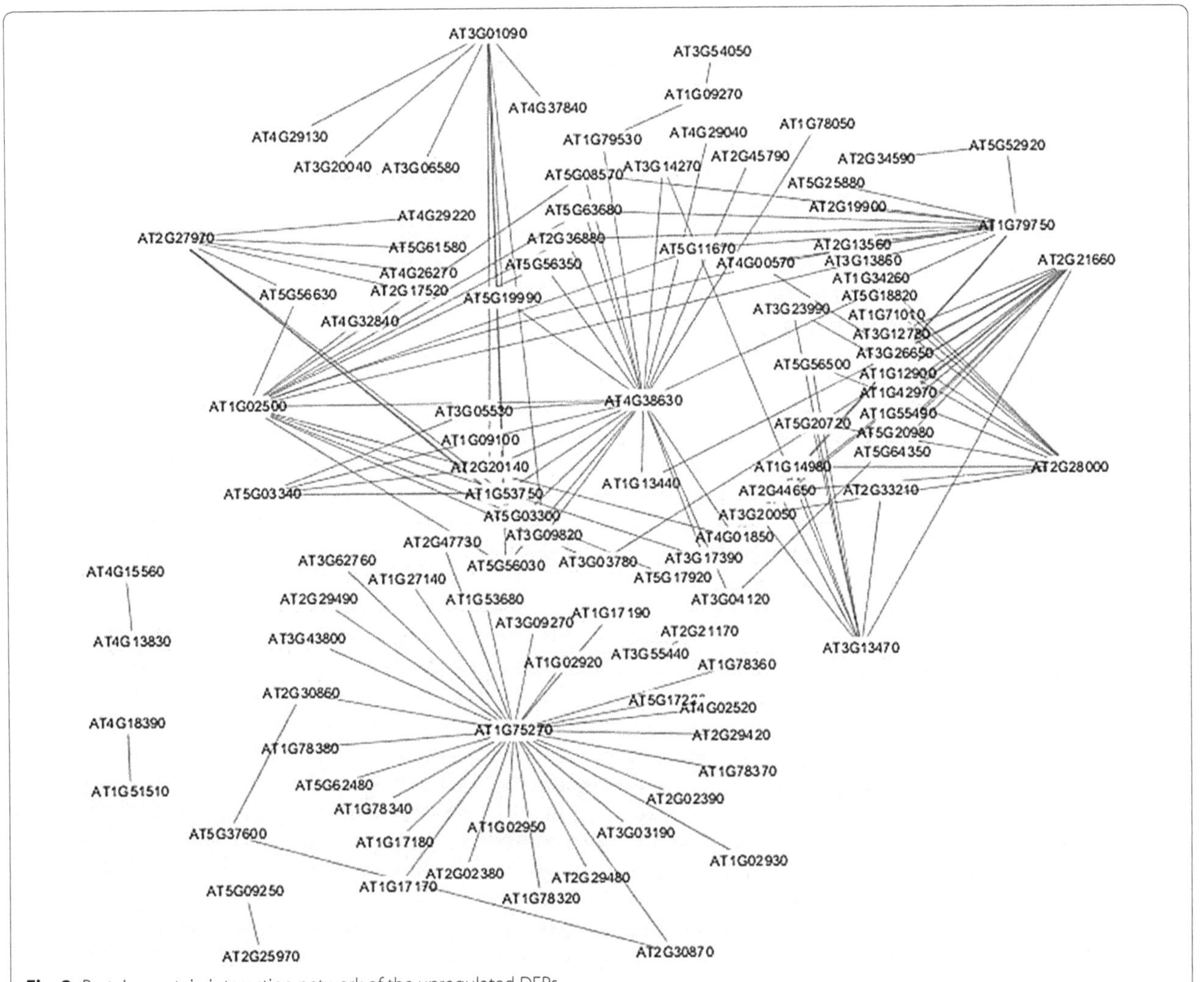

Fig. 2 Protein–protein interaction network of the upregulated DEPs

ROS production and repression of photosynthesis-related genes orchestrate the trade-off between plant growth and stress tolerance, which can improve plant tolerance of environmental stress (Su et al. 2018).

Photosynthetic genes are encoded by nulear genome and chloroplast genome together. We compared the chloroplast genes expression at both transcriptional level and translational level. Although most of the genes encoded by chloroplast genome are repressed, the correlation between transcript level and protein level is poor. It is reported that translation regulation plays more important roles in chloroplast genes expression than transcription regulation because protein levels vary considerably in response to environmental factors even the corresponding mRNA remain constant (Eberhard et al. 2002). In our study we find that several ribosome genes are both repressed at transcriptional and translational level, which represents a key limit factors for the translational repression for the chloroplast genes (Beligni et al. 2004). Different factors regulate translation of chloroplast genes, so how TERF1 regulates this process deserves future research.

TERF1 may regulate the interaction between chloroplast and mitochondria

It has been recognized that there exists communication between chloroplast and mitochondria (Raghavendra and Padmasree 2003). From the results we find some important clues for the interaction between chloroplast and mitochondria in TERF1 tobacco that may improve plant tolerance of drought stress. (1) Upregulation of the biosynthesis of valine, leucine and isoleucine not only provide important osmolytes under abiotic stress condition but also can promote oxidization of them in mitochondria, which can provide ATP for photosynthetic carbon fixation under environmental stress conditions (Joshi

Table 3 Function analysis of the proteins with more than 10 interacting partners in the PPI network

Protein node	Gene name	Enriched GO Term	Degree	Function	Reference
AT1G75270	Dehydroascorbate reductase 2 (DHAR2)	GO:0009407	28	activating the salicylic acid pathway; modulating the redox states of ascorbate–glutathione cycle	Ball et al. (2004); Noshi et al. (2017); Rahantaniaina et al. (2017)
AT4G38630	Multiubiquitin chain binding protein 1 (MBP1)	GO:0006458; GO:0006096; GO:0006090; GO:0046835; GO:0046500; GO:0046086; GO:0019318	23	autophagic receptor; promoting the ubiquitin-mediated degradation of the oxidized proteins degradation	Kurepa et al. (2008)
AT2G28000	Chaperonin 60 Alpha (CPN60A)	GO:0006986; GO:0006458 GO:0051084; GO:0061077	17	chloroplast development; plant response to abiotic and biotic stress	Ke et al. (2017); Suzuki et al. (2009)
AT3G13470	Chaperonin 60 Beta (CPN60B)		14		
AT2G21660	Glycine-rich RNA-binding protein 7 (GR-RBP7)	GO:0006458; GO:0006096; GO:0051084; GO:0050667 GO:0046500; GO:0019318 GO:0061077	14	Stress response; innate immune response; circadian rhythm; flower time	Fu et al. (2007); Kwak et al. (2013); Meyer et al. (2017)
AT1G02500	S-adenosylmethionine synthetase 1 (SAM1)	GO:0006096; GO:0006090; GO:0050667; GO:0046500; GO:0061077	14	Ethylene biosynthesis; lignin deposition, alkaloid biosynthesis	Belbahri et al. (2000); Jin et al. (2017)
AT1G79750	NADP-malic enzyme 4 (NADP-ME4)	GO:0006458; GO:0006096 GO:0006090; GO:0046500; GO:0061077	12	Fatty acid biosynthesis	Wheeler et al. (2005)

et al. 2010). As TERF1 significantly repressed ATPase (Additional file 6: Table S4), upregulation of valine, leucine and isoleucine biosynthesis may provide an alternative way to supplement the energy for carbon fixation and cytosol demand under drought stress condition. (2) A key pathway "oxoacid metabolic process" is enriched for the upregulated DEPs, which has been proven to be correlated with a modified TCA cycle in mitochondrial and provides carbon skeleton for N assimilation in chloroplast under environmental stress condition (Padmasree et al. 2002). Then how TERF1 regulates mitochondrial function and the communication between chloroplast and mitochondrial deserves future research.

TERF1 regulates plant nutrition utilization to improve plant tolerance of drought stress

Our study has shown that TERF1 can regulate plant utilization of nutrition, which can contribute the improvement of plant tolerance of environmental stress. Firstly, the upregulation of the genes related with carbon fixation positively regulates plant response to different environmental stresses and promotes plant growth in the following darkness (Prasch et al. 2015; Fernandez et al. 2017). Secondly, the enhancement of N utilization can improve plant tolerance of environmental stresses and maintain cell membrane stability and leaf tissue integrality (Chang et al. 2016). Finally, the reduction of Pi utilization can repress the expression of PhANGs, which may finally lead to the upregulation of stress-related genes (Su et al. 2018).

TERF1 can regulate nuclear gene expression in more diversified ways

Previously TERF1 was found to regulate nuclear gene expression through binding the *cis*-element of DRE and GCC box. Our study reveals new potential mechanisms of TERF1 in regulating nuclear gene expression. Firstly, TERF1 can regulate nuclear gene expression through different retrograde signals derived from chloroplast. We have confirmed the interaction between ethylene signaling and retrograde signaling mediated by TERF1 through Q-PCR method, which regulates the expression of PhANGs, plastid redox-associated nuclear genes (PRANGs), singlet oxygen responsive genes (SORGs) (Wu et al. 2018). These retrograde signals not only regulate nuclear gene expression but also connect with different signaling pathways in response to environmental stress as described above. Secondly, TERF1 can regulate gene expression not only at transcriptional level but also at posttranscriptional level. Our analysis shows that TERF1 can regulate Pi uptake at posttranscriptional level through regulating different miRNA. The detailed mechanism of TERF1 regulating nuclear gene expression at posttranscriptional level deserves detailed research in the future, which can enrich our knowledge about ethylene signaling pathway.

Conclusions

In this study we compared the chloroplast proteome between WT and TERF1 tobacco under drought stress condition when the RWC of WT tobacco was significantly lower than the transgenic tobacco. Our results showed significant difference on the proteins related with photosynthesis, carbon fixation, nutrition utilization between WT and TERF1 tobacco, which provide some new insights into the function of TERF1 and ethylene signaling pathway. This proteome analysis also partially confirmed our previous report about the retrograde signaling mediated by TERF1 (Wu et al. 2018), which is a new mechanism for ERF1 protein regulating nuclear gene expression. We also provide new clues for further research on ethylene signaling and ERF1 protein.

Materials and methods

Plant materials and drought stress treatment

Nicotiana tabacum L. cv. NC89 was used for *Agrobacterium*-mediated transformation with construct of PROKII containing the ORF of TERF1 driven by 35S promoter (Wu et al. 2018).

Transgenic and WT seeds were first germinated in Petri dish with a moist filter paper at the bottom. After germination seedlings with unanimous growth status were selected and transferred to the soil. The growth condition for the plant is as followings: 28 °C during day time (300 μmol m^{-2} s^{-1}) for 16 h, 23 °C at night time for 8 h. All WT and transgenic tobaccos were kept in the same pot to keep unanimous growth condition. Natural dehydration stress was applied for WT and transgenic tobacco at 30 days after germination and relative water content was measured. After 20 days there was significant difference between WT and transgenic tobacco (Additional file 1: Figure S1) and leaves and roots were harvested for further proteome and gene expression analysis. The latest growing leaves were harvested for analysis to minimize the growth difference owing to the drought stress. WT tobaccos under drought stress condition were utilized as control. In this study the proteome and gene expression analysis was applied with three replicates for WT and transgenic tobacco.

Chloroplast isolation and protein extraction

About 10 g leaves were grinded in 100 mL ice cold grinding medium [50 m MHEPES–KOH (pH 8.0), 330 mM sorbitol, 2 mM EDTA-Na_2 (pH 8.0), 5 mM ascorbic acid, 5 mM cysteine, 0.05% bovine serum albumin (BSA)]. Filter the homogenate through a double layer of nylon cloth (22 μm). Collect crude chloroplasts by centrifugation for 3 min at 1300*g*. Resuspend the crude chloroplast pellet in wash medium [50 mM HEPES–KOH (pH 8.0), 330 mM sorbitol, 2 mM EDTA-Na_2 (pH 8.0)] by swirling the suspension. Load the resuspended chloroplast on a Percoll step cushion [PF Percoll (40% or 85%), 0.5 mM EDTA, 50 mM HEPES-KOH (pH 8.0), 330 mM sorbitol] and spin for 10 min at 3750 g in a swing-out rotor. Collect the intact chloroplasts from the 40/85% Percoll interface using a pipet and add wash medium to dilute the Percoll. Spin the chloroplasts for 3 min at 1200 g. Remove the supernatant. The pellets are intact chloroplasts.

Chloroplast protein extraction

Suspend the chloroplast samples in protein extraction buffer (1.52 g thiourea, 0.3 g 3-[(cholamidopropyl) dimethylammonio]-1-propanesulfonate (CHAPS), 2 mL water, 1 mL 1 M Tris–HCl (pH 8.0), 2 mL 50% glycerol, 0.2 mL Triton X-100, 0.2 mL 1 M DTT, 3.75 mL 8 M urea) and sonicate. Centrifuge the homogenate at 12,000*g* for 5 min at 4 °C and transfer the supernatant to a new tube. Add 1/10 volume of 100% trichloroacetic acid (TCA) to the supernatant and mix. Incubate the solution on ice for 15 min. Centrifuge the homogenate at 10,000*g* for 5 min at 4 °C and discard the supernatant. Add 1 mL ice-cold acetone to the pellet and mix. Centrifuge the homogenate at 10,000*g* for 10 min at 4 °C. Discard the supernatant and resuspend it in 20 μL dissolution buffer and 1 μL denaturant (2% SDS) supplied by iTRAQ kit.

iTRAQ labeling and strong fractionation by cation exchange chromatography

Add iTRAQ labeling reagents to each protein sample and incubate for 1 h at 37 °C. A 100 μL aliquot of water was added to stop the labeling reaction. A 1 μL aliquot of sample was removed from each group to test labeling and extraction efficiency, and the sample was subjected to a matrix assisted laser desorption ionization procedure after Ziptip desalting. The six sample groups were pooled and vacuum-dried. Each pool of mixed peptides was lyophilized and dissolved in solution A [2% acetonitrile and 20 mM ammonium formate (pH 10.0)]. Then, the samples were loaded onto a reverse-phase column (Luna C18, 4.6 × 150 mm; Phenomenex, Torrance, CA, USA) and eluted using a step linear elution program: 0–10% buffer B [500 mM KCl, 10 mM KH_2PO_4 in 25% acetonitrile, (pH 2.7)] for 10 min, 10–20% buffer B for 25 min, 20–45% buffer B for 5 min, and 50–100% buffer B for 5 min at a flow rate of 0.8 mL/min. The samples were collected each min and centrifuged for 5–45 min. The fractions (about 40) collected were finally combined into 10 pools and desalted on C18 Cartridges (Empore™ standard density

SPE C18 Cartridges, bed I.D. 7 mm, 3 mL volume; Sigma, St. Louis, MO, USA).

LC-electrospray ionization-MS/MS analysis

The reconstituted peptides were analyzed with the Q-Exactive mass spectrometer (Thermo Fisher Scientific, Waltham, MA, USA) coupled with a nano high-performance liquid chromatography system (UltiMate 3000 LC Dionex, Thermo Fisher Scientific). The peptides were loaded onto a C18-reversed phase column (3 μm-C18 resin, 75 μm × 15 cm) and separated on an analytical column (5 μm C18 resin, 150 μm × 2 cm; Dr. Maisch GmbH, Ammerbuch, Germany) using mobile phase A: 0.5% formic acid [FA]/H_2O and B: 0.5% FA/ACN at a flow rate of 300 nL/min, using a 150 min gradient. Spectra were acquired in data-dependent mode. The 10 most intense ions selected for MS scanning (300–1800 m/z, 60,000 resolution at m/z 400, accumulation of 1 × 106 ions for a maximum of 500 ms, 1 microscan). The isolation window was 1.3 m/z, and the MS/MS spectra were accumulated for 150 ms using an Orbitrap. MS/MS spectra were measured at resolution of 15,000 at m/z 400. Dynamic precursor exclusion was allowed for 120 s after each MS/MS spectrum measurement and was set to 17,500 at m/z 200. Normalized collision energy was 30 eV and the underfill ratio, which specifies the minimum percentage of the target value likely to be reached at the maximum fill time, was defined as 0.1%. The instrument was run with peptide recognition mode enabled.

Sequence database search

The raw mass data were processed for the peptide data analysis using Proteome Discoverer 1.4 (ver.1.4.0.288; Thermo Fisher Scientific) with a false discovery rate [FDR = N(decoy)*2/((N(decoy) + N(target))) < 1% and expected cutoff or ion score < 0.05 (with 95% confidence] for searching the Uniprot Nicotiana tobacco Complete Proteome database. Protein probabilities were assigned using the Protein Prophet algorithm, and proteins with at least two unique peptides were identified. The following options were used to identify the proteins: peptide mass tolerance = ±15 ppm, fragment mass tolerance = 20 mmu, enzyme = trypsin, max missed cleavage = 2, fixed modification: carbamidomethyl (C), iTRAQ8plex (N-term), iTRAQ8plex (K), iTRAQ8plex (Y), variable modification: Oxidation (M), Acetyl (Protein N-term).

RNA isolation and reverse transcription

Total RNA and miRNA were extracted from about 200 mg leaf and 500 mg root using RNAiso Plus Kit and RNAiso for small RNA kit (Takara Biomedical Technology, Japan), respectively. RNA integrity was confirmed by electrophoresis and RNA quantity was determined by NanoDrop *ND-1000* spectrophotometer (NanoDrop Technologies, Wilmington, DE, USA). Total RNA and miRNA were utilized for reverse transcription using Prime Script™ RT reagent Kit with gDNA Eraser and Mir-X™ miRNA First Strand Synthesis Kit (Takara Biomedical Technology, Japan), respectively.

Quantitative real-time polymerase chain reaction (qRT-PCR)

qRT-PCR was performed with 3 replicates by *CFX96* real-time PCR detection system (Bio-Rad, Hercules, USA). SYBR® Premix Ex Taq™ II (Tli RNaseH Plus) kit and Mir-X™ miRNA qRT-PCR SYBR® Kit (Takara Biomedical Technology, Japan) was used for mRNA and miRNA qRT-PCR, respectively. Tobacco *L25* (GenBank accession number L18908) and *ERF1α* (GenBank accession number AF120093) were utilized to normalize the mRNA genes expression value. 5S RNA was utilized to normalize the miRNA genes expression value. *Beacon designer 8.0* was utilized to design the gene specific primers (Additional file 12: Table S1). The qPCR reaction was performed as following: 95 °C for 5 min, followed by 40 cycles (95 °C for 30 s, 60 °C for 30 s, and 72 °C for 30 s), followed by melting curve analysis: at 50 °C for 30 s and then at 65–95 °C (0.5 °C increments, 5 s for each). The chloroplast genome sequences of tobacco were retrieved from NCBI database (Accession number: NC_001879.2). The nuclear-encoded genes sequences of tobacco were retrieved from *Sol Genomics Network* database according to the latest genome annotation (https://solgenomics.net/).

Relative water content analysis

Relative water content (RWC) was calculated by the formula: RWC (%) = (FW − DW)/(TW − DW) × 100. FW: fresh weight, TW: turgid weight, DW: dry weight.

Additional files

Additional file 1: Figure S1. Comparison of relative water content (RWC) under drought stress condition.

Additional file 2: Table S9. Chloroplast proteins identified in WT and TERF1 tobacco by iTRAQ method.

Additional file 3: Table S2. Identification and quantification of chloroplast proteins in WT and TERF1 tobacco.

Additional file 4: Figure S2. Principal component analysis of the chloroplast proteome in WT and TERF1 tobacco.

Additional file 5: Table S3. Proteins up regulated by TERF1 under drought stress condition.

Additional file 6: Table S4. Proteins downregulated by TERF1 under drought stress condition.

Additional file 7: Table S5. Gene ontology enrichment analysis for the upregulated proteins by TERF1.

Additional file 8: Table S6. Gene ontology analysis for the downregulated proteins by TERF1.

Additional file 9: Table S7. Identification and quantification of the chloroplast genome encoded proteins in WT and TERF1 tobacco.

Additional file 10: Table S8. Expression analysis of the chloroplast genome encoded genes.

Additional file 11: Figure S3. Expression analysis of MPK6 between WT and TERF1 tobacco. Means ± SDs, n = 3; * - significantly different at 5% level of probability.

Additional file 12: Table S1. Primers Used in This Research.

Abbreviations
ABA: abscisic acid; DEPs: differentially expressed proteins; ERF: ethylene response factor; GO: gene ontology; GS: glutamine synthetase; GSLA: gene set linkage analysis; N: nitrogen; NR: nitrate reductase; Pi: phosphorus; PPI: protein–protein interaction; RWC: relative water content; TERF1: tomato ethylene responsive factor 1; Trx: thioredoxin; WT: wild type.

Authors' contributions
YY conceived the concept, designed the experiment and provide revisions for the manuscript. WW performed the experiments, analyzed the data and wrote the manuscript. Both authors read and approved the final manuscript.

Acknowledgements
We are grateful to Dr. Rongfeng Huang for providing the construct of TERF1.

Competing interests
The authors declare that they have no competing interests.

Funding
This study was supported by grants from Chinese Academy of Agricultural Sciences Basal Research Fund (No. 1610042018006).

References
Ball L, Accotto GP, Bechtold U, Creissen G, Funck D, Jimenez A, Kular B, Leyland N, Mejia-Carranza J, Reynolds H, Karpinski S, Mullineaux PM (2004) Evidence for a direct link between glutathione biosynthesis and stress defense gene expression in Arabidopsis. Plant Cell 16:2448–2462

Belbahri L, Chevalier L, Bensaddek L, Gillet F, Fliniaux MA, Boerjan W, Inze D, Thomas D, Thomasset B (2000) Different expression of an S-adenosylmethionine synthetase gene in transgenic tobacco callus modifies alkaloid biosynthesis. Biotechnol Bioeng 69:11–20

Beligni MV, Yamaguchi K, Mayfield SP (2004) The translational apparatus of *Chlamydomonas reinhardtii* chloroplast. Photosynth Res 82(3):315–325

Broin M, Cuine S, Eymery F, Rey P (2002) The plastidic 2-cysteine peroxiredoxin is a target for a thioredoxin involved in the protection of the photosynthetic apparatus against oxidative damage. Plant Cell 14:1417–1432

Brunkard JO, Runkel AM, Zambryski PC (2015) Chloroplasts extend stromules independently and in response to internal redox signals. Proc Natl Acad Sci USA 112:10044–10049

Chan KX, Phua SY, Crisp P, Mcquinn R, Pogson BJ (2016) Learning the languages of the chloroplast: retrograde signaling and beyond. Annu Rev Plant Biol 67:25–53

Chang Z, Liu Y, Dong H, Teng K, Han L, Zhang X (2016) Effects of cytokinin and nitrogen on drought tolerance of creeping bentgrass. PLoS ONE 11(4):e154005

Chien PS, Chiang CB, Wang Z, Chiou TJ (2017) MicroRNA-mediated signaling and regulation of nutrient transport and utilization. Curr Opin Plant Biol 39:73–79

Daniell H, Lin CS, Yu M, Chang WJ (2016) Chloroplast genomes: diversity, evolution, and applications in genetic engineering. Genome Biol 17:134

de Souza A, Wang JZ, Dehesh K (2017) Retrograde signals: integrators of interorganellar communication and orchestrators of plant development. Annu Rev Plant Biol 68:85

Eberhard S, Drapier D, Wollman FA (2002) Searching limiting steps in the expression of chloroplast-encoded proteins: relations between gene copy number, transcription, transcript abundance and translation rate in the chloroplast of *Chlamydomonas reinhardtii*. Plant J 31(2):149–160

Fernandez O, Ishihara H, George GM, Mengin V, Flis A, Sumner D, Arrivault S, Feil R, Lunn JE, Zeeman SC, Smith AM, Stitt M (2017) Leaf starch turnover occurs in long days and in falling light at the end of the day. Plant Physiol 174:2199–2212

Filleur S, Dorbe MF, Cerezo M, Orsel M, Granier F, Gojon A, Daniel-Vedele F (2001) An arabidopsis T-DNA mutant affected in Nrt2 genes is impaired in nitrate uptake. FEBS Lett 489:220–224

Fu ZQ, Guo M, Jeong BR, Tian F, Elthon TE, Cerny RL, Staiger D, Alfano JR (2007) A type III effector ADP-ribosylates RNA-binding proteins and quells plant immunity. Nature 447:284–288

Gabilly ST, Dreyfuss BW, Karamoko M, Corvest V, Kropat J, Page MD, Merchant SS, Hamel PP (2010) CCS5, a thioredoxin-like protein involved in the assembly of plastid c-type cytochromes. J Biol Chem 285:29738–29749

Galvez S, Lancien M, Hodges M (1999) Are isocitrate dehydrogenases and 2-oxoglutarate involved in the regulation of glutamate synthesis? Trends Plant Sci 4:484–490

Glass F, Hartel B, Zehrmann A, Verbitskiy D, Takenaka M (2015) MEF13 requires MORF3 and MORF8 for RNA editing at eight targets in mitochondrial mRNAs in Arabidopsis thaliana. Molecular Plant 8:1466–1477

Hausler RE, Heinrichs L, Schmitz J, Flugge UI (2014) How sugars might coordinate chloroplast and nuclear gene expression during acclimation to high light intensities. Mol Plant 7:1121–1137

Holmgren A (1985) Thioredoxin. Annu Rev Biochem 54:237–271

Huang Z, Zhang Z, Zhang X, Zhang H, Huang D, Huang R (2004) Tomato TERF1 modulates ethylene response and enhances osmotic stress tolerance by activating expression of downstream genes. FEBS Lett 573:110–116

Huang C, Yu QB, Li ZR, Ye LS, Xu L, Yang ZN (2017) Porphobilinogen deaminase HEMC interacts with the PPR-protein AtECB2 for chloroplast RNA editing. Plant J 92:546–556

Jin Y, Ye N, Zhu F, Li H, Wang J, Jiang L, Zhan J (2017) Calcium-dependent protein kinase CPK28 targets the methionine adenosyltransferases for degradation by the 26S proteasome and affects ethylene biosynthesis and lignin deposition in Arabidopsis. Plant J 90:304–318

Joshi V, Joung JG, Fei Z, Jander G (2010) Interdependence of threonine, methionine and isoleucine metabolism in plants: accumulation and transcriptional regulation under abiotic stress. Amino Acids 39:933–947

Kato Y, Miura E, Ido K, Ifuku K, Sakamoto W (2009) The variegated mutants lacking chloroplastic FtsHs are defective in D1 degradation and accumulate reactive oxygen species. Plant Physiol 151:1790–1801

Kavas M, Kizildogan A, Gokdemir G, Baloglu MC (2015) Genome-wide investigation and expression analysis of AP2-ERF gene family in salt tolerant common bean. Excli J 14:1187–1206

Ke X, Zou W, Ren Y, Wang Z, Li J, Wu X, Zhao J (2017) Functional divergence of chloroplast Cpn60alpha subunits during Arabidopsis embryo development. PLoS Genet 13:e1007036

Kmiecik P, Leonardelli M, Teige M (2016) Novel connections in plant organellar signalling link different stress responses and signalling pathways. J Exp Bot 67:3793–3807

Kosmala A, Perlikowski D, Pawlowicz I, Rapacz M (2012) Changes in the chloroplast proteome following water deficit and subsequent watering in a high- and a low-drought-tolerant genotype of Festuca arundinacea. J Exp Bot 63:6161–6172

Kurepa J, Toh-E A, Smalle JA (2008) 26S proteasome regulatory particle mutants have increased oxidative stress tolerance. Plant Journal 53:102–114

Kusano M, Fukushima A, Redestig H, Kobayashi M, Otsuki H, Onouchi H, Naito S, Hirai MY, Saito K (2010) Comparative metabolomics charts the impact

of genotype-dependent methionine accumulation in Arabidopsis thaliana. Amino Acids 39:1013–1021
Kwak KJ, Kang H, Han KH, Ahn SJ (2013) Molecular cloning, characterization, and stress-responsive expression of genes encoding glycine-rich RNA-binding proteins in *Camelina sativa* L. Plant Physiol Biochem 68:44–51
Li W, Xiang F, Zhong M, Zhou L, Liu H, Li S, Wang X (2017) Transcriptome and metabolite analysis identifies nitrogen utilization genes in tea plant (*Camellia sinensis*). Sci Rep 7(1):1693
Liu W, Sun Q, Wang K, Du Q, Li WX (2017a) Nitrogen limitation adaptation (NLA) is involved in source-to-sink remobilization of nitrate by mediating the degradation of NRT1.7 in Arabidopsis. New Phytol 214:734–744
Liu X, Liu R, Li Y, Shen X, Zhong S, Shi H (2017b) EIN3 and PIF3 Form an interdependent module that represses chloroplast development in buried seedlings. Plant Cell 29:3051–3067
Martinis J, Gas-Pascual E, Szydlowski N, Crevecoeur M, Gisler A, Burkle L, Fitzpatrick TB (2016) Long-distance transport of thiamine (Vitamin B1) is concomitant with that of polyamines. Plant Physiol 171:542–553
Meyer K, Koster T, Nolte C, Weinholdt C, Lewinski M, Grosse I, Staiger D (2017) Adaptation of iCLIP to plants determines the binding landscape of the clock-regulated RNA-binding protein AtGRP7. Genome Biol 18:204
Miller AJ, Fan XR, Shen QR, Smith SJ (2008) Amino acids and nitrate as signals for the regulation of nitrogen acquisition. J Exp Bot 59(1):111–119
Nikolau BJ, Ohlrogge JB, Wurtele ES (2003) Plant biotin-containing carboxylases. Arch Biochem Biophys 414:211–222
Noshi M, Yamada H, Hatanaka R, Tanabe N, Tamoi M, Shigeoka S (2017) Arabidopsis dehydroascorbate reductase 1 and 2 modulate redox states of ascorbate-glutathione cycle in the cytosol in response to photooxidative stress. Biosci Biotechnol Biochem 81:523–533
Nott A, Jung HS, Koussevitzky S, Chory J (2006) Plastid-to-nucleus retrograde signaling. Annu Rev Plant Biol 57:739–759
Padmasree K, Padmavathi L, Raghavendra AS (2002) Essentiality of mitochondrial oxidative metabolism for photosynthesis: optimization of carbon assimilation and protection against photoinhibition. Crit Rev Biochem Mol Biol 37(2):71–119
Prasch CM, Ott KV, Bauer H, Ache P, Hedrich R, Sonnewald U (2015) ss-amylase1 mutant Arabidopsis plants show improved drought tolerance due to reduced starch breakdown in guard cells. J Exp Bot 66:6059–6067
Raghavendra AS, Padmasree K (2003) Beneficial interactions of mitochondrial metabolism with photosynthetic carbon assimilation. Trends Plant Sci 8(11):546–553
Rahantaniaina MS, Li S, Chatel-Innocenti G, Tuzet A, Issakidis-Bourguet E, Mhamdi A, Noctor G (2017) cytosolic and chloroplastic DHARs cooperate in oxidative stress-driven activation of the salicylic acid Pathway. Plant Physiol 174:956–971
Seung D, Soyk S, Coiro M, Maier BA, Eicke S, Zeeman SC (2015) Protein targeting to starch is required for localising granule-bound starch synthase to starch granules and for normal amylose synthesis in Arabidopsis. PLoS Biol 13:e1002080
Shimada H, Sugiura M (1991) Fine structural features of the chloroplast genome: comparison of the sequenced chloroplast genomes. Nucleic Acids Res 19:983–995
Su J, Yang L, Zhu Q, Wu H, He Y, Liu Y, Xu J, Jiang D, Zhang S (2018) Active photosynthetic inhibition mediated by MPK3/MPK6 is critical to effector-triggered immunity. PLoS Biol 16:e2004122
Sulmon C, Gouesbet G, Ramel F, Cabello-Hurtado F, Penno C, Bechtold N, Couee I, El AA (2011) Carbon dynamics, development and stress responses in Arabidopsis: involvement of the APL4 subunit of ADP-glucose pyrophosphorylase (starch synthesis). PLoS ONE 6:e26855
Suzuki K, Nakanishi H, Bower J, Yoder DW, Osteryoung KW, Miyagishima SY (2009) Plastid chaperonin proteins Cpn60 alpha and Cpn60 beta are required for plastid division in *Arabidopsis thaliana*. BMC Plant Biol 9:38
Taniguchi M, Taniguchi Y, Kawasaki M, Takeda S, Kato T, Sato S, Tabata S, Miyake H, Sugiyama T (2002) Identifying and characterizing plastidic 2-oxoglutarate/malate and dicarboxylate transporters in *Arabidopsis thaliana*. Plant Cell Physiol 43:706–717
Thormahlen I, Ruber J, von Roepenack-Lahaye E, Ehrlich SM, Massot V, Hummer C, Tezycka J, Issakidis-Bourguet E, Geigenberger P (2013) Inactivation of thioredoxin f1 leads to decreased light activation of ADP-glucose pyrophosphorylase and altered diurnal starch turnover in leaves of Arabidopsis plants. Plant Cell Environ 36:16–29
Tiller N, Bock R (2014) The translational apparatus of plastids and its role in plant development. Mol Plant 7:1105–1120
Uberegui E, Hall M, Lorenzo O, Schroder WP, Balsera M (2015) An Arabidopsis soluble chloroplast proteomic analysis reveals the participation of the executer pathway in response to increased light conditions. J Exp Bot 66:2067–2077
Wang L, Liang W, Xing J, Tan F, Chen Y, Huang L, Cheng CL, Chen W (2013) Dynamics of chloroplast proteome in salt-stressed mangrove *Kandelia candel* (L.) Druce. J Proteome Res 12:5124–5136
Wang YY, Cheng YH, Chen KE, Tsay YF (2018) Nitrate transport, signaling, and use efficiency. Annu Rev Plant Biol. 69:85–122
Wheeler MC, Tronconi MA, Drincovich MF, Andreo CS, Flugge UI, Maurino VG (2005) A comprehensive analysis of the NADP-malic enzyme gene family of Arabidopsis. Plant Physiol 139:39–51
Wu W, Liu LL, Yang T, Wang JH, Wang JY, Lv P, Yan YC (2018) Gene expression analysis reveals function of TERF1 in plastid-nucleus retrograde signaling under drought stress conditions. Biol Plant. https://doi.org/10.1007/s10535-018-0771-x
Xiao Y, Savchenko T, Baidoo EE, Chehab WE, Hayden DM, Tolstikov V, Corwin JA, Kliebenstein DJ, Keasling JD, Dehesh K (2012) Retrograde signaling by the plastidial metabolite MEcPP regulates expression of nuclear stress-response genes. Cell 149:1525–1535
Yamburenko MV, Zubo YO, Borner T (2015) Abscisic acid affects transcription of chloroplast genes via protein phosphatase 2C-dependent activation of nuclear genes: repression by guanosine-3′-5′-bisdiphosphate and activation by sigma factor 5. Plant J 82:1030–1041
Yao H, Wang X, Chen P, Hai L, Jin K, Yao L, Mao C, Chen X (2018) Predicted Arabidopsis interactome resource and gene set linkage analysis: a transcriptomic analysis resource. Plant Physiol 177(1):422–433
Zhang X, Zhang Z, Chen J, Chen Q, Wang XC, Huang R (2005) Expressing TERF1 in tobacco enhances drought tolerance and abscisic acid sensitivity during seedling development. Planta 222(3):494–501
Zhang X, Szydlowski N, Delvalle D, D'Hulst C, James MG, Myers AM (2008) Overlapping functions of the starch synthases SSII and SSIII in amylopectin biosynthesis in Arabidopsis. BMC Plant Biol 8:96
Zhang Y, Sun K, Sandoval FJ, Santiago K, Roje S (2010) One-carbon metabolism in plants: characterization of a plastid serine hydroxymethyltransferase. Biochem J 430(1):97–105
Zhang K, Liu H, Tao P, Chen H (2014) Comparative proteomic analyses provide new insights into low phosphorus stress responses in maize leaves. PLoS ONE 9(5):e98215
Zubo YO, Yamburenko MV, Kusnetsov VV, Borner T (2011) Methyl jasmonate, gibberellic acid, and auxin affect transcription and transcript accumulation of chloroplast genes in barley. J Plant Physiol 168:1335–1344

20

Characterization of *LhSorP5CS*, a gene catalyzing proline synthesis in Oriental hybrid lily Sorbonne: molecular modelling and expression analysis

Le Wang[1,2], Zhihong Guo[1], Yubao Zhang[1], Yajun Wang[1], Guo Yang[1], Liu Yang[1,2], Ruoyu Wang[1] and Zhongkui Xie[1*]

Abstract

Background: Abiotic stresses negatively affect plant growth and flower production. In plants, P5CS proteins are key enzymes that catalyzed the rate-limiting steps of proline synthesis, and proline is a well-known osmoprotectant that is closely related to abiotic stress tolerance. However, information about the P5CS genes, their effects on proline accumulation, and their role in abiotic stress tolerance in *Lilium* is still lacking.

Results: We isolated and characterized a novel gene (*LhSorP5CS*) from Oriental hybrid lily cultivar Sorbonne. Phylogenetic analysis indicated that *LhSorP5CS* is a member of the *P5CS* family. The three-dimensional structure of LhSorP5CS predicted by homology modeling showed high similarity to its correspondant human P5CS template. Further gene expression analysis revealed that *LhSorP5CS* expression was up-regulated by NaCl, mannitol, and ABA, and that stress-exposed plants accumulated proline at a significantly higher level than in the control.

Conclusions: *LhSorP5CS* characterized in this study is involved in proline synthesis in lily, and that it might play an important role in abiotic stress tolerance. However, there should be other *P5CS* homologues in the lily genome, and some of them could be highly stress-induced and more important for proline accumulation. Future studies on *P5CS* family genes would be of great importance to proline-related stress tolerance in lily.

Keywords: *LhSorP5CS* gene, Homology modeling, Proline, Abiotic stress, *Lilium* spp.

Background

Abiotic stresses such as salinity, drought, heat, and water-logging negatively affect plant growth. These environmental constraints further restrict the range of sites that are suitable for cultivation, and cause decreased agricultural productivity around the world (Zhu 2001). To protect themselves against adverse conditions, plants have evolved many physiological, cellular, and molecular mechanisms (Tan et al. 2013). One of the most well-studied protective mechanisms relates to proline metabolism. Stress responses in plants are often accompanied by the accumulation of proline in different tissues (Verbruggen and Hermans 2008). The accumulated proline can function as an intracellular osmolyte (Chinnusamy et al. 2005), a scavenger for reactive oxygen species (ROS) (Matysik et al. 2002), a maintainer of cell structure (Verslues et al. 2006), and a signaling molecule that primes multiple stress response pathways (Maggio et al. 2002). More recent findings have connected proline to redox status (Sharma et al. 2011; Shinde et al. 2016), and proline was considered as a storage of both energy and reducing potential (Szabados and Savoure 2010). Besides, proline metabolism was also involved in programmed cell death and plant-pathogen interaction. Accumulation of proline

*Correspondence: wxhcas@lzb.ac.cn
[1] Gaolan Station of Agricultural and Ecological Experiment, Northwest Institute of Eco-Environment and Resources, Chinese Academy of Sciences, Lanzhou 730000, China

was observed at the site of infection during incompatible interaction in Arabidopsis (Fabro et al. 2004).

The metabolic pathways for proline synthesis have been deciphered and proven to be evolutionarily conserved from bacteria to higher organisms (Rai and Penna 2013). In plants, two routes for proline synthesis are known to exist, a glutamate pathway and an ornithine pathway (Adams and Frank 1980; Delauney and Verma 1993). However, the glutamate-derived pathway is the major route for proline synthesis under stress conditions (Trovato et al. 2008). In this pathway, glutamate is reduced by Δ1-pyrroline-5-carboxylate synthetase (P5CS) to form γ-glutamate-semialdehyde (GSA) in a manner that depends on ATP and NADPH, then GSA cyclized itself to form P5C (Liang et al. 2013). These reactions are rate-limiting steps in proline synthesis and are catalyzed by P5CS; thus, *P5CS* genes play an important role in stress tolerance.

Given that P5CS plays such an important role in proline synthesis, many attempts to improve stress tolerance by increasing proline production have centered on *P5CS* genes. For example, overexpression of *P5CSF129A* resulted in increased proline production and enhanced salt tolerance in *indica* rice (*Oryza sativa* L.) (Kumar et al. 2010). In citrus species, plants overproduced proline exhibited superior osmotic adjustment and higher photosynthetic rates under drought stress (Molinari et al. 2004).

In the present study, we cloned the full-length cDNA sequence of a *P5CS* gene (*LhSorP5CS*) from the Oriental hybrid lily cv. Sorbonne (*Lilium* spp.) by rapid amplification of cDNA ends (RACE). Phylogenetic analysis of the deduced LhSorP5CS amino acid sequence by comparison with other P5CS proteins in plants indicated that LhSorP5CS was a member of the P5CS family. We investigated the expression of *LhSorP5CS* in different tissues and in response to salt, drought, and abscisic acid (ABA) stresses by means of real-time quantitative polymerase chain reaction (qPCR). To our knowledge, *LhSorP5CS* is the first *P5CS* gene that has been functionally characterized in *Lilium*.

Methods

Plant growth conditions

The Oriental hybrid lily cultivar Sorbonne was planted in the greenhouse of the Gaolan Agricultural and Ecological Experiment Station, in northern China. Bulbs 10–12 cm in diameter were grown in plastic pots containing peat and kept under a 16 h light/8 h dark photoperiod at 22 °C.

Total RNA extraction and first-strand cDNA synthesis

Total RNA was extracted with RNAprep Pure Plant kits (TIANGEN Corporation, Beijing, China) following the manufacturer's instructions. We used 1 μg of total RNA for first-strand cDNA synthesis. First-strand cDNA was synthesized using the PrimeScript first-strand cDNA synthesis kit (Takara, Dalian, China) or the SMARTer RACE cDNA Amplification Kit (for RACE cloning; Clontech, USA). Synthesized cDNA was subsequently diluted to a final concentration of 20 ng μL^{-1} with nuclease-free water for use in cloning.

Full length cloning of *LhSorP5CS*

We designed a degenerate primer set (P5CS-Deg-F: 5′-GGNATHTTYTGGGAYAAYGA-3′; P5CS-Deg-R: 5′-GTYTCCATNGCRTTRCANGC-3′) based on conserved regions for plant P5CS enzymes. PCR was performed in a 25-μL reaction mixture containing 20 ng of template cDNA, 200 μmol L^{-1} of dNTPs, 1.0 μmol L^{-1} of each primer, and 1.25 U of TransTaq HiFi DNA polymerase (TransGen, Beijing, China). PCR products were gel-purified, ligated into the pMD-18T vector (Takara, Dalian, China), and sequenced (Sangon, Shanghai, China). Based on the partial coding sequence that we obtained, we designed gene-specific primers for 5′-RACE (5RACE-GSP-1: 5′-GGTCGGCCTTCAACTC CATCGCC-3′; 5RACE-GSP-2: 5′-AGCCCCGCTAA GCTGTCATTGTCCC-3′) and 3′-RACE (3RACE-GSP-1: 5′-GATCTCGTGATTCCGAGGGGTAGC-3′; 3RACE-GSP-2: 5′-CGCCAAGACAGATTACCCAGCAGC-3′). The full-length sequence was obtained by assembling the 5′-RACE sequence, the partial coding sequence, and the 3′-RACE sequence.

In silico analysis of *LhSorP5CS*

The open reading frame (ORF) of the full-length cDNA of the *LhSorP5CS* sequence was predicted using the ORF finder. Homology of the deduced P5CS protein with other members of this family was verified by querying the NCBI database using the BLAST software. Multiple alignment of sequences was performed using the Clustal W program. A phylogenetic tree of the P5CS proteins was constructed using the Mega 4.1 software (Tamura et al. 2007) using the neighbor-joining method.

Three-dimensional structure prediction by homology modeling

Prediction of 3D structure was carried out via accessing Swiss-Model(Schwede et al. 2003), a web-based protein structure prediction tool. Superposition of the built LhSorP5CS model on the template was achieved by Swiss PDB viewer programme. The root mean square deviation (RMSD),which measure the average distance between the backbone atoms of superposed proteins, was calculated using the PDBeFold programme (Krissinel 2007). Evaluation of stereochemical quality of the LhSorP5CS model was carried out by using Procheck V.3.5.4 (Laskowski et al. 1993).

Expression analysis of *LhSorP5CS* in different tissues and under various stress treatments

Tissues (root, stem, leaf, petal, and scale) were sampled from 2-month-old fully bloomed plants and used as the materials for tissue-specific expression analysis. For exogenous stress treatments, the bulbs of 30-day-old seedlings were drenched with 500 mL solutions containing 500 mM mannitol, 500 mM NaCl, or 2 μM ABA. Leaf samples were taken 0, 2, 4, 8, 12, and 24 h after the stress treatments. All samples were frozen immediately in liquid nitrogen and kept in a refrigerator at −80 °C until use.

After RNA extraction, first strand cDNA synthesis of templates for qPCR was carried out using the HiScript II Q RT SuperMix for qPCR kits (Vazyme Biotech, Nanjing, China), and contamination with genomic DNA was removed by using the gDNA wiper following the manufacturer's protocol. Real-time qPCR was performed using an MX3000P qPCR thermocycler (Stratagene, USA). The Oriental hybrid lily cv. Sorbonne *polyubiquitin4* gene (GenBank accession no. DW718023) was used as the reference gene (Yamagishi 2011). The P5CS-specific primer set (P5CS-q-F: 5′-GCTTTCATCAGGGACGCCAAG-3′; P5CS-q-R: 5′-CAACACCAACCGCACCAGAAG-3′) was designed using OligoArchitect (Fourie et al. 2014), online software developed by Sigma-Aldrich. qPCR was performed using the AceQ qPCR SYBR Green Master Mix kit (Vazyme Biotech, Nanjing, China) according to the manufacturer's instructions. The following protocol was used for amplification: pre-denaturation for 10 min at 95 °C, followed by 40 cycles of 94 °C for 15 s and 60 °C for 30 s. Relative changes in gene expression were analyzed by using the $2^{-\Delta\Delta CT}$ method, and the C_T values used were from three independent biological replicates.

Proline content determination

We determined the free proline content according to the method proposed by Bates et al. (1973). In summary, 0.5 g fresh weight (FW) of leaf tissue was homogenized in 3% sulfosalicylic acid solution using a mortar and pestle. The homogenate was then filtered through Whatman 42 filter paper. We then added 2 mL of 2.5% acidic ninhydrin and 2 mL of acetic acid to 2 mL of the filtrate. After boiling for 1 h, the mixture was placed on ice and proline was extracted using 4 mL of toluene. The proline content per unit FW was calculated based on the absorbance readings at 520 nm using a UV-1200 spectrometer (Macy, Shanghai, China). Pure proline was used as the analytical standard.

Statistical analysis

All experiments were repeated three times, with three replicates each time. We used version 8 of the SAS software for statistical analysis and the Origin Pro 8.1 software to plot the data.

Results

Molecular cloning and sequence analysis of *LhSorP5CS*

The cloned full-length *LhSorP5CS* cDNA contained 2645 bp, comprising a 5′ untranslated region (UTR) of 170 bp, a predicted ORF of 2139 bp, and a 3′ UTR of 196 bp. The ORF encoded a protein of 712 amino acids with a theoretical molecular mass of 77.3 kDa and a predicted isoelectric point (pI) of 5.71. The *LhSorP5CS* cDNA sequence has been submitted to Genbank and deposited with the accession number KU057356.

Multiple alignment showed that the deduced LhSorP5CS amino acid sequence exhibited high homology with other P5CS proteins from plants, such as MaAAAP5CS (83%), TaP5CS (77%), OsP5CS1 (79%), and ZmP5CS (79%). The LhSorP5CS protein sequence was compared with homologues from *Oryza sativa* and *Arabidopsis thaliana* (Additional file 1: Figure S1), and conserved regions included the ATP binding site, two Leu-rich domains, the Glu-5-kinase domain, the NAD(P)H binding domain, and the GSA-DH domain. A conserved phenylalanine residue (Phe, at position 125), which functioned in proline feedback inhibition, was also found in LhSorP5CS.

A phylogenetic tree was built to analyze the evolutionary relationship of the LhSorP5CS protein with P5CS proteins from other species. Phylogenetic analysis revealed that the P5CS proteins in plants clustered into two major groups: the monocot group and the dicot group. LhSorP5CS was most closely related to MaAAAP5CS in the monocot group (Fig. 1). These findings suggest that the *LhSorP5CS* that we cloned is a new gene in the *P5CS* family, and that the putative LhSorP5CS might be involved in proline synthesis in lily.

Predicted three-dimensional structure of the LhSorP5CS protein

The deduced LhSorP5CS protein sequence was submitted to server to BLAST against the Swiss-Model template library to obtain the best template. The searching result revealed that human P5CS (2h5g.1.A) which share 50.12% identity with LhSorP5CS was the best alignment. Thus, the human P5CS protein was used as template (Fig. 2a) to build the LhSorP5CS 3D model (Fig. 2b). Superposition (Fig. 2c) of the built LhSorP5CS model with the human P5CS template showed that 3D structure of these two proteins were highly similar, and the

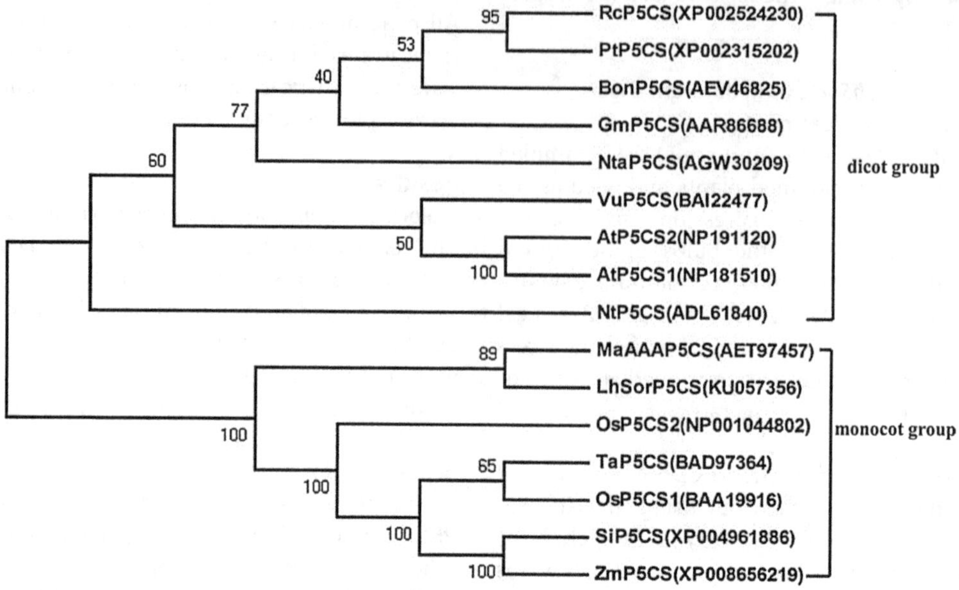

Fig. 1 Phylogram of LhSorP5CS and other P5CSs from different species. Phylogenetic relationship of LhSorP5CS with AtP5CS1 (*Arabidopsis thaliana* P5CS1), AtP5CS2 (*A. thaliana* P5CS2), OsP5CS1 (*Oryza sativa* P5CS1), OsP5CS2 (*O. sativa* P5CS2), RcP5CS (*Ricinus communis* P5CS), PtP5CS (*Populus trichocarpa* P5CS), BonP5CS (*Boehmeria nivea* P5CS), GmP5CS (*Glycine max* P5CS), NtaP5CS (*Nitraria tangutorum* P5CS), VuP5CS (*Vigna unguiculata* P5CS), NtP5CS (*Nicotiana tabacum* P5CS), MaAAAP5CS (*Musa acuminata* AAA Group P5CS), TaP5CS (*Triticum aestivum* P5CS), SiP5CS (*Setaria italica* P5CS), and ZmP5CS (*Zea mays* P5CS). *Numbers in brackets* refer to the sequences accession number in GenBank

calculated RMSD of the superposed proteins was 0.45. Ramachandran plot (Additional file 2: Figure S2) showed that 90.3% of residues in the LhSorP5CS model were in the most favored regions, 8.7% in the additional allowed regions, 0. 5% in the generously allowed regions; while only 0.5% of residues were in the disallowed regions. The result that over 90% of the residues were in the most favored regions indicated that the LhSorPR5 model presented was well built.

Expression analysis

Tissue-specific expression of LhSorP5CS

Figure 3 shows that *LhSorP5CS* expression was detected in all five tissues (root, stem, leaf, petal, and scale), but the abundance of transcripts was significantly higher in the root and stem tissues than in the other tissues. The levels of *LhSorP5CS* transcripts in root and stem tissues were 5.08 and 5.03 times, respectively, that in the petal tissue, which had the lowest expression. The abundance of *LhSorP5CS* transcripts in petal tissue was also significantly lower than that in leaf and scale tissues (which were 2.77 and 2.41 times the petal level, respectively).

Response of *LhSorP5CS* expression to abiotic stresses

We used the levels of *LhSorP5CS* transcripts to investigate whether *LhSorP5CS* expression was induced by salt, drought, or ABA stress. Figure 4a shows that *LhSorP5CS* expression was strongly and significantly induced by NaCl 2 h after treatment, reaching a level 6.92 times that in the control. Similarly, significant up-regulation of *LhSorP5CS* also occurred 2 h after mannitol treatment, to 1.88 times the control level (Fig. 4b); that is, the response was weaker than that to the NaCl treatment. Under ABA treatment, *LhSorP5CS* expression was significantly up-regulated by 2 h after treatment, reaching 1.77 times the control level, and then increased to 2.66 times the control level by 4 h after treatment (Fig. 4c); this was also weaker than the NaCl response. The rapid induction of *LhSorP5CS* in the presence of salt, drought, and ABA stress suggests that *LhSorP5CS* is an abiotic stress-response gene in lily.

Proline accumulation

We also measured proline accumulation in leaves 12 h after the abiotic stress treatments, using the same samples used for qPCR. The NaCl, mannitol, and ABA treatments all significantly increased proline accumulation in the leaves compared with the control, with values reaching 3.39, 1.81, and 1.34 times the control value, respectively (Fig. 5). These results suggest that proline accumulation was significantly induced by salt, drought, and ABA stress treatments in lily and that the rapid increase in *LhSorP5CS* expression may be involved in proline production as part of lily's stress response.

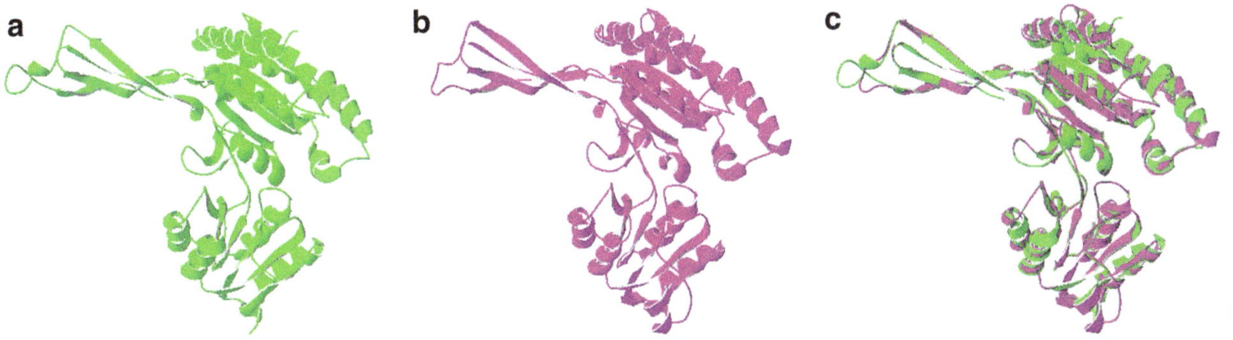

Fig. 2 Three dimensional structure of LhSorP5CS generated by homology modeling with the human P5CS as template. **a** Structure of the human P5CS used as template. **b** Predicted structure of LhSorP5CS. **c** Superposition of LhSorP5CS and the human P5CS protein shows highly similar structure between the two homologs

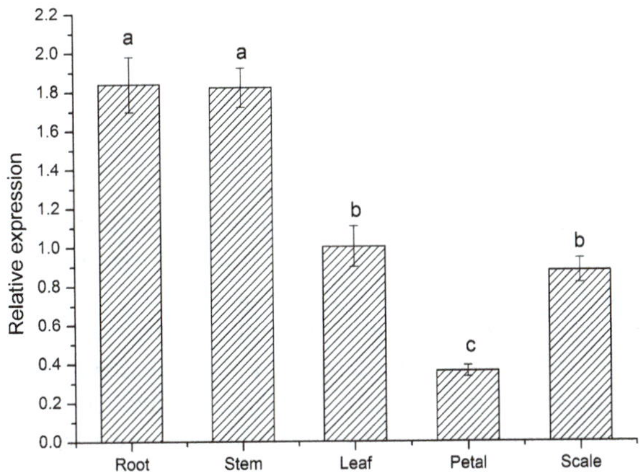

Fig. 3 Tissue-specific expression of *LhSorP5CS* from Oriental hybrid lily cv. Sorbonne (*Lilium* spp.). Transcript levels of the target gene were normalized against the *polyubiquitin4* internal control. Values are mean ± standard deviation (SD) for three replicates. *Bars* labeled with *different letters* differ significantly ($P < 0.05$, *t* test)

Discussion

Proline, an amino acid known to function as an osmoprotectant, a ROS scavenger, a cell structure maintainer, and a signaling molecule, plays an active role in plant stress responses. In plants, the rate-limiting steps in proline synthesis are catalyzed by P5CS. The function of the first plant *P5CS* gene from *Vigna aconitifolia* was revealed by its complementation of the *proA* and *proB* mutations in *E. coli* (Hu et al. 1992). Since then, *P5CS* homologues have been cloned from *A. thaliana* (Strizhov et al. 1997), *O. sativa* (Igarashi et al. 1997), *M. truncatula* (Ginzberg et al. 1998) and other species (Chen et al. 2010; Fujita et al. 1998; Zhu et al. 2012). In this study, we cloned the full-length cDNA sequence of the *LhSorP5CS* gene, the first *P5CS* gene reported in *Lilium*. Phylogenetic analysis revealed that LhSorP5CS was closely related to P5CS proteins from monocot plants, and had 83% identity with MaAAAP5CS. The *LhSorP5CS* transcript level was significantly up-regulated by salt, drought, and ABA stress.

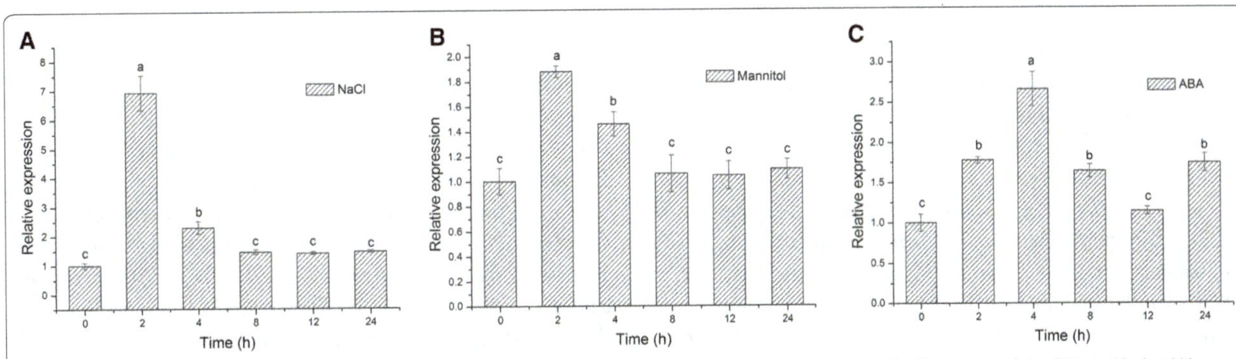

Fig. 4 *LhSorP5CS* expression pattern in response to abiotic stresses. *LhSorP5CS* transcript level was quantified in leaves of the Oriental hybrid lily cv. Sorbonne (*Lilium* spp.) under the **a** NaCl, **b** mannitol, and **c** ABA treatments. Values were determined by real-time qPCR, and normalized using transcript levels of *LhSorP5CS* in the leaves of unstressed plants (the control), which were set to a value of 1. Values represent the mean ± standard deviation (SD) for three replicates. *Bars* labeled with *different letters* differ significantly ($P < 0.05$, *t* test). Note that the y-axis scale differs greatly between graphs

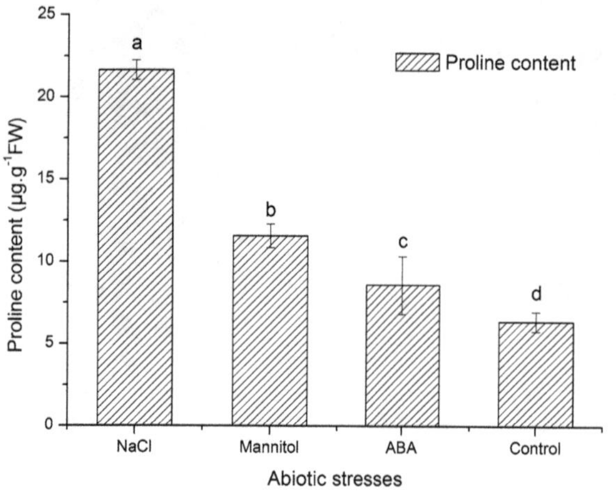

Fig. 5 Proline contents in leaves of Oriental hybrid lily cv. Sorbonne (*Lilium* spp.) sampled 12 h after treating lily plants with different abiotic stresses. Values represent mean ± standard deviation (SD) for three replicates. *Bars* labeled with *different letters* differ significantly ($P < 0.05$, *t* test)

By 12 h after stress treatments, a significantly higher proline content was detected in leaf tissue, revealing proline accumulation induced by salt and drought stress in lily. This agrees with previous research that showed the accumulation of *P5CS* transcripts and proline in many different plant species under stress. In sweet sorghum (*Sorghum bicolor*) (Su et al. 2011), induction of *P5CS* expression was observed before proline accumulation under salt and drought stress. Similarly, rapid up-regulation of *PvP5CS* was observed prior to proline accumulation in *Phaseolus vulgaris* (Chen et al. 2009). Hence, it appears that *LhSorP5CS* is involved in proline synthesis and that the accumulation of proline might result from up-regulation of *LhSorP5CS* expression in lily.

Because *P5CS* genes appear to be the most impor tant genes involved in catalyzing proline synthesis, they have been selected as candidate genes to improve abiotic stress tolerance by elevating in vivo proline levels. Increased proline production and enhanced tolerance to multiple stresses have been reported in several transgenic plants, including potato (*Solanum tuberosum*) (Hmida-Sayari et al. 2005), Arabidopsis (Chen et al. 2013), chickpea (*Cicer arietinum*) (Ghanti et al. 2011), and wheat (*Triticum aestivum*) (Vendruscolo et al. 2007). The *LhSorP5CS* gene that we cloned in the present study is therefore also a candidate gene for future transgenic research in lily.

In higher plants, stress-induced proline accumulation has been observed in many species, and halophytes generally accumulated more proline than glycophytes under adverse conditions (Verbruggen and Hermans 2008). Therefore, the proline level was often correlated with stress tolerance, especially for salt and drought tolerance (Chakraborty et al. 2012; Chauhan et al. 1980; Putnik-Delic et al. 2013). However, this correlation was not always consistent. A recent study of rice seedlings suggested that the proline accumulation level was not a reliable parameter for assessing tolerance of saline alkaline conditions (Lv et al. 2015). Similarly, a study by Deng (2013) suggested that there was no correlation between proline accumulation and drought tolerance in ten genotypes of Tibetan hull-less barley (*Hordeum vulgare* var. *nudum*). Hence, it is worth noting that even though proline plays a multifaceted role in plant stress adaption, its accumulation level may not be significantly associated with stress tolerance and its role may differ among species. It will be necessary to confirm this association in future research on other lily accessions.

The non-consistent correlation between proline level and stress tolerance can be ascribed to the fact that proline turnover, rather that just proline accumulation, was more important to plant growth maintaining under stress. Although many early attempts have proven that improvement of stress tolerance through elevating proline level was feasible, recent evidence (Bhaskara 2015) showed that mutation of *proline dehydrogenase 1* (*PDH1*), the key enzyme catalyze proline catabolism, resulted in similar reduction in growth as the *p5cs1* mutant in spite of elevated proline production. Such findings emphasize the importance of proline catabolism to plant growth maintaining under stress, and underlie that the flux and turnover of proline, rather than just proline accumulation, might be more important to stress adaption for plants.

Conclusions

LhSorP5CS, the first *P5CS* gene cloned from *Lilium*, encoded a protein with 712 amino acids and a pI of 5.71. Expression of *LhSorP5CS* was significantly up-regulated when the plants were exposed to salt, drought, and ABA stresses. Proline accumulation increased significantly 12 h after the stress treatments. These findings suggest that *LhSorP5CS* plays an important role in proline synthesis and stress tolerance in lily. However, there should be other *P5CS* homologues in the lily genome, and some of them could be highly stress-induced and more important for proline accumulation. Future studies on *P5CS* family genes would be of great importance to proline-related stress tolerance in lily.

Abbreviations

P5CS: Δ1-pyrroline-5-carboxylate synthetases; GSA: γ-glutamate-semialdehyde; RACE: rapid amplification of cDNA ends; ABA: abscisic acid; ORF: open reading frame; UTR: untranslated region.

Authors' contributions

ZX and YW conceived and designed the experiments; LW and ZG performed the experiments; YZ and LY analyzed the data; LW, GY and RW wrote the manuscript. All authors read and approved the final manuscript.

Author details

[1] Gaolan Station of Agricultural and Ecological Experiment, Northwest Institute of Eco-Environment and Resources, Chinese Academy of Sciences, Lanzhou 730000, China. [2] University of Chinese Academy of Sciences, Beijing 100049, China.

Acknowledgements

This research was supported by the Ningxia Agricultural Comprehensive Development Office (NTKJ2015-05-01), the National Natural Science Foundation of China (Grant No. 31201651) and the "100 Talents Programme" of the Chinese Academy of Sciences (No. 27Y127L41002).

Competing interests

The authors declare that they have no competing interests.

References

Adams E, Frank L (1980) Metabolism of proline and the hydroxyprolines. Annu Rev Biochem 49:1005–1061

Bates LS, Waldren RP, Teare ID (1973) Rapid determination of free proline for water-stress studies. Plant Soil 39:205–207

Bhaskara GB, Yang TH, Verslues PE (2015) Dynamic proline metabolism: importance and regulation in water limited environments. Front Plant Sci 6:484

Chakraborty K, Sairam RK, Bhattacharya RC (2012) Salinity-induced expression of pyrrolline-5-carboxylate synthetase determine salinity tolerance in Brassica spp. Acta Physiol Plant 34:1935–1941

Chauhan RPS, Chauhan CPS, Kumar D (1980) Free proline accumulation in cereals in relation to salt tolerance. Plant Soil 57:167–175

Chen JB, Wang SM, Jing RL, Mao XG (2009) Cloning the PvP5CS gene from common bean (*Phaseolus vulgaris*) and its expression patterns under abiotic stresses. J Plant Physiol 166:12–19

Chen JB, Zhang XY, Jing RL, Blair MW, Mao XG, Wang SM (2010) Cloning and genetic diversity analysis of a new P5CS gene from common bean (*Phaseolus vulgaris* L.). Theor Appl Genet 120:1393–1404

Chen JB, Yang JW, Zhang ZY, Feng XF, Wang SM (2013) Two P5CS genes from common bean exhibiting different tolerance to salt stress in transgenic *Arabidopsis*. J Genet 92:461–469

Chinnusamy V, Jagendorf A, Zhu JK (2005) Understanding and improving salt tolerance in plants. Crop Sci 45:437–448

Delauney AJ, Verma DPS (1993) Proline biosynthesis and osmoregulation in plants. Plant J 4:215–223

Deng G, Liang J, Xu D, Long H, Pan Z, Yu M (2013) The relationship between proline content, the expression level of P5CS (delta(1)-pyrroline-5-carboxylate synthetase), and drought tolerance in Tibetan hulless barley (*Hordeum vulgare* var. nudum). Russ J Plant Physiol 60:693–700

Fabro G, Kovacs I, Pavet V, Szabados L, Alvarez ME (2004) Proline accumulation and AtP5CS2 gene activation are induced by plant-pathogen incompatible interactions in *Arabidopsis*. Mol Plant Microbe Interact 17:343–350

Fourie JJ, Joubert A, Labuschagne M, Beugnet F (2014) New method using quantitative PCR to follow the tick blood meal and to assess the anti-feeding effect of topical acaricide against *Rhipicephalus sanguineus* on dogs. Comp Immunol Microbiol Infect Dis 37:181–187

Fujita T, Maggio A, Garcia-Rios M, Bressan RA, Csonka LN (1998) Comparative analysis of the regulation of expression and structures of two evolutionarily divergent genes for delta(1)-pyrroline-5-carboxylate synthetase from tomato. Plant Physiol 118:661–674

Ghanti SKK, Sujata KG, Kumar BMV, Karba NN, Reddy KJ, Rao MS, Kishor PBK (2011) Heterologous expression of P5CS gene in chickpea enhances salt tolerance without affecting yield. Biol Plant 55:634–640

Ginzberg I, Stein H, Kapulnik Y, Szabados L, Strizhov N, Schell J, Koncz C, Zilberstein A (1998) Isolation and characterization of two different cDNAs of delta(1)-pyrroline-5-carboxylate synthase in alfalfa, transcriptionally induced upon salt stress. Plant Mol Biol 38:755–764

Hmida-Sayari A, Gargouri-Bouzid R, Bidani A, Jaoua L, Savoure A, Jaoua S (2005) Overexpression of delta(1)-pyrroline-5-carboxylate synthetase increases proline production and confers salt tolerance in transgenic potato plants. Plant Sci 169:746–752

Hu CAA, Delauney AJ, Verma DPS (1992) A bifunctional enzyme (delta-1-pyrroline-5-carboxylate synthetase) catalyzes the 1st 2 steps in proline biosynthesis in plants. Proc Natl Acad Sci USA 89:9354–9358

Igarashi Y, Yoshiba Y, Sanada Y, YamaguchiShinozaki K, Wada K, Shinozaki K (1997) Characterization of the gene for delta(1)-pyrroline-5-carboxylate synthetase and correlation between the expression of the gene and salt tolerance in *Oryza sativa* L. Plant Mol Biol 33:857–865

Krissinel E (2007) On the relationship between sequence and structure similarities in proteomics. Bioinformatics 23:717–723

Kumar V, Shriram V, Kishor PBK, Jawali N, Shitole MG (2010) Enhanced proline accumulation and salt stress tolerance of transgenic indica rice by over-expressing P5CSF129A gene. Plant Biotechnol Rep 4:37–48

Laskowski RA, Macarthur MW, Moss DS, Thornton JM (1993) Procheck—a program to check the stereochemical quality of protein structures. J Appl Crystallogr 26:283–291

Liang XW, Zhang L, Natarajan SK, Becker DF (2013) Proline mechanisms of stress survival. Antioxid Redox Signal 19:998–1011

Lv BS, Ma HY, Li XW, Wei LX, Lv HY, Yang HY, Jiang CJ, Liang ZW (2015) Proline accumulation is not correlated with saline–alkaline stress tolerance in rice seedlings. Agron J 107:51–60

Maggio A, Miyazaki S, Veronese P, Fujita T, Ibeas JI, Damsz B, Narasimhan ML, Hasegawa PM, Joly RJ, Bressan RA (2002) Does proline accumulation play an active role in stress-induced growth reduction? Plant J 31:699–712

Matysik J, Alia B Bhalu, Mohanty P (2002) Molecular mechanisms of quenching of reactive oxygen species by proline under stress in plants. Curr Sci India 82:525–532

Molinari HBC, Marur CJ, Bespalhok JC, Kobayashi AK, Pileggi M, Leite RP, Pereira LFP, Vieira LGE (2004) Osmotic adjustment in transgenic citrus rootstock Carrizo citrange (*Citrus sinensis* Osb. × *Poncirus trifoliata* L. Raf.) overproducing proline. Plant Sci 167:1375–1381

Putnik-Delic M, Maksimovic I, Venezia A, Nagl N (2013) Free proline accumulation in young sugar beet plants and in tissue culture explants under water deficiency as tools for assessment of drought tolerance. Rom Agric Res 30:141–148

Rai AN, Penna S (2013) Molecular evolution of plant P5CS gene involved in proline biosynthesis. Mol Biol Rep 40:6429–6435

Schwede T, Kopp J, Guex N, Peitsch MC (2003) SWISS-MODEL: an automated protein homology-modeling server. Nucleic Acids Res 31:3381–3385

Sharma S, Villamor JG, Verslues PE (2011) Essential role of tissue-specific proline synthesis and catabolism in growth and redox balance at low water potential. Plant Physiol 157:292–304

Shinde S, Villamor JG, Lin W, Sharma S, Verslues PE (2016) Proline coordination with fatty acid synthesis and redox metabolism of chloroplast and mitochondria. Plant Physiol 172:1074–1088

Strizhov N, Abraham E, Okresz L, Blickling S, Zilberstein A, Schell J, Koncz C, Szabados L (1997) Differential expression of two P5CS genes controlling proline accumulation during salt-stress requires ABA and is regulated by ABA1, ABI1 and AXR2 in *Arabidopsis*. Plant J 12:557–569

Su M, Li XF, Ma XY, Peng XJ, Zhao AG, Cheng LQ, Chen SY, Liu GS (2011) Cloning two P5CS genes from bioenergy sorghum and their expression

profiles under abiotic stresses and MeJA treatment. Plant Sci 181:652–659
Szabados L, Savoure A (2010) Proline: a multifunctional amino acid. Trends Plant Sci 15:89–97
Tamura K, Dudley J, Nei M, Kumar S (2007) MEGA4: molecular evolutionary genetics analysis (MEGA) software version 4.0. Mol Biol Evol 24:1596–1599
Tan CM, Chen RJ, Zhang JH, Gao XL, Li LH, Wang PR, Deng XJ, Xu ZJ (2013) OsPOP5, a prolyl oligopeptidase family gene from rice confers abiotic stress tolerance in *Escherichia coli*. Int J Mol Sci 14:20204–20219
Trovato M, Mattioli R, Costantino P (2008) Multiple roles of proline in plant stress tolerance and development. Rend Lincei 19:325–346
Vendruscolo ECG, Schuster I, Pileggi M, Scapim CA, Correa Molinari HB, Marur CJ, Esteves Vieira LG (2007) Stress-induced synthesis of proline confers tolerance to water deficit in transgenic wheat. J Plant Physiol 164:1367–1376
Verbruggen N, Hermans C (2008) Proline accumulation in plants: a review. Amino Acids 35:753–759
Verslues PE, Agarwal M, Katiyar-Agarwal S, Zhu J, Zhu JK (2006) Methods and concepts in quantifying resistance to drought, salt and freezing, abiotic stresses that affect plant water status. Plant J 45:523–539
Yamagishi M (2011) Oriental hybrid lily Sorbonne homologue of LhMYB12 regulates anthocyanin biosyntheses in flower tepals and tepal spots. Mol Breed 28:381–389
Zhu JK (2001) Plant salt tolerance. Trends Plant Sci 6:66–71
Zhu XY, Li XP, Zou Y, Chen WX, Lu WJ (2012) Cloning, characterization and expression analysis of delta(1)-pyrroline-5-carboxylate synthetase (P5CS) gene in harvested papaya (*Carica papaya*) fruit under temperature stress. Food Res Int 49:272–279

21

Range extension of *Christisonia scortechinii* from mainland Southeast Asia into Borneo and notes on the distinction between *Aeginetia* and *Christisonia* (Orobanchaceae)

Antony van der Ent[1,2] and K. M. Wong[3*]

Abstract

Background: *Christisonia* is a little-documented and poorly studied root-parasitic genus in the Orobanchaceae occurring in India, China, Indochina and part of the Malesian region. Recent collection of a *Christisonia* taxon in Kinabalu Park in Sabah, Borneo, taxonomically identical to earlier Sabah collections that have hitherto not been recorded in the literature, led to an assessment of the taxonomic identity of the species against *Christisonia scortechinii*, *C. siamensis*, *C. sinensis* and related species.

Results: Some taxa in China, Indochina, the Malay Peninsula, and the Philippines are morphologically identical to the Borneo taxon except in the number of calyx lobes, but differ by several distinctive characters from other well-distinguished species in the region. Studies of dried herbarium specimens, augmented by photographic images of different stages of fresh flowering material and a scrutiny of available descriptions confirmed that the calyx has two primary lobes in the bud that may separate into 3–5 portions, giving a variable number of apparent lobes in specimen material collected at different localities. This new scrutiny of the calyx also permitted an improved description of the calyx differences that separate *Christisonia* and the closely related *Aeginetia*, which have not been clearly elucidated in the past.

Conclusions: *Christisonia scortechinii* Prain (Orobanchaceae), the only species that was described as having an initially spathaceous calyx among species of this root-parasitic genus, is newly recorded for Borneo (including Kinabalu Park, where its presence has been overlooked). The range of the species in mainland Southeast Asia, previously extended from Peninsular Malaysia to Thailand and Vietnam, is here further extended to Laos and China. *Christisonia wightii* Elmer (relevant to the Philippines) and *C. sinensis* Beck (China) are reduced to synonymy.

Keywords: *Aeginetia*, Borneo, China, *Christisonia*, Kinabalu, Laos, New record, Orobanchaceae, Philippines, Ultramafics

Background

The root-parasitic genus *Christisonia* Gardn. has ca. 20 species distributed from India to China and the Malay Peninsula (Parnell 2001, 2008), Philippines, SW Celebes and Flores (Van Steenis 1967). *Christisonia* has hitherto not been documented for Borneo Island, in spite of pre-existing collections, although the closely related *Aeginetia* L. is recorded there (Beaman and Anderson 2004; under Scrophulariaceae). Both *Christisonia* (Quisumbing 1940) and *Aeginetia* (Kusano 1903, Lee and Goseco 1932) have been reported to parasitize the roots of sugarcane crops. In nature, *Christisonia* is also a root-parasite on other grasses, including bamboos (Barlow 1982; Parnell 2001), and other plant families, including Acanthaceae

*Correspondence: wkm2000@gmail.com
[3] Singapore Botanic Gardens, 1 Cluny Road, Singapore 259569, Singapore
Full list of author information is available at the end of the article

(Nandikar et al. 2013) and Vitaceae (Benniamin et al. 2012).

Christisonia and *Aeginetia* are resolved in the same clade, within the tribe Buchnereae, in the general molecular phylogeny of the Orobanchaceae (McNeal et al. 2013). However, a closer study of the phylogenetic relationship between these two genera with adequate taxon sampling is still elusive. *Christisonia* species have been distinguished by their distinctly lobed calyx from *Aeginetia*, also a root-parasitic genus, which has spatheate calyx (Parnell 2001). Prain (1904) had, however, also described the calyx of the Malayan *C. scortechinii* as "spathaceous rupturing into usually 2, occasionally 3–4 lobes in the fully opened flower", so causing uncertainty if the spatheate calyx of *Aeginetia* is a consistent character for distinguishing it from *Christisonia* (Parnell et al. 2014). Nevertheless, the presence of a covering of colourless slime over *Christisonia* flower buds (Fig. 1a) seems to provide another consistent distinction from *Aeginetia*, in which this slime is absent (Parnell et al. 2014). Also, most *Christisonia* spp. have largely white corollas, and most *Aeginetia* spp. lilac, purple, red or yellow corollas, although this is not mutually exclusive and a purplish corolla form has been reported for *C. scortechinii* (Yong 1989) (Fig. 1d) and species such as *A. indica* also have white corolla forms (Parnell 2012). Another general difference is that *Christisonia* flowers are typically on short pedicels borne on short or obscure stems that do not protrude much from the ground level, whereas *Aeginetia* species often have taller stems 10–40 cm long that carry the flowers clearly above ground level.

More recently, the distribution of several *Christisonia* species has become better documented. For example, *C. siamensis* Craib, once thought to be endemic to Thailand

Fig. 1 Flowering material of *Christisonia scortechinii* from Sabah (Kinabalu Park) (**a–c**) and a *purple*-tinged variant from Peninsular Malaysia (Sungei Wi, Cameron Highlands) (**d**). **a** Side view of flower showing the transparent slime covering. **b** Cluster of fresh flowers with short calyx lobes and a yellow band on the lip that continues down into the corolla tube. **c** Open flowers, with one showing the peltate discoid stigma in the throat. **d** *Christisonia* with unusual *purple*-tinged corolla. Photos: Sukaibin Sumail (**a**), A. van der Ent (**b, d**) and R. van Vugt (**c**)

(Parnell 2001) has been documented in Nagaland, NE India (Benniamin et al. 2012) and SW China (Tan 2013). Nevertheless, the identification of these root-parasites, the flowers of which are delicate and difficult to preserve (spirit collections are essential), remains challenging. Comparatively few collections of the various species and the poor state of preservation of key reference specimens of earlier documented species hamper comparison and reliable identification (Parnell 2001; Parnell et al. 2014). Only in a number of cases where photographic images of sufficient quantity are available can we elucidate the morphology and variation. This, fortunately, is becoming the case as amateur and professional botanists make their images more readily available on the World Wide Web.

This paper formally records *Christisonia* for Borneo through specimens collected recently from Kinabalu Park by Antony van der Ent (Fig. 1a–c), which began this investigation, as well as previous collections from there and elsewhere in Sabah.

Methods

Herbarium material of *Christisonia* for Borneo was studied at the Singapore Herbarium (SING) and the Sabah Parks Herbarium (SNP), and through images on JStor for the Global Plants Initiative, the latter including material from K and WU; herbarium acronyms used follow Thiers (2015). The SNP specimens collected by A. van der Ent in Kinabalu Park were preserved in FAA mixture. Photographic images of different stages of fresh flowering material taken by A. van der Ent and colleagues or which were available from various internet sources (as respectively acknowledged below) aided the interpretation of floral structures and development, especially that of the calyx.

Results and discussion

Christisonia newly recorded for Borneo

Cedric E. Carr, better known for his orchid collecting (Kiew 2003), was apparently the first person to collect a *Christisonia* from Mount Kinabalu and Borneo, although this was never recorded in any publication, including the recently completed *Plants of Mount Kinabalu* where Beaman and Anderson (2004) had missed Carr's number *SFN 26457* of this taxon in the Singapore Herbarium. Carr's locality was the Kinataki River (below the famous Marai Parai spur) at around 1400 m on the western flank of Mount Kinabalu.

Christisonia scortechinii (Fig. 1) was recorded by Antony van der Ent from two localities inside Kinabalu Park, the first at Marai Parai (near the Kinataki River where also Carr collected this species on Mount Kinabalu over 80 years earlier), and the second on Mount Tambuyukon in the north of Kinabalu Park. Both localities sit at 1500–1800 m asl in dense montane cloud forest on ultramafic soils and have remarkably similar vegetation in aspect and composition. Specimens were collected only from the Marai Parai site on Mount Kinabalu, while the Mount Tambuyukon occurrences were documented with photographs.

The *Christisonia* is root-parasitic on the bamboo *Racemobambus gibbsiae* (Poaceae) which is abundant at both sites. The parasitic plants occur as scattered typically "clumped" individuals in the permanently moist humus-rich soil. It appears to be flowering year-round as it has been recorded on a number of occasions during fieldwork. The Marai Parai site is a steep slope along the Kinataki River below the famed spur that is the type-locality for many of Kinabalu's enigmatic plant species, such as *Nepenthes rajah* and *N. edwardsiana* (Nepenthaceae). Mount Tambuyukon (2579 m asl) is the third highest mountain on the island of Borneo and the highest ultramafic mountain in the region. The vegetation on this mountain ranges from tall mixed dipterocarp forest on the foot slopes to stunted graminoid scrub near the summit. The *Christisonia* occurs between Musang Camp and General Camp on the east flank roughly 400 m below the summit.

Voucher material for *C. scortechinii* recorded on Mount Kinabalu include the following:

SABAH, Ranau, Mount Kinabalu, Kinataki River, ca. 4500 ft (1370 m), 7 Mar. 1933, *C.E. Carr SFN 26457* (SING), Marai Parai, near Kinataki River, 1630 m, 31 Mar. 2011, *A. van der Ent, S. Sumail & R.M. Karim SNP 23658* (SNP).

Christisonia scortechinii is also known from elsewhere in Sabah, also on ultramafic soils:

SABAH, Kinabatangan district. Telupid, Bukit Tawai, near summit, 1 Sep. 1999, *R. Kiew RK 4817* (SING), Sungai Ruku-Ruku Forest Reserve, 3 Sep. 1999, *Ali Ibrahim & R. Kiew AI 540* (SING).

The flora of Kinabalu Park is the richest globally in terms of the number of plant species per unit area with more than 5500 taxa recorded from an area roughly 1200 km^2 in extent (Beaman 2005, Van der Ent unpublished). Plants with alternative metabolism or nutrient-acquisition strategies in Kinabalu Park include insect and/or rodent scat trapping ('carnivorous') plants (Nepenthaceae, Droseraceae, Lentibulariaceae), myrmecophytes (Rubiaceae: *Hydnophytum*, *Myrmecodia*), myco-heterotrophic ('saprophytic') plants (mainly Burmanniaceae, Orchidaceae, Triuridaceae), and parasitic plants (principally Loranthaceae and Rafflesiaceae). Most of the hemi-parasitic plants (those that have an active photosynthesis system) in Kinabalu Park are in the order Santalales represented by approximately 60 species predominantly in the families Loranthaceae and Santalaceae (Beaman and Anderson 2004). Apart from the newly recorded *Christisonia*, holo-parasitic plants (which derive all carbohydrates exclusively from the host) include three species of *Balanophora* (*B. lowii, B. reflexa* and *B. papuana*) (Balanophoraceae), one species of *Rhizanthes* (*R. lowii*), one species of *Mitrastema* (*M. yamamotoi*) and two species of *Rafflesia* (*R. keithii* and *R. pricei*) (the latter three genera in Rafflesiaceae).

The local occurrence of ultramafic soils has been suggested as a driving force for speciation, in combination with recent uprising of the Mount Kinabalu Massif, isolation and other factors (Beaman and Beaman 1990). Soils developed from ultramafic (often called 'serpentine' by botanists and ecologists) bedrock induce an edaphic filter on the vegetation that includes nutrient deficiencies, major cation imbalances and metal toxicities (Proctor 2003). As a consequence, ultramafic vegetation in Sabah is often of lower stature, with distinct species composition compared to mixed dipterocarp forest on sedimentary soils. Ultramafic habitats in Borneo are known to harbour a multitude of rare and localised plant species (Wong 1998). Although ultramafic host a number of endemic taxonomic entities that could be evolutionarily derived from closely related taxa from adjacent environments, so-called "neo-endemics" (Van der Ent et al. 2014, 2015), we could not detect any consistent morphological differences between the *C. scortechinii* from the Sabah ultramafic sites and that from elsewhere.

Comparing Sabah *Christisonia* material with closely related taxa

At the outset we were able to determine that the Sabah taxon (Fig. 1a–c) is most related to *C. scortechinii* Prain, *C. siamensis* Craib and *C. sinensis* Beck, in that all these typically or frequently have glabrous white corollas with a single yellow patch that runs from the base of the lip to within the throat and peltate stigmas.

Christisonia scortechinii

An unnumbered collection of *C. scortechinii* at Kew (K000899781) represents potential type material against which to interpret the species in its original Malay Peninsula provenance, given the type sheet bearing Scortechini's number 2121 at Calcutta is in poor condition (Parnell et al. 2014). But even on the Kew sheet, it is difficult to make out intact calyx lobes and verify other characters because the dried material is in poor condition and mounted on a sheet. Thus the interpretation of *C. scortechinii* has depended in large part on direct visual assessment of fresh material from Peninsular Malaysia (Fig. 1d), other herbarium material available (see below), the original description by Prain (1904), as well as the sketch in Henderson (1954) and available photographic images, such as in http://www.parasiticplants.siu.edu/Orobanchaceae/images/Christisonia1.jpg and http://www.parasiticplants.siu.edu/Orobanchaceae/images/Christisonia3.jpg.

Christisonia siamensis

Christisonia siamensis is well distinguished from *C. scortechinii, C. sinensis* and the Sabah taxon in frequently having violet-purplish corolla coloration and a distinct hair tuft at the insertion of the anther that obstructs the corolla throat (Parnell et al. 2014), which is absent in the latter three taxa, which have quite glabrous filaments.

Christisonia sinensis

Christisonia sinensis, found in SW and SE China, was reduced to the synonymy of *C. hookeri* C.B.Clarke ex Hook. in the Flora of China (Zhang and Tzvelev 1998). However this may not be justified because, as pointed out by Parnell et al. (2014), the stigma was described as unequally bilobed in the type description of *C. hookeri* (Hooker 1885) whereas the Flora of China treatment describes a discoid stigma 4–6 mm in diameter, which fits the peltate stigma illustrated in photographic images from Chinese material, as shown in Fig. 2, for example.

Although Parnell et al. (2014) expressed doubt that Chinese material (with peltate stigmas) had been correctly identified as *C. hookeri* (described as having bilobed stigmas), they did not comment on the status of the name *C. sinensis*.

Identification of the Sabah *Christisonia*

Table 1 compares a number of characters among the Sabah taxon, *C. scortechinii, C. siamensis* and *C. sinensis*, as well as *C. calcarata* and *C. hookeri*, both of which are clearly differentiated from the others by their bilobed stigmas. Other very conspicuous and consistent characters also help differentiate consistently recognised species in Table 1, such as the slender pedicellate flowers of *C. calcarata*; a distinct hair tuft at the anther insertion in *C. siamensis*; and the densely pubescent corolla in *C. calcarata*. It would appear that while the predominantly white corolla is also important, occasional violet or purplish tinges do occur. The single band of yellow marking the lip and down into the corolla throat in *C. scortechinii* (Fig. 1), *C. siamensis* and *C. sinensis* (Fig. 2a) also distinguishes these species from *C. calcarata*, where the entire throat is yellow. These provide an idea of various specialised characters that clearly distinguish species.

It can be seen from Table 1 that, while *C. calcarata, C. hookeri* and *C. siamensis* are immediately distinguished from the rest as stated above, *C. scortechinii, C. sinensis* and the Sabah taxon remain rather similar, including in the variable number of calyx lobes, a character that had caused much hesitation in deciding whether or not *C.*

Fig. 2 *Christisonia sinensis* from Chinese material. **a** Flowers, with slime over young buds and (*center-right*) suturing of calyx lobes that incompletely separate. **b** Corolla cut open to show the peltate discoid stigma and two pairs of stamens; in the shorter stamens, the anther appendage has a short acute projection on the dorsal side. (Reproduced with the kind permission of Ming-I Weng, from: https://www.flickr.com/photos/mingi-weng/4927639081/; http://www.flickriver.com/photos/mingiweng/4928219216/)

scortechinii from the type provenance (Peninsular Malaysia) and other provenances in Thailand and Vietnam should be conspecific.

In order to understand the nature of the calyx lobes better, we examined specimen material of *C. scortechinii* from Peninsular Malaysia (the type provenance) to see if they agreed with the original description of Prain (1904); and both dried preserved material (mostly expected to be in poor condition) and photographic images that could assist in interpreting this character for other similar taxa in China (as *C. sinensis* Beck or *C. hookeri* sensu Zhang & Tzvelev, non (C.B.Clarke) Hook.), Vietnam, and Thailand.

For material from the type provenance of *C. scortechinii*, the diagram given by Henderson (Henderson 1954: 335, Fig. 317) clearly shows calyces with 2 or 3 lobes. Earlier, Ridley (1923) had also described "Calyx spathaceous splitting into 2–4 lobes...", which is not different from what Prain (1904) stated. We also studied the following material:

PENINSULAR MALAYSIA, Perak, Bukit Kinta Forest Reserve, 28 Apr. 1987, *R. Kiew RK 2583* (KEP, SING); Perak, Ulu Batang Padang, no date, *Ridley 13693* (SING); Selangor, Ginting Simpah, top of pass, ca. 2000 ft (609 m), 20 Sep. 1947, *J. Reid s.n.* (SING).

In fact, Reid's specimen had helped the description in Henderson (1954), as evident from his letter to Henderson pasted on the specimen, which, among other things, gave the occurrence of the species as "sparsely distributed in small localised patches at moderate altitudes." All five calyces of young as well as open flowers on the Reid specimen sheet show 2 lobes; one young bud shows a single deep suturing across the top of the calyx (implying two lobes from a younger stage). The Ridley specimen has only one 'good' flower in the packet showing clearly two main lobes, each with a short median tear (which can be expected to give at least four lobes perceptible at a later stage). The Kiew specimen shows 2, 3, 4, or 5 calyx lobes, unequal in width as if from tears rather than from preformed equivalent calyx members.

Christisonia sinensis is very similar to *C. scortechinii*, including some Thai and Vietnamese (Central Mountains Region) material that was identified as the latter species (Parnell et al. 2014). The wider concept of *C. scortechinii* (Parnell et al. 2014) accepted that the calyx may be spatheate in the bud but becoming 2–4(–5)-lobed. Although herbarium material for the genus is admittedly poor (Parnell et al. 2014), the type of *C. sinensis*, *Handel-Mazzetti 9595*, Yunnan, 2850 m, 31 July 1916 (WU), clearly shows the short thick pedicels, short-lobed calyces and peltate stigmas that accord with the images of material from China as well as the Thai and Vietnam material mentioned. Moreover, the type also includes flowers where the calyx is variably 2–3-lobed or even 5-lobed, indicating that the calyx in *C. sinensis* does not always part into 5 lobes.

Table 1 The Sabah *Christisonia* compared with five other taxa

	Sabah *Christisonia*	*C. scortechinii*	*C. siamensis*	*C. sinensis*	*C. calcarata*	*C. hookeri*
Pedicel	Short-thick, ca. 1 cm long	Short-thick, ca. 1 cm long	Short-thick, ca. 1 cm long	Short-thick, ca. 1 cm long	*Slender, 2–3.5 cm long*	Short-thick, ca. 1 cm long
Calyx colour	White	White to purplish	White, yellowish to dark purple	White	Pale yellow	Violet
Calyx surface	Glabrous	Glabrous	Glabrous	Glabrous	Glabrous	Glabrous[a]
Calyx lobes/teeth	2–4 (–5)	2–4 (–5)	(2–) 3 (–4)	(2–) 3–4 (–5)	5	(4–) 5
Enclosure of corolla tube by calyx (herbarium material)	About half	Less than half	Less than half	About half	Less than half, to half	About half
Enclosure of corolla tube by calyx (fresh material)	About a quarter	A quarter to a third	A third to half	A quarter to a third	A quarter to a third	?
Corolla colour	White	White to tinged with violet	White tinged with violet	White	White	White with violet seams
Corolla pubescence	Glabrous	Glabrous	Glabrous	Glabrous	*Tube outer surface and both sides of lobes white-pubescent*	Glabrous
Corolla lobe & throat colour	White, with yellow-orange patch from base of lip down to inside the tube	White, with dark yellow patch from base of lip down to inside the tube	Bluish to violet, with yellow patch from base of lip down to inside the tube	White to slightly lilac, with yellow-orange patch from base of lip down to inside the tube	White to greyish or bluish, *bright yellow patch on all lobes to down inside the tube*	?
Hairs at point of attachment of anthers	Glabrous	Glabrous	*Distinct hair tuft* obstructing corolla throat	Glabrous	*Sparse scattered short hairs on filaments*	Glabrous
Anther appendage, dorsal side	With short acute projection	With "projecting acute posterior process" (Prain 1904)	With short acute projection	With short acute projection	*With a slender spur-like projection*	*Rounded, without acute projection*
Stigma	Peltate	Peltate	Peltate	Peltate	*Bilobed*	*Bilobed*
Known distribution	N Borneo (Sabah)	Peninsular Malaysia, Thailand, Vietnam[b]	Thailand (N NE, SW), India (Nagaland)	SW and SE China	India, Pakistan	India
Habitat	Montane sites over 2000 m	Montane sites ca. 650–1700 m	Mixed deciduous forest at 300–800 m	Montane forest, 1500–2000 m	Montane sites	Montane sites

[a] Not pubescent as stated in Benniamin et al. (2012)

[b] Material from Thailand (N, NE, SE, Peninsular, from ca. 1200–1655 m) and a more northerly latitude in C Vietnam (642 m) placed in *C. scortechinii* by Parnell et al. (2014)

Prain (1904) described that the lower pair of stamens of *C. scortechinii* had "a projecting acute posterior process", and *C. sinensis* material from China, as well as Thai and Vietnam material ascribed to *C. scortechinii* in Parnell et al. (2014), also develop a small acute or conical projection on the anther appendage. This can be seen in Chinese material (Fig. 2b) as well as the Sabah material (Fig. 3b). These vary slightly in prominence but are instantly distinguished from the slender elongate, spur-like projection from the dorsal side of the anther appendage developing in another species, *C. calcarata*.

Fresh Chinese material from Taiwan referable to *C. sinensis* beginning with two calyx lobes that sustain visible apical tears can be seen in images from the website: https://www.flickr.com/photos/mingiweng/4928220164/ and examples of calyx tearing into 3–5 lobes can be seen in: http://www.plant.csdb.cn/details?guid=photo:cfh@5c7f74b1-31df-4c72-a872-a2d446a8fa22.

Incompletely separating calyx lobes are clearly seen in Fig. 2a.

Even more instructive, Chinese (Taiwan) material showing a bud calyx with only a small apical opening, and other calyces tearing irregularly into several lobes are obvious in: http://blog.xuite.net/bibibear599/twblog/137870776.

Likewise, the variable number of calyx lobes that can be seen in web-images of fresh material from mainland China, Vietnam and Thailand has been discussed in detail by Parnell et al. (2014), including their Figs. 1c and 2a. Vietnam material with calyx tearing into 5 or more lobes is also available in: https://www.flickr.com/photos/biodivn/14351717859/.

To these we wish only to add a new record of the same taxon for Laos:

LAOS, Houa Phan Province, Vieng Thong District, Nam Et-Phou Louey Natural Protected Area, Nam Et side, near Ban Phon Song, 20°20′46″N, 103°22′04″E, 953 m, 8 Jun. 2013, *J. Leong-Skornickova, M. Rodda, Aung Thame, U. Souvannakhoummane, K. Phoutthavong, M. Norsaengsri & S. Sangvirotjanapat JLS 2407* (E, P, QBG, SING).

Most calyces from this specimen are torn, but one clearly shows 2 lobes, another in bud has the calyx showing only a single small median tear (or slit).

These observations imply that there is no consistent difference in the number of calyx lobes between *C. scortechinii* and *C. sinensis*, supporting a key premise on which Parnell et al. (2014) based their extension of the range of *C. scortechinii* to include sites in Thailand and Vietnam. Although for some other species, e.g., *C. calcarata*, the number of calyx lobes seems evenly formed and are consistently 5 in number, for other species such as *C. scortechinii* and *C. sinensis*, there is generally a variable number of calyx lobes. It would seem that the basic condition representative of the latter two taxa is a basically 2-lobed calyx, where the lobes are fused to only slightly separated in the bud stage, later separating into two clear

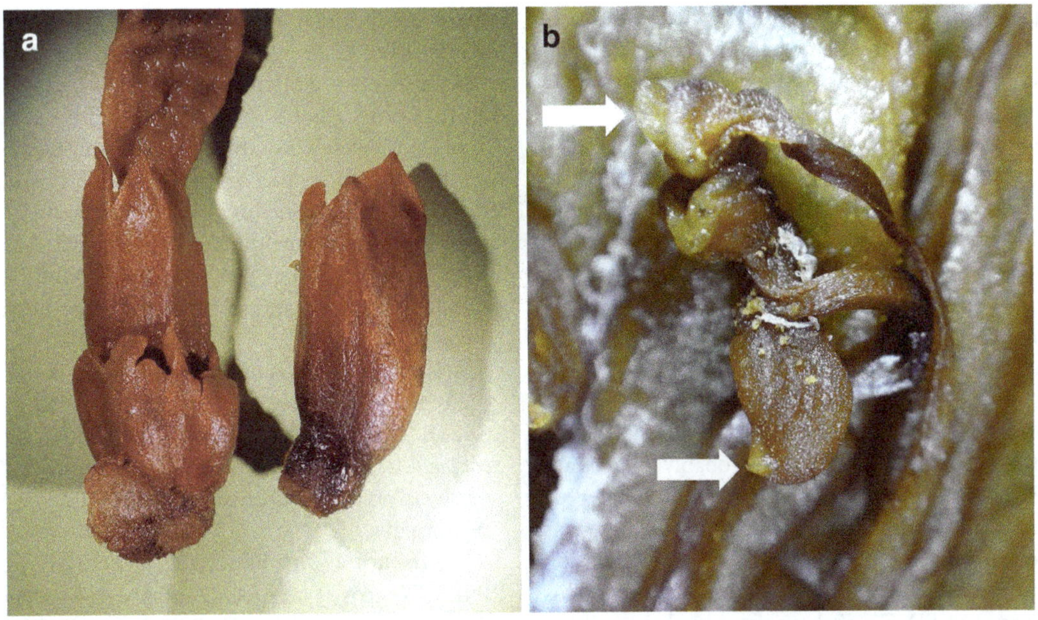

Fig. 3 *Christisonia scortechinii* from Kinabalu Park, showing calyces separating into 3 or more lobes (**a**) and detail of one set of stamens (**b**) showing anther of longer stamen (*upper arrow*) close to the anther of shorter stamen, which has a dorsal lobe with an acute process at its tip (*lower arrow*). From spirit-preserved material of *Van der Ent* et al. *SNP 23658* (SNP)

lobes, each of which can sustain a secondary median tear, in all giving 2, 3, or 4 lobes (as was recorded by Prain for typical *C. scortechinii* and not authoritatively recorded for *C. sinensis*), or even 5 lobes (as common in and recorded for *C. sinensis* and some material from Vietnam and Thailand). Other distinct species may also show a similar calyx character, e.g., the painting of *Christisonia thwaitesii* Trimen by H. DeAlwis in Trimen (1893–1898), t. 69, depicts calyces with 2 lobes, 3 lobes and irregularly tearing lobes: http://plantillustrations.org/illustration.php?id_illustration=157328&mobile=0.

In fact, the description of the calyx by Trimen (1895: 263) of *C. thwaitesii* is not only accurate but also describes the condition in *C. scortechinii*: "...quite closed in the bud (which is beaked), deeply 2-lipped, the lips irregularly or faintly 2- and 3-toothed...".

The material from Sabah also amply displays the same condition of basically two calyx lobes that may further divide (tear) to give 2, 3, or 4, even 5 lobes. Carr's material is rather well preserved, with one calyx clearly showing division into two main lobes, and another having two main lobes with apical tears. The recent material collected by van der Ent and colleagues also shows 3–4 calyx lobes (Fig. 3a).

A wider geographical range for *Christisonia scortechinii*

The impression from descriptions, dried herbarium specimens and photographic images of different stages of fresh flowering material in *C. scortechinii*, *C. sinensis* and the Sabah taxon is that the calyx is basically composed of two lobes that slightly to mostly cohere in the young flower bud, with an apical slit along which the main calyx lobes part and through which the corolla protrudes. In summary, there seem to be no significant differences between *C. scortechinii* and *C. sinensis* and our Sabah taxon. The Malayan name would take precedence.

Christisonia scortechinii is probably also distributed in the Philippines. Images by Leonardo L. Co identified as *C. wightii* Elmer from the Sierra Madre Mountain Range, Cagayan Province, Luzon Island, and the (ultramafic) Mount Victoria massif on Palawan Island (http://www.philippineplants.org/Families/Orobanchaceae.html), both show *Christisonia* flowers with a peltate discoid stigma, and appear to be the same as *C. scortechinii*. *Christisonia wightii* Elmer, described based on material from Negros, was the only species described for the Philippines, and of which the type is not found and probably not extant (likely at PNH and destroyed during WWII). Of the type specimen *Elmer 9510*, Elmer (1915) says "A specimen was however sent...to *Dr. Ridley* [italics his] for determination and who pronounced it a species of *Christisonia*." This collection is not at SING. We have examined the following material, however most calyces are fragmented:

PHILIPPINES, Mindanao, Camp Keelthler, Lake Loamao, Apr. 1907, *M.S. Clemens, s.n.* (SING).

Elmer (1915) described the stigma as "composed of 2 broad and fleshy lobes", which differs from the clearly peltate discoid stigma in *C. scortechinii*, otherwise the details he gives do not vary specially from the characters of *C. scortechinii* or *C. sinensis*. We regard that Elmer's description of the stigma may have been inaccurate and consider that his species is likely the same as *C. scortechinii*.

The material from China, Vietnam and Thailand (as detailed by Parnell et al. 2014), Laos, Peninsular Malaysia (Prain 1904; Ridley 1923) and Borneo (Sabah), probably also that from the Philippines which we identify as *C. scortechinii* all share the following characters: short thick flower pedicels barely 1 cm long; a calyx with 2 primary lobes in the bud that may separate into 3–5 lobes; a typically white, glabrous corolla with a distinctive yellow–orange patch on the lip that continues down into the corolla tube; also additionally, an anther appendage with a short acute-conical projection on the dorsal side (Figs. 2b, 3b). Furthermore, as with the Peninsular Malaysia, Sabah and Philippine records, the ecological distribution of the taxa here included into *C. scortechinii* is clearly montane: the Thai records were from 1200 to 1655 m, the Laos record was from 953 m, the Vietnam record from 642 m at Bi Doup in the Central Mountains Region (Parnell et al. 2014), and the Chinese account generally recorded 1500–2000 m for the species (as '*C. hookeri*') (Zhang and Tzvelev 1998).

A corollary by way of conclusion: *Christisonia* vs *Aeginetia*

Our current understanding of how the calyx structure varies within *C. scortechinii* allows us to recognise it as a species with the main part of its range from the Malay Peninsula to parts of Thailand, Vietnam, Laos and south China, and which is now recorded for Borneo (Sabah) and probably found also in the Philippines. We thus interpret the species to be vicariant across the South China Sea, between mainland Southeast Asia and the Philippine islands and Borneo. Correspondingly, *C. sinensis* and *C. wightii* should be regarded as synonyms of *S. scortechinii*:

Christisonia scortechinii Prain (1904: 205); Ridley (1923: 489); Henderson (1954: 335); Parnell et al. (2014: 16–23). Type: MALAY PENINSULA, Perak, no date, *Scortechini 2121* (holotype CAL, barcode 0000025048; possible isotype K, barcode K000899781).

Synonyms:

Christisonia wightii Elmer (1915: 2793). Type: PHILIPPINES, Negros, Oriental, Dumaguete (Cuernos Mountains), Mar. 1908, *Elmer 9510* (holotype PNH, not found, probably destroyed).

Christisonia sinensis Beck (1930: 314). Type: CHINA, Yunnan, 2850 m, 31 Jul. 1916, *Handel-Mazzetti 9595* (holotype WU, barcode WU0060576).

Christisonia hookeri sensu Zhang and Tzvelev (1998: 242), non Hooker (1885).

It was perhaps unfortunate that Prain (1904) had chosen the term "spathaceous" to describe the calyx of *C. scortechinii.* Use of this term has also led to some concern that an initially "spathaceous" calyx in *C. scortechinii,* the sole example of this in a genus where generally the species were described as having several (up to 5) discrete calyx lobes, would weaken the distinction between *Christisonia* and *Aeginetia* by their otherwise different calyx forms.

In fact, in *C. scortechinii* the two primary calyx lobes (which may remain intact or further tear to give more lobes) are subequal and so ensheath the emerging corolla tube at the beginning; this character is therefore rather different from the conspicuously spathaceous calyx in *Aeginetia.* In *Aeginetia,* there is only one (incomplete) split developing along the ventral median in the conspicuously curved calyx, through which the corolla emerges. Thus, the genera *Aeginetia* and *Christisonia* are still readily distinguished from their calyx form and the way it ruptures during floral development. We venture a distinction between these two otherwise very closely related genera as follows:

(1a) Flower buds covered in a translucent jelly-like slime; flowers typically on short pedicels borne on short or obscure stems that do not protrude much above ground; calyx not or only slightly curved, with 2-several distinct primary lobes, if only 2 then the lobes subequal but possibly dividing further through apical tears; corollas largely white outside, sometimes infused with pink or blue-violet to various degrees ..... *Christisonia*

(1b) Flower buds without a slimy covering; flowers carried on taller stems typically 10–40 cm above ground level; calyx markedly curved or unequally distended, developing only a single split along the ventral median and therefore effectively without distinct lobes; corollas lilac, purple, pink, red or yellow, less frequently white ..... *Aeginetia*

Authors' contributions

AVDE collected and photographed specimen material from Kinabalu Park and Peninsular Malaysia, and compiled the information on vegetation, habitat and host plants. KMW conducted the morphological studies and taxonomic comparisons, and initiated the manuscript draft. Both authors read and approved the final manuscript.

Author details

[1] Centre for Mined Land Rehabilitation, Sustainable Minerals Institute, The University of Queensland, St. Lucia, QLD 4072, Australia. [2] Laboratoire Sols et Environnement, Université de Lorraine-INRA, UMR 1120, 54000 Nancy, France. [3] Singapore Botanic Gardens, 1 Cluny Road, Singapore 259569, Singapore.

Acknowledgements

We thank Sabah Parks, especially Dr Jamili Nais and Rimi Repin, for research support and for the loan of the herbarium material. AVDE is grateful to Rositti Karim and Sukaibin Sumail for help during the fieldwork in Kinabalu Park, and Anthea Phillipps for an earlier identification of *Christisonia*. KMW acknowledges the support of the National Parks Board, Singapore and the Singapore Botanic Gardens.

Compliance with ethical guidelines

Competing interests

The authors declare that they have no competing interests.

References

Barlow HS (1982) A note on *Christisonia scortechinii*, a parasitic flowering plant of the Broom-Rape family (Orobanchaceae). Malay Nat J 36:140

Beaman JH (2005) Mount Kinabalu: Hotspot of plant diversity in Borneo. Biol Skr 55:103–127

Beaman JH, Anderson C (2004) The Plants of Mount Kinabalu. 5. Dicotyledon Families Magnoliaceae to Winteraceae. Natural History Publications (Borneo), Kota Kinabalu

Beaman JH, Beaman RS (1990) Diversity and distribution patterns in the flora of Mount Kinabalu. In: Baas P, Kalkman K, Geesink R (eds) The Plant Diversity of Malesia. luwer Academic Publishers: Dordrecht, The Netherlands

Beck von Mannagetta G (1930) Orobanchaceae. In: Engler A (ed) Das Pflanzenreich. regni vegetabilis conspectus, vol. 261, Part 96. W Engelmann, Leipzig

Benniamin A, Chaturvedi SK, Dey S, Moaakum (2012) Supplements to the root parasitic plant in India. A new recorded species *Christisonia siamensis* Craib (Orobanchaceae). Taiwania 57: 217–221

Elmer ADE (1915) Two Hundred Twenty Six New Species—II. Leafl Philipp Bot 8:2719–2884

Henderson MR (1954) Malayan wild flowers. Malay Nat Soc, Kuala Lumpur

Hooker JD (1885) Orobanchaceae. Flora of British India, vol 4. L. Reeve & Company, London, pp 319–328

Kiew R (2003) Kinabalu diary and orchid determinations. C.E. Carr's Kinabalu field diary. Gardenwise 20:8–12

Kusano S (1903) Notes on *Aeginetia indica*. Linn Bot Mag 17:81–84

Lee A, Goseco F (1932) Studies of the sugar-cane root parasite *Aeginetia indica*. Int Soc Sugar Cane Tech Congr San Juan Bull 101:1–12

McNeal JR, Bennett JR, Wolfe AD, Mathews S (2013) Phylogeny and origins of holoparasitism in Orobanchaceae. Am J Bot 100:971–983

Nandikar M, Singh R, Gurav R, Shinde R (2013) Lectotypification and notes on the identity of *Christisonia calcarata* (Orobanchaceae). Rheedea 23:26–29

Parnell J (2001) A revision of Orobanchaceae in Thailand. Thai For Bull (Bot) 29:72–80

Parnell J (2008) Orobanchaceae. In: Santisuk T, Larsen K (eds) Flora of Thailand, vol 9(2). The Forest Herbarium, National Park, Wildlife and Conservation Department, Bangkok, pp 142–147

Parnell J (2012) *Aeginetia flava*: a new and remarkable species of *Aeginetia*: Orobanchaceae from South-eastern Thailand. Kew Bull 67:81–84

Parnell J, Strijk JS, Maknoi C, Luu HT, Pooma R, Sirimongkol S, Rueangruea S, Tichai W, Ritphet N (2014) New records expand the ranges of *Christisonia siamensis* and *Christisonia scortechinii*, the latter species being new to Thailand. Thai For Bull (Bot) 42:16–23

Prain D (1904) Noviciae Indicae XXIV. Some new Indian plants. J As Soc Bengal 73(II, Natural History): 192–206

Proctor J (2003) Vegetation and soil and plant chemistry on ultramafic rocks in the tropical Far East. Perspect Plant Ecol Evol Syst 6:105–124

Quisumbing E (1940) On *Christisonia wightii*, Elmer, a parasite of sugarcane. Philipp J Plant Ind 11:397–400

Ridley HN (1923) The Flora of the Malay Peninsula: Gamopetalae, vol II. L. Reeve & Co., London

Tan Y-H (2013) *Christisonia siamensis* Craib, a newly recorded species of *Christisonia* (Orobanchaceae) from China. Guihaia 33:521–522

Thiers B (2015). Index Herbariorum: a global directory of public herbaria and associated staff. New York Botanical Garden's Virtual Herbarium. Available from: http://sweetgum.nybg.org/ih/. Accessed 1 July 2015 (continuously updated)
Trimen H (1893–1898). Plates in Illustration of A Hand-Book to the Flora of Ceylon. Dulau & Co., London
Trimen H (1895) A Hand-Book to the Flora of Ceylon, vol 3. Dulau & Co., London
Van der Ent A, Repin R, Sugau J, Wong KM (2014) The Ultramafic Flora of Sabah. An introduction to the plant diversity on ultramafic soils. Natural History Publications (Borneo). Kota Kinabalu
Van der Ent A, Repin R, Sugau J, Wong KM (2015) Plant diversity and ecology of ultramafic outcrops in Sabah (Malaysia). Aust J Bot 63:204–215
Van Steenis CGGJ (1967) Miscellaneous botanical notes XVIII. Blumea 15:145–155
Wong KM (1998) Patterns of plant endemism and rarity in Borneo and the Malay Peninsula. In: Peng C-I, Lowry PP II (eds) Rare, Threatened and Endangered Floras of Asia and the Pacific Rim. Institute of Botany, Taiwan
Yong HS (1989) Colour polymorphism in the flowers of the parasitic plant Malayan Bamboorape (*Christisonia scortechinii*). Nat Malays 14:100–103
Zhang Z-Y, Tzvelev NN (1998) Orobanchaceae. Flora of China vol 18. Science Press, Beijing & Missouri Botanical Garden, St. Louis, pp 229–243

PERMISSIONS

All chapters in this book were first published in BS, by Springer; hereby published with permission under the Creative Commons Attribution License or equivalent. Every chapter published in this book has been scrutinized by our experts. Their significance has been extensively debated. The topics covered herein carry significant findings which will fuel the growth of the discipline. They may even be implemented as practical applications or may be referred to as a beginning point for another development.

The contributors of this book come from diverse backgrounds, making this book a truly international effort. This book will bring forth new frontiers with its revolutionizing research information and detailed analysis of the nascent developments around the world.

We would like to thank all the contributing authors for lending their expertise to make the book truly unique. They have played a crucial role in the development of this book. Without their invaluable contributions this book wouldn't have been possible. They have made vital efforts to compile up to date information on the varied aspects of this subject to make this book a valuable addition to the collection of many professionals and students.

This book was conceptualized with the vision of imparting up-to-date information and advanced data in this field. To ensure the same, a matchless editorial board was set up. Every individual on the board went through rigorous rounds of assessment to prove their worth. After which they invested a large part of their time researching and compiling the most relevant data for our readers.

The editorial board has been involved in producing this book since its inception. They have spent rigorous hours researching and exploring the diverse topics which have resulted in the successful publishing of this book. They have passed on their knowledge of decades through this book. To expedite this challenging task, the publisher supported the team at every step. A small team of assistant editors was also appointed to further simplify the editing procedure and attain best results for the readers.

Apart from the editorial board, the designing team has also invested a significant amount of their time in understanding the subject and creating the most relevant covers. They scrutinized every image to scout for the most suitable representation of the subject and create an appropriate cover for the book.

The publishing team has been an ardent support to the editorial, designing and production team. Their endless efforts to recruit the best for this project, has resulted in the accomplishment of this book. They are a veteran in the field of academics and their pool of knowledge is as vast as their experience in printing. Their expertise and guidance has proved useful at every step. Their uncompromising quality standards have made this book an exceptional effort. Their encouragement from time to time has been an inspiration for everyone.

The publisher and the editorial board hope that this book will prove to be a valuable piece of knowledge for researchers, students, practitioners and scholars across the globe.

LIST OF CONTRIBUTORS

Guanhua Liu
Tea Research Institute, Chinese Academy of Agricultural Sciences, Hangzhou 310008, People's Republic of China
Key Laboratory of Tea Quality and Safety Control, Ministry of Agriculture and Rural Affairs, Hangzhou, 310008, People's Republic of China
Graduate School of Chinese Academy of Agricultural Sciences, Beijing 100081, People's Republic of China

Jianyu Fu
Tea Research Institute, Chinese Academy of Agricultural Sciences, Hangzhou 310008, People's Republic of China
Key Laboratory of Tea Quality and Safety Control, Ministry of Agriculture and Rural Affairs, Hangzhou, 310008, People's Republic of China

Paweena Traiperm and Sasivimon C. Swangpol
Department of Plant Science, Faculty of Science, Mahidol University, Rama VI Road, Ratchathewi, Bangkok 10400, Thailand

Janene Chow and Possathorn Nopun
Department of Plant Science, Faculty of Science, Mahidol University, Rama VI Road, Ratchathewi, Bangkok 10400, Thailand
Department of Pharmaceutical Botany, Faculty of Pharmacy, Mahidol University, Bangkok 10400, Thailand

G. Staples
HRA, Herbarium, Royal Botanic Gardens, Kew, Richmond, Surrey TW9 3AE, UK

Peng Li and Wangzhen Guo
State Key Laboratory of Crop Genetics and Germplasm Enhancement, Nanjing Agricultural University, Nanjing 210095, Jiangsu Province, People's Republic of China

Martin Schnittler, Carl Barnick and Anja Klahr
Institute of Botany and Landscape Ecology, Ernst Moritz Arndt University Greifswald, Soldmannstr. 15, 17487 Greifswald, Germany

Angela Peterson
Institute of Biology, Martin-Luther-University of Halle-Wittenberg, Weinbergweg 10, 06120 Halle/Saale, Germany

Jens Peterson
State Office for Environmental Protection of Saxony-Anhalt, Reideburger Str. 47, 06116 Halle/Saale, Germany

Akmaral Nursafina
L.N. Gumilyov Eurasian National University, 5 Munaytpassov Str, 010008 Astana, Kazakhstan

Shunlei Peng
Key Laboratory of Ecosystem Network Observation and Modeling, Institute of Geographic Sciences and Natural Resources Research, CAS, Beijing 100101, People's Republic of China
Key Laboratory of Ecological Restoration in the Hilly Area, Pingdingshan University, Pingdingshan 467000, Henan, People's Republic of China

Nianpeng He, Guirui Yu and Qiufeng Wang
Key Laboratory of Ecosystem Network Observation and Modeling, Institute of Geographic Sciences and Natural Resources Research, CAS, Beijing 100101, People's Republic of China
College of Resources and Environment, University of Chinese Academy of Sciences, Beijing 100049, People's Republic of China

Wei Chang, Wei Huang, Shi-Bao Zhang and Hong Hu
Key Laboratory of Economic Plants and Biotechnology, Kunming Institute of Botany, Chinese Academy of Sciences, 132 Lanhei Road, Heilongtan, Kunming 650201, Yunnan, China
Yunnan Key Laboratory for Wild Plant Resources, 132 Lanhei Road, Heilongtan, Kunming 650201, Yunnan, China

Andreas Kempe, Christoph Neinhuis and Thea Lautenschläger
Department of Biology, Faculty of Science, Institute of Botany, Technische Universität Dresden, 01062 Dresden, Germany

Ying-Jie Yang
Key Laboratory of Economic Plants and Biotechnology, Kunming Institute of Botany, Chinese Academy of Sciences, 132 Lanhei Road, Heilongtan, Kunming 650201, Yunnan, China
Yunnan Key Laboratory for Wild Plant Resources, 132 Lanhei Road, Heilongtan, Kunming 650201, Yunnan, China
University of Chinese Academy of Sciences, 19 A Yuquan Rd, Shijingshan District, Beijing 100049, People's Republic of China

Michael F. Fay
Royal Botanic Gardens, Kew, Richmond, Surrey TW9 3AB, UK
School of Biological Sciences, University of Western Australia, Crawley, WA 6009, Australia

Kulsumpun Krobanan and Wei-Chiang Shen
Department of Plant Pathology and Microbiology, National Taiwan University, No. 1 Roosevelt Road, Sec. 4, Taipei 10617, Taiwan

Yu-Chen Chuang, Ming-Chi Lee and Ya-Lan Chang
Department of Life Sciences, National Cheng Kung University, Tainan 701, Taiwan

Wen-Huei Chen
Orchid Research and Development Center, National Cheng Kung University, Tainan 701, Taiwan

Hong-Hwa Chen
Department of Life Sciences, National Cheng Kung University, Tainan 701, Taiwan
Orchid Research and Development Center, National Cheng Kung University, Tainan 701, Taiwan

Yu-Hsuan Huang, Chih-Yu Hung, I-Rhy Lin, Tomonori Kume and Chih-Hsin Cheng
School of Forestry and Resource Conservation, National Taiwan University, Taipei 106, Taiwan

Oleg V. Menyailo
Institute of Forest Research RAI SR, Krasnoyarsk 660036, Russia

Wen-Chieh Tsai and Hong-Hwa Chen
Institute of Tropical Plant Sciences, National Cheng Kung University, Tainan 701, Taiwan
Orchid Research and Development Center, National Cheng Kung University, Tainan 701, Taiwan
Department of Life Sciences, National Cheng Kung University, Tainan 701, Taiwan

Yu-Yun Hsiao
Orchid Research and Development Center, National Cheng Kung University, Tainan 701, Taiwan
Department of Life Sciences, National Cheng Kung University, Tainan 701, Taiwan

Chia-Chi Hsu, Shang-Yi Chiou and Hsin Huang
Department of Life Sciences, National Cheng Kung University, Tainan 701, Taiwan

Anne Dievart
CIRAD, UMR AGAP, TA A 108/03, Avenue Agropolis, 34398 Montpellier, France
School of Life Sciences and Biotechnology, Shanghai Jiao Tong University, 800 Dongchuan Road, Life Sciences Building, Room 3-117, Shanghai 200240, People's Republic of China

Hui-Ju Shen and Wei-Chin Chang
Institute of Plant and Microbial Biology, Academia Sinica, Taipei 115, Taiwan, ROC

Jen-Tsung Chen
Department of Life Sciences, National University of Kaohsiung, Kaohsiung 811, Taiwan, ROC

Hsiao-Hang Chung
Department of Horticulture, National Ilan University, Yilan 260, Taiwan, ROC

Hisayoshi Nozaki and Shota Yamashita
Department of Biological Sciences, Graduate School of Science, University of Tokyo, Hongo, Bunkyo-ku, Tokyo 113-0033, Japan

Noriko Ueki
Department of Biology, Brooklyn College, City University of New York, Brooklyn, NY 11210, USA

Mari Takusagawa
Department of Botany, Graduate School of Science, Kyoto University, Sakyo-ku, Kyoto 606-8502, Japan

Osami Misumi
Department of Biological Science and Chemistry, Faculty of Science, Graduate School of Medicine, Yamaguchi University, Yoshida, Yamaguchi 753-8512, Japan

Ryo Matsuzaki and Masanobu Kawachi
Center for Environmental Biology and Ecosystem Studies, National Institute for Environmental Studies, Onogawa, Tsukuba, Ibaraki 305-8506, Japan

Yin-Ru Chiang and Jiunn-Tzong Wu
Biodiversity Research Center, Academia Sinica, Nankang, Taipei 11529, Taiwan

Kuan-Ting Hsin
Institute of Ecology and Evolutionary Biology, National Taiwan University, Taipei, Taiwan

Chun-Neng Wang
Institute of Ecology and Evolutionary Biology, National Taiwan University, Taipei, Taiwan
Department of Life Science, National Taiwan University, Taipei, Taiwan

Heng-An Lin, Szu-Yu Chen, Yi-Chia Chen, Wei-Chiang Shen and Chia-Lin Chung
Department of Plant Pathology and Microbiology, National Taiwan University, No. 1, Sec. 4, Roosevelt Rd., Taipei 10617, Taiwan

Fang-Yu Chang and Chih-Wen Wu
Kaohsiung District Agricultural Research and Extension Station, No. 2-6, Dehe Rd., Pingtung County 90846, Taiwan

Chih-Wei Tung
Department of Agronomy, National Taiwan University, No. 1, Sec. 4, Roosevelt Rd., Taipei 10617, Taiwan

Ruey-Shyang Chen
Department of Biochemical Science and Technology, National Chiayi University, No. 300, Syuefu Rd., Chiayi City 60004, Taiwan

Yu-Hsin Tseng, Ching-I Peng, Yoshiko Kono and Kuo-Fang Chung
Research Museum and Herbarium (HAST), Biodiversity Research Center, Academia Sinica, Taipei 11542, Taiwan

Young-Dong Kim
Department of Life Science, Hallym University, Chuncheon, Gangwon 24252, Republic of Korea

Khin Myo Htwe
Popa Mountain Park, Nature and Wildlife Conservation Division, Ministry of Natural Resources and Environmental Conservation, Kyaukpadaung Township, Mandalay Region, Myanmar

Seong-Hyun Cho
International Biological Material Research Center, Korea Research Institute of Bioscience and Biotechnology, Daejeon 34141, Republic of Korea

Muhammad Khalid, Danial Hassani and Danfeng Huang
School of Agriculture and Biology, Shanghai Jiao Tong University, Shanghai 200240, China

Muhammad Bilal
State Key Laboratory of Microbial Metabolism, and School of Life Sciences and Biotechnology, Shanghai Jiao Tong University, Shanghai 200240, China

Hafiz M. N. Iqbal
School of Engineering and Science, Tecnologico de Monterrey, Campus Monterrey, Ave. Eugenio Garza Sada 2501, Monterrey, N.L. 64849, Mexico

Hang Wang
Instrumental Analysis Center, Shanghai Jiao Tong University, Shanghai 200240, China

Wei Wu and Yanchun Yan
Graduate School of Chinese Academy of Agricultural Sciences, No. 12 Zhongguancun South St., Haidian District, Beijing 100081, People's Republic of China

Le Wang and Liu Yang
Gaolan Station of Agricultural and Ecological Experiment, Northwest Institute of Eco-Environment and Resources, Chinese Academy of Sciences, Lanzhou 730000, China
University of Chinese Academy of Sciences, Beijing 100049, China

Zhihong Guo, Yubao Zhang, Yajun Wang, Guo Yang, Ruoyu Wang and Zhongkui Xie
Gaolan Station of Agricultural and Ecological Experiment, Northwest Institute of Eco-Environment and Resources, Chinese Academy of Sciences, Lanzhou 730000, China

Antony van der Ent
Centre for Mined Land Rehabilitation, Sustainable Minerals Institute, The University of Queensland, St. Lucia, QLD 4072, Australia
Laboratoire Sols et Environnement, Université de Lorraine-INRA, UMR 1120, 54000 Nancy, France

K. M. Wong
Singapore Botanic Gardens, 1 Cluny Road, Singapore 259569, Singapore

Sheng-Zehn Yang and Po-Hao Chen
National Pingtung University of Science and Technology, No. 1, Shuefu Rd., Neipu, Pingtung 912, Taiwan

Gang-cai Liu
Key Laboratory of Mountain Surface Processes and Ecological Regulation, Institute of Mountain Hazards and Environment, Chinese Academy of Sciences and Ministry of Water Resources, Chengdu 610041, China

Xue-mei Wang and Bang-guo Yan
Key Laboratory of Mountain Surface Processes and Ecological Regulation, Institute of Mountain Hazards and Environment, Chinese Academy of Sciences and Ministry of Water Resources, Chengdu 610041, China
University of Chinese Academy of Sciences, Beijing 100049, China

Li Zhao
College of Environment and Resource Science, Southwest University of Science and Technology, Mianyang 621010, China

Liang-tao Shi
Institute of Tropical Eco-agricultural Sciences, Yunnan Academy of Agricultural Sciences, Yuanmou 651300, Yunnan Province, China

Yu-xiao He
Institute of Resources and Environment, He'nan Polytechnic University, Jiaozuo 454000, He'nan, China

Index

www.ingramcontent.com/pod-product-compliance
Lightning Source LLC
LaVergne TN
LVHW080717210926
848485LV00003B/480